Textbook of

Remote Sensing and Geographical Information Systems

Fourth Edition

Textbook of

Remote Sensing and Geographical Information Systems

Fourth Edition

M. ANJI REDDY

Professor of Environmental Science and Technology,
Jawaharlal Nehru Technological University Hyderabad,
Kukatpally, Hyderabad-72 (A.P.) India.
mareddyanjireddy@rediffmail.com

BS Publications
A unit of **BSP Books Pvt., Ltd.**

4-4-309/316, Giriraj Lane, Sultan Bazar,
Hyderabad - 500 095
Phone : 040 - 23445605, 23445688

Published by :

BSP BS Publications
A unit of **BSP Books Pvt., Ltd.**

4-4-309/316, Giriraj Lane, Sultan Bazar,
Hyderabad - 500 095
Phone : 040 - 23445605, 23445688
e-mail : info@bspbooks.net

ISBN : 978-93-85433-35-1 (HB)

To the cause of human welfare and elevation
which are ever so dear to my heart...

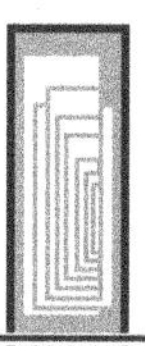

Preface to Fourth Edition

The fourth edition is the upgradation of the third edition with notable chapters on various applications. The new chapters, namely Photogrammery with very few topics of digital photogrammetry, Global Positioning System (GPS) and the chapters of applications like, Forest Resources Management, Watershed Management and Natural Disaster Management: Landslides are added. Since, the users of geomatics technology for various applications have been using high resolution image data, the photogrammetry with fundamental concepts are included for better understanding of the student community. The application chapters are the output of the sponsored research projects of the author. These chapters are very much useful to the students who focussed their research on geospatial technologies.

I am deeply grateful to many students, faculty and reviewers who have suggested changes for the fourth edition. In addition to those who suggested changes, I wish to thank individuals and organisations for their support to the second and third editions: a support that helped in making this edition a reality. In this regard, I thank Sri C.K.P. Unni, guest faculty, JNTU for his technical support in reviewing the Chapters 2 and 6.

Special thanks to my research students Dr. Santhosh Kumar, Dr. Kavita Singh, Mr. Vamshi Mohan Reddy, Mr. Vamsi Krishna Reddy, U. Kezia, Smt. Hema Sailaja, Swaraj and other studnets for their continuous support.

Finally, special thanks to my wife Jaya and children Chandu and Rajiv for their support and cooperation in writing this book.

Hyderabad **M. Anji Reddy**
2012

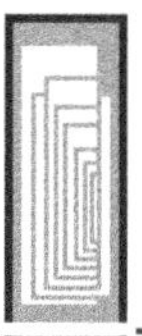

Contents

2 Photogrammetry

4 Microwave Remote Sensing

5 Remote Sensing Platforms and Sensors

6 Visual Image Interpretation

19 Natural Disaster Management : Landslides

20 Creation of Information System: A Case Study

1

Map Language

1.1 Introduction

The collection of data about the spatial distribution of the significant features of the earth's surface has long been an important part of the activities of organized societies. From the earliest civilizations to modern times, spatial data have been collected by navigators, geographers and surveyors, and rendered into pictorial form by map makers or cartographers. Originally, maps were used to describe far-off places, as an aid for navigation and military strategies (Hodgkiss 1961). During the eighteenth century, many governments realised the advantages of systematic mapping of their lands, and commissioned national government institutions to prepare topographical maps. These institutions are still continuing the mapping work. Many of the developing countries are making all attempts to obtain the status of a developed country. These attempts are based on certain strategies relating to areas like natural resources management and development,information technology, tourism development, infrastructure development,rural development, environmental management, facility management, and e-governance. In order to make an effective study of these thrust and emerging fields, new and innovative technologies have been developed.

In the last two decades innovative technologies have been greatly applied to experimental and operational activities. These technologies have their historical antecedents. For instance, Remote Sensing and GIS have been developed from earlier technologies such as surveying, photogrammetry, cartography, mathematics, and statistics. Laurini and Thompson (1992) adopted the umbrella term “Geomatics” to cover all these disciplines. They stated that the different aspects of each of these areas are necessary for formulating and understanding spatial information systems. The traditional method of storing, analysing and presenting spatial data is the map. The map or spatial language, like any other language, functions as a filter for necessary information to pass through (Witthuhn et al, 1974). It modifies the way we think, observe, and make decisions. Maps are thus the starting point in any analysis and are used in the presentation of results of any operational project. Whether it is remote sensing, photogrammetry, cartography, or GIS, the ultimate output will be the production of high quality, more accurate and clearer map, so that the user finds it easy to make appropriate decisions. Therefore, maps and their production using modern technologies is an essential starting point and they are the necessary tools to explore the characteristics of spatial phenomena. This Chapter is exclusively devoted to providing the fundamental concepts of a map, map scale, various terms used in mapping, review of map projections, and map symbolism.

1.2 Map as Model

The map constitutes the language of simple geography as well as automated geography. As a graphic form of spatial data abstraction it is composed of different grid systems, projections, symbol libraries, methods of simplification and generalisation, and scale.

A map is the representation of the features of the earth drawn to scale. The surface of the map is a reduction of the real scenario. The map is a tool of communication and it has been in use since the days of the primitive man who had to move about constantly in search of food and shelter. A map from any local planning agency provides different kinds of information about the area. This map focuses on the infrastructure and legal descriptions of the property boundaries, existing and planned roadways, the locations of existing and planned utilities such as potable water, electric and gas supplies, and the sanitary sewer system. The planning map may not be of the same scale as the topographic map, the former probably being drawn to a larger scale than the latter. Further, the two may not be necessarily based on the same map projection. For a small area, the approximate scale of the data is probably more important than the details of the map projection. As Robinson et.al (1984) observe, "A map is a very powerful tool and maps are typical reductions which are smaller than the areas they portray". As such, each map must have a defined relationship between what exists in the area and its mapped representation. The scale of a map sets limits on both the type and manner of information that can be portrayed on a map.

1.2.1 Spatial Elements

Spatial objects in the real world can be thought of as occurring in four easily identifiable types namely, points, lines, areas and surfaces (Fig. 1.1). Collectively, these four features, or various permutations and combinations of these four spatial features

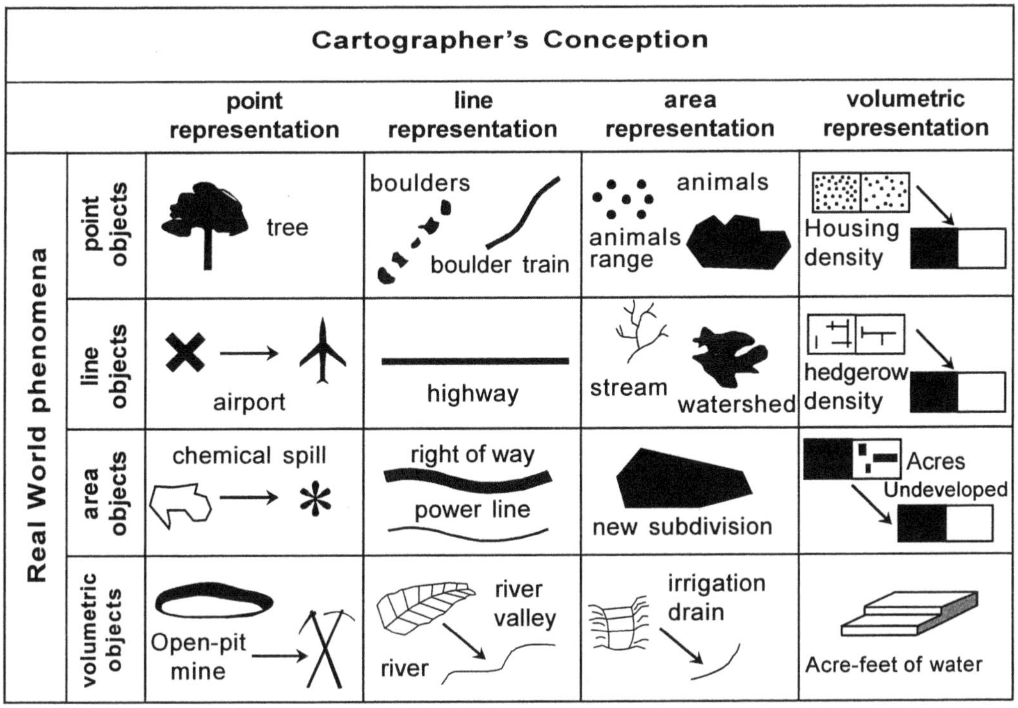

Real World phenomena	Cartographer's Conception: point representation	line representation	area representation	volumetric representation
point objects	tree	boulders; boulder train	animals; animals range	Housing density
line objects	airport	highway	stream; watershed	hedgerow density
area objects	chemical spill	right of way; power line	new subdivision	Acres Undeveloped
volumetric objects	Open-pit mine	river valley; river	irrigation drain	Acre-feet of water

Fig. 1.1 Comparison of real-world phenomena and the cartographer's conception. Point, line, area, and surface features with examples (Source : Demers, 1998).

can form the human phenomena or the spatial real world. Points, lines and areas can be represented by using symbols to depict the real world. Surfaces are represented by any combination of these spatial entities. In general, all the geographic surfaces are in two tangible forms, namely, discrete and continuous. Trees, houses, road intersections and similar items are discrete spatial features. A feature can be termed as discrete, if it occupies a given point in space and time, that is, each feature can be referenced by its locational coordinates. All discrete features are said to have a zero dimensionality but have some spatial dimension. Fig. 1.2 shows the continuous spatial feature. The earth's surface occurs all around as natural features like hills, ridges, cliffs, and trenches, which can be described by citing their locations, the area they occupy, and how they are oriented with the addition of the third dimension. All these are considered continuous surface features. These features are composed of an infinite number of possible height values distributed without interruption across the surface.

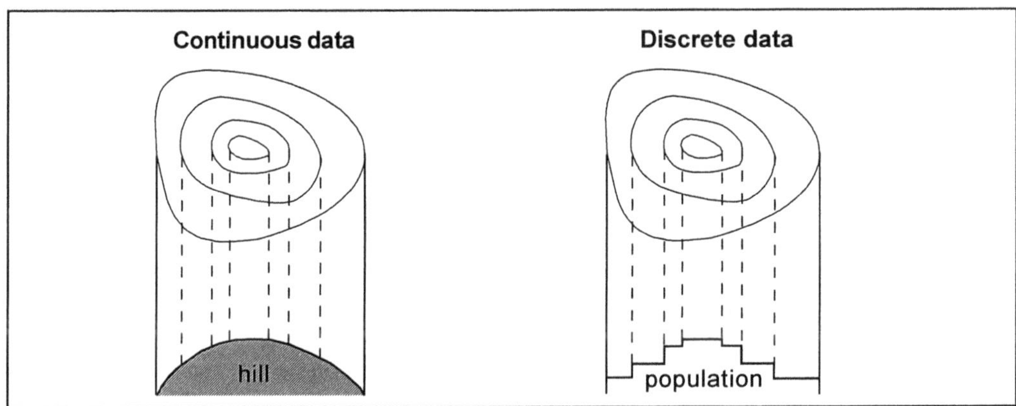

Fig. 1.2 Continuous versus discrete surface.

A topographic map depicts several kinds of information both discrete and continuous. Elevation on the site is portrayed as a series of contour lines. These contour lines provide us with a limited amount of information about the shape of the terrain. Different kinds of man-made features including structures and roadways are typically indicated by lines and shapes. In many cases, the information on this map is five to fifteen years out of date, a common situation resulting from the rate of change of land cover in the area and the cycle of map updates. Each of these different kinds of information, which we may decide to store in various ways, is called a theme.

A topographic map describes the shape, size , position, and relation of the physical features of an area. In addition to mountains, hills, valleys, and rivers, most topographic maps also show the culture of a region, that is, political boundaries, towns, houses, roads, and similar features.

1.2.2 Terminology

The terminology used in describing any kind of spatial/geographic features is discussed in this section. Elevation or altitude is the vertical distance between a given point and the datum plane. Datum plane is the reference surface from which all altitudes on a map are measured. This is usually mean sea level. The height is defined as the vertical difference in elevation between an object and its immediate surroundings. The difference in elevation of an area between tops of hills and bottoms of valleys is known as relief of the terrain. A point of known elevation and position usually indicated on a map by the letters B. M. with the altitude given to the nearest foot is termed as bench mark. On some maps the altitude is given along the contour line (Survey of India Maps).

A map line connecting points representing places on the earth's surface that have the same elevation is called contour line. It thus locates the intersection with the earth's surface of a plane at any arbitrary elevation parallel to the datum plane. Contours represent the vertical or third dimension on a map which otherwise has only two dimensions. They show the shape and size of physical features such as hills and valleys. A depression is indicated by an ordinary contour line except that hachures or short dashes are used on one side and point toward the center of the depression. The difference in elevation represented by adjacent contour lines is termed as contour interval. (Monmonier and Schnell, 1988)

Maps are a very important form of input to a geographical information system, as well as a common means to portray the results of an analysis from a GIS. Like GIS, maps are concerned with two fundamental aspects of reality, locations and attributes. Location represents the position of a point in a two-dimensional space. Attributes at a location are some measure of a qualitative or quantitative characteristic such as land cover, ownership, or precipitation. From these fundamental properties a variety of topology and metric properties of relationship may be identified including distance, direction, connectivity, and proximity.

1.3 Classification of Maps

Maps are thus the cartographer's representation of an area and a graphic representation of selected natural and man-made features of the whole or a part of the earths surface on a flat sheet of paper on a definite scale. Even though there are many different types of maps, all the maps are broadly classified on the basis of two criteria, namely, scale and, contents and purpose. On the basis of the scale, the map may be classified as either a small scale map or a large scale map. Some of large scale maps, are cadastral or revenue maps, utility maps, urban plan maps, transportation or network maps. On the basis of the content, maps are classified either as physical maps considered as small scale maps, or cultural maps. In the process of preparing a map one should remember that inside a GIS, one is likely to encounter a greater variety of maps than one might have expected on the basis of the subject matter. In addition, based on the thematic content of GIS coverages, the maps can be termed as thematic maps like vegetation maps, transportation maps, land use land cover maps, and remotely sensed imagery. These thematic maps will be portrayed as prism maps, choropleth maps (Fig. 1.3), point distribution maps, surface maps, graduated circle maps, and a host of other types. Plate 1 shows land use/land cover hydrogeomorphological map of the Shivanngudem watershed, which is a classical example of a thematic map.

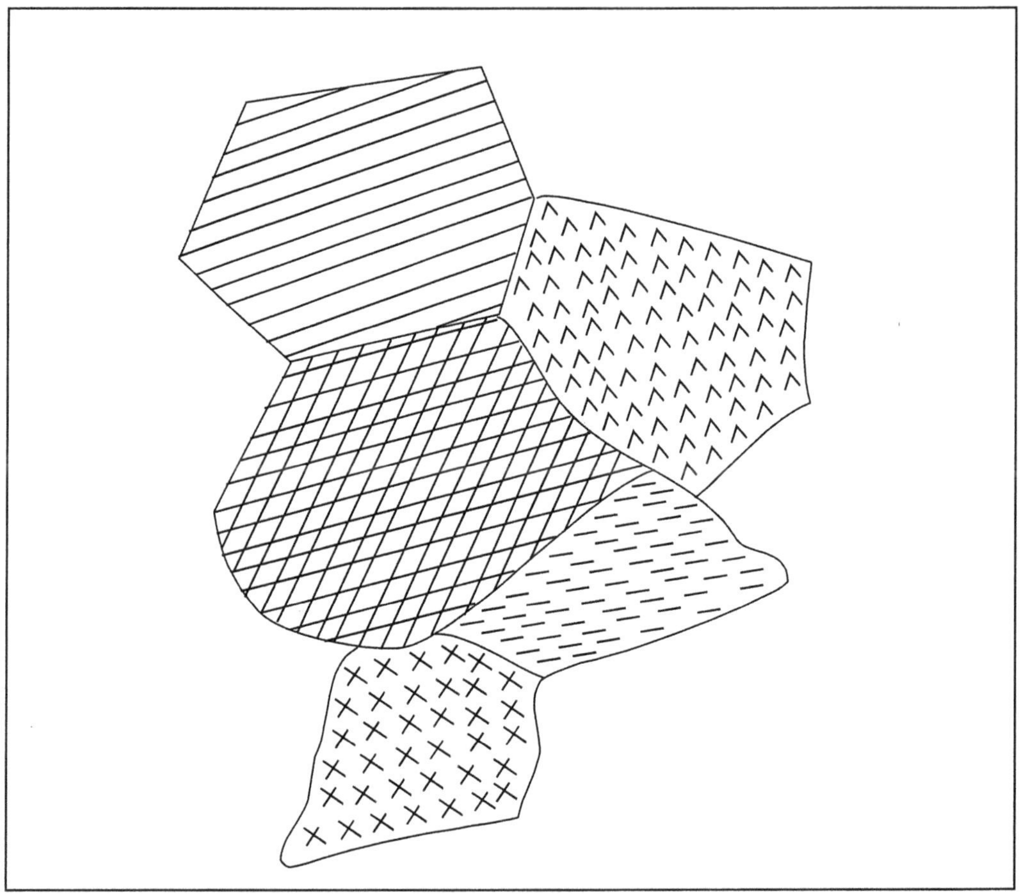

Fig. 1.3 Choropleth map.

1.4 Map Scale

By necessity, the process of representing geographic features on a sheet of paper involves the reduction of these features. The ratio between the reduced depiction on the map and the geographical features in the real world is known as the map scale, that is the ratio of the distance between two points on the map and the corresponding distance on the ground. The scale may be expressed in three ways and the pictorial representation of these three types is shown in Fig. 1.4.

Fractional scale : If two points are 1 km apart in the field, they may be represented on the map as separated by some fraction of that distance, say 1 cm. In this instance, the scale is 1 cm to a kilometer. There are 100,000 cm in 1 km; so this scale can be expressed as the fraction or ratio of 1:100,000. Many topographic maps of the United States Geological Survey have a scale of 1:62,500; and many recent maps have a scale of 1:31,250, and others of 1:24,000. In India, commonly used fractional map scales are 1:1,00,000,00; 1:250,000, 1:50,000; 1:25,000 and 1:10,000. The method of representing this type of scale is called Representation Fraction (RF) method.

Graphic scale : This scale is a line printed on the map and divided into units that are equivalent to some distance such as 1 km or 1 mile. The measured ground distance appears directly on the map in graphical representation.

Verbal scale : This is an expression in common speech, such as, "four centimeters to the kilometer", "an inch to a mile". This common method of expressing a scale has the advantage of being easily understood by most map users.

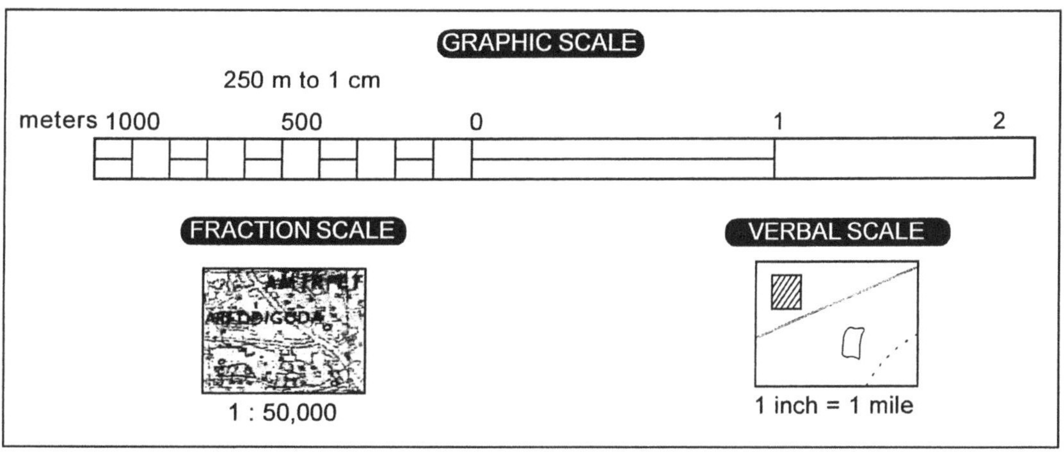

Fig. 1.4 Three different types of scale representation.

The ratio and map scale are inversely proportional. Therefore, 1:1,00,000 (large ratio) map is considered a small scale. In this instance, the scale is 1 cm to a kilometer. As there are 100,000 cm in 1 km, this scale can be expressed as the fraction or ratio of 1:10,000 (small ratio) map is considered a large scale. The small scale maps depict large tracts of lands (such as continents or countries) usually with a limited level of detail and a simple symbology. Large scale maps can depict small areas (such as cities) with a richness of detail and a complex symbology.

The terms 'small scale' and 'large scale' are in common use. A simple example helps illustrate the difference. Consider a field of 100 meters on a side. On a map of 1:10000 scale, the field is drawn 1 centimeter on a side. On a map of 1:1,000,000 scale, the field is drawn 0.1 millimeter on a side. The field appears larger on the 1:10000 scale map; we call this a large-scale map. Conversely, the field appears smaller on the 1:1,000,000 scale map, and we call this a small-scale map. Alternately, if we have a small area of the earth's surface on a page, we have a large-scale map; if we have a large area of the earth's surface on a page we have a small-scale map. Fig. 1.5 shows the effect of scale. On 1:50,000 scale map (large), even small knicks can be mapped where as on 1:250,000 scale the mapping of knicks is practically impossible.

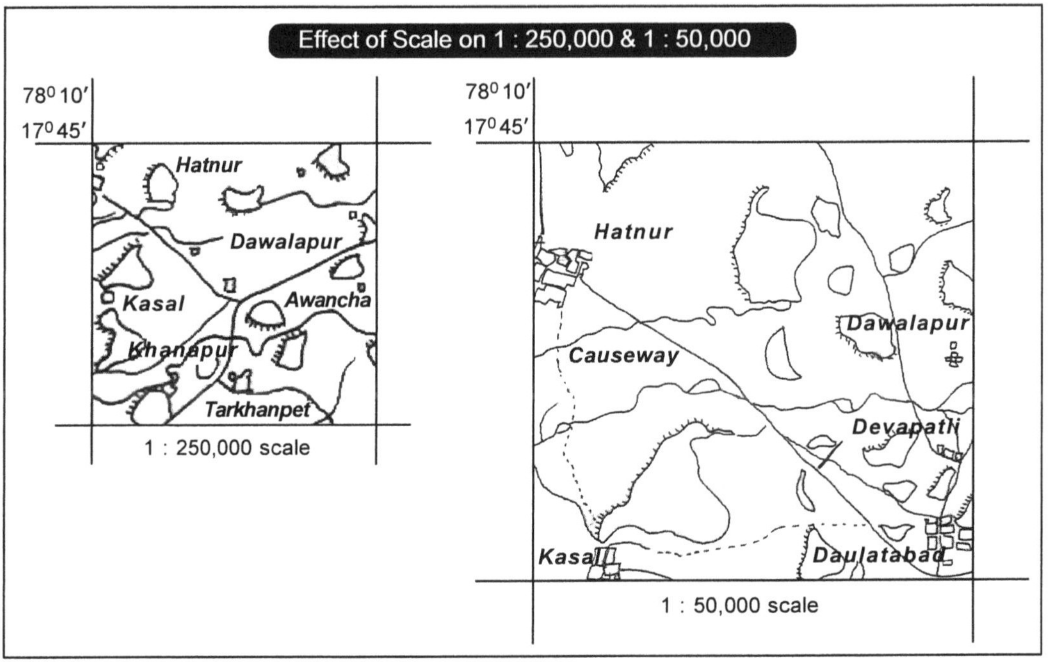

Fig. 1.5 Effect of Scale-1:250,000 vs 1:50,000

Scale values are normally written as dimensionless numbers, indicating that the measurements on the map and the earth are in the same units. A scale of 1:25,000, pronounced one to twenty five thousand, indicates that one unit of distance on a map corresponds to 25,000 of the same units on the ground. Thus, one centimeter on the map refers to 25,000 centimeters (or 250 meters) on the earth. This is exactly the same as one inch on the map corresponding to 25,000 inches (or approximately 2,080 feet) on the earth. Scale always refers to linear horizontal distances, and not measurements of area or elevation,

An explanation of the symbols used on topographic maps is printed on the bottom of each topographic sheet along the margin, and for other maps on a separate legend sheet. In general, culture (artificial works) is shown in black. All water features, such as streams, swamps and glaciers are shown in blue. Relief is shown by contours in brown. Red may be used to indicate main highways, and green overprints may be used to designate areas of woods, orchards, vineyards, and scrub.

1.5 Spatial Referencing System

The first important spatial concept in mapping technology is to locate objects with respect to some reference system. The system must have a structured mechanism to communicate the location of each object under study. The characteristics that a referencing system should possess include stability, the ability to show points, lines and areas and the ability to measure length, size and shape (Dale and McLaughlin, 1988). There are several methods of spatial referencing systems and they can be grouped into three categories, namely, geographic coordinate systems, rectangular coordinate systems and non-coordinate systems. In geographic coordinate systems, the coordinates of any location on the earth surface can be defined by latitude and longitude. Lines of longitude, called meridians are, drawn from pole to pole. The starting point for these lines called the prime meridian runs through Greenwich (Fig. 1.6).

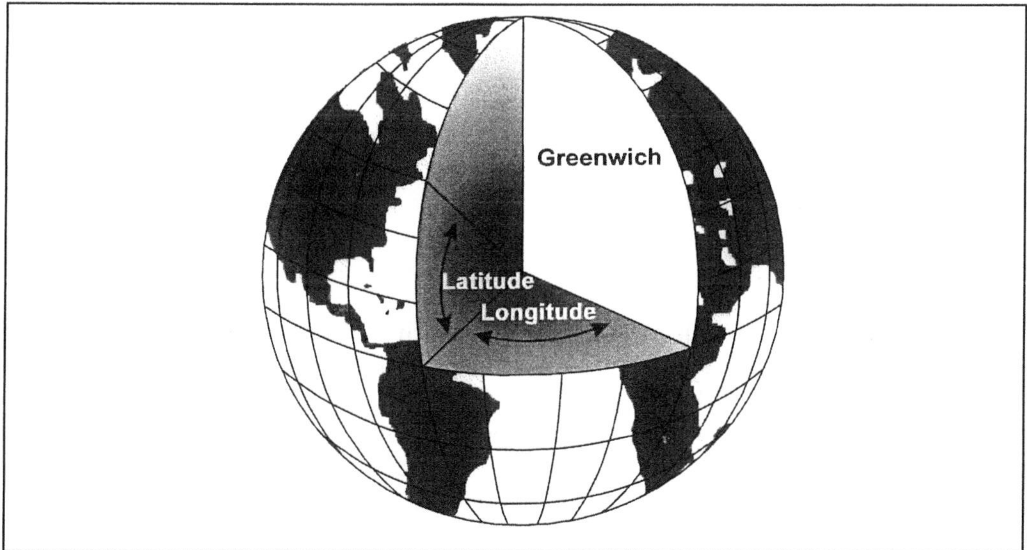

Fig. 1.6 The geographic grid.

The prime meridian is the starting or zero point for angular measurements, east and west. The lines of latitude lie at right angles to lines of longitude, and run parallel to one another. That is, each line of latitude represents the circle rounding the globe. Each circle will have a definite circumference and area depending on where it lies relative to the two poles. The circle of greatest circumference is called the equator and will be at equidistant from the poles. This type of location is called absolute location and it gives a definitive, measurable, and fixed point in space. This system is also called spherical grid system. This spherical grid system is produced by slicing the entire globe and placing two sets of imaginary lines around the earth. The first set of lines starts at the middle of the earth or equator called parallels, and circle the globe from east to west. These parallels are called latitudes. The second set of lines, called meridians, are drawn from pole to pole. These are called longitudes. Simply, it can be stated that the system of angular measurements allows us to state the absolute position of any location on the surface of the earth while calculating the degrees of latitude north or south of the equator, and the degrees of longitude east and west of the prime meridian.

Most of the spatial data available by means of remote sensing systems or any other sources of data for use in GIS are in two-dimensional form. This coordinate referencing system to locate any object point is called rectangular coordinate system. The location of any point on the earth's surface with reference to rectangular coordinate system is generally termed as relative position. The third coordinate system, namely, non-coordinate system, provides spatial references using a descriptive code rather than a coordinate, such as, postal codes, which are numeric in nature. Some countries use the non-coordinate system which is alpha numeric as in the case of UK and Canadian postal codes. This type of reference system is of particular importance for GIS users. Public land survey systems of Western United states is a classical example of this non coordinate-referencing system (Heywood et. al 2000).

For instance, in relation to prominent features a point may be referred to as being so many kilometers in a given direction from a city, mountain, lake, river mouth, or any other easily located feature on the map.

1.6 Map Projections

Map projection is a basic principle of map making in that when projected on to a flat map, objects on the earth's surface are distorted in some way, either in size, shape or in relative location (Maling, 1980). When the information is digitised from a map, the recorded locations will be often based on a rectangular coordinate system determined by the position of the map on the digitising table (star and Eastes, 1999). In order to determine the true earth locations of these digitised entities, it is necessary to devise the mathematical transformation required to convert these rectangular coordinates into the positions on the curved surface of the earth as represented on the map. Mathematical formulae to convert map units into latitude and longitude are available for most common projections (Snyder, 1987). Such mathematical transformation functions are normally built into projection as it is mathematically produced and is a two-fold process. First by, an obvious scale change converts the actual globe to a reference globe based on the desired scale. Secondly, the reference globe is mathematically projected on to the flat surface (Robinson et. al,1995). In this process of projection there is a change in scale. The representative fraction for the reference globe called the principle scale, can be calculated by dividing the earth's radius by the radius of the globe. The scale divided by the principle scale, is by definition 1.0 at every location on the reference globe. The process of transformation of three-dimensional space into a two dimensional map inevitably distorts at least one of the properties, namely, shape, area, distance or direction, and often more than one. Therefore, the scale factor will differ in different places on the map (Robinson et. al, 1995).

Map projection properties can be evaluated by means of applying three principal cartographic criteria, namely, conformity orthomorphic projections, equivalent projections, and equidistant projections (peuckar and chrisman, 1975). The projection that retains the property of maintaining correct angular correspondence can be preserved, and this is called angular conformity, conformal, or orthomorphic projection. The conformal type of projection results in distortions of areas leading to incorrect measurements. The projections by which areas can be preserved are called equal area or equivalent projections, the scale factor being equal to 1.0 mm. The projections by which the distances are preserved are known as equidistant projections. These three criteria are basic and mutually exclusive and other properties have only a peripheral importance. In fact, there is no ideal map projection, but only the best representation for a given purpose can be achieved.

A special emphasis is laid on transforming the satellite data on to a map. One of the requirements of the remotely sensed data is its ability to process an image from a generic coordinate system on to a projected coordinate system. Projecting imagery from line and pixel coordinates to the Universal Transverse Mercator (UTM) is an example in this regard. This is particularly critical when different trends of information from a Geographical Information Systems (GIS) are to be combined. The imagery must be accurately projected and rectified. By applying relevant map projections, a few basic concepts essential to the understanding of a map projections (Fig. 1.7) are reviewed in this section.

Fig. 1.7 Concept of Map Projections.

To transfer the image of the earth and its irregularities on to the plane surface of a map, three factors are involved, namely, a geoid, an ellipsoid or a datum with ellipsoid, and a projection. The geoid is a rendition of an irregular spheroidal shape. The variations in gravity are accounted for at this level. The observations made on the geoid are then transfered to a regular geometric reference surface, the ellipsoid. Many countries and organisations have calculated a variety of ellipsoids over the years. Variation in ellipsoid calculations are in part due to different observations on the geoid from different points upon the earth. The geographical relationships of the ellipsoid, still in a three-dimensional form, are transformed into two-dimensional plane of a map by a process called 'map projection' or simply projection. As illustrated in Fig. 1.8, the vast majority of projections are based upon cones, cylinders and planes. Each of these formats has advantages and disadvantages in terms of distortions and accuracy. Every flat map misrepresents the surface of the earth in some way. No map can rival a globe in truly representing the surface of the entire earth. However , a map or parts of a map can show one or more, but never all-of the following: True shapes, true directions, true distances, true areas.

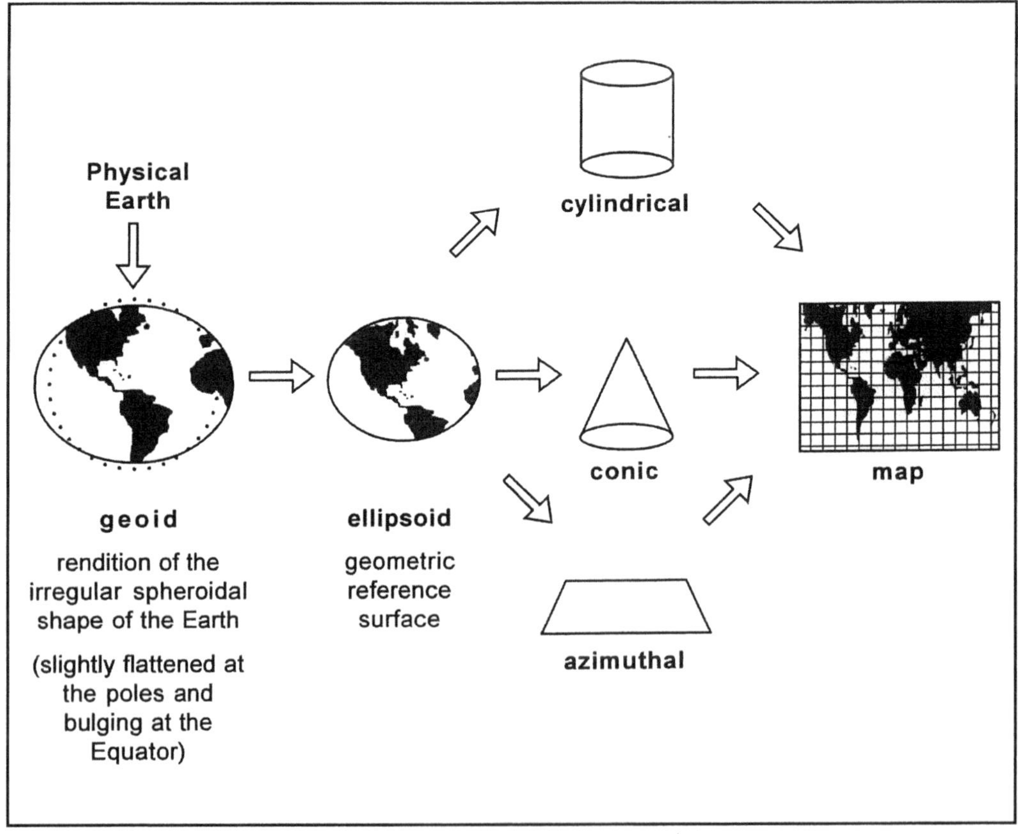

Fig 1.8 Geoid Ellipsoid - Projection : Relationship.

1.6.1 Grouping of Map Projections

All the map projections are grouped into four main families. They are,

(i) the family of planar projections

(ii) the family of cylindrical projections

(iii) the family of conical projections, and

(iv) the family of azimuthal projections.

If we wrap a sheet of paper round the globe in the form of a cylinder, transfer the geographical features of the globe on to it, and then unroll the sheet and lay of on a flat surface, we would achieve a cylindrical projection. The resulting graticule would be rectangular. In conic projection, if we repeat the above process, by wrapping the sheet of paper round the globe in the form of a cone, the resulting graticule would be fan shaped. The cone can be either tangent to a chosen parallel or it may intersect the plane along two parallels. Conic projections are specially suited to mapping areas having east-west extents, such as, Canada, USA and China. If a sheet of paper is laid tangent to a point on the globe and transfer the geographical features of the globe on to it, we would achieve azimuthal projections that appear as straight lines intersecting at the designated centre point, and parallels that appear as concentric circles round the centre point. A combination of any two of the above projections forms an hybrid projection.

The classification of map projections should follow a standard pattern so that any regular projection can be described by a set of criteria, and, conversely, a set of criteria will define a regular projection. Thus a classification scheme may follow a number of criteria subdivided into classes. Adler (1968) have named five basic criteria, as follows :

(i) Nature of the projection surface as defined by geometry,

(ii) Coincidence or contact of the projection surface with the datum surface,

(iii) Position or alignment of the projection surface with relation to the datum surface,

(iv) Properties of cartographic requirements, and

(v) Mode of generation of datum surface and coordinate systems.

The datum surface of the earth is usually an ellipsoid of revolution, but sometimes it is also approximated by a sphere. Although this assumption that the earth is a sphere can be used for small-scale maps to maintain the accuracy, for large scale maps the earth must be treated as a spheroid. Coordinate systems are necessary for the expression of position of points upon the surface, be it on an ellipsoid or a sphere or a plane. For the ellipsoid or the sphere the system of longitude and latitude is expressed in degrees, minutes and seconds of arc. For the plane a system of rectangular X and Y coordinates, sometimes referred to as Northings and Eastings, is usually applicable. Some of the commonly used map projections are described below.

1.7 Commonly used Map Projections and their Comparison

Though several conventional and non conventional map projections exist and are used, only a few projections with specific advantages are considered here.

1.7.1 Mercator

This is used for navigation for maps of equatorial regions. Any straight line on map is a thumb line (line of constant direction). Directions along a thumb line are true between any two points on a map, but a thumb line usually is not the shortest distance between points. Distances are true only along equator and are reasonably correct. Special scales can be used to measure distances along other parallels. Two particular parallels can be made correct in scale instead of the equator. Areas and shapes of large areas are distorted. Distortion increases as distance increases from the equator and is extreme in polar regions. The map, however, is conformal in that angles and shapes within any small area (such as that shown by a USGS topographic map) are essentially true.

1.7.2 Transverse Mercator

This is used by USGS for many quadrangle maps at scales from 1:24,000 to 1:250,000. Such maps can be joined at their edges only if they are in the same zone with one central meridian. Transverse Mercator is also used for mapping large areas that are mainly north-south in extent. Distances are true only along the central meridian selected by the map maker, or else along two lines parallel to it, and all distances, directions, shapes, and areas are reasonably accurate. Distortion of distances, directions, and size or area increases rapidly outside the defined distance. Since the map is conformal, shapes and angles within any small area are essentially true. Graticule spacing increases away from central meridian. Equator is straight. Other parallels are complex curves concave towards the nearest pole. The Central meridian and each meridian 90^0 from it are straight. Other meridians are complex curves concave toward central meridian.

This projection is a transverse cylindrical case, in which the scale will be kept exact along the central meridian and along the equator. The mapped area may be extended without limit in the north-south direction. This projection is also an orthomorphic projection with small shapes and angles maintained accurately. The scale distortions are systematic and can be predetermined.

1.7.3 Oblique Mercator

This is used to show regions along a great circle other than the equator or a meridian. These regions have their general extent oblique to the equator. This kind of map can be made to show as a straight line, the shortest distance between any two pre-selected points along the selected great circle.

Distances are true only along the great circle (the line of tangency for this projection), or along two lines parallel to it. Distances, directions, areas, and shapes are fairly accurate within 15^0 of the great circle. Distortion of areas, distances, and shapes increases away from the great circle. It is excessive toward the edges of a world map except near the path of the great circle. The map is conformal, but not perspective, of equal area, or equidistant. Graticule spacing increases away from the great circle, but conformity is retained. Both poles can be shown. Equator and other parallels are complex curves concave towards the nearest pole. Two meridians 180 apart are straight lines, all others being complex curves concave towards the great circle.

1.7.4 Polyconic Projection

This projection has generally been accepted as the best for a small, regular shaped area, such as, the standard quadrangle. Survey of India uses this projection for making topographical maps of Scale 1:250,000 and more. Although this projection is not conformal, the scale is not uniform, shapes and areas not being retained exactly. It comes closer to compliance with most of these projections. It cannot be used on a large area without noticeable distortion, and although two or three adjacent sheets can be matched continuously in one direction, two or more strips cannot be matched for any great distance without developing gaps along the abuting edges.

1.7.5 Lambert Conical Orthomorphic Projection

This projection portrays a portion of the earth's surface on the developed surface of a secant cone. It is used along the parallel of latitude at orthomorphic projection with two standard parallels by countries having predominant east-west directions for topographical mapping. In India we use this projection for geographical maps. When using two standard parallels, in the area between these parallels, the map scale will be too small and in the area outside the parallels it will be too great. The mapped area may be extended without limit in the east-west direction, but restricted within narrow limits in the north-south direction. The principal advantages of this projection is that the scale distortions are systematic and can be predetermined, and that a map sheet of any portion of a zone can be matched perfectly with an adjoining sheet in the same zone. However, a map of one zone cannot be matched with a map of an adjoining zone.

1.8 Grid Systems

Though the map projections discussed in the foregoing paragraphs are generally used by the National Mapping Agencies for preparation of various kinds of maps and charts, grid maps are also used mainly in large scale maps. A Grid system has certain advantages over the use of geographical coordinates. Firstly, every grid square is of the same size and shape. Secondly, linear values can be used rather than angular values. A rectangular grid system is required for the army, and for engineering and cadastral maps, as the computations of reference azimuth and survey coordinates distance can be easily done in this system, and a brief, precise, unique and convenient map references can be given. Two of the important grid systems that are essential to our studies are Lambert Grid system used in India and Universal Transverse Mercator (UTM) coordinate system used in more than a hundred countries the world over.

1.8.1 Lambert Grid System for India

The grid system adopted for India is the Lambert Grid system. The grid squares in this system can be superimposed on any conformal projection. However, it is only on the Lambert conformal conical projection that they represent equal and perfect squares. On the polyconic maps of Survey of India, the grid squares appear as squares, but they may differ by as much as 1 in 824.

A series of eight Lambert Grids cover India and Burma which are named as I, IIA, IIB, IIIA, IIIB, IVA, IVB and O. The N -S extent of the grid is limited to 8 degrees in order to limit the maximum scale error to 1 in 824 and the E - W extent limited to 16 degrees owing to limitations on the grid convergence -- the deviation of the grid north from true (geographical) north. The grid origins are spaced 7 degrees apart in latitude with 1 degree overlap at borders, and are assigned arbitrary grid coordinates of 1 million grid meters north and 3 million grid meters East. The scale factor is 823/824 along the central parallel and 824/823 along the edges. (Fig. 1.9)

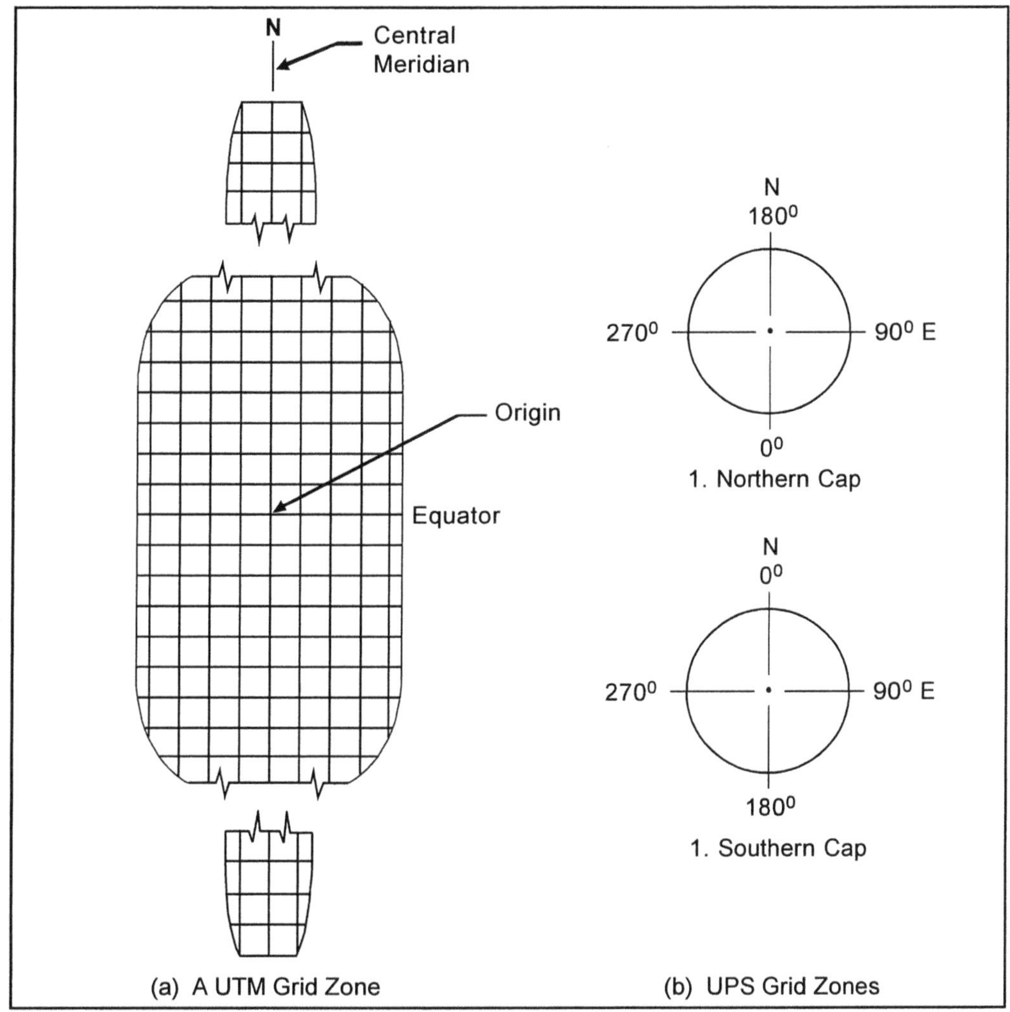

Fig 1.9 Coordinate system for UTM grid.

1.8.2 Universal Transverse Mercator (UTM) Grid

For the UTM grid the world is divided into 60 north to south zones. Each zone covers a 6^0 - wide strip of Longitude. The maximum extent of the zone was chosen to minimise distortion. The zones are numbered consecutively, beginning at zone 1 between 180^0 and 174^0 west to zone 60 between 174^0 and 180^0 east longitude. Each zone is then divided into 19 segments with an 8^0 difference in latitude plus an additional segment at the extreme north with a 12^0 difference in latitude. The rows of these segments are lettered from south to north by the letters C through X (with omission of the letters I and O to avoid confusion). By specifying a letter and a number, each element in the UTM system is uniquely identified.

Within each zone , the location of a point is made by specifying its X coordinate, the easting and its Y coordinate, the northing. The unit of the coordinates is meters. The central meridian of each zone is assigned the false easting value of 500,000 m. This is done to avoid negative numbers at the western edge of the zone. The easting numbers increase from west to east. For north-south values in the north hemisphere, the equator has a false northing value of 0 m, and the northing decrease towards the south pole. The coordinate system for the UTM grid is illustrated in (Fig.1.9). Of course, the scale of the grid must be consistent with the scale of the base transverse mercator projection.

For the UTM grid the mapping cylinder is secant to the earth. The local scale factor for the central meridian, halfway between zone boundary meridians is 0.9996. This limits maximum variations to 1 in 1000. No less than 100 countries in the world have accepted this projection, as they must be finding it very satisfactory for both mapping and rectangular referencing. A few of the advantage are :

(i) It is a conformal projection and hence the direction errors are minimal within a selected UTM zone.

(ii) It has continuity over sizable areas coupled with a minimal number of zones. In case of India it involves only 6 UTM zones.

(iii) Scale errors caused by the projections do not exceed the specified tolerance, that is, 1/2,500.

(iv) Unique referencing is possible in a plane coordinate system for all zones.

(v) Transformation formulae from one zone to another are uniform throughout the system (assuming one reference ellipsoid). This helps in easy programming for the computer eliminating the need for auxillary tables based on manual computations.

(vi) Meridionial convergence does not exceed five degrees.

In addition to the above quoted advantages, modern satellite based-geodetic measurement techniques like Global Positioning System provide directly the coordinate in UTM projection avoiding the hustles of conversion from one coordinate system to another. In the area of digital mapping also, most of the GIS software today handles the cartographic data in a UTM projection with the possibility of converting the coordinates from one projection system to another.

The basic disadvantage of the UTM lies in the maximum projections scale distortion which reaches 4 parts in 10,000. The main reason for this is the choice of zone width, 6 degrees in the case of UTM.

1.9 Computer in Map Production

By 1977, however, the experience of using computers in map making considerably advanced. Rhind (1977) was able to present a list of reasons for using computers in the process of making maps. These reasons are listed below.

(i) To make existing maps more at a quicker pace.

(ii) To make existing maps at a cheaper rate.

(iii) To make maps for specific user needs.

(iv) To make map production possible in situations where skilled staff are unavailable.

(v) To allow experimentation with different graphical representations of the same data.

(vi) To facilitate map making and updating when the data are already in digital form.

(vii) To facilitate analysis of data that demand interaction between statistical analysis and mapping .

(viii) To minimize the use of the printed map as a data store and thereby to minimize the effects of classification and generalization on the quality of the data.

(ix) To create maps like the 3-D type.

(x) To create maps in which selection and generalization procedures are explicitly defined and consistently executed.

(xi) Introduction of automation can lead to a review of the whole map-making process, which can also lead to savings and improvements.

During the 1960's and 1970's there were two main trends in the application of computer methods for mapping. One was the automation of existing tasks, with an accent on cartographic accuracy and visual quality, and the other with the accent on spatial analysis but at the expense of good graphical results.

1.10 Digital Database in a GIS

The essential component of a GIS is its information database. This database is developed by capturing information from different sources like topographical maps, thematic maps, and cadastral maps, in analog form, remotely sensed images in digital form and so on. Whatever may be the source of information, they need to be integrated on to a common projection and registration.

The general technique of converting the analog map to digital form is either by manual digitisation or by scanning. Manual digitisation is the commonly used technique for including analog data in a digital database. The use of remotely sensed data which is available in digital form for maintaining the dynamicity of the database for GIS, has become very common. Map projection aspects involved in these two techniques are discussed in the following paragraphs.

1.10.1 Digitiser Units Vs Real-world Coordinates

After determining the projection of the data source and deciding upon a projection in which to store the database, we need to project data into a real-world coordinate system. Registering all layers to common coordinate system ensures data integrity during spatial joins.

Coverages may be digitised in digitiser units or real-world units. The digitiser is based on a type of rectangular coordinate system with its origin in the lower-left corner. On the digitiser surface, moving one inch up or down covers the same distance as moving one inch left or right. Anywhere on the table's surface, an inch is an inch. Whether the unit of measurement be inches or centimeters, when associated with digitising, it is called a digitiser unit.

Location and distance are the key features in mapping geographic information. The real world has a curved surface and is often measured in feet or meters. Both feet and meters are standard, and the system of spherical coordinates (Global Reference System) is used to reference system location on the surface of the globe with points on a map. The units of this reference system are degrees of latitude and longitude. But the distance represented by a degree depends upon its location on the globe which is the global reference system and not a rectangular coordinate system. Thus, a coverage can be digitised in meters, but not in degrees. The advantages and disadvantages of digitising a coverage in either digitiser units or in real-world units are outlined in Table 1.1.

Table 1.1 Advantages and disadvantages of digitiser units and Real-world units

S.No.	Digitiser units	Real-world units
(i)	Easy to create edit plots at scale of source map	Maps need to be plotted at a precise scale to overlay edit plots
(ii)	Digitising staff has less to learn and understand	Digitising staff should understand transformation and projection concepts
(iii)	Coverages are not spatially referenced and cannot be displayed simultaneously	Allow multiple coverages to be shown, such as background or adjacent coverages.
(iv)	Inconvenient for update	Usually used for update
(v)	Less concern over bad information	Must have correct projection parameters
(vi)	Don't know whether initial digitising of tics was accurate	RMS error indicates actual tic accuracy in real-world units.

A common step in quality assurance when initially developing a database is to compare a digitised file with its source map. This is most commonly done by creating an edit plot and overlaying it on a light table. If a map is digitised in real-world units, it may have been stretched and scaled so that it will no longer register accurately with the source map even if the file was digitised accurately. If coverages are to be digitised in real-world units, the digitising staff should understand how to project and transform coverages. This naturally requires some knowledge of projection concepts.

The ability to display adjacent and background coverage can be quite helpful, particularly for updating maps. This is not possible when maps are digitised in digitiser units. Coverages should not be edited, cleaned, built, or buffered, nor should any spatial analysis be performed when they are stored in reference units (latitude-longitude). The algorithms that perform snapping functions using a measurement of length or area are based upon Cartesian coordinates. The length of a line of latitude between two meridians varies with latitude , and the area is confusing when measured in square degrees.

1.11 Linkage of GIS to Remote Sensing

This book aims to provide an introduction to the theoretical and technical principles that need to be understood to work effectively and critically with GIS. Today maps are not just made using GIS, but the infrastructure of utilities in the streets of our towns will be held in a GIS. Your taxis and emergency services may be guided to their destination using satellite-linked spatial systems; the foresters and farmers will be monitoring their standing crops with spatial information systems; natural resources management developments, and environmental management strategies may be compared with the integration of satellite data while processing results. To use GIS technology, for the above allied application areas, a huge GIS database on geographical features is to be created. Creating such a database is a complex operation which may involve data capture, verification and structuring process. Because raw geographical data are available in many different analogue or digital forms like maps, aerial photographs, and satellite images, a spatial database can be built in several, not mutually exclusive ways, such as acquiring data in digital form from a data supplier, digitising existing analogue data, carrying out field survey of geographic entities, and interpolating from point observations to continuous surfaces.

Images derived from optical and digital remote sensing systems mounted in aircraft and satellites provide much spatial information and major data as an input to GIS. Remote sensing data are a major source of data for the mapping of resources like geology, forestry, water resources, land use and land cover. Integration of the two technologies, remote sensing and GIS, can be used to develop decision support systems for a planner or decision maker. Remotely sensed images can be used for two purposes, as a source of spatial data withinGIS and using the functionality of GIS in processing remotely sensed data in both pictorial and digital modes.

Since digital remote sensing images are collected in a raster format, digital images are inherently compatible spatially with other sources of information in a raster domain. Because of this, "raw" images can be directly and easily included as layers in a raster-based GIS. Similarly, such image processing procedures as automated land cover classification result in the creation of interpreted or derived data files in a raster format. These derived data are again inherently compatible with the other sources of data represented in a raster format.

Remote sensing images need not be digital in format to be of any value in a GIS environment. Visual interpretation of hardcopy images is used extensively to locate specific features and conditions, which are then subsequently geocoded for inclusion in a GIS. At the same time, the information resident in a GIS can also be used to aid in a visual or digital image interpretation process. For example, GIS information on elevation, slope, and aspect might be used to aid in the classification of forest types appearing in images acquired over areas of high relief. Thus, the interaction between remote sensing and GIS techniques is two-way in nature.

Remote sensing images including the information extracted from such images, along with GPS data, have become primary data sources for modern GIS. Indeed, the boundaries between remote sensing, GIS, and GPS technology have become blurred, and these combined fields will continue to revolutionise the inventory, monitoring, and managing natural resources on a day-to-day basis. Similarly, these technologies are assisting us in modeling and understanding biophysical process at all scales. They are also permitting us to develop and communicate cause-and-effect "what-if" scenarios in a spatial context in ways never before possible.

The importance of remote sensing, GIS, GPS, and related information technologies in the professional careers of today's students involved in measuring, studying, and managing earth resources cannot be over-stated.

Hence, in recent years, remote sensing has become a powerful source of spatial data as an input for GIS through which a detailed map can be generated with the help of other collateral data derived from several other sources. There are, two methods of extracting data for GIS from the remote sensing data. They are, visual interpretation of satellite imageries in pictorial format, and computer processing of remotely sensed digital data. The output of either of these analysis methods can be considered an input for GIS for any kind of application. Fig. 1.10 shows an overview of the linkage of remote sensing and GIS.

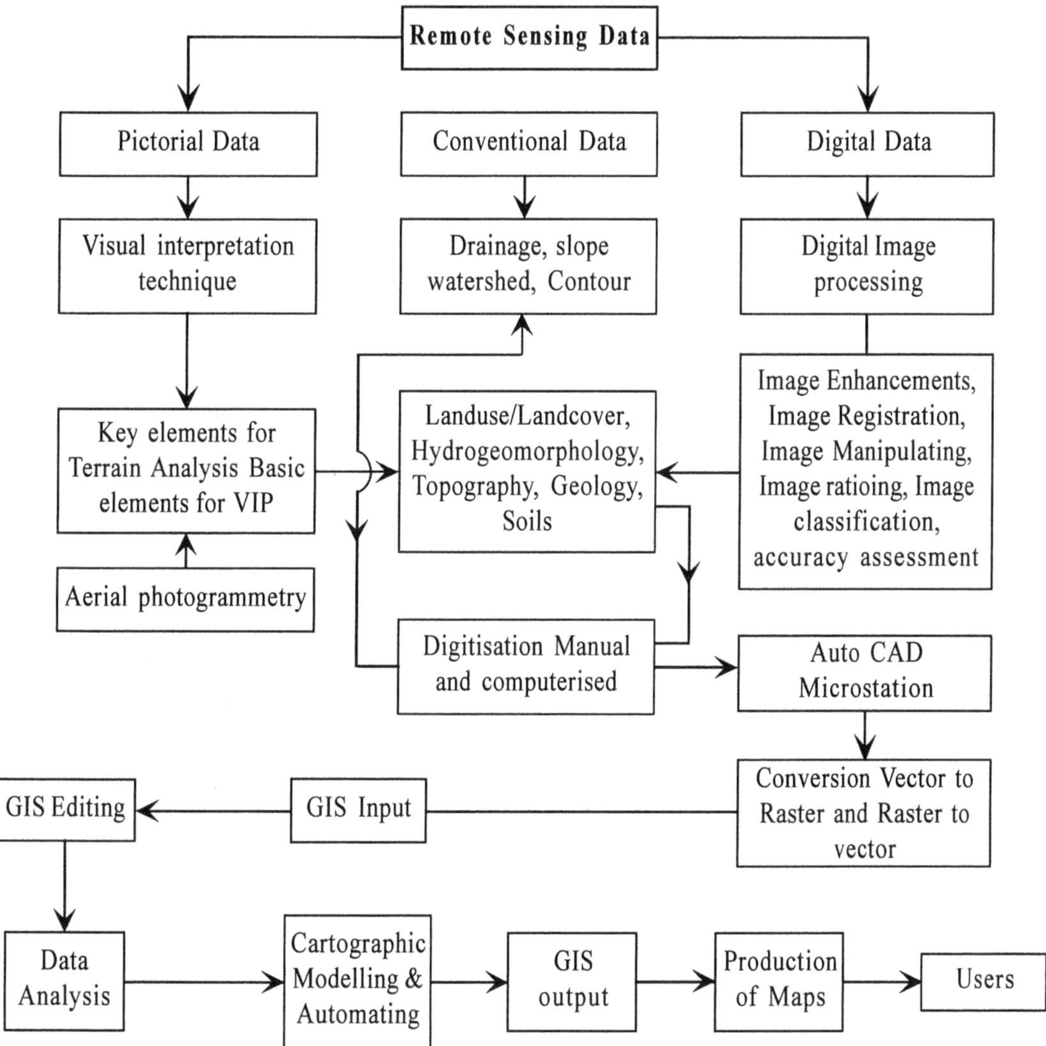

Fig 1.10 Overview of the linkage of remote sensing and GIS.

2

Photogrammetry

2.1 Introduction

Photogrammetry is defined by American Society of Photogrammetric Engineering and Remote Sensing (ASPE & RS) as an art, science and technology of obtaining reliable information about objects, environment through the process of recording, measuring and interpreting photographic images and patterns of recorded electromagnetic radiation through a device. The device is a camera for acquiring a photograph. The science or art of obtaining reliable measurements is by means of photographs. Photogrammetry is frequently decided based on the type of photographs or images used, location of the camera from which they have taken and the purpose of using these photographs. Accordingly, in general, four types of photogrammetry are available today, to the resource scientists, engineers and managers. They are 1. Terrestrial Photogrammetry, 2. Close Range Photogrammetry, 3. Aerial or Space Photogrammetry and (4) Stereo Photogrammetry. The terrestrial photogrammetry deals with the photographs taken from the ground surface as platform for the photographic instruments, like camera. In this photographic system, the distance between the object and the photographer should be 300 mts or more than 300 mts. The terrestrial photogrammetry is called as close range photogrammetry if this distance is less than 300 mts.

Aerial photogrammetry deals with the photographs taken by an aerial camera onboard aircraft. Aerial or space photogrammetry connotes the use of traditional analog photographs (Colwell, 1997) or digital photographs (Hinz 1999) that have been taken from any airborne/space borne platform. These photographs may be either vertical or oblique images based on its geometry and characteristics. The geometry of this type of photographic system is illustrated in Fig. 2.1.

Stereophotogrammetry deals with the overlapping pairs of photographs are obtained, measured and interpreted in a stereoscopic viewing device. This provides a three dimensional view and creates the illusion that the observer or photogrammetrist is viewing a relief model of the terrain. Stereophotogrammetry is essential for creating a digital elevation model of real world by measuring the elevations of 'h - measurements'. This can be performed by using specially designed instruments like stereometer or by computers digitally.

In practice, airborne/space borne photogrammetry is embraced into two broad categories, namely, metric photogrammetry and photo interpretation. In the category of metric photogrammetry, the quantitative measurements are made, from which ground co-ordinates, elevations, distances, areas, volumes and angles can be computed. This is also used to prepare planimetric or topographic maps. This category is often called as quantitative photogrammetry. The most common applications are, (1) The production of topographic maps that contain contours, transportation network and other baseline information of the terrain, (2) The extraction of digital elevation model, (3) The 3-D extraction of planimetric details and contours, (4) Production of orthophoto images.

The second category is photo interpretation in which the photographs are evaluated or interpreted for the extraction of qualitative data like water pollution, classification of soils, geological formations, crops identification and military intelligence etc. (Colwell 1982). This type of photogrammetry is also called as interpretative photogrammetry. Photo-interpretation coupled with fundamental photogrammetric analysis results in the creation of thematic maps of landuse/landcover, hydrogeomorphological and biophysical information maps, which can be used as data sources and data input to GIS.

2.2 Brief history of Photogrammetry

As early as in eighteen century itself the principles of perspective for preparation of maps was suggested by J. H. Lambert. Terrestrial photographs were tried and used for topographic mapping from 1859 to 1909, in France, Canada and Germany. The airplane was first used for obtaining aerial photographs for mapping in 1913.

Instrumentation and techniques contributed during nineth and twentyth century has made photogrammetry the most accurate, efficient and advantageous tool for preparation of topographic maps and also for other applications. With the advances in electronics and computer science, the new concept in photogrammetry called

"Soft copy photogrammetry" or " Digital photogrammetry" has already proven its efficiency in practice. The age old classical analogue and analytical photogrammetric plotters are replaced by computer based "Digital Photogrammetric Work Stations".

2.3 Classification of Aerial Photographs

Broadly, photographs are classified based on the nature of camera and the platform on which the camera is located into two types : terrestrial and aerial photographs.

Terrestrial photographs are the photographs taken with ground based cameras. Camera axis is horizontal photo theodolites are used for taking terrestrial photographs.

Aerial photographs are those photographs taken from an air craft. Camera axis in this case is vertical.

In this chapter, aerial photogrammetric principles are presented as it is more relevant to geoinformatics.

Aerial photographs taken by an aerial camera on board aircraft can be further classified into two categories, namely, vertical and oblique. The vertical photograph for which the camera axis is truly vertical. It can also happen that the camera axis may not remain truly vertical : Normal tilt for preparation of topographical maps can vary from 1^0 to 3^0 such photographs are called nearly vertical. In other words, if the axis of the aerial camera makes an angle 1^0 to 3^0 with the plumbline, such photographs are called as vertical photographs.

In the case of oblique photography, the camera axis is intentionally tilted from the vertical. In this category, there are two types of oblique photographs, namely high oblique and low oblique.

If the camera axis is tilted in such a way that horizon appears on the photograph, such photograph is called as high oblique photograph.

If the amount of tilt is made in such a way that horizon does not appear on photograph, such photograph is called as low oblique photograph. Fig. 2.1 shows the schematic representation of classification of aerial photograph. Fig. 2.2 shows the types of aerial photographs.

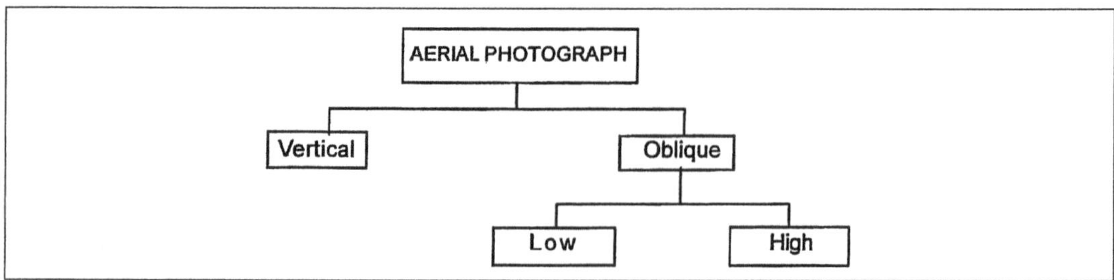

Fig. 2.1 Classification of Aerial Photograph

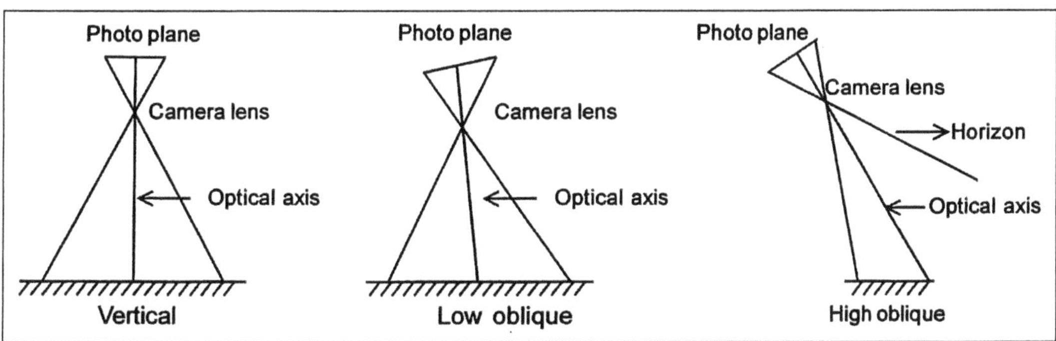

Fig. 2.2 Vertical, Low oblique and High Oblique Photographs

2.4 Organisation of Aerial Photography

Aerial Photography covering a particular area is organised as follows for mapping purpose (Fig. 2.3, Fig. 2.4 and Fig. 2.5).

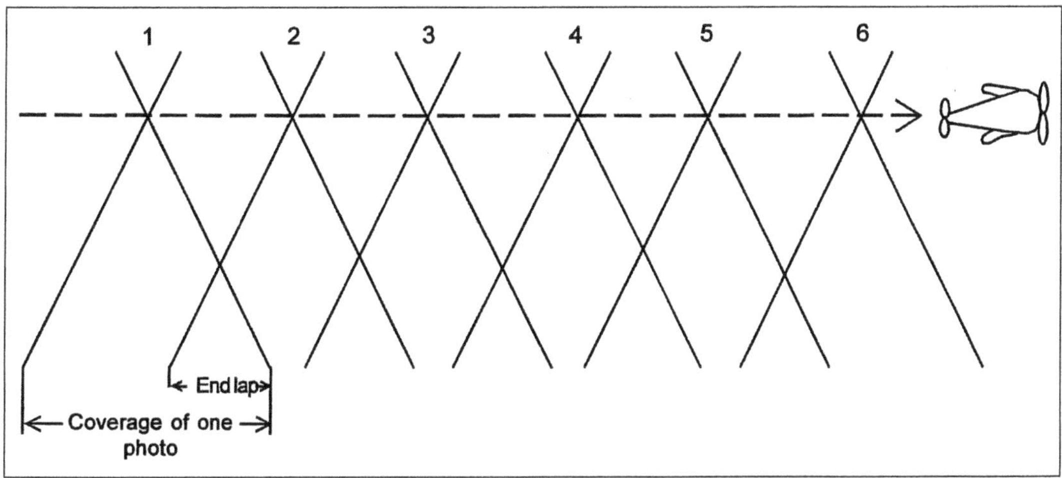

Fig. 2.3 Overlap between two photographs (End Lap)

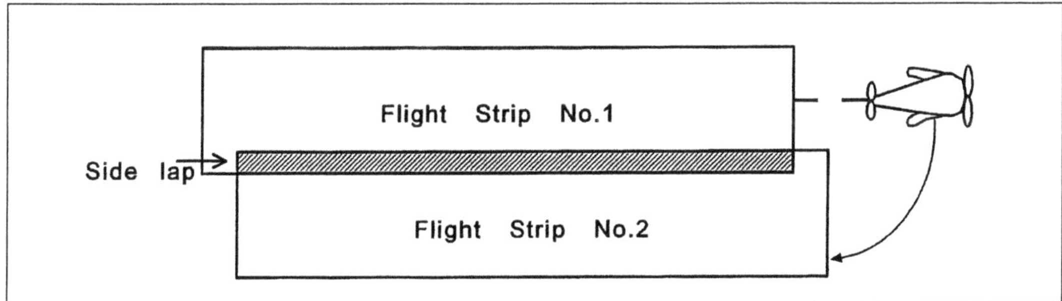

Fig. 2.4 Over lap between two strips - Side Lap

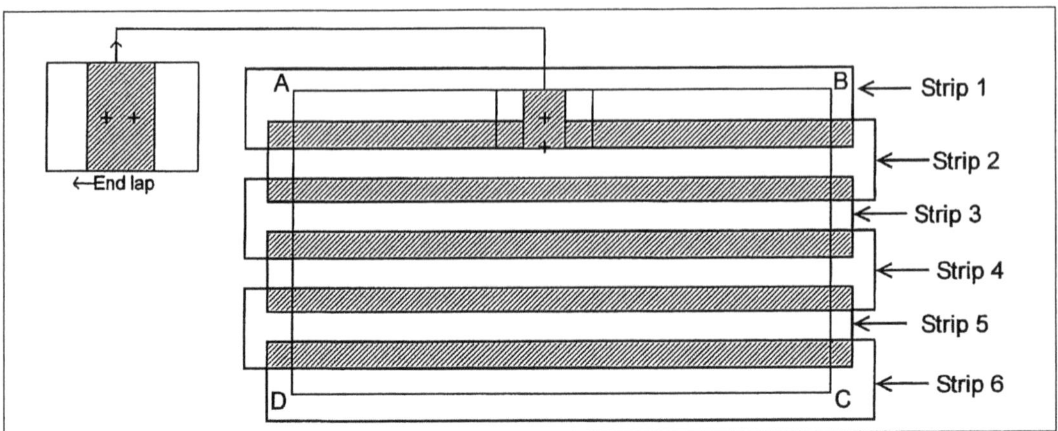

Fig. 2.5 Aerial Photography Block Covering Area ABCD consisting of 6 Strips

Aerial Photograph : One photograph taken from the air at one specified exposure station by camera fitted to air craft.

Aerial Photography STRIP : Number of aerial Photographs along one flight line with specified over laps between the adjoining photographs, normally for the purpose of mapping the area. The over lapping area between the adjoining photographs is kept at 60% or more for mapping. This over lap between adjoining photographs in a strip in called end lap.

Aerial Photography BLOCK : Number of strips covering an area is called block. For the purpose of mapping overlap between each strip is kept at 25% to 30%. This over lap between each strip is called side lap.

2.5 Aerial Cameras

Aerial camera is one of the most important instruments in photogrammetry. Aerial camera is used to obtain aerial photography upon which photogrammetry depends.

Requirements of Aerial Camera are :

- Lens of high geometric quality
- Capable of exposing in rapid succession a great number of photographs to exacting specifications.
- Short cycling time, fast lenses and efficient shutters.
- Capable of functioning under most extreme weather conditions, in spite of air craft vibrations.
- Uses of roll films-capacities ranging from 200 to 400 feet.

Camera mount is a device is used to attach camera to aircraft. The camera mount is designed to carryout the three functions, They are (i) to prevent aircraft vibrations from being transmitted to the camera (ii) to correct the crab and (iii) keeping properly oriented for aerial photos.

2.5.1 Types of Cameras

Broadly aerial cameras are classified in three types.

They are Single lens frame camera, Multi lens frame camera and Convergent camera. Scope of discussions will be single lens frame camera, as they are extensively used for obtaining photographs for mapping purpose.

The features of single lens cameras are related to lens assembly and format of the photographs.

Lens is held fixed to the focal plane to preserve the geometric property and entire format of the film is exposed simultaneously with single click of shutter. The single lens frame cameras are classified into 4 types based on the angular field of view which is an angle subtend by the focal plane of the camera system. If the angular field of view (Fig. 2.6) is less than 60^0, than such cameras are called as narrow angle cameras. The normal angle cameras are those cameras whose angular field of view is upto 75^0. If the angular field of view is from 75^0 to 100^0, then such cameras are wide angle. The last category is super wide angle cameras for which the angular field of view is more than 100^0, some brands of single lens cameras are

RMK 15/23 : f = 150 mm, format 23 × 23cm (f = focal length of the lens).

RC 10 : f = 90 mm, 150 mm and 300 mm format 23 × 23 cm

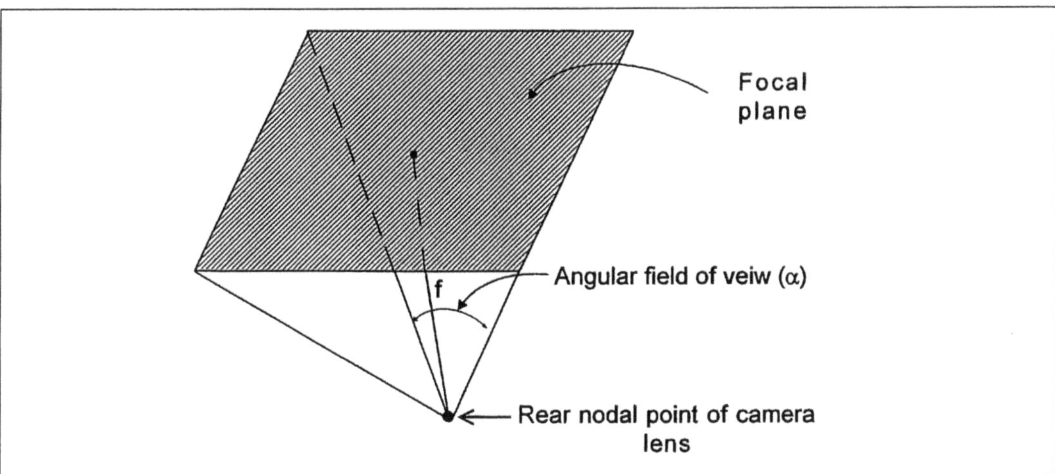

Fig. 2.6 Showing angular field of view (α).

2.5.2 Components of Aerial Camera

All cameras which are used for aerial photography required for mapping purposes, are the metric cameras. The format size, that is, the actual area covered by photographic emulsion and focal length are known as metric cameras. The basic components or assemblies of a frame metric aerial camera are the magazine, the camera body and the lens cone assembly (Fig. 2.7). The functions of these components in aerial photography are described below.

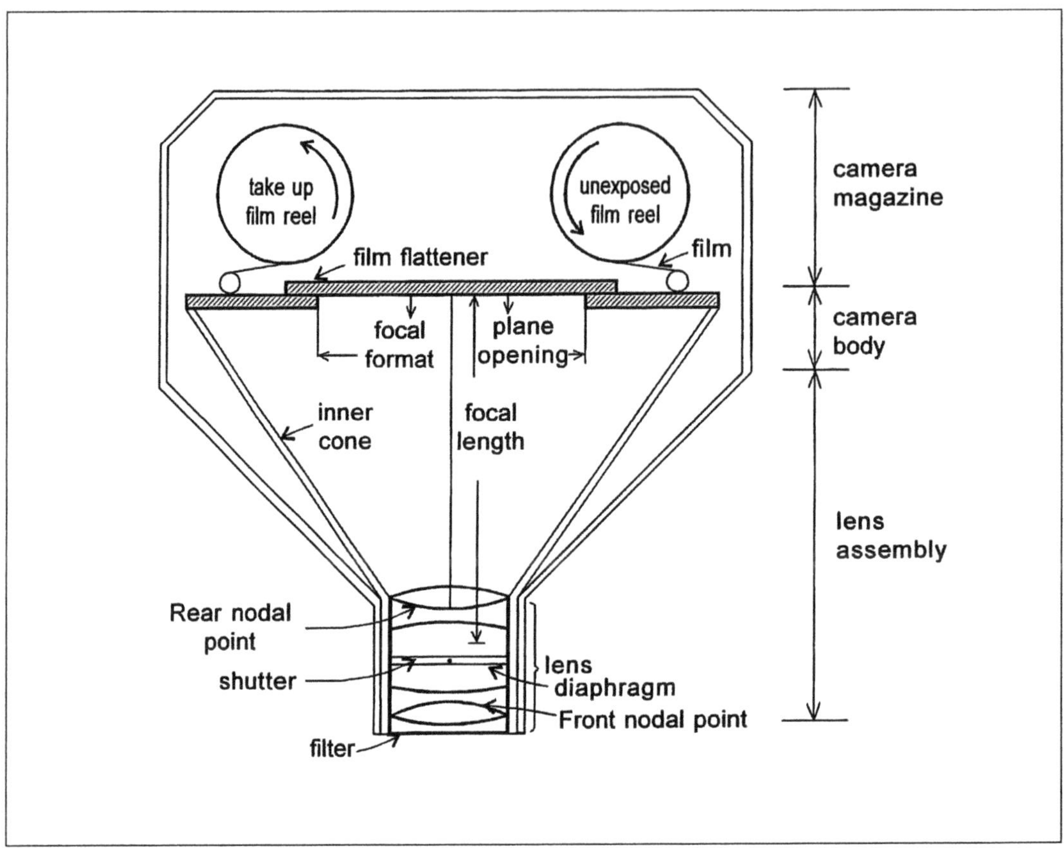

Fig. 2.7 Generalised cross section of a frame aerial mapping camera.

2.5.3 Camera Magazine

The camera magazine houses the reels which hold exposed film and unexposed film contains the film - advancing and film flattening mechanisms. Film flattening may be accomplished in any of the following ways.

(i) By applying tension to the film during exposure.

(ii) By pressing the film firmly against a flat focal - plane glass, which lies in front of the film.

(iii) By applying air pressure into the air tight camera cone, there by film will flat leasing behind the focal plane.

(iv) By drawing the film tightly up against a vaccum plate whose surface lies in the focal plane.

2.5.4 Camera Body

Camera body contains the drive mechanism. Drive mechanism operates the camera through it cycle. The cycle consisting of

(i) Advancing the film
(ii) Flattening the film
(iii) Cocking the shutter
(iv) Tripping the shutter.

Power for the drive mechanism provided by an electric motor.

2.5.5 Lens Cone Assembly

It contains a number of parts and serves several functions. The parts are lens, filter, shutter and diaphragm. It also contains an inner cone of spider.

The camera lens gathers light rays from the object space and brings them to focus in the focal plane behind the lens. The filter serves three purposes, (i) Introduces the effect of atmospheric haze, (ii) It helps to provide light distribution over the entire format, (iii) It protects the lens from damage and dust. The shutter and diaphragm together regulates the length of time, a given amount of light is allowed to pass through the lens to make the exposure. The shutter controls the length of time that light is permitted to pass through the lens. The diaphragm regulates the f - stops of the camera by varying the size of the aperture to control the amount of light passing through the lens.

Camera Mount : The function of camera mount is to absorb the vibration set-up by the aeroplane and maintain vertically of the optic axis at all times during photographic operation. The amount of exposure controls the density of the negative and the quality of details. The recording instrument consists of an assembly of instruments to record relevant data on one of the margins of the negative at the time of exposure.

2.5.6 Resolution of a Lens System

The resolution of a lens is an important characteristic feature of aerial photographic system. The resolution or resolving power of lens is defined as the ability of the lens to show detail. One common method of measuring lens resolution is to count the number of line pairs, i.e., black lines separated by white line spaces of equal thickness which can be distinguished within a width of 1mm in aerial image produced by the lens.

Good resolution is important in aerial cameras because photo images must be sharp and clearly defined for precise measurements and accurate interpretative work. Photographic resolution is not just a function of the lens but also depends on film quality and processing.

Resolution is influenced by a host of parameters, such as the resolving power of the film and camera lens used to obtain an image, any uncompensated image exposure and the atmospheric conditions present at the time of image exposure. An alternative method of determining the film resolution is Modulation Transfer Function (MTF).

The focal length (f) of a lens is the distance from the centre of the lens and the image plane that is the distance from the lens at which parallel light rays are focused to a point.

It is not applicable to a photographic lens. In this, distance is taken from the rear nodal point to the focal plane. For an aerial camera the focal length is defined as the distance between the centre of the lens system and the film plane. It is at this fixed distance away from the camera cone to the focus on the film. For mapping purposes, 152 mm focal length lenses are widely used. Longer focal lengths are used for very high altitude applications. The focal length of a lens, in other words can be defined as the intercept on the optical axis between the rear node of lens and the point at which a sharp image of a distant object is formed. The calibrated focal length of lens is adjusted so as to distribute the effect of lens distortion over the entire field used in aerial camber. This produces an overall mean distribution of radial lens distortion. It is also called as principle distance. The principal distance is the perpendicular distance from the internal perspective centre to the plane of finished negative or print.

2.5.7 Format of a Film

Format of a film is defined as the measurements of a film for one exposure. Various sizes of formats are available such as 70mm by 70mm, 140mm by 140mm or 180 by 180mm. These are used in the earlier cameras but now the format size in most modern cameras has been almost standardised to be as 230mm by 230mm.

Since different companies are manufacturing the aerial cameras differently, it is very important to standardize the format for the production of same size of photographs. The measurements that are in common use are :

Name of Camera	Focal Length (mm)	Format size (mm)	Manufacturer
RMK - A	150 mm	230 *230 mm	Zeiss. W. Germany
WILDRC - 8	150 mm	230 *230 mm	Wild Heebrugh Switzerland
WILDRC - 10	153 mm	230 *230 mm	Wild Heebrugh Switzerland

F - Stop

The diaphragm, in any camera system, regulates the amount of light passing through the camera lens, normally operated automatically and defined by its relative aperture and is equal to focal length / Diameter of diaphragm i.e., relative aperture = f/d = focal length / Diameter of diaphragm. It is also called f-Stop number. The nominal f - Stop setting is 1, 1.4, 2.0, 2.8... and 32. Not all cameras will have all these, but the most explosive cameras will have most of them. In making photographic exposures, the correct amounts of illuminance and time may be regulated by varying f-Stop setting on the camera, while time of exposure is set by varying shutter speed. Variations in f-Stop settings are actually variations in the diameter of the aperture, which can be controlled with a diaphragm a circular shield that enlarges or contracts, changing the diameter of the opening of the lens and this regulating the amount of light that is allowed to pass through it. As the f-Stop increased, the diameter of the lens opening decreases and accordingly the film exposure decreases. Because the area of the lens opening varies as the square of the diameter the change in exposure with f-Stop is proportional to the square root of the f-Stop.

It is not possible to take the photographs by using one camera with fixed focal length and cone angle in order to get all details of the terrain. In aerial photography focal length and angular fields of view play on important role for getting photographs with all required details. Hence it is need to classify the cameras on the basis of these two measurements. For example, for buildup areas, normal angle cameras are used whereas super wide angle cameras are used for the study of hilly terrains. Some time wide - angle cameras are also used for different purpose.

Usually wide-angle camera is used for photographs both types areas, unless the photographs typically required only for a specific purpose.

2.5.8 Photographic Bases

The *photographic bases* are three types, (1) Glass Plates, (2) Film and (3) Paper.

Glassplates are used almost exclusively as base for light - sensitive emulsion. The glass plate can be made almost as plane and constitutes a stable and a non-shrinkable base for the image. This can produce the documentary quality of photographs. But there are disadvantages also. The glass plate is heavily brittle, clumsy to handily and comparatively expensive, changing the negative material becomes complicated which easily courses trouble. And lastly the accommodation problem in the room of the platform.

Now, in aerial photogrammetry, uses film as base for the emulsion. This is non-inflammable and easy to store and handle. It has improved dimensional stability. No accommodation problem. The main draw back of film is, it dimensions variations, of course this change is uniform in all directions. A serious defect is the irregular shrinkage, which occurs in spite of all efforts to improve the film base.

Paper base, it has much lower degree of dimensional stability. For this reason paper prints are not used in photogrammetric work. The distortion in the x - direction is different from distortion in y - direction and this type of differential distortion will be the serious consequences if we neglected. The two types of film are one layer black and white and three layer colour, both of which can be sensitized to either visible only or visible and near infrared wavelengths. The resultant film types are (1) black and white, (2) black and white near infrared, (3) colour film: (a) false colour near IR (b) true colour.

Uses and Merits

1. ***Black and White Film*** : This is orthochromatic emulsion and is sensitive to blue and green. This film is sensitive to broad waveband of visible light is called panchomatic film. The two types of panchromatic black and white film are mapping film which has equal sensitivity of all visible wavelengths. Second one is equal sensitivity to all visible wavelengths. Second one is reconnaissance film which has reduced sensitivity to blue wavelengths to minimise the effect to atmospheric scatter. Mapping film is used in the environmental sciences.

   A Black and White aerial photographs have been used for many applications in the earth sciences, for

   (i) Geological mapping

   (ii) Hydrological investigations and

   (iii) Terrain analysis

   B These are used in agriculture for the identification of crop types, crop disease and soil erosion.

   C Also used is a soil surveys and vegetation survey.

   D The majority of planning authorities through out the world hold these photographs for regional planning, urban planning etc.

2. **Black and white near infrared film** : This film is similar to the above film, but the main difference is its spectral sensitivity which extends beyond visible wavelengths to a wavelength of around 1.0mm, in the near IR region of EMS.

   A Used in studies of vegetation, it has been used for the location of crop diseases, crop type and mapping of forest and semi-national vegetation.

   B It has been used for knowing the soil moisture states of agricultural fields.

   C It has been used for locating the geological boundaries.

2.6 Geometry of Vertical Aerial Photograph

ABCD is the terrain which is photographed. a'b'c'd' is the *negative* obtained during the photography. The negative is a reversal of both tone and geometry of the object space ABCD. The negative is situated at a distance equal to the *focal length* (f) 0'L, of the camera. The *positive* is obtained by contact process. Geometrically it is situated at a distance (OL) equal to the focal length f(O'L) of the camera O'P is the optical axis, which is along the direction of the focal length and perpendicular to both negative and positive planes. The basic geometry of vertical photograph is shown in Fig. 2.8.

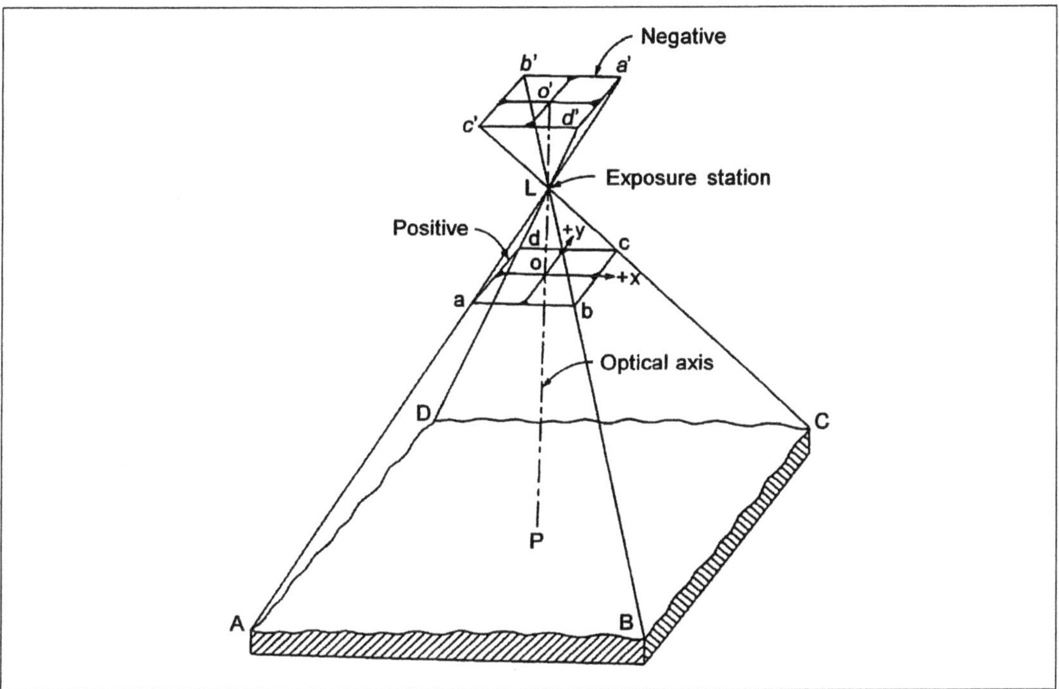

Fig. 2.8 Geometry of vertical photograph.

The first thing that a photogrammetrist/photointerpreter should understand is the geometry of aerial photogrpahy. The aerial photograph is taken by means of an aerial camera on board aircraft / space craft at an altitude based on the scale needed for the given application. The geometry of the photograph is often controlled by the application, scale, size and the condition of the terrain. In general, the geometry of the photography can be known during flight planning. Let us assume that the aircraft is flying at an altitude of 'H' from the Mean Sea Level (MSL). The focal length of the lens used in camera system is 'f'. The focal plane of an aerial camera is the plane in which all incident light rays are brought to focus. In aerial photography the image distances are less than the object distances. Aerial cameras are therefore have their focus fixed for infinite object distances. This is done by setting the focal plane as exactly as possible at a distance equal to the focal length behind the rear nodal point of the camera lens. The focal plane is defined by the upper surface of the focal plane frame. This is the surface upon which the film emulsion rests when an exposure is made. Camera fiducial marks, impressions of holdings, appear on the photograph in the shape of 'V' or 'W' are usually four or eight in number. These fiducial marks are situated in the middle of the sides of the focal plane resulting an impressions on the negative when the picture is taken.The line joining opposite fiducial pints is called fiducial axes. The point of intersection of these fiducial axes is called the principal point. The location of this principal point is based on the position of fiducial points. The principal point may or may not appear at the centre of the photograph. The fudicial axes provide a rectangular coordinate axes which are taken as a photocoordinate system for photogrammetric measurements. The Fig. 2.9 and Fig. 2.10 illustrates the photographic coordinate system.

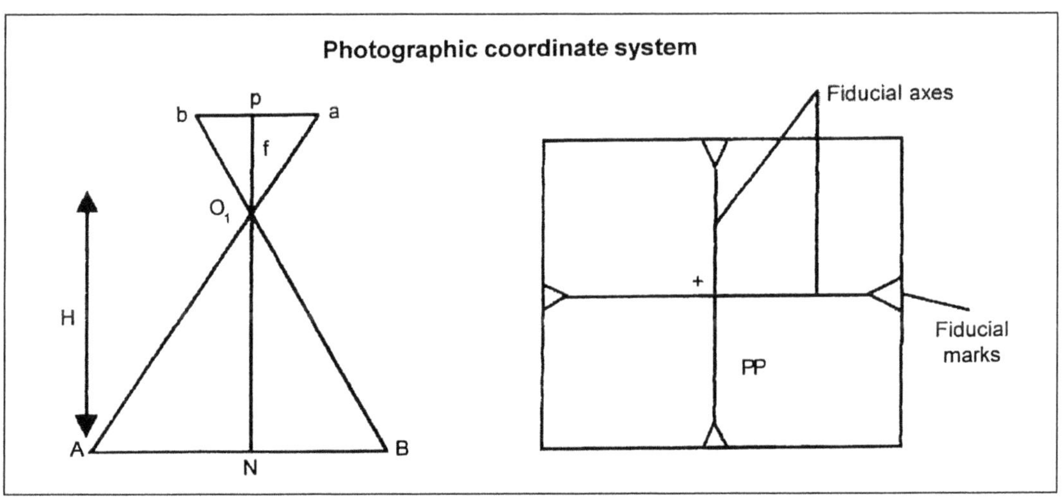

Fig. 2.9 Fig. 2.10

2.7 Scale of a Vertical Aerial Photograph

Based on the geometry of vertical aerial photograph, scale of an aerial photograph can be determined. Scale of a map is different from the scale of an aerial photograph as the map is symbolic representation of earth surface features to some scale and it is uniform throughout the map. The scale of an aerial photograph is different for different points as the photograph is having central projections and the real features are appeared on the photograph. The scale of the photograph can be dealt in two ways based on the nature of the topography. One for flat terrain and the other is for variable or rugged terrain.

2.7.1 Scale of a Vertical Aerial Photograph Over of a Flat Terrain

Let 'A' and 'B' are any two points on the ground and 'a' and 'b' are the corresponding image points on the aerial photograph. 'f' is the focal length of the lens used in aerial camera and 'H' is the flying height of the aircraft by which the photograph be taken. Then the scale of aerial photograph for flat terrain is determined as follows.

From Fig. 2.11.

Δ^{les} aob and AoB are similar triangles.

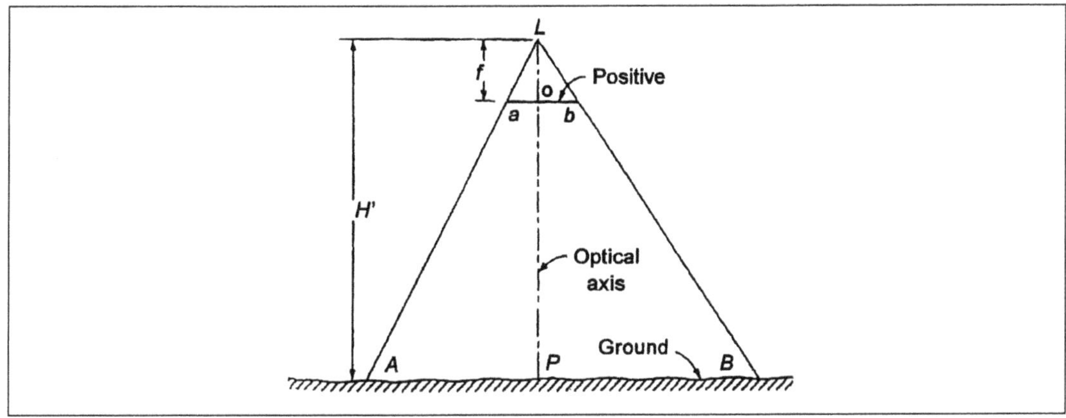

Fig. 2.11. Scale of flat terrain

Scale, $$S = \frac{\text{Photo Distance}}{\text{Corresponding Ground Distance}} = \frac{ab}{AB} = f/H$$

i.e., $$S = \frac{\text{focal length of the lens}}{\text{flying height of aircraft above the terrain}}$$

2.7.2 Scale of Vertical Photograph Over Undulating Terrain

Fig. 2.12 scale of vertical photograph-variable terrain camera station-O, flight height H, object point AB image point ab, and elevation of AB from datum h. Ground points AB is imaged ab, on the positive plane; elevation of AB above datum is h AB

The scale at AB = $\frac{ab}{AB}$

from similar triangles ab and AB, scale = $\frac{f}{H-h}$ considering AB to be infinitesimal, the scale at any point over an undulating terrain with average terrain height h is given by $\frac{f}{H-h}$.

Scale of Aerial Photograph for Variable / rugged terrain at the points A, B and C are

$S_A = f / H - h_A$

$S_B = f / H - h_B$ and

$S_C = f / H + h_C$

Where A, B and C are ground points with respect to Mean Sea Level (MSL) (FIg. 2.12) hA, hB are the elevations of A and B where s 'hC' is the depression. Then the scale of points A, B and C can be calculated as given in the above equations.

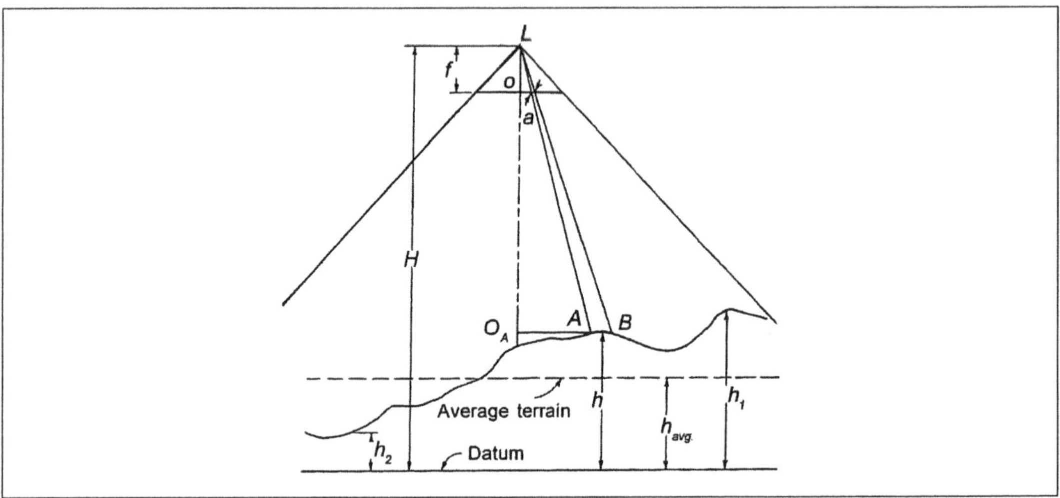

Fig. 2.12 Scale of aerial photograph over a undulating terrain.

2.7.3 Other Methods for Determining Scale of Vertical Aerial Photograph

By measuring ground distance and corresponding distance on photo, between two points, preferably at the same height.

$$\text{Scale} = \frac{\text{Photo Distance}}{\text{Ground Distance}}$$

The exercise can be repeated at 2 to 3 place, for average scale .

By comparison with existing map covering the photographs, measure the distance between two well defined points on photo and corresponding points on maps. Then the scale of aerial photograph is given by

$$\text{Scale} = \frac{\text{Distance on Photo}}{\text{Distance on Map}} \times \text{Scale of map}$$

The exercise can be repeated at two to tree locations and the average scale of the photo can be determined.

2.8 Relief Displacement

In Fig. 2.13 'h' height of the building with A is top point and B is bottom point 'a' and 'b' are corresponding image points on the aerial photos. 'f' and H are focal length of the lens and flying height of the aricarft, respectively. The relief displacement is governed by Vertical exaggeration (V) of the photographic system.

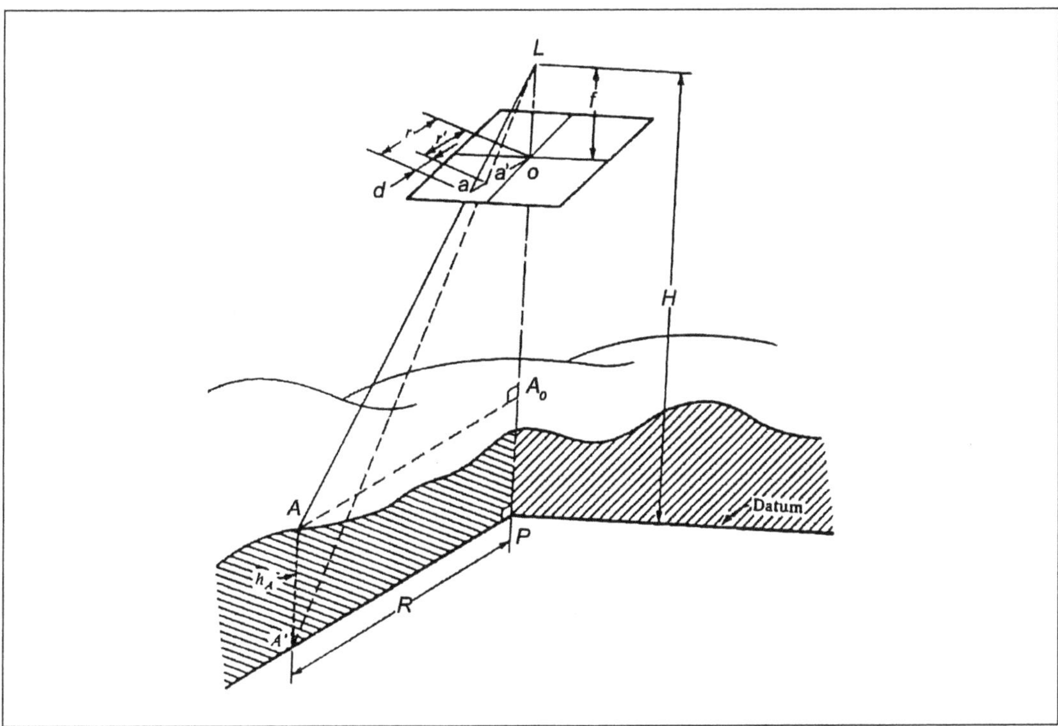

Fig. 2.13 Determination of Relief Displacement.

The displacement of the image due to the height of the object is known as height distortion. The distortion is radial from the principal/plumb point. This means that the distortion or displacement is radial from the plumb point and it increases from the plumb point. Then we can derive the following relationship between h, d (displacement) and 'r' (radial distance) which are depicted in the above figure.

- As the height 'h' of the building increases, the displacement 'd' increases,

  i.e., $d \alpha h$
- As the height 'h' of the building increases, the radial distance 'r' increases,

  i.e., $h \alpha r$
- As the 'H' increases the displacement 'd' decreases, i.e., $d \alpha 1/H$

  Therefore, we have, $d = r \cdot h/H$

$$\Rightarrow \quad h = d \cdot h/r$$

That is, height of the building is determined by displacement 'd', radial distance 'r' whcih can be measured from the photographs and 'H' can be obtained from the suppliers of photographs. This equation can be used to find the height of any object with the help of a single aerial photograph.

2.9 Stereo Photogrammetry-

2.9.1 Stereoscopic Depth Perception

Methods of judging or measuring depths or distances of objects around us is called as depth perception. Depth perception can be classified as : relative sizes of objects, hidden objects, shadows and differences in eye focussing as per varying distances. Stereoscopic depth perception is much greater degree of accuracy than monoscopic, hence it is of extreme importance in photogrammetry.

Stereoscopic vision is binocular vision. When the left and right eyes are focussed on a certain point, the optical axes of the two eyes converge on that point intersecting at an angle called the parallactic angle. Nearer the object, greater the *parallactic angle* and vice versa. In the Fig. 2.14 L and R are the left and right eyes. The distance between the two eyes is called the *eye base be.* Where the eyes are focussed at point A the optical axes converge, forming parallactic angle ϕ a. Similarly where sighting at point B, the optical axis converge at 'B' forming parallactic angle ϕ b.

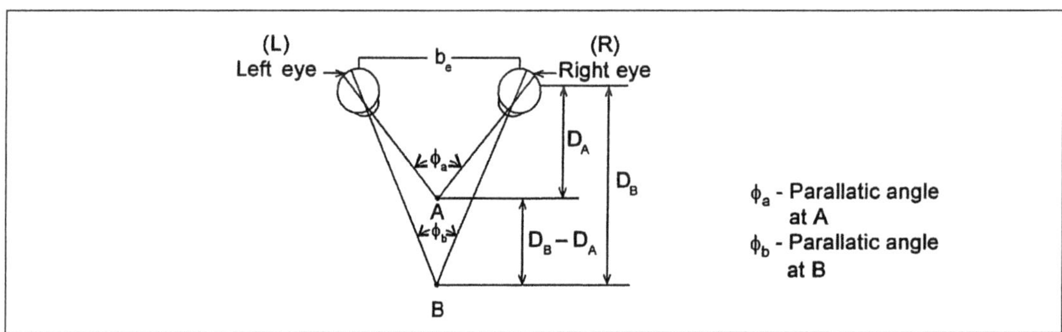

Fig. 2.14 Stereoscopic depth perception as a function of parallatic angle.

The brain associates the distance D_A and D_B with the corresponding parallactic angles ϕ a and ϕ b. The depth between the objects A and B is perceived as the difference between these two parallactic angles. Shortest distance for clear stereoscopic depth perception is about 10 inches for an average adult. Eye base of an average adult is about 6.3 to 6.9 cms or 2.6 inches. Hence the maximum parallactic angle can be calculated by the formula :

$$\phi = 2 \tan^{-1}\left(\frac{1.3}{10}\right)$$

$$= 15^{o} \text{ (app.)}$$

Maximum distance for distinct stereo - for distinct stereoscopic depth perception is not well defined. But for average adult, it is about 2000 ft. Beyond this, parallactic angles are extremely small and changes cannot be differentiated.

2.9.2 Stereoscopic Viewing of Overlapping Aerial Photographs

Fig. 2.15 (a) photographs taken from two exposure stations with building AB in the common overlap area.

Aerial photos are taken from exposure stations L1, L2 so that the building appears on both photos as shown in Fig. 2.15(a). The object point A and B of the top and the ground. This is the situation when aerial photographs of over lapping areas are taken during the flight.

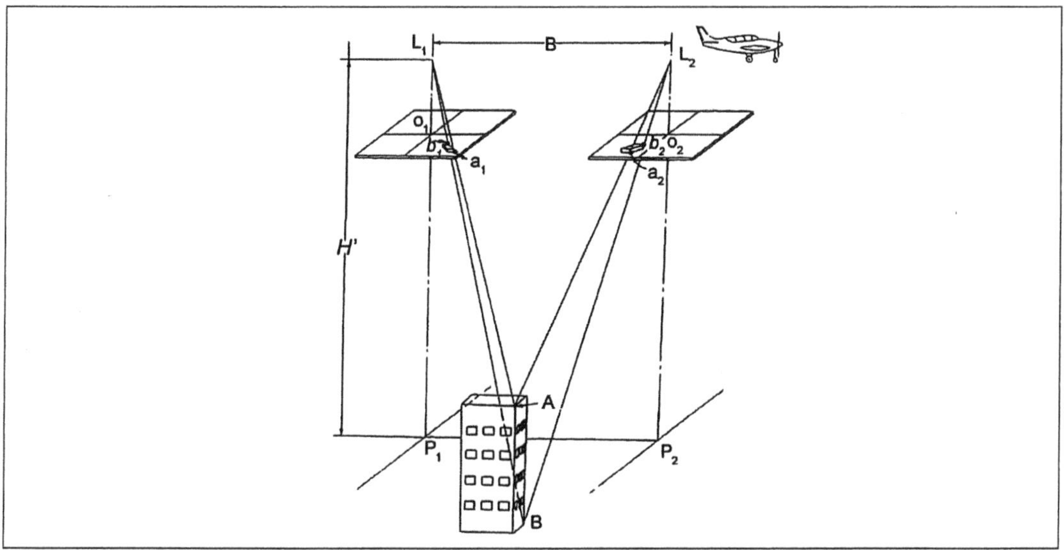

Fig. 2.15 (a) Photographs taken from two exposure stations with builging AB in the common overlap area.

The two photos (left and right) are laid on the table and viewed in such a way that left eye sees only left photo and right eye sees only right photo as shown in Fig. 2.15 (b); the three dimensional impression of the building is obtained. The three dimesional impression appears to be below the table top at a distance 'h' from the eyes. The brain judges the height of the building by associating the depths to point A and B with parallactic angles a and b.

When the eyes view the entire overlap area, the brain receives a continuous three dimesional impression of the terrain. This is achieved by the continuous perception of changing parallactic angles of the infinite number of image points which make up the terrain. The three dimensional model thus formed is called a "*Stereoscopic Model*" and the over lapping pair of photographs is called a "Stereo Pair".

It is difficult to view photographs stereoscopically, without the aid of optical devices. The difficulties can be overcome by an instrument called *Stereoscope.* The most commonly used stereoscopes are Pocket stereoscope and Mirror stereoscope.

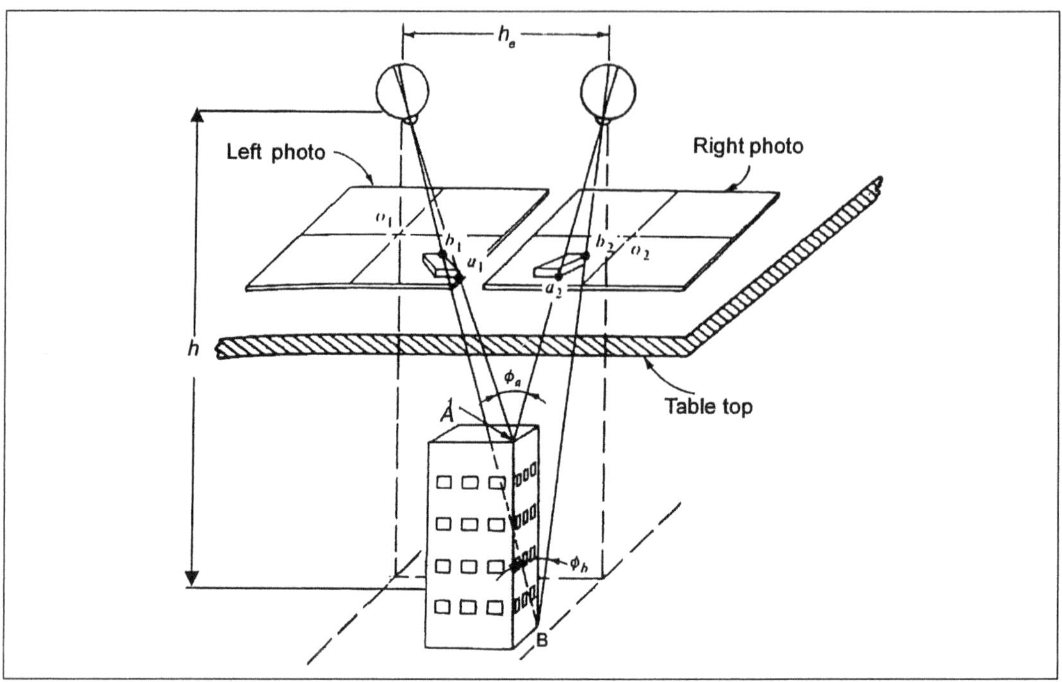

Fig. 2.15 (b) Viewing the building stereoscopically.

Pocket stereoscope are most commonly used and simple construction and consists of two simple convex lenses mounted or a frame. The spacing between the lenses can be varied to accommodate various eye bases.

The Mirror stereoscope has two large wing mirrors and two small eye piece mirrors. Light rays from the photo points a1, a2 are reflected from the mirror surfaces and according to the principles of reflection are received at the eyes, forming the parallactic angle ϕ a. Similarly, for point b1, b2 also forming parallaetic angle ϕ b. The brain automatically associates the depth of the points A and B with their respective parallactic angles ϕ a and ϕ b. This happens for the images of left and right photo, which ultimately generates the 3D stereoscopic viewing, which appear to be below the surface of the photographs.

2.9.3 Procedure for use of Stereoscopes

The following points are to be noted for stereoscopic viewing of aerial photos:

- Place the photographs in such a way that the left photo is seen by the left eye and right by the right eye.
- Reversal will give pseudoscopic view ups and downs and down are reversed in viewing.
- For comfortable (without strain to eye) stereoscopic viewing; the eye base, the line joining the centres of the stereoscopic pairs and flight lines are to be mode parallel.

For this the following procedures are to be carried out :

- Marking of the conjugate principal points of the two over lapping stereopair (The conjugate Principal point for a photograph is the principal pointf of the adjoining photograph transferred to the over lapping photo).

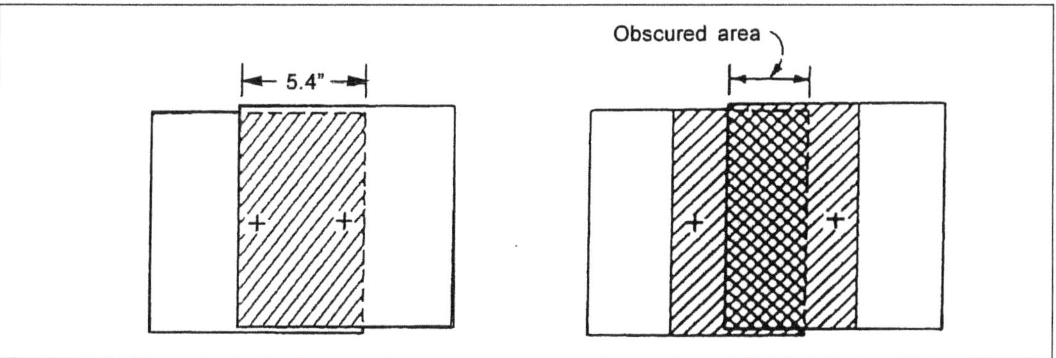

Fig. 2.16 Overlapping area of a stereopair.

2.9.4 Steps for Stereoveiwing by Stereoscope

- Drawing a straight line one a plane (paper fixed) indicating the flight line direction (f f').
- Fix the left photo along the flight line direction (ff') so that line O_1, O_2' direction coincides with the direction ff '.
- Orient the right photo such that line O_1 O_2 O_1' O_2 lie on the straight line ff ';
- Move the right photographs along this direction for getting comfortable stereoscopic view.

Y-parallax

The essential condition for perfect stereoscopic viewing is that the line joining the two corresponding image points must be parallel to the direction of the flight line. If this condition is not satisfied "Y-parallax" exists. Y-parallax causes, strain to eye during stereoscopic veiwing. The more the Y-parallax, the more the strain and vice-versa.

Y-parallax can occur due to

- Improper orientation of photos.
- When flight height H varies at L_1, L_2 of adjoining photographs.
- Hue to lilt in adjoining photograph.

2.10 Stereoscopic Parallax

Apparent displacements in the position of an object, with respect to a frame of reference, by shift in the position of observation. Closer the object, greater the movement. The change in position of an image from one photograph to next caused by the Air Craft's motion is termed as stereoscopic parallax or x-parallax.

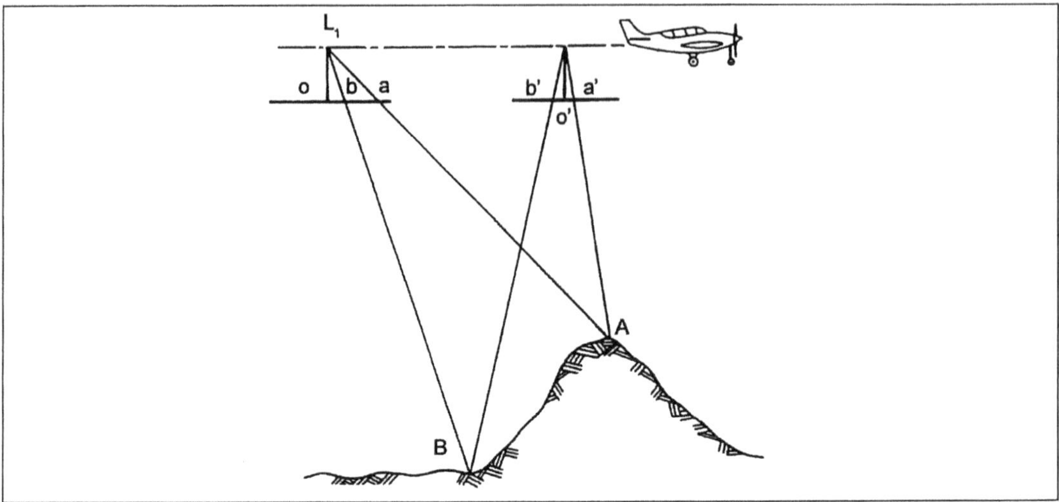

Fig. 2.17 (a) Stereoscopic parallax of vertical over lapping aerial photographs.

A, B ground point are imaged at a, b on left photo and a', b' on right photo parallax of point A is greater than that of B.

Importance of parallax measurements :

- Parallax of any point is directly related to elevation of that point.
- Parallax is greater for higher point and less for lower points.
- Parallax measurements provides basis for determination of elevation of points from photographic measurements.

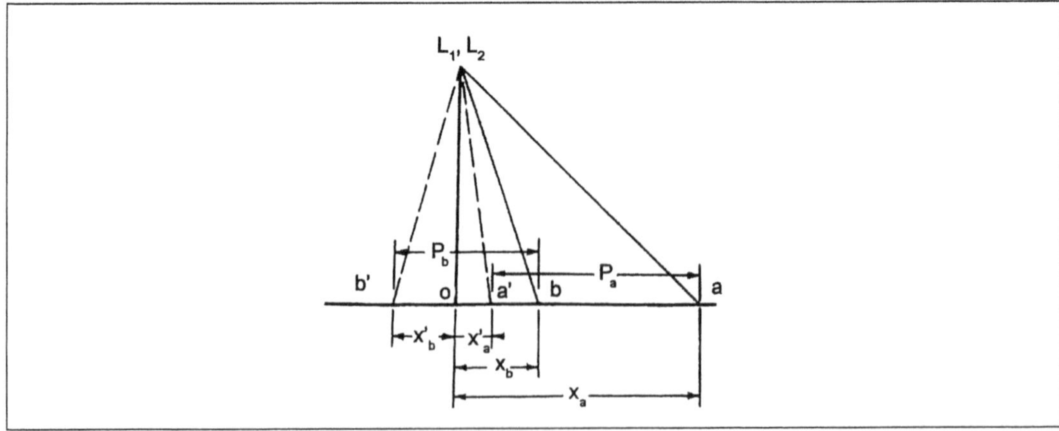

Fig 2.17(b) L, R photographs in superimposition.

Parallax of point A is P_a, parallax of point B is P_b. Parallax expressed in terms of flight line coordinates.

$$P_a = xa - xa'$$

$$P_b = xb - xb'$$

2.10.1 Parallax measurement - Monoscopic

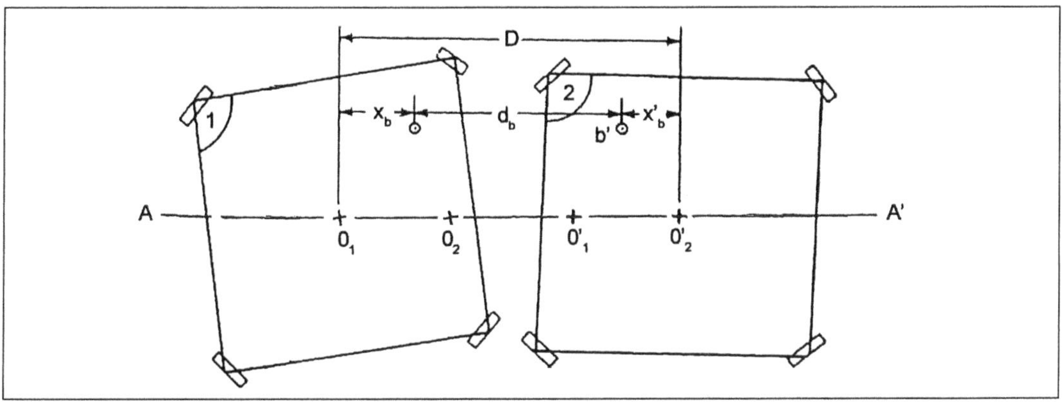

Fig 2.18. Photos 1,2 oriented along flight line axes for stereovision and parallax measurement.

01, 02 - Principal points of photos 1, 2

01', 02' - Conjugate principal points.

xa, x' a - x – Coordinates of point a, a' on photos 1, 2

D - Distance between principal points

da - Distance between image points a, a'

Direct measurement

Parallax of point a = xa-(-x'a)

By measurement of D, da

D, the distance between the principal points O_1, O_2 of the photos is a constant and can be measured. da, measure the distance between the image points a, a'. Parallax Pa is given by **Pa = D - da**

2.10.2 Principal of Floating Mark

Two small identical marks (.) are etched on clear glass. Place these marks on the left and right photographs. Fuse the marks over the stereomodel in such a way that, they fuse into a single mark and appear to rest on the stereomodel at some elevation. If the marks are moved apart and fused, the parallax is increased and the point appears to move down. If the marks are moved closer, the parallax is increased and appears to rise in the stereomodel. This apparent variation in elevation of the point by increasing or decreasing the distances between them is the basis of giving the name "FLOATING MARK" to the point.

The principal of FLOATING MARK in stereo model is extremely important. This is used for point transfer, mapping of terrain and elevation measurement by placing the mark on the terrain at required places and moving along the terrain.

2.10.3 Stereoscopic Methods of Parallax Measurement

By using the principal of floating mark, parallax of stereomodel points are measured. The instrument used along with stereoscope is called "*Parallax bar*".

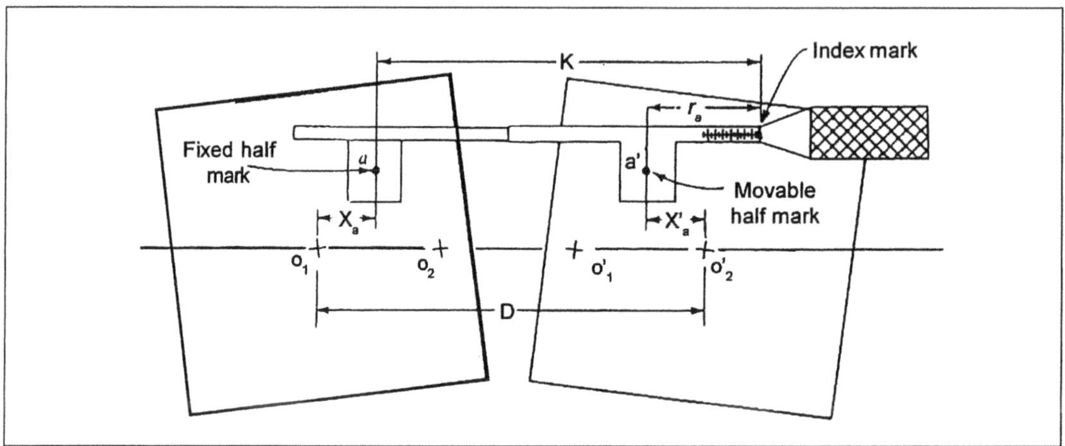

Fig. 2.19

Schematic diagram of parallax bar with measuring principal of parallax of image points a, and a stereo pair of photographs.

Steps

1. Orient the photographs (1) and (2) along flight axis line A A' for comfortable stereovision and fix both photographs.
2. Choose an average elevation point on the terrain, fixed the left hand floating mark of the parallax bar, such that the micrometer drum indicator is at the middle.
3. Left hand mark will be fixed measurements in this particular stereo model.
4. Constant D-Distance between the two principal points.
5. Constant K-Distance between the left hand fixed point of parallax bar and index reading of the micrometer of parallax bar.
6. Adjust the floating mark, such that it rests on the elevation of the fused point a a' in the Stereo model, and take the micro meter reading (ra)

Parallax of point $Pa = Xa - X'a$

$$= D - (K - ra)$$

$$= (D - K) + ra$$

D – K is called the parallax bar Constant (C) for the set up. Substituting,

$$\mathbf{Pa = C + ra}$$

2.10.4 Determination of parallax bar constant

1. Measure the micrometer reading of any point, by setting the floating mark.
2. Calculate the parallax of the point by monoscopic method.

   Then substituting in this equation.

   $$Px = C + rx$$

   $$C = Px - rx \qquad \text{..........(1)}$$

3. Similarly take the micrometer of another point by setting the floating mark on the point under Stereofusion, derreading be ry

   than $$C = py - ry \qquad \text{..........(2)}$$

The average value of C from equation (1) and (2) can be taken for final determination of the parallax bar, constant C has to be determined before each observation.

C can also be determined from the parallax reading of the principal point.

Advantage of stereoscope determination by parallax bar are accuracy and speed. Once C is obtained, all parallax can be measured by one measurement.

2.10.5 Parallax Equations

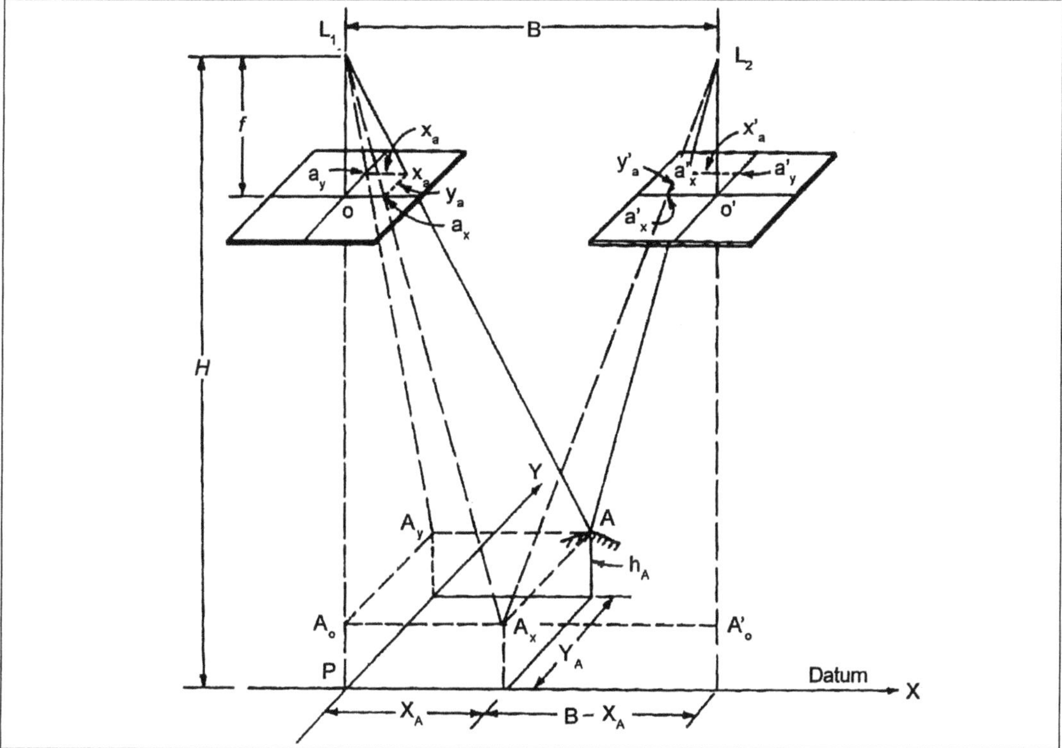

Fig. 2.20 Geometry of overlapping vertical photographs.

B - air base

L_1, L_2 - exposure stations

H - flight height, f - focal length.

A - object on the ground, a, a' images of A or L and R photos.

(XA, YA) ground co-ordinates of point A

(xa, ya), (x'a, y'a) co-ordinates of the image points or left photo and Right photo of ground object A.

Every stereopair of photos have its own unique coordinate system.

Derivation of h_A (ground height of object point A) and (X_A, Y_A)

from lll^r Δ^S $L_1O_1\ a_y$, $L_1\ A_O\ A_y$, $\dfrac{Y_A}{H-h_A} = \dfrac{y_a}{f}$

$$Y_A = \frac{y_a}{f}(H - h_B) \qquad \text{..........(1)}$$

from lll^r Δ^S $L_1O_1\ a_x$, $L\ A_O\ A_x$, $\dfrac{X_A}{H-h_a} = \dfrac{x_a}{f}$

$$X_A = \frac{x_a}{f}(H - h_A) \qquad \text{..........(2)}$$

from lll^r Δ^S $L_2O_2\ a'_x$, $L_2\ A'_O\ A_x$, $\dfrac{B-X_A}{H-h_A} = \dfrac{-x_a}{f}$

$$X_A = B + \frac{x'_a}{f}(H - h_A) \qquad \text{..........(3)}$$

Equating (2) and (3)

$$\frac{x_a}{f}(H - h_A) = B + \frac{x'_a}{f}(H - h_A)$$

$$(H - h_A)\left(\frac{x_a}{f} - \frac{x'_a}{f}\right) = B$$

$$H - h_A\left(\frac{x_a - x'_a}{f}\right) = B$$

$$H - h_A = \frac{Bf}{x_a - x'_a}$$

Substituting $P_a = x_a - x'_{a'}$

$$h_A = H - \frac{Bf}{P_a} \text{ or } P_a = \frac{fB}{H - h_A} \qquad \text{..........(4)}$$

Substituting eqn. (4) in eqns. (2) and (3)

$$X_A = \frac{x_a}{f}\left(H - \left(H - \frac{Bf}{x_a}\right)\right)$$

$$= \frac{x_a}{f}\left(\frac{Bf}{P_a}\right)$$

$$X_A = B\left(\frac{x_a}{P_a}\right) \qquad \text{..........(5)}$$

|||^{rly} $$Y_A = B\left(\frac{Y_a}{P_a}\right) \qquad \text{..........(6)}$$

Equations (4), (5) and (6) are called *Parallax equations.*

h_A - height of point A above datum.

H - flying height

B - air base, f - focal length of camera.

P_a - parallax of image point A.

(x_A, y_A) ground co ordinates of point A.

(x_a, y_a) photo co ordinates of point a.,

(XA,YA) based or unique co ordinate system with origin at datum principal point of O_1 at P.

Parallax equations are the most useful for a photogrammetrist. The equations enable to complete survey of the stereo overlap area, provided focal length f, and sufficient number of ground controls are available for calculation of B and H.

Conversion of unique arbitrary co-ordinate system of a stereo pair to tree ground co-ordinates system is possible, by two dimensional co-ordinate transformation.

2.10.6 Elevation by Parallax Differences

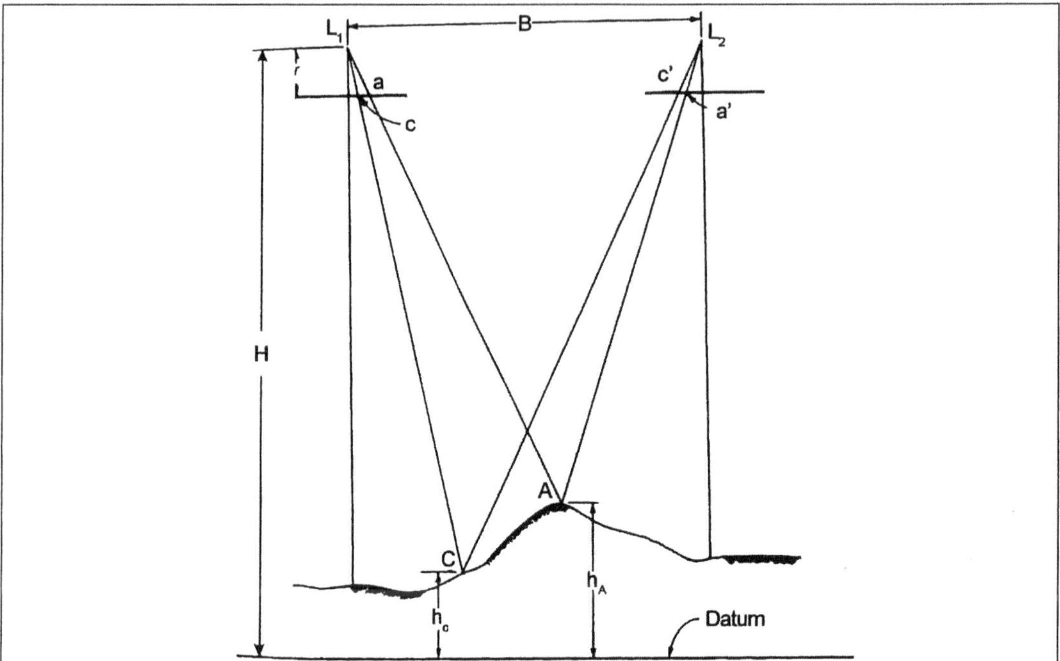

Fig. 2.21 Elevation by parallax differences.

Parallax differences are caused by difference in elevation of the two points. If elevation of one point is known; the parallax differences of the known and unknown point can be applied to determine the elevation of the other point.

In the figure if h_C is known, h_A can be calculated by the difference of parallaxes of A, and C.

$$P_C = \frac{fB}{H - h_C} \qquad \text{..........(7)}$$

$$P_a = \frac{fB}{H - h_A} \qquad \text{..........(8)}$$

$$(8) - (7) = P_a - P_C = \frac{fB}{H - h_A} - \frac{fB}{H - h_C}$$

$$= \frac{fB\{(H - h_C) - (H - h_A)\}}{(H - h_A)(H - h_C)}$$

$$= \frac{fB(h_A - h_C)}{(H - h_A)(H - h_C)}$$

Let $\Delta_P = P_a - P_C$

and substitute $H - h_A = \frac{fB}{P_a}$

Then $$\Delta_P = f\,B\,\frac{(h_A - h_C)}{\frac{fB}{P_a}(H - h_C)}$$

$$\Delta_P = \frac{P_a(h_A - h_C)}{(H - h_C)} \qquad \text{..........(9)}$$

$$\Delta_P = (H - h_C) = P_a\,(h_A - h_C)$$

$$h_A - h_C = \frac{\Delta_P(H - h_C)}{P_a}$$

$$\therefore \quad h_A = h_C + \frac{\Delta_P(H - h_C)}{P_a} \qquad \text{..........(10)}$$

From equation (10), if the height of a given point is known (say c), height of and other point (A) can be determined, by finding P_a, P_c, Δ_p and knowing the flight height H.

2.11 Stereoscopic Plotting Instruments-Analogue and Analytical

Stereoscopic plotting instruments, commonly called as Stereo plotters are widely used for preparation of topo-graphical maps. The subject of stereo plotters is most important in the study of photogrammetry.

Basic concept : An over lapping pair of aerial photos, (stereo pair) is exposed and transparencies (diapositives) are carefully prepared. After preparation of the diapositives these are placed on the projectors of the stereo plotter in such a way that left photo is put on left project and right photo on right projector. Before plotting of the details and contours, the following orientations are carried out :

2.11.1 Interior Orientation

This stage is for recreating the geometry of the projected rays to duplicate exactly the geometry of the orientation photos. The procedures involved in inner orientation are :

- Preparation of diapositives
- Compensation for image distortions
- Centreing diapositives in the projectors.
- Setting of proper principal distance in the projects.

2.11.2 Relative Orientation

The condition that is fulfilled in relative orientation is that each model point and the two projection centres form a plane in miniature, just like the plane that existed for the corresponding ground point and the two exposure stations. This ensures that the parallactic angle for any point of the stereo model is equal to the original parallactic angle of the corresponding object point at the time of taking photograph. This implies that the projected rays of each the corresponding point on the left and right diapositives must intersect at a point.

Since the X-component Px, of the mismatch of the images can be removed by raising or lowering the plate carriers (Px being a function of elevation of the point), the remaining Y-component Py, called Y-parallax has to be cleared for all points in the stereo model in order to obtain a relatively oriented model.

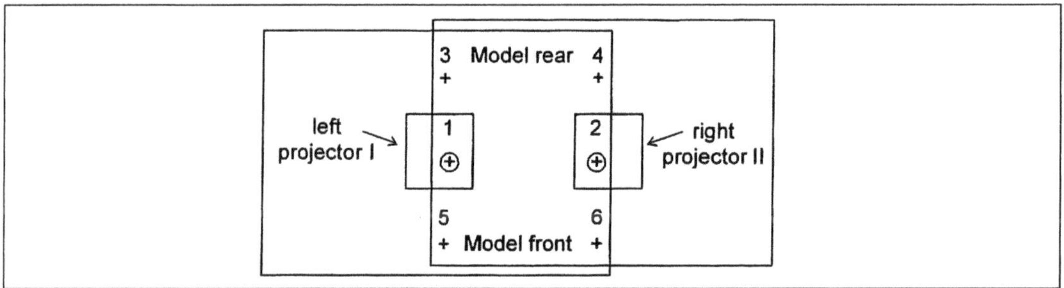

Fig. 2.22(a) Stereo model locations of six points conventionally used for Relative Orientation.

Conventional procedure for clearing Y-parallax at five standard points (plus sixth for check), located in a model, is cleared, then the entire model will be cleared of Y-parallax.

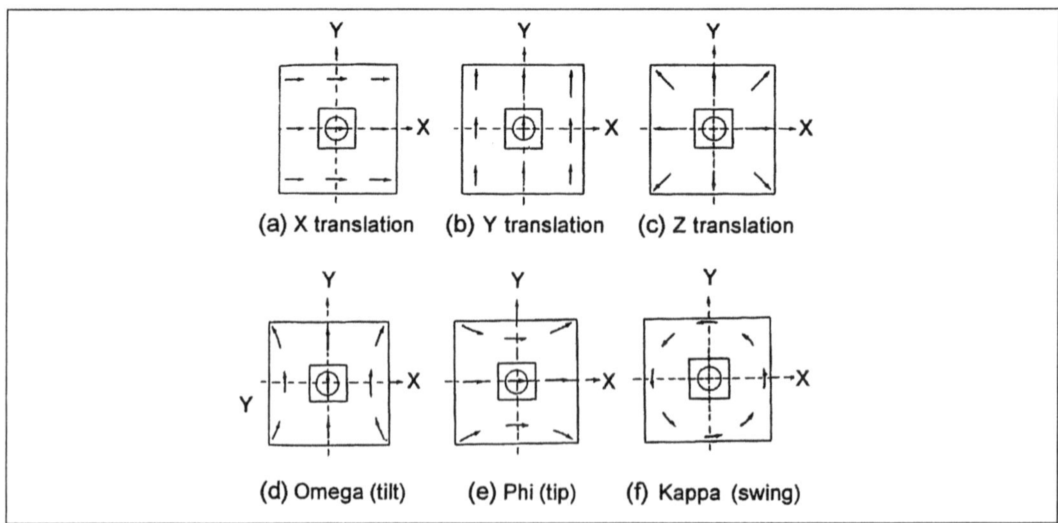

Fig .2.22(b) Stereo model locations of six points conventionally used in relative orientation.

The image movements caused by the six projector motions are given above. The movements are as follows:

X-translation : Only X movement to all images hence no Y-parallax can be removed.

Y- translation : Y movement of all projected images. Hence Y - parallax can be cleared at any point.

Z-translation : Radially outward for Z translation upward and inward forZ translation downword. Y-parallax could be cleared at any point, other than along X-axis.

Omega (Ω) rotation (tilt) : causes Y-movement of all projected images, Hence Y-parallax can be cleared anywhere in the model.

Phi-rotation (tip) : Causes X movement on all projected images, but also imparts Y- Component for points not on X and Y axis.

Kappa - rotation (swing) : Y components to all points everywhere in the model except along Y axis,

Procedure have been detailed after understanding the above image movements for removal of Y-parallax. Relative orientation consists of procedure for clearing Y - parallax for the entire model based on the above movements.

2.11.3 Absolute Orientation

After relative orientation, a true three dimensional model of the terrain in generated. *Model scale fixation* and also *levelling* the model for vertical scale is called Absolute Orientation.

Fixing model scale

Scaling carried out by the formula.

$$Sm = \frac{b}{B} = \frac{h}{H'}$$

also

$$Sp = \frac{f}{H'}$$

$\therefore$

$$Sm = \frac{h}{f'} . Sp$$

Model scale can be calculated from photo scale by the above formula where f in the principal distance, h is the optimum projection distance.

Levelling the model: Final step in absolute orientation. Minimum requirement of height points are three which should form a triangle. Four height Control situated at the four corners of the model is preferred along with one at the centre of the model. The model is rotated about X - axis and Y-axis till the heights agree in the triangular set up. After agreement, the fourth and final points are checked and fine corrections to the projectors are carried out.

After the scaling and levelling is carried out, the model is really for plotting of details and contours as per laid out specifications. (Three dimensional data capture).

Map Compilation: These are number of stereo plotters available for map compilation. These can be classified as :

- Optical (analogue)
- Optical Mechanical (analogue)
- Semi-analytical
- Analitical

Different plotters have different methods of procedure for plotting and map compilation.

2.12 Aerial Photography - Planning and Execution

For successful execution of any project based on aerial photography, proper planning and execution of aerial photography mission is required. The basic requirements are the specifications and the preparation of the flight map. The specification for Aerial photography is based on the following item variables:

- Purposed of photography
- Camera and film
- Scale of photography
- End lap and side lap
- Flying height.
- Ground coverage
- Weather conditions
- Season of the year

Purpose of photography : The first and foremost consideration is the purpose for which the photography is taken. For topographic mapping, photos having good metric qualities are needed. For this calibrated cameras, films having fine grained and high resolution emulsion are selected.

Scale of photography : Average photo scale is one of the most important variables that must be selected in Aerial photography planning. For topographic mapping photo scale depends upon the final map scale, contour interval and the Photogrammetric instruments for map preparation. In most of the Photogrammetric instruments the enlargement ratio from photo scale to map scale is five. Hence photo scale for mapping can be fixed at one fifth of the map scale.

End lap and side lap : For topographic mapping and Photogrammetric control extension, the end lap requirement is 60% and the side lap is 30%. The side lap is required for tieing the strips and making the block for the specified area.

Flying height : Once the selection of the camera and scale of photography is finalised, the flying height for the project is determined

Ground coverage : Once the photo scale and camera format dimensions are selected the ground surface area covered by a single photograph can be calculated. In addition, if the end lap and side lap are known the area covered by a *stereoscopic neat model* can be calculated. Neat model is the stereoscopic area between adjacent principal points and up to the middle of the side lap in both directions, extending side ways.

Weather conditions : The weather for aerial photography is an important consideration cloud less days, have sky and atmosphere clear without haze, smoke are chosen for flight.

Season of the year : For topographic mapping the season to be chosen in based on alitude of sun and the trees having leaves fallen or intact. The ground should not be obscured by tree leaves. Normally start of the winter season is chosen in for aerial photo missions.

Flight map : Based on the above specifications and consideration, a flight map is prepared. The flight map gives the project boundaries and flight lines, the pilot must fly to obtain the desired coverage. Flight maps are prepared on existing maps covering the projects Area. (Fig. 2.23).

A transparent templete showing neat model positions based on the scale, end lap, side lap and base is prepared for the project area and super imposed for preparation of the flight map.

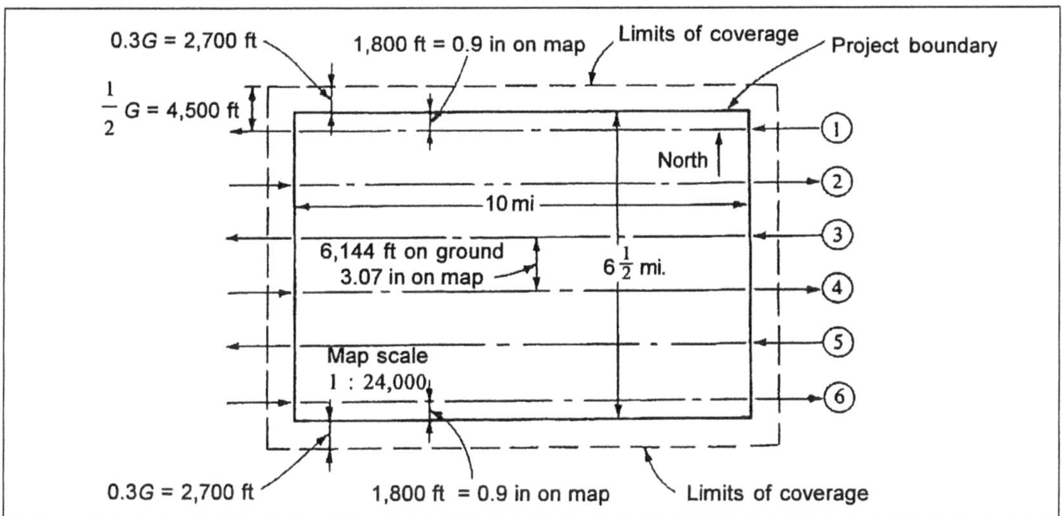

Fig. 2.23 Flight map for project area ABCD

2.13 Control for Aerial Photography

Control points based on some coordinate system are required for any type of mapping project. Similarly, for topographic mapping projects based on Aerial photography also requires ground control. Control selected in this case are sharp well defined points in the object space (ground) so that they are clearly identifiable on the photographs. Ground control for photogrammetry also known as photogrammetric control is classified into two different categories:

- ***Horizontal control :*** Position of point on ground is known with reference to a Horizontal datum or origin.
- ***Vertical control :*** Elevation of a point is known with reference to a vertical datum.

2.13.1 Requirement of Photogrammetric Control

In photogrammetric mapping projects, control are required for orienting stereo models. Further ground surveyed photo-control points are needed for bridging (densifying or supplementing) photo control. The ground surveyed photo-control requirement varies depending on the methods, instruments and personnel. The procedure for densifying control is called Aerial Triangulation. Both horizontal and vertical control are required for Aerial Triangulation.

2.13.2 Methods of Providing Control on Aerial Photographs

Post - pointing : Control points as per requirements are provided on ground on well defined points which are easily identifiable on aerial photograph. The photographs are taken on the ground and the control point locations are transferred on to the photos by identifying them and under stereofusion. This method of transferring, existing fresh control already provided on ground to aerial photographs is called post-pointing.

Pre-pointing : In this methods locations where the points are required are first identified as per planning and targets are mode on the ground at specified locations. The dimension of the target are calculated from the proposed scale of photography. Indicators to the point which are five times bigger than the target mark is laid for quick identification. The back ground contrast is generated by making the target pure white and the back ground colour dark. During aerial photography all such marks are imaged on to the aerial photographs, Although expensive, this methods is much more accurate than post-pointing.

2.14 Aerial Triangulation

For large mapping projects the number of control points needed is extensive and the cost of establishing them by ground methods can be very high. With the help of electronic computers and improved photogrammetric equipment and techniques much of the needed control can be established by photogrammetric control extension often called “Aerial Triangulation”.

2.14.1 Aerial Triangulation can be classified into two broad categories

- *Analoguge Method*, where procedure involves, manual relative orientation and measurement of model co-ordinates by using stereoscopic plotting instrument.
- *Analytical Method*, consists of photo coordinate measurements followed by numerical relative orientation and Computation of model coordinates using electronic computer.

2.14.2 Aerial Triangulation - Analogue Method

For analogue method of aerial triangulation, Universal Instruments are used, where "base-in" and "base-out" facility is available. This enables the strip to be formed on the machine itself. The procedure for base-in base-out is as follows.

1. In base-in mode 1-2 photos are kept in L and R projector and observation carried out.
2. After observation photo (1) is removed and 3 is kept at (L) projector and base-out position is obtained. Observation of coordinates is carried out with 3-2 photos at L and R projectors.
3. After this photo 2 in removed and replaced by photo (4). This 3-4 photo is now in L and R projectors and base-in position in model observations are carried out.

The same is Continued next by (3) replaced by (5) and 5-4 Base out position and so-on, till the strip is completed.

After completion, the error curves are plotted on graph paper and all the co-ordinate adjusted and used for photogrammetric plotting.

2.14.3 Independent Model Triangulation

Also referred to as semi-analytical aerial triangulation. In this method each stereopair of a strip is relatively oriented in the plotter, Model coordinates of all control points and pass points are then read and recorded for each stereo model in turn. The coordinate system of each model is independent from others. By means of pass points common to each adjacent models, three dimensional co-ordinate transformation is carried out to tie each successive model to the pervious one. The coordinates of the perspective centres are also determined in each independent model and included as common points for gaining needed geometrical strength (Fig. 2.24). The right exposure station of model 1-2 is the same as the left exposure station of model 2-3.

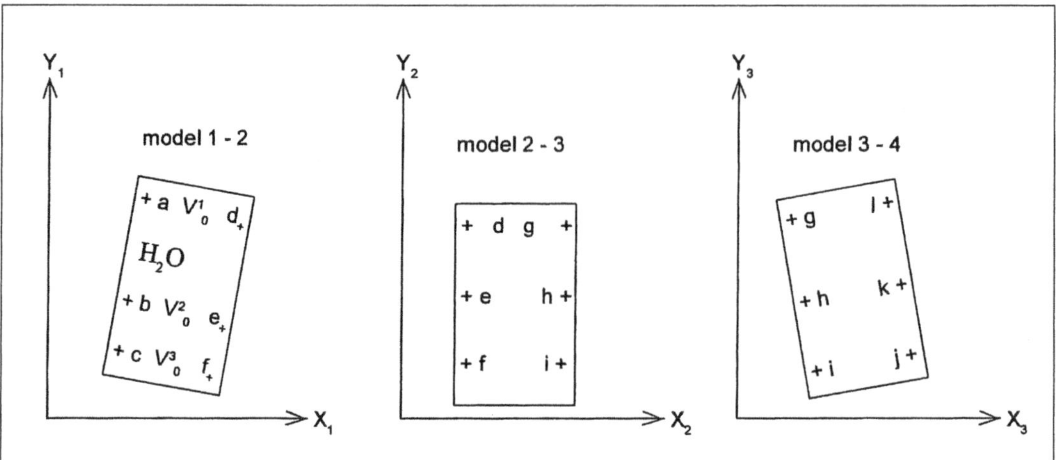

Fig. 2.24. Independent model triangulation observation

To transform model 2-3 into model 1-2 the co-ordinates of common points d, e, f and O_2 of model 2-3 are made to coincide with corresponding model 1-2 coordinates.

By applying coordinate transformations, a continuous strip of stereomodels are formed. This strip may then be adjusted to ground control using polynomial methods to obtain corrected coordinates for all pass points.

For obtaining model coordinates of perspective centres number of methods are available. The grid plate method is most commonly used.

2.14.4 Aerial Triangulation - Analytical Method

Analytical procedure in Aerial Triangulation also involves the basic steps:

- Relative orientation of each model
- Connection of adjacent model to form a continuous strip.
- Adjustment of strip to ground co-ordinates.

In analytical method, the basic input consists of precisely measured photo coordinates of control points and pass points. Relative Orientation is then performed analytically, based upon the measured coordinates and known camera constants.

Basically, analytical method consists of writing condition equations which relate to unknown elements of exterior orientation of each photo to camera constants and measured and refined coordinates. The equations are solved for unknown Orientation parameters and either simultaneously or subsequently the coordinates of pass-points are calculated.

2.14.5 Collinearity or Coplanarity

Collinearity Condition : The collinearity condition specifies that the exposure station, any object point and its photo image all lie along a straight line. (L, a and A) lie along a straight line. Two equations express collinearity conditions of any point. One for x-photo coordinate and other for y-photo coordinate. The basic equations are:

$$x = f\left[\frac{m_{11}(X - X_L) + m_{12}(Y - Y_L) + m_{13}(Z - Z_L)}{m_{31}(X - X_L) + m_{32}(Y - Y_L) + m_{33}(Z - Z_L)}\right]$$

$$y = f\left[\frac{m_{21}(X - X_L) + m_{22}(Y - Y_L) + m_{23}(Z - Z_L)}{m_{31}(X - X_L) + m_{32}(Y - Y_L) + m_{33}(Z - Z_L)}\right]$$

Where x, y are the photo co-ordinates, x, y, z are the ground coordinates, y_L, Y_L, Z_L are the ground co-ordinates of the exposure station, camera focal length and the ms are the functions of rotation angles omega, phi, and kappa.

The collinearity condition equations are non-linear and are linearised by using. Taylor's theorem.

Coplananity Condition : The condition stipulates that the two exposure stations of the stereo pair, any object point and corresponding image points on the two photos, all lie in a common plane. The condition equation is

$$O = B_X (D_1F_2 - D_2F_1) + B_Y (E_2F_1 - E_1E_2) + B_2 (E_1D_2 - E_2D_1)$$

Where :

$$B_X = X_{L2} - X_{L1}$$
$$B_Y = Y_{L2} - Y_{L1}$$
$$B_Z = Z_{L2} - Z_{L1}$$
$$D = (m_{12})\, x + (m_{22})\, y + (m_{32})\, f$$
$$E = (m_{11})\, x + (m_{21})\, y + (m_{31})\, f$$
$$D = (m_{13})\, x + (m_{23})\, y + (m_{33})\, f$$

The coplanarity equation are non-linear and must be linearised using Taylor's theorem.

2.14.6 Blocks of Photos (Block Adjustment)

Control extension can also be made through blocks of photos (two or more side lapping strips). In dealing with blocks number of pass points must be carefully selected so that they appear in the side lap area of adjacent strips.

Using numerical methods, strips are successively joined by matching common tie points and the block is adjusted to available horizontal and vertical field control. Experience has shown that strongest solution is obtained when horizontal control points are located near the periphery of the block. Vertical control must be uniformly distributed throughout the block for block adjustment.

2.14.7 Bundles Adjustment

With the emergence of faster and larger capacity computers, another technique called "*Simultaneous Bundle Block Adjustment*" or "*Bundle Adjustment*" has become practical in analytical photogrammetry. In the procedure all the possible collinearity equations for the entire block of photographs are formulated and solved simultaneously. By solving both control and pass points collinearity equations simultaneously a numerical relative and absolute orientation of all stereo models in the block is achieved.

The symultaneous solutions of all collinearity equations that is highly redundant and is solved by method of least squares, which is mathematically more rigorous than polynomial methods.

2.15 Digital or Soft Copy Photogrammetry

Digital photogrammetry or soft copy photogrammetry has already proven its efficiency in practice. The production of the photos, aerial triangulation, derivation of digital terrain model or photogrammetric plotting are increasingly carried out on "*digital photogrammetric workstation*". The assistance afforded to the operator by image processing tools and the easy interaction with GIS data bases, distinguish digital photogrammetric work stations from the classical analogue analytical plotters.

2.15.1 Scanning Of Aerial Photographs

Digital photogrammetry is now widely used. But the imaging technique have not been changed. Most of the aerial photos are still taken with film cameras. The information content of an aerial photograph requires an average storage space on disc of 1.2 Giga byte, without image compression. The present technology does not permit one to reach such performances with reasonable means under flight conditions. Consequently, film cameras, will remain standard technique for image registration for several years and it will be necessary to convent the analog images to digital format by scanning process. The process in semiconductor field for digital applications is the basis of rapid advances in the image sensor field

2.15.2 Application of GPS in Photogrammetry

Kinematic GPS positioning of air borne platforms coupled with aerial photogrammetry has become a reality. Real time GPS positioning is used for navigation, while post processed carrier phased differencing is employed for exposure station locations. Since it is not possible for exposure station event to control GPS collection, the coordinates are derived by interpolation method. By employing combination of GPS and ground control, the demand for ground control has been reduced. The operational efficiency has increased as the need for expensive and labour intensive ground control has diminished. The technology is rapidly maturing, such that, it will become the most commercially viable approach for aerial mapping.

Airborne GPS offers a flexible and cost-effective alternative to land based methods for obtaining geodetic control needed in aerotriangulation block adjustments.

2.15.3 Integrated GPS Flight Management System

- ***System Details :*** 33 M Hz P. C Wild Re 20 metric mapping camera, GPS receiver and associated peripherals.
- ***Functions :***
  - *GPS receiver Functions :* Operates in continuous kinematic mode (second).
  - *Mapping camera Functions* : For recording of exposure instant location by GPS. The parameters of each exposure event is recorded (all). In Flight Management System the aircraft can be navigated exactly as per the Flight planning Chart.

2.15.4 Aerial Triangulation using GPS Data

Air borne GPS data is used for carrying out Aerial Triangulation Block Adjustments. Application of GPS positioning data in photogrammetry, (GPS photogrammetry) can be carried out in the following ways;

- Determination of ground control points by terrestrial GPS
- GPS flight navigation
- GPS Camera positioning for Aerial Triangulation
- Positioning of Other airborne Sensors.

2.15.5 Storage and Data Compression in Digital Photogrammetry

Storage and retrieving large volumes of high resolution digital imagery in soft copy photogrammetry system require huge amounts of data space and advanced data transfer techniques. The difference in hardware between general purpose systems and photogrammetric work stations is characterized by two features.

- 3D display
- large Capacity short and long term storage subsystems.

Compression is an integral part of any large volume information processing and archiving system. Images usually exhibit quite high degrees of redundancy, they compress well. Handling very large digital images is already a daily practice. Most of the commercial soft copy systems use Industry standard JPEG still, continuous tone lossy image compression scheme. To achieve high performance the compression is typically implemented in hardware.

2.15.6 DEM and DTM in Soft Copy Photogrammetry

The term DEM (Digital Elevation Model) Is some times used generically to mean the digital cartographic representation of earth in any form, rectangular grids, or lattices, triangular net works or irregular spot heights and break lines. In this context, a DEM is a digital cartographic representation of the elevation of land (Z value) at regularly spaced intervals in x and y directions.

The term DTM (Digital Terrain Model) is similar to DEMs, but incorporate the elevation of significant topographic features on land and change points (break lines), which are irregularly placed so as to better characterize the shape of the terrain. Break lines are linear features that describe a change in the smoothness or continuity of the surface.

Advantages of Soft Copy Photogrammetry on DEM/DTM generation : Soft copy photogrammetry has significant benefits over hard copy analytical plotters for the extraction, editing and quality control of DEMs. Soft copy offers the possibility of manual assistance and or automated extraction. In addition, high quality accurate stereo superimposition provides tremendous benefits in effective DTM collection, real time stereo displays and also unmatched capabilities in quality control of DEMS/DTMs.

Soft Copy photogrammetric Work stations : The most essential components of a Digital photogrammetric work station can be identified in the Fig. 2.25.

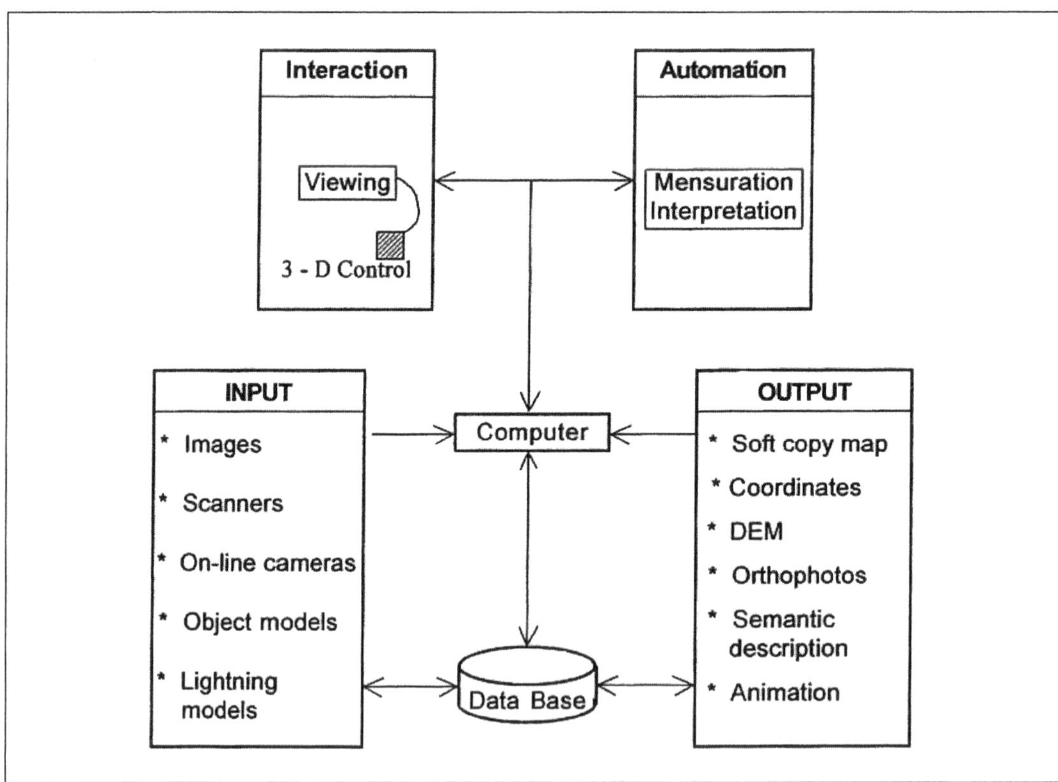

Fig. 2.25 The fundamental components of a soft copy photogrammetric Work station and the related input-output devices, tools and products.

Computer, Data base and Graphics Subsystem : One of the core elements of a soft copy photogrammetry Work Station in the computer.

All basic software and soft copy application software, the user interface, the data transfer and communication are controlled by one or more processing units.

Image and vector data base : In the data base the images, image parameters, preprocessed images, project parameters, lighting models, object models semantic information, resulting vector data, DTM, orthophotos etc. are archived. There are high requirements on the data management, lie, archiving and retrival of the data.

Network : The computer, the data base, peripheral input and output devices are typically connected via very fast bases (eg : VME bus, S bus) or a fast net work like Ethernet.

Interaction : The latest trend in design of soft copy photogrammetric Workstation is to develop the human interface following the principle of uniform interface design. The interaction requires a veiwing system and 2D or 3D control devices.

Veiwing : The display of digital images and super imposed vector information requires high resolution monitor (e.g., 1280 × 1024 pixels or more). In 8 bit colour or 24 bit true colour with a non-destructive overlay. Images are displayed in monomodel or in stereo mode using special techniques. Use of an additional control monitor with input menus and process information is very common.

Stereo Veiwing : For many applications stereo veiwing is essential. The stereo veiwing requires the separation of the two images of a stereo pair. The separation is based on temporal separation (alternate display of two images) or radiometric separation (polarization or anaglyph principle) or on spatial separation (split screen and stereoscope) or on combination of these principles (Fig. 2.26(a) & (b)).

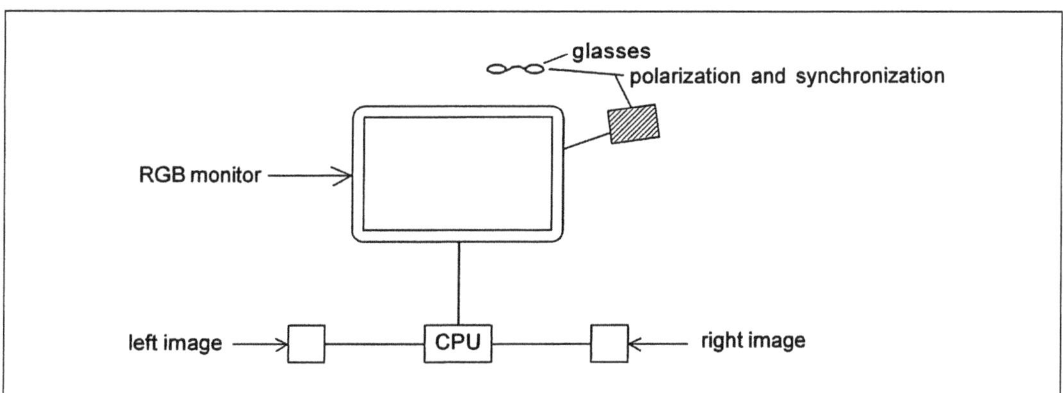

Fig. 2.26(a) Stereoveiwing by polarization. Left and right image at 120 Hz and synchronized polarization for veiwing by polarized glance.

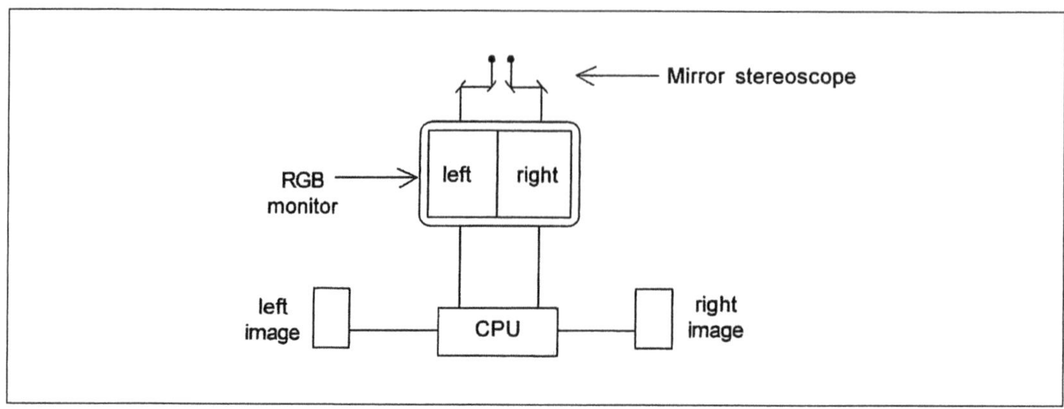

Fig. 2.26(b) Stereoveiwing by optical separation. Left and righ image are displayed on split screen veiwing in performed through a mirror stereoscope.

Super-Imposition : The graphical display of object data on the image is called superimposition. In the case of stereo it is called stereographic superimposition. For 3D superimposition in required for high quality overlay.

Control : Subpixel positioning is an essential prerequisite of a soft copy photogrammetric Workstation. The cursor (floating mark) can be easily defined in any desired shape and size. These are solutions specially for 2D and 3D control. A mouse is most often used for the positioning. Some systems use hand wheels and foot and wheel to control movement of floating mark. Advanced 3D cursors allow the control of 3D movement and after programmable function buttons for special tasks.

2 D	3D
- Mouse - Track ball - Key board keys - Hand wheels	- 3D Mouse - Track ball - key board keys - 3 D cursor with function buttons - Hand wheels & foot wheel / foot disc

3

Remote Sensing - Basic Principles

3.1 Introduction

Remote sensing is the science and art of obtaining information about an object, area or phenomenon through an analysis of the data acquired by a device which is not in contact with the object, area or phenomenon under investigation. In the present context, the definition of remote sensing is restricted to mean the process of acquiring information about any object without physically contacting it in anyway regardless of whether the observer is immediately adjacent to the object or millions of miles away. It is further required that such sensing may be achieved in the absence of any matter in the intervening space between the object and the observer. Consequently, the information about the object, area or any phenomenon must be available in a form that can be impressed on a carrier vacuum. The information carrier, or communication link, is electromagnetic energy. Remote sensing data basically consists of wavelength intensity information acquired by collecting the electromagnetic radiation leaving the object at specific wavelength and measuring its intensity.

Remote sensing of earth's environment comprises measuring and recording of electromagnetic energy reflected from or emitted by the planet's surface and atmosphere from a vantage point above the surface, and relating of such measurements to the nature and distribution of surface materials and atmospheric conditions. Sensors mounted on aircraft or satellite platforms measure the amounts of energy reflected from or emitted by the earth's surface. These measurements are made at a large number of points distributed either along a one-dimensional profile on the ground below the platform or over a two-dimensional area on either side of the ground track of the platform. The sensors scan the ground below the satellite or aircraft platform and as the platform moves forward, an image of the earth's surface is built up. Fig. 3.1 shows how a sensor on board satellite scans along line AB. Each scan line of a remotely sensed image is a digital or numerical record of radiance measurements made at regular intervals along the line. A set of consecutive scan lines forms an image (Mather, 1987). Two-dimensional image data can be collected by means of two types of imaging sensors, namely, nadir looking or side looking sensor.

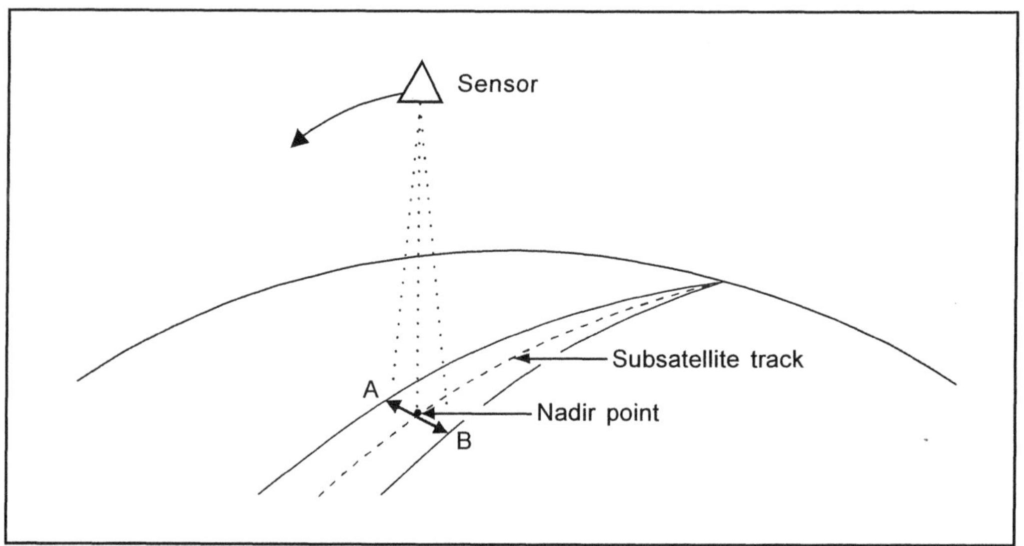

Fig. 3.1 Sensor on-board satellite scans along line AB. As the platform moves forward, an image of the swath region is built up.

In the case of nadir looking, the ground area to either side of the satellite or aircraft platform is imaged, whereas an area of the earth's surface lying to one side of satellite track is imaged by means of side looking sensor. Spatial patterns evident in remotely sensed images are interpreted in terms of geographical variation in the nature of material forming the surface of the earth. Such materials may be vegetation, exposed soil and rock, or water. These materials are not themselves detected directly by remote sensing, and their nature inferred from the measurements made. The characteristic of digital image data is that they can be adjusted so as to provide an estimate of physical measurements of properties of the targets, such as, radiance or reflectivity (Mather, 1987).

Broadly, there are two types of sensing systems to record the information about any target. They are active sensing system and passive sensing system. An active sensing system generates and uses its own energy to illuminate the target and records the reflected energy which carries the information content or entropy. Synthetic aperture radar (SAR) is one of the best examples of active sensing systems. These sensing systems operate in the microwave region of electromagnetic spectrum and include radiation with wavelengths longer than 1 mm. These systems do not rely on the detection of solar or terrestrial emissions as the solar irradiance in microwave region is negligible. The active remote sensing operation principles and the general details of latest imaging radar systems are described in the following chapter. The second type of remote sensing systems are passive systems mainly depending on the solar radiation operates in visible and infrared region of electromagnetic spectrum. The nature and properties of the target materials can be inferred from incident electromagnetic energy that is reflected, scattered or emitted by these materials on the earth's surface and recorded by the passive sensor (for example, a camera without flash). The remote sensing system that uses electromagnetic energy can be termed as electromagnetic remote sensing.

3.2 Electromagnetic Remote Sensing Process

The generalised processes involved in electromagnetic remote sensing system or passive remote sensing system, namely, data acquisition and data analysis are outlined below and a schematic diagram of electro-magnetic remote sensing process is shown in Fig. 3.2. The data acquisition process comprises distinct elements, namely, (i) energy sources, (ii) propagation of energy through the atmosphere, (iii) energy interactions with earth's surface features (iv) airborne/spaceborne sensors to record the reflected energy and (v) generation of sensor data in the form of pictures or digital information. These elements are described in detail further in this chapter.

The data analysis process involves examining the data using various viewing instruments to analyse pictorial data which is called the 'visual image interpretation techniques. Use of computers to analyse digital data through a process is known as digital image processing techniques. The analysis of a data utilising visual image interpretation involves use of the fundamental picture elements, namely tone, texture, pattern, size and shape in order to detect and identify various objects. Aerial or satellite imagery are seen through stereoscopic instruments to obtain three-dimensional images. There are many photogrammetric instruments today for visual-interpretation and for transferring the details on to base maps. If the data is available in digital form, it can be analysed on interactive computer systems for extracting statistical data or classified to obtain thematic information about resources. The scene is interactively analysed using computers by comparing with the actual "signature" of the object collected through field visits. This system of classification of objects is quite accurate and depends on the dispersion of training data sets over the area of the scene.

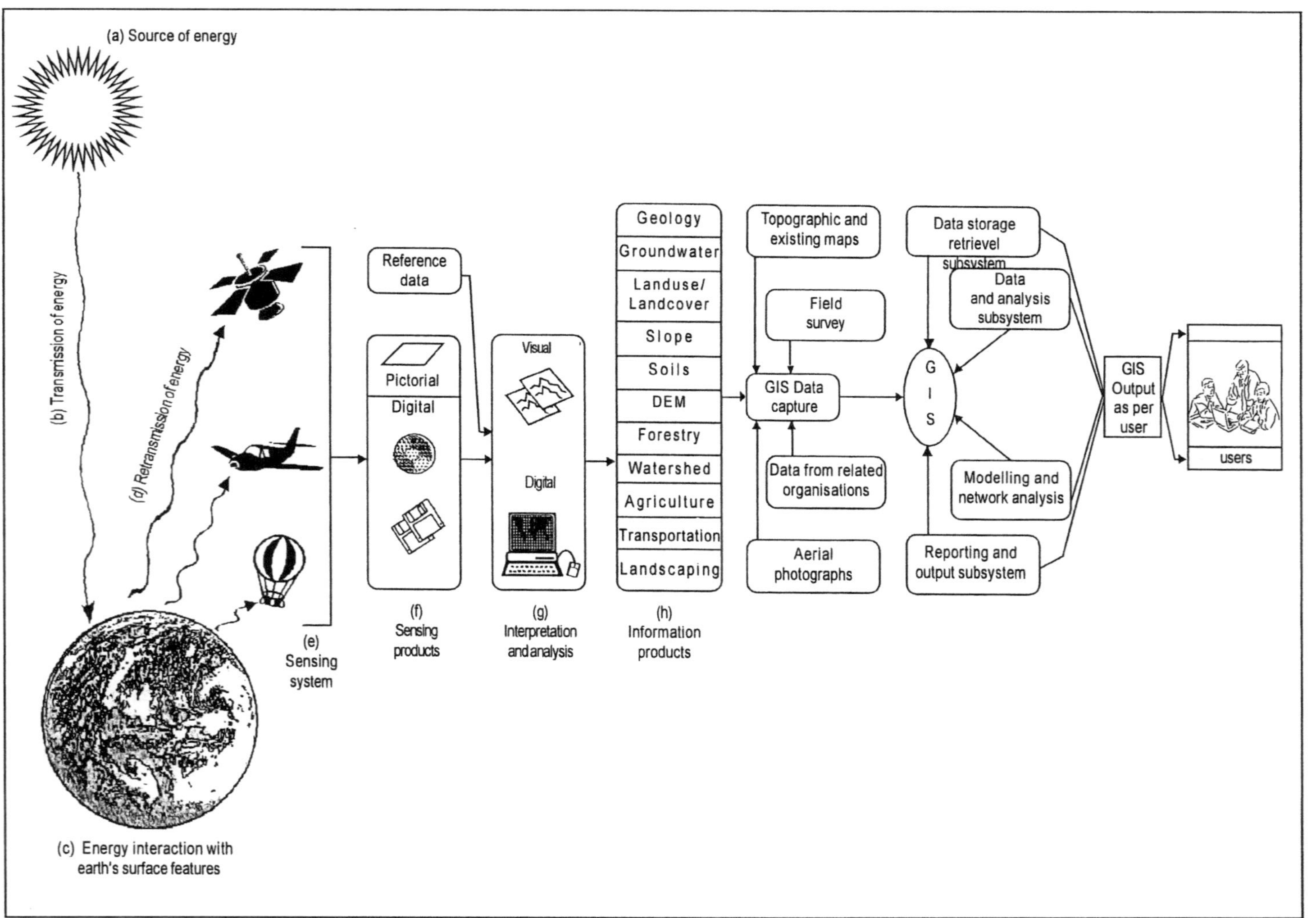

Fig. 3.2 Electromagnetic Remote Sensing Process with overview on GIS.

Digital data can also be used for certain functions like edge detection and height extraction, using specialised image processing techniques. Computer analysis of remotely sensed images offers several advantages like quick processing of large volumes of data and special image processing possibilities, such as geometrical and other types of corrections, scale changing, band ratioing, contrast enhancement, edge enhancement, and feature enhancement. Besides the standard peripherals, the digital system consists of high resolution display terminal for interaction with the resource scientist and output devices like plotters. Sample reference ground data, called training data, is collected and used conjunctively in order to obtain a more accurate thematic information by means of image classification. This thematic information layers are then used as input data for a GIS.

Whether it is the pictorial data analysis or digital data analysis, reference data should be generated as supporting data for the entire remote sensing data analysis. Reference data also called as ground truth or field check, is an essential part of remote sensing data processing. Reference data is used to serve any or all of the following purposes: (i) to analyse and interpret remotely sensed data, (ii) to calibrate a sensor, and (iii) to verify information extracted from remote sensing data. (Lillesand, T.M. and keifer, R.W, 1994). Collection of reference data consists of either time-critical or time-stable measurements. Time critical measurements are those made in cases where ground conditions, such as, vegetation conditions or water pollutants which change rapidly with time. Time stable measurements like the geology of the area of interest are those involving the materials under observation, which do not change with time.

3.3 Physics of Radiant Energy

Radiant energy, which like all other energies expressed in Joules, is the energy associated with electromagnetic radiation. The rate of transfer of radiant energy is called the flux and has watts as the units of power. Density implies distribution over the surface on which the radiant energy falls. If radiant energy falls upon a surface then the term irradiance (E) is used in place of radiant flux density. If the flow of energy is away from the surface, as in the case of thermal energy emitted by the earth or incoming solar energy which is reflected by the earth, then the term radiant exitance or radiant emittance as measured in units of Wm^{-2} is used (Mather, 1987).

Radiance (L) is defined as the radiant flux density transmitted from a small area on the earth's surface and viewed through a unit solid angle. It is measured in watts per square meter per steradian ($Wm^{-2}\ Sr^{-1}$). The concepts of the radian and steradian are illustrated in Fig. 3.3. The other important terms we come across remote sensing technology is 'reflectance' denoted by e. It is defined as the ratio between the irradiance and the radiant emittance of an object. When remotely sensed images collected over a time period are to be compared, it is most appropriate to convert the radiance values recorded by the sensor into reflectance in order to eliminate the effects of variable irradiance over the seasons of the year.

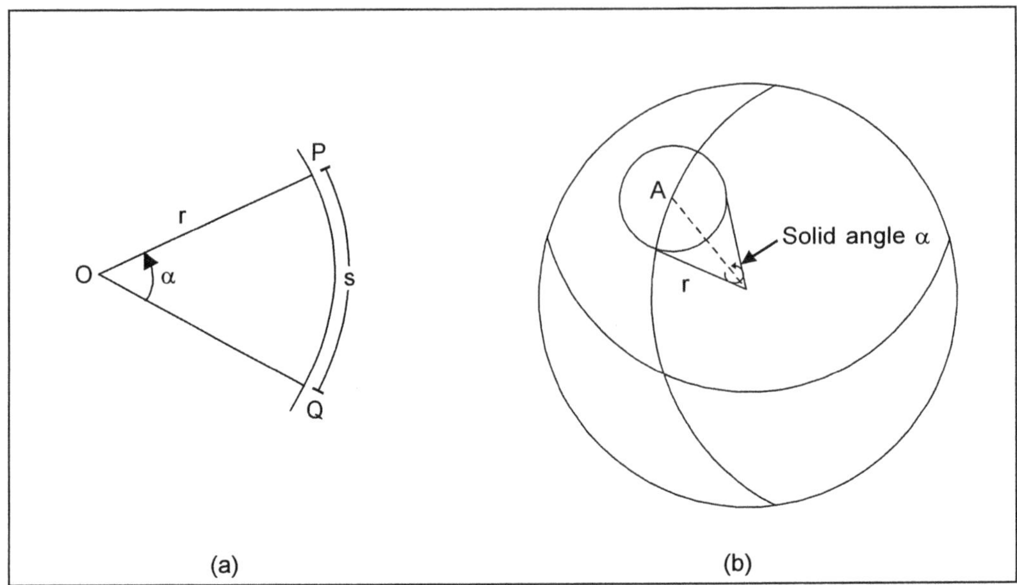

Fig. 3.3 (a) The angle formed when arc length s equals r, the radius of the circle, is equal to 1 radian. Thus, angle α = s/r radians. There are 2π radians (360 degrees) in a circle.

(b) A steradian is the solid (three-dimensional) angle formed when the area A delimited on the surface of the sphere is equal to the square of the radius of the sphere. A need not refer to a uniform shape. Thus angle a = A/r^2 steradians (sr). There are 4π radians in a sphere. (Mather, 1987)

The reflectance characteristic of earth's surface features may be quantified by measuring the portion of incident energy that is reflected. It is a dimensionless quantity. The quantities described above are very often used to refer to particular narrow wavebands rather than to the whole spectrum. The terms are then preceded by the word 'spectral', as in 'spectral radiance for a given waveband is the radiant flux density in the waveband per unit solid angle per unit wavelength' (Curran, 1988). The sun's light is the form of electromagnetic radiation most familiar to human beings. The light as reflected by physical objects travels in a straight line to the observer's eye. On reaching the retina, it generates electrical signals which are transmitted to the brain by the optic nerve. These signals are used by the brain to construct an image of the viewer's surroundings. This is the process of vision and it is closely analogous to the process of remote sensing; indeed, vision itself is a form of remote sensing. The set of all electromagnetic waves is called the electromagnetic spectrum, which includes the range from the long radio waves, through the microwave and infrared wavelengths to visible light waves and beyond them to the ultraviolet and to the short wave X-and gamma rays (Fig. 3.4).

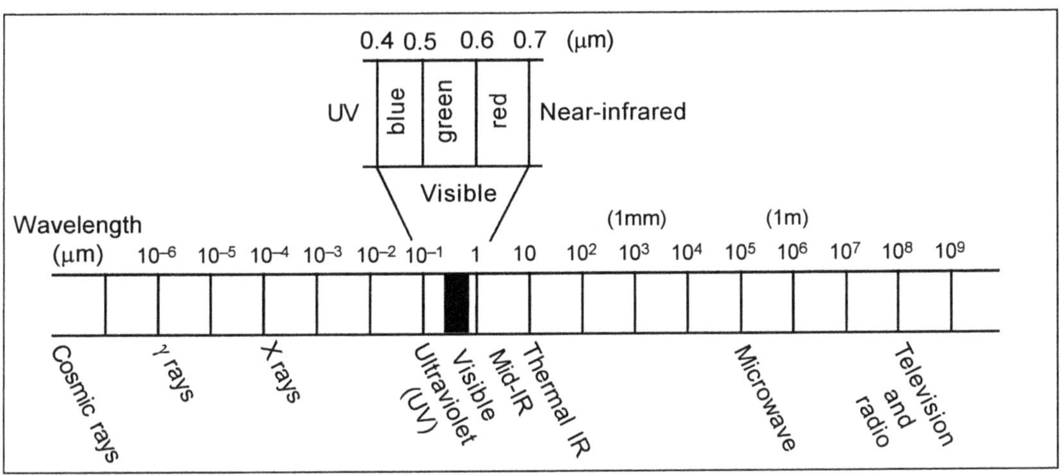

Fig.3.4 Electromagnetic Spectrum.

3.3.1 Nature of Electromagnetic Radiation

A basic difficulty encountered in analysing electromagnetic energy and interaction with matter derives from the dual nature of the behavior of electromagnetic energy. To be precise, in some situations electromagnetic energy behaves like waves, while in others it displays the properties of particles. This has been a very controversial point for the past 250 years concerned the nature of electromagnetic energy. This controversy has been sufficient to explain the nature of visible light, even though originally it was not realised that light is a form of electromagnetic energy.

Electromagnetic wave theory formulated by Maxwell in 1962, succeeded in characterising the electric and magnetic fields and their relation to charges and current, and expressing these relationships in a set of partial differential equations now known generally, as Maxwell's equations. Maxwell demonstrated that it was possible to have wave-like configurations of electric and magnetic fields. Maxwell's equations explain a great variety of phenomena relating to propagation, dispersion, reflection, refraction, and interference of electromagnetic waves; but they do not explain the interaction of electromagnetic energy with matter on an atomic and molecular level. In 1990 Planck found that, in order to calculate the correct distribution of energy emitted by a black body, he could not assume that the constituent oscillators gain and lose energy continuously. He was rather forced to assume that a particular oscillator of frequency 'v' is able to exist only in discrete states whose energies are separated by the interval hv, where 'h' is known as the Planck's constant. Planck's ideas were applied and extended shortly afterwards.

The decisive step was provided in 1925 by De Broglie who proposed that waves accompanied particles, or particles had a wave nature such that l = h/mv, where l is the wavelength and mv (mass x velocity) is the momentum. However, it was Schrodinger in 1926 who formulated wave mechanics in terms of a wave equation. The Schrodinger wave equation for atomic-molecular scale problems is not really derivable, and should be regraded as the counterpart of Newton's laws of motion for macroscopic bodies. It is used and accepted, not because of its derivation showing validity, but because when properly applied it yields correct results consistent with observation and experiment. The Schrodinger wave equation directly yields the allowed energy levels of an atomic or molecular systems.

Based on the historical development of understanding the nature of electromagnetic energy, it is presently possible to furnish a consistent and unambiguous theoretical explanation for all optical phenomena using a combination of Maxwell's electromagnetic wave theory and the modern quantum theory. Maxwell's theory deals primarily with the propagation and macroscopic optical effects of electromagnetic energy, while quantum theory is concerned with the atomic molecular absorption and emission aspects of radiation.

Maxwells Theory

The four differential equations that form the basis of electromagnetic theory are generally referred to as "Maxwells's equations," and they are expressed in mathematical terms. The electric and magnetic fields may exist in regions where no electric charges are present. When the fields at one point in space vary with time, then some variation of the fields must occur at every other point in space at some other time, and consequently, changes in the fields propagate throughout space. The propagation of such a disturbance is called an electromagnetic wave. According to Maxwell the electromagnetic state at a point in a vacuum can be specified by two vectors : E, the electric field in volts per meter and H, the magnetic field in ampere turns per meter. These vector quantities are completely independent of each other in the static case, and are determined by the distribution of all charges and currents in space. In the dynamic case, however, the fields are not independent, but rather their space and time derivatives are interrelated as expressed by the curl (∇) equations.

$$\nabla \times \mathbf{E} = \mu_0 \frac{\partial \mathbf{H}}{\partial t} \text{ and } \nabla \times \mathbf{H} = \varepsilon_0 \frac{\partial \mathbf{E}}{\partial t} \qquad ....(3.1)$$

where $\mu_0 \cong$ permeability of the vacuum = 4 $\Delta \times 10^{-7}$ h/m and $\varepsilon_0 \cong$ permitivity of the vacuum = 8.85×10^{-12} farads/m. and

$$\nabla . \mathbf{E} = 0 \quad \text{and} \quad \nabla . \mathbf{H} = 0 \qquad ....(3.2)$$

Eq. (3.2) indicate that there is no charge at the point in question, and this is true in both the static and the dynamic cases. These four equations are "Maxwell's equations" for a vacuum. The four equations or both the fields satisfy the same partial differential equation,

$$\nabla^2(X) = \frac{1}{c^2} \times \frac{\partial^2(X)}{\partial t^2} \qquad \text{..........(3.3)}$$

where X = E or H, and c = $\frac{1}{\mu_0 \varepsilon_0}$ and this is called the wave equation, which occurs in connection with many different kinds of physical phenomena. The major implication of the equation is that changes in the fields E and H propagate through space at a speed equal to the constant value c, which is known as the speed of light, with a measured value of 2.9979×10^8 m/s. The Maxwell curl equations are precisely the same for isotropic nonconduction media as they are for vacuum, except that the vacuum constants μ_0 and ε_0 are replaced by corresponding constants for the medium, denoted μ and ε. It can be shown, for the case where the spatial variation occurs in the z direction, that the function $E_z t = E_0 \cos(k_z - wt)$ is a solution to the wave equation. This is the fundamental solution to the wave equation, representing a plane harmonic wave for which the solution is the same as that for the magnetic field. It can be shown that the magnetic and electric components are perpendicular to each other and that these plane waves are both perpendicular to the direction of propagation (Fig. 3.5). In summary, it can be seen that all electromagnetic radiation is energy in

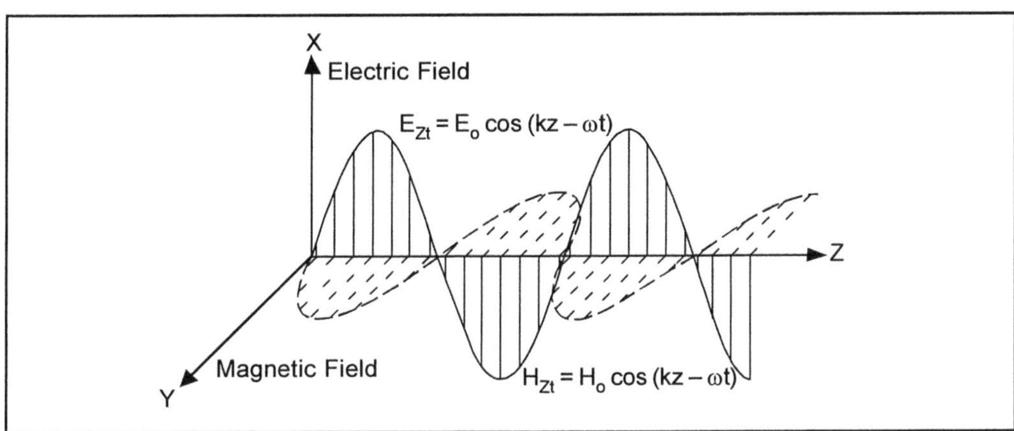

Fig. 3.5 Electromagnetic wave : the electric (vertical) and magnetic (horizontal) components are indicated.

transit and can be regarded as a wave motion. It consists of inseparable oscillating electric and magnetic fields that are always mutually perpendicular to each other and to the direction of propagation, the rate of propagation being constant in a vacuum.

Quantum Theory

The basic idea of quantum theory is that radiant energy is transmitted in indivisible packets whose energy is given in integral parts, of size hν, where **h** is Planck's constant = 6.6252 × 10^{-34} J - s, and ν is the frequency of the radiation. These are called quanta or photons.

The dilemma of the simultaneous wave and particle waves of electromagnetic energy may be conceptually resolved by considering that energy is not supplied continuously throughout a wave, but rather that it is carried by photons. The classical wave theory does not give the intensity of energy at a point in space, but gives the probability of finding a photon at that point. Thus the classical concept of a wave yields to the idea that a wave simply describes the probability path for the motion of the individual photons.

The particular importance of the quantum approach for remote sensing is that it provides the concept of discrete energy levels in materials.The values and arrangement of these levels are different for different materials. Information about a given material is thus available in electromagnetic radiation as a consequence of transitions between these energy levels. A transition to a higher energy level is caused by the absorption of energy, or from a higher to a lower energy level is caused by the emission of energy. The amounts of energy either absorbed or emitted correspond precisely to the energy difference between the two levels involved in the transition. Because the energy levels are different for each material, the amount of energy a particular substance can absorb or emit is different for that material from any other materials. Consequently, the position and intensities of the bands in the spectrum of a given material are characteristic to that material.

3.3.2 Electromagnetic Spectrum

The electromagnetic spectrum may be defined as the ordering of the radiation according to wavelength, frequency, or energy. The wavelength, denoted by λ, is the distance between adjacent intensity maximum (for example) of the electromagnetic wave, and consequently, it may be expressed in any unit of length. Most commonly wavelength is expressed in meters (m) or centimeters (cm); microns or micrometers (μ or μm = 10^{-4} cm); nanometers (nm = 10^{-7} cm); or Angstrom units (Å = 10^{-8} cm). The frequency denoted by ν, is the number of maxima of the electromagnetic wave that passes a fixed point in a given time. Its relationship to wavelength is simply, $\nu = C/\lambda$

where, 'c' is the speed of light. Frequency is commonly expressed in reciprocal centimeters, also called wave numbers (cm^{-1}) or cycles per second (cps) which are also called Hertz (Hz). The wavelengths may assume any value, although for most practical purposes the spectrum is usually presented between 10^{-16} and 10^{7} m, or from the cosmic ray to the audio range. However, wavelengths as long as 10^{11}m have been detected by sensitive magnetometers.

No matter what the wavelength of the electromagnetic radiation, it is all generated by electrically charged matter. However, there is no universal radiation generator that provides a useful intensity of radiation at all wavelengths for practical purposes, and there is no universal wavelength resolving instrument or universal detector. Consequently, the spectrum has been divided into regions that bear names related to the sources that produce it, such as, the "ray" regions, or as extensions from the visible range such as, the ultraviolet and the infrared regions, or according to the way in which wavelengths in a range are used such as, radio and television. The extent of the wavelength ranges corresponding to these names were made arbitrarily, and the decision as to where the divisions should be was made mostly on the basis of the limits imposed by the range of the human eye (visible), the properties of optical materials, and the response limits of various sources and detectors.

In brief, the electromagnetic spectrum is the continuum of energy that ranges from meters to nano-meters in wave length, travels at the speed of light, and propagates through a vacuum like the outer space (Sabins 1986). All matter radiates a range of electromagnetic energy, with the peak intensity shifting toward progressively shorter wave lengths at an increasing temperature of the matter. In general, the wavelengths and frequencies vary from shorter wave length high frequency cosmic waves to long wavelength low frequency radio waves. The wave lengths of greatest interest in remote sensing are visible and near-infrared radiation in the wave band 0.4 - 3 micrometers, infrared radiation in the wave band 3 - 14 micrometers and microwave radiation in the wave band 5 - 500mm.

Spectral Wave Bands

Visible light is electromagnetic radiation with wavelengths between 0.4 μm and 0.7 μm. The eye is not uniformly sensitive to light within this range and has its peak sensitivity at 0.55 μm (Fig. 3.6). This peak in the response function of the human eye corresponds closely to the peak in the sun's radiation emittance distribution. Electromagnetic radiation with wavelengths shorter than those of visible light (0.4μm) is divided into three spectral regions, namely, the gamma rays, X-rays, and ultraviolet rays. Because of the effect of scattering and absorption, none of these bands is used in satellite remote sensing.

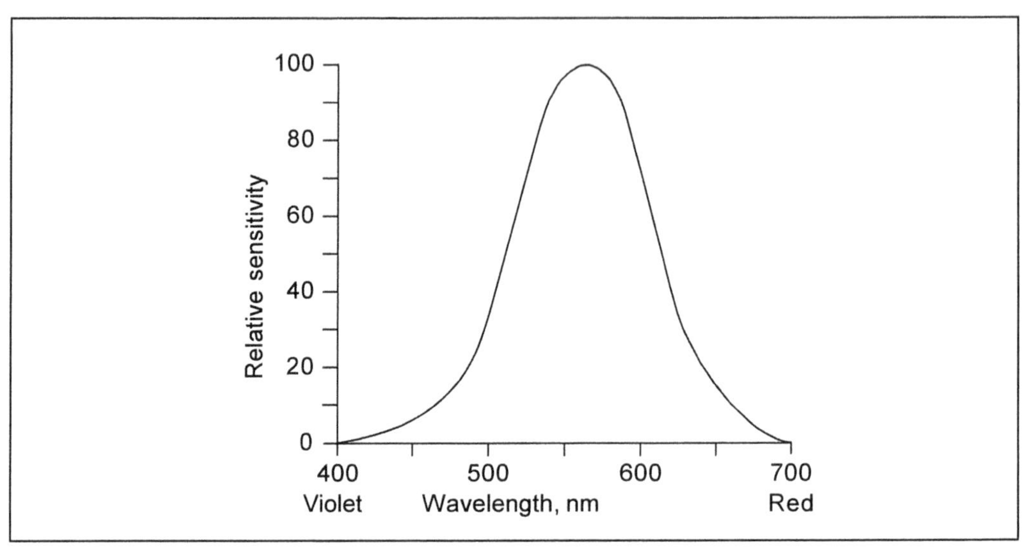

Fig. 3.6 Response Function of Human Eye.

Wavelengths longer than the visible red are sub-divided into the infrared (IR), microwave and radio frequency wavebands. The infrared waveband, extending from 0.7 μm to 1 μm is not a uniform region. Short wavelength or near - IR between 0.7 μm and 0.9 μm behaves like visible light and can be detected by special photographic film. Infrared radiation with a wavelength upto 3 μm is reflected by the surface of the earth. Beyond a wavelength of 3 μm, IR radiation emitted by the earth's surface can be sensed in the form of heat. The region of the spectrum composed of electromagnetic radiation with wavelengths between 1 mm and 300 cm is called the microwave band and radiation at these wavelengths can penetrate the clouds. The microwave band is thus a valuable region for remote sensing. Beyond the microwave region is the radioband of very long wavelengths used in certain radar applications. The electromagnetic wavebands with their utility in remote sensing are described in Table 3.1.

3.4 Energy Source and its Characteristics

All objects whose temperature is greater than an absolute zero (273°k), emit radiation. All stars and planets emit radiation. Our chief star, the sun is almost a spherical body with a diameter of 1.39×10^6 km at a mean distance from the earth equal to 1.5×10^8 km. The continuous conversion of hydrogen to helium which is the main constituents of the sun, generates the energy that is radiated from the outer layers. If the energy received at the edge of earth's atmosphere were distributed evenly over the earth, it would give an average incident flux density of 1367 w/m^2. This is known

Table 3.1 Electromagnetic spectral regions (Sabines, 1987)

Region	Wavelength	Remarks
Gamma ray	<0.03 nm	Incoming radiation is completely absorbed by the upper atmosphere and is not available for remote sensing.
X-ray	0.03 to 3.0 nm	Completely absorbed by atmosphere. Not employed in remote sensing.
Ultraviolet	0.3 to 0.4 μm	Incoming wavelengths less than 0.3 μm are completely absorbed by ozone in the upper atmosphere.
Photographic UV band	0.3 to 0.4 μm	Transmitted through atmosphere. Detectable with film and photodetectors,but atmospheric scattering is severe
Visible	0.4 to 0.7 μm	Imaged with film and photodetectors. Includes reflected energy peak of earth at 0.5 μm.
Infrared	0.7 to 1.00 μm	Interaction with matter varies with wave length. Atmospheric transmission windows are separated.
Reflected IR band	0.7 to 3.0 μm	Reflected solar radiation that contains information about thermal properties of materials. The band from 0.7 to 0.9 μm is detectable with film and is called the photographic IR band.
Thermal IR	3 to 5 μm band	Principal atmospheric windows in the 8 to 14 μm thermal region. Images at these wavelengths are acquired by optical mechanical scanners and special vidicon systems but not by film. Microwave 0.1 to 30 cm longer wavelengths can penetrate clouds, fog, and rain. Images may be acquired in the active or passive mode.
Radar	0.1 to 30 cm	Active form of microwave remote sensing. Radar images are acquired at various wavelength bands.
Radio	>30 cm	Longest wave length portion of electromagnetic spectrum. Some classified radars with very long wavelengths operate in this region.

as the solar constant. 35 percent of the incident radiant flux is reflected back by the earth. This includes the energy reflected by clouds and atmosphere. 17 percent of it is absorbed by atmosphere and 48 percent is absorbed by the earth's surface materials (Mather, 1987).

If the sun were a perfect emitter, it would be an example of an ideal black body. A black body transforms heat energy into radiant energy at the possible maximum rate consistent with Planck's law which defines the spectral exitance of a black body as follows (Henderson, 1970):

$$M_\lambda = \frac{C_1}{\lambda^5[\exp(C_2/\lambda T)-I]} \qquad \text{..........(3.4)}$$

Where, $C_1 = 3.742 \times 10^{-16}\, Wm^{-2}$, $C_2 = 1.4388 \times 10^{-2}\, m^0k$

λ = Wavelength (μm)

T = Temperature (in ^{0}k)

M_λ = Spectral exitance per unit wavelength.

Curves showing the variation of spectral exitance for black bodies at different temperatures are shown in Fig. 3.7. The wavelength at which the maximum spectral exitance is achieved is reduced as the temperature increases. The dashed line in Fig. 3.7 which joins the peaks of the spectral exitance curves, is described by

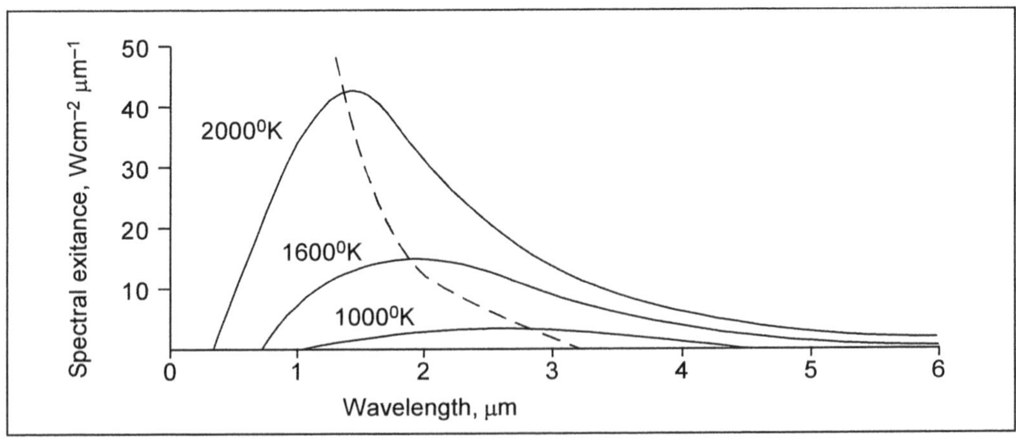

Fig. 3.7 Spectral Exitance curves for black bodies.

Wien's displacement law. The law gives the wavelength of maximum spectral exitance (λ_m) in the following form :

$$\lambda_m = C_3 / T \qquad \text{.........(3.5)}$$

where, $C_3 = 2.898 \times 10^{-3}$ mk, T = temperature of body

The total spectral exitance of a black body is given by the following Stefan Boltzman law:

$$M = \sigma T^4 \qquad \text{..........(3.6)}$$

in which, $\sigma = 5.6697 \times 10^{-8}\ Wm^{-2}\ k^{-4}$ (Stefan - Boltzman constant)

The distribution of the spectral exitance for a black body at 5900^{0}k closely approximates the sun's spectral exitance curve (Mather, 1987), while the earth can be considered to act like a black body with a temperature of 290^{0}k (Fig. 3.8.) The solar radiation, maximum of which occurs at 0.47 μm, is within the visible spectrum. Wavelength dependent mechanisms of atmospheric absorption alter the solar irradiance that actually reaches the surface of the earth. Fig. 3.9 shows the spectral irradiance from the sun at the edge of the atmosphere (solid curve) and at the earth's surface (dashed line).

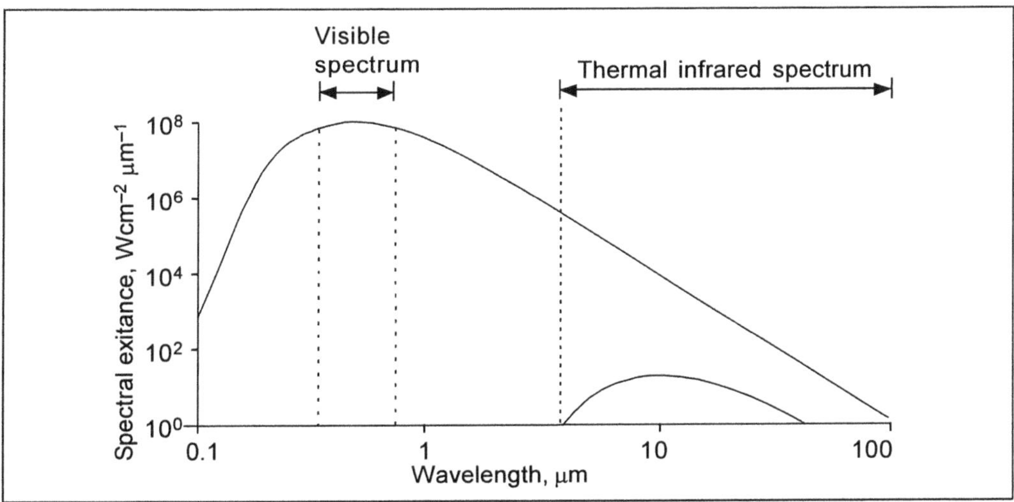

Fig. 3.8 Spectral exitance curves at temperatures of Sun and Earth.

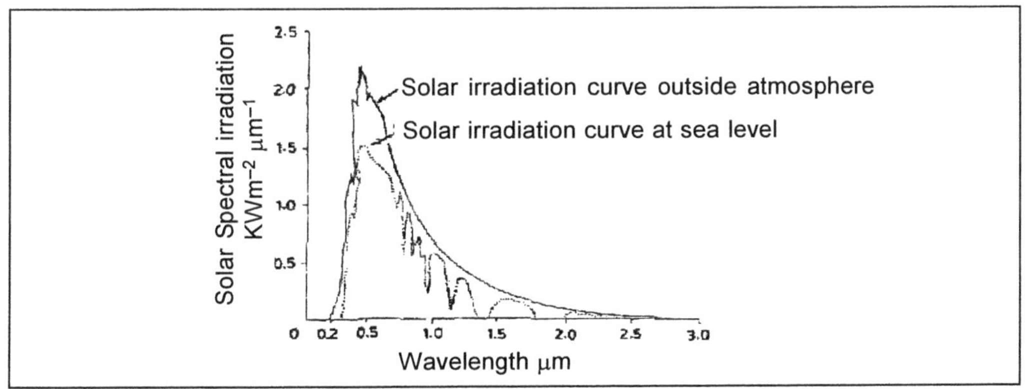

Fig. 3.9 Solar irradiance curves.

The characteristics of radiation sources impose some limitations on the range of wavebands that can be utilized in remote sensing. Generally, the selection of wavebands for use depends on (a) the characteristics of the radiation source, (b) the effects of atmospheric absorption and scattering, and (c) the nature of the target.

3.5 Atmospheric Interactions with Electromagnetic Radiation

All electromagnetic radiation detected by a remote sensor has to pass through the atmosphere twice, before and after its interaction with earth's atmosphere. This passage will alter the speed, frequency, intensity, spectral distribution, and direction of the radiation. As a result atmospheric scattering and absorption occur (Curran, 1988). These effects are most severe in visible and infrared wavelengths, the range very crucial in remote sensing.

During the transmission of energy through the atmosphere, light interacts with gases and particulate matter in a process called *atmospheric scattering*. The two major processes in scattering are selective scattering and non-selective scattering. Rayleigh, Mie and Raman scattering are of selective type. Non selective scattering is independent of wavelength. It is produced by particles whose radii exceed 10 μm, such as, water droplets and ice fragments present the clouds. This type of scattering reduces the contrast of the image. While passing through the atmosphere, electromagnetic radiation is scattered and absorbed by gasses and particulates. Besides the major gaseous components like molecular nitrogen and oxygen, other constituents like water vapour, methane, hydrogen, helium and nitrogen compounds play an important role in modifying the incident radiation and reflected radiation. This causes a reduction in the image contrast and introduces radiometric errors. Regions of the electromagnetic spectrum in which the atmosphere is transparent are called *atmospheric windows.* The atmosphere is practically transparent in the visible region of the electromagnetic spectrum and therefore many of the satellite based remote sensing sensors are designed to collect data in this region. Some of the commonly used atmospheric windows are 0.38 – 0.72 μm (visible), 0.72 –3.00 μm (near infrared and middle infra-red) and 8.00 –14.00 μm (thermal infrared). Fig. 3.10. Shows relative scatter as

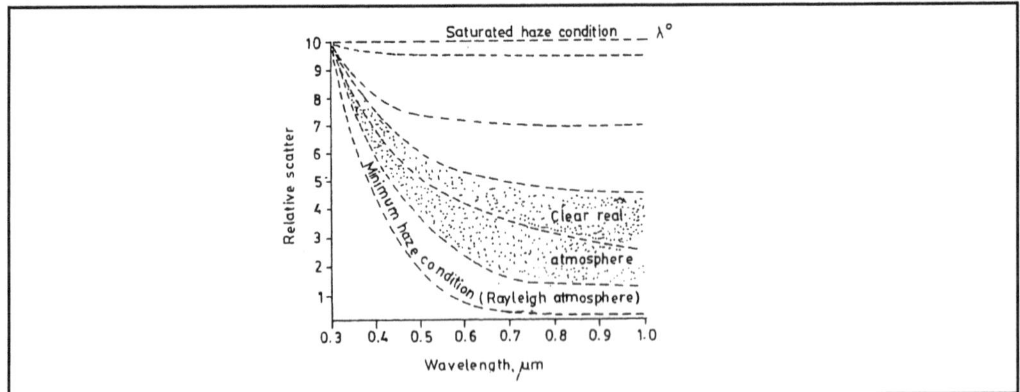

Fig. 3.10 Relative scatter for various levels of atmospheric haze.

a function of wavelength from 0.3 to 1 μm of the spectrum for various levels of atmospheric haze. The characteristics of all the four types of scattering in the order of their importance in remote sensing are given in Table 3.2.

TABLE 3.2

Types of Atmospheric Scatter in order of importance (Curran. 1988)

Type of Scatter of particles	Size of effective atmospheric particles	Type of effective atmospheric particles	Scatter of particles	Effect of scatter on visible and near visible wavelength
Rayleigh	Smaller than the wavelength of radiation.	Gas molecules	Molecule absorbs high energy radiation and re-emits. skylight scatter is inversely proportional to fourth power of wave length.	Affects short visible wave lengths,resulting in haze in photography, and blue skies.
Mie	Same size as the wavelength of radiation.	Spherical particles, fumes and dust	Physical scattering under overcast skies.	Affects all visible wave lengths
Non-selective	Larger than the wavelength of radiation.	Water droplets and dust.	Physical scattering by fog and clouds.	After all visible wave lengths equally, resulting white fog and clouds
Raman	Any	Any	Photon has elastic collision with molecule resulting in a loss or a gain in energy; this can decrease or increase wave length.	Variable

3.5.1 Atmospheric properties

The main part of the radiance measured from high flying aircraft or satellite stems form multiple scattering in the atmosphere. Therefore, the remaining signal can be interpreted in terms of suspensions only after a careful correction for the atmospheric contribution. For this reason the varying optical parameters of atmosphere must enter the radiative transfer calculations (Fischer J, 1989). Before we study the effects of solar radiation and atmospheric properties, we shall consider the mass quantities which determine the spectral upward radiance. The source of the shortwave radiation field in atmosphere is the Sun emitting in a broad spectral range. The extraterrestrial irradiance at the top of the atmosphere, the solar constant, depends on the black body emission of the Sun's photosphere and on the scattering and absorption process in the Sun's chromosphere. Important Fraunhofer lines caused by the strong absorption in the Sun's chromosphere show some prominent drops in the spectral distribution of the solar radiation. Fig. 3.11 shows the solar irradiance at the top of the earth's atmosphere to be between 0.4 and 0.8 μm as determined by Necked and Labs.

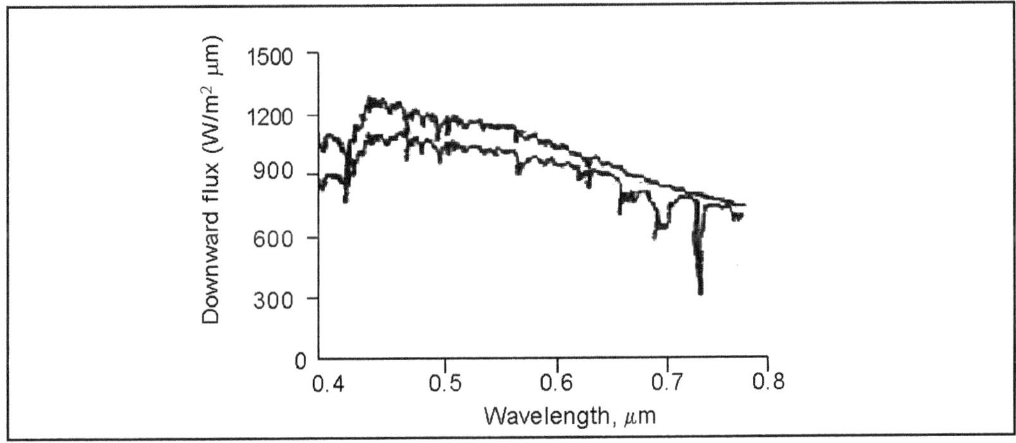

Fig. 3.11 Solar irradiance at the top of the atmosphere illuminating the Earth between 0.4 μm – 0.8 μm.

3.5.2 Absorption of Ozone

Ozone is a trace gas in the atmosphere mainly confined to stratospheric heights between 20 and 40 km with a maximum concentration near 25 km. At these levels, ozone dominates the short wave radiation budget, while at other heights its influence is nearly negligible. The Chappuis band of ozone in the visible spectrum is the only ozone band used to detect the oceanic constituents from space.

The transmission of the chlorophyll fluorescence to the top of the atmosphere is hindered through the absorption by water vapour and molecular oxygen in their vibration action bands. In order to study the selective gaseous absorption in the radiative transfer calculations the transmission functions of O_2 and H_2O are computed from absorption line parameters by explored through areas of Lorentz's theory of collision broadening. The contribution from resonance broadening is negligible in the spectral region considered. Also the Doppler line broadening, which is small when compared with Lorentz line widths, is neglected since the area absorption takes place in the atmosphere below 40 km (Barrow 1962). The transmission functions are averaged over 1nm wavelength intervals. The reduction in the solar flux due to absorbtion and scattering by a clear mid-latitude summer atmosphere. Response studies for the temperature and pressure dependence of the transmission function have been performed and show only a weak influence for the temperature effect. The pressure impact is not negligible and has to be accounted for. Air molecules are small compared to the wavelength of the incoming sunlight. Hence, the extinction through molecular scattering can be determined with Rayleigh theory. The necessary property for the determination of the scattering coefficient of the vertical profile of the atmospheric pressure has been estimated (Hunt, et. al., 1973). Since molecular scattering within the atmosphere depends mainly on pressure, the scattering coefficient can be estimated by climatological measurement.

Atmospheric spectral turbidity variations are caused by variations in aerosol concentration, composition and size distribution. The vertical distribution of the aerosols is taken from (Adler and Ken, 1963). The phase functions of aerosols are nearly wavelength independent within the visible and near infrared. For the radiative transfer calculations the scattering functions are estimated by Mie theory. The range of atmospheric turbidity values used to study the effects of aerosol scattering on the measured spectral radiances correspond to horizontal visibilities at the surface between 6 and 88 km.

3.5.3 Atmospheric effects on Spectral Response Patterns

The energy recorded by a sensor is always modified by the atmosphere between the sensor and the ground. As shown in Fig. 3.12, the atmosphere influences the radiance recorded by a sensor in two ways, namely, (a) it attenuates or reduces the energy illuminating a ground object and (b) the atmosphere acts as a reflector itself adding the path radiance to the signal detected by the sensor. These two atmospheric effects are expressed mathematically as follows (Lillesand, and Kiefer 1979):

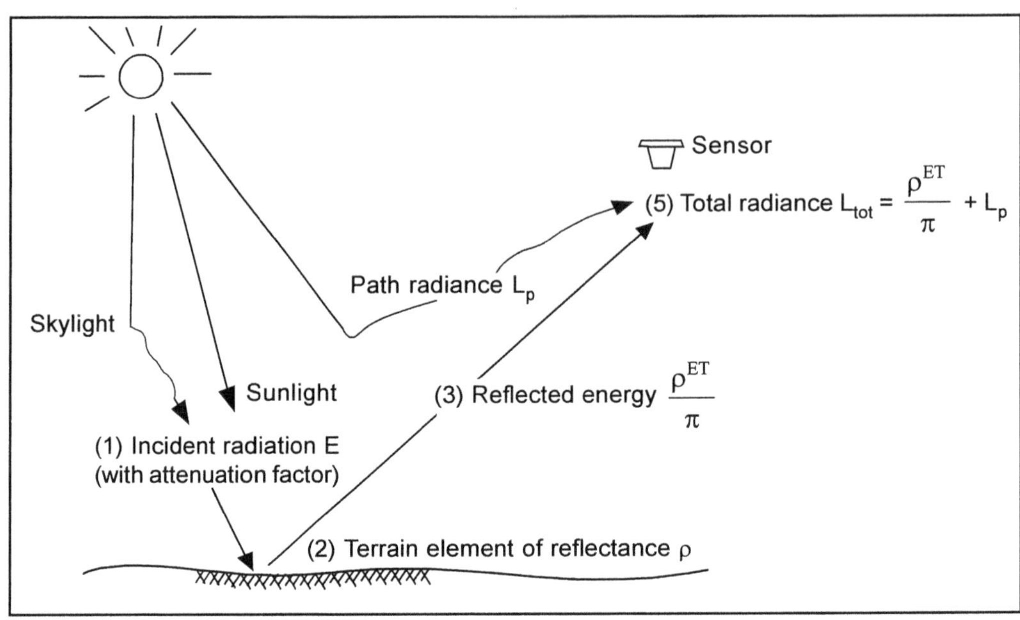

Fig. 3.12 Atmospheric effects influencing the Spectral Radiance.

$$L_{tot} = \frac{\rho ET}{\pi} + L_P \qquad \text{..........(3.7)}$$

Where, L_{tot} = Total spectral radiance measured by the sensor

L_P = Path radiance from the atmosphere

ρ = Reflectance of object

E = Irradiance on object or incoming energy.

T = Transmitted energy

The irradiance (E) is caused by directly reflected 'sunlight' and diffused 'skylight', which is the sunlight scattered by the atmosphere. The amount of irradiance depends on seasonal changes, solar elevation angle, and distance between the earth and sun.

3.6 Energy interactions with Earth's surface materials

When electromagnetic energy is incident on any feature of earth's surface, such as a water body, various fractions of energy get reflected, absorbed, and transmitted as shown in Fig. 3.13. Applying the principle of conservation of energy, the relationship can be expressed as:

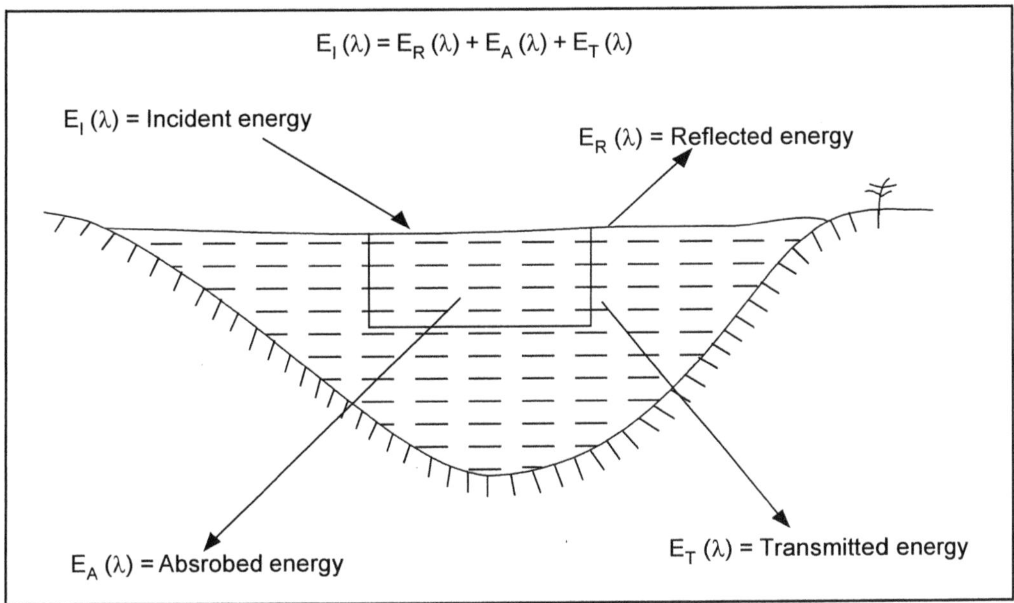

Fig. 3.13 Basic interactions between Electromagnetic Energy and a water body.

$$E_I(\lambda) = E_R(\lambda) + E_A(\lambda) + E_T(\lambda) \qquad \text{..........(3.8)}$$

Where, E_I = Incident energy

E_R = Reflected energy

E_A = Absorbed energy

and, E_T = Transmitted energy

All energy components are functions of wavelength, (l). In remote sensing, the amount of reflected energy $E_R(\lambda)$ is more important than the absorbed and transmitted energies. Therefore, it is more convenient to rearrange these terms like

$$E_R(\lambda) = E_I(\lambda) - [E_A(\lambda) + E_T(\lambda)] \qquad \text{..........(3.9)}$$

Eq. (3.9) is called balance equation. From this mathematical equation, two important points can be drawn. Firstly,

$$\frac{E_R(\lambda)}{E_I(\lambda)} = \frac{E_I(\lambda)}{E_I(\lambda)} - \left[\frac{E_A(\lambda)}{E_I(\lambda)} + \frac{E_T(\lambda)}{E_I(\lambda)}\right] \qquad \text{..........(3.10)}$$

According to principles of physics, it is known that

$\frac{E_R(\lambda)}{E_I(\lambda)}$; $\frac{E_A(\lambda)}{E_I(\lambda)}$ and $\frac{E_T(\lambda)}{E_I(\lambda)}$ are called reflectance, absorbance and transmittance and can be denoted as $\rho(\lambda)$, $\alpha(\lambda)$ and $\gamma(\lambda)$. Simply, it can be understood that, the measure of how much electromagnetic radiation is reflected off a surface is called its reflectance. The reflectance range lies between 0 and 1. A measure of 1.0 means that 100% of the incident radiation is reflected off the surface, and a measure '0' means that 0% is reflected. The reflectance characteristics are quantified by "spectral reflectance, $\rho(\lambda)$ which is expressed as the following ratio :

$$\rho(\lambda) = \frac{E_R(\lambda)}{E_I(\lambda)}$$

$$= \frac{\text{energy of wavelength '}\lambda\text{' reflected from the object}}{\text{energy of wavelength '}\lambda\text{' incident upon the object}} \quad \text{..........(3.11)}$$

Eq. (3.10) can be written as

$$\rho(\lambda) = 1 - [\alpha(\lambda) + \gamma(\lambda)] \quad \text{..........(3.12)}$$

Since, almost all earth surface features are very opaque in nature, the transmitance $\gamma(\lambda)$ can be neglected. According to Kirchoff's law of physics, the absorbance is taken as emissivity (ξ). Therefore Eq. (3.12) becomes :

$$\rho(\lambda) = 1 - \xi(\lambda) \quad \text{..........(3.13)}$$

Eq. (3.13) is the fundamental equation by which the conceptual design of remote sensing technology is built. If $\xi(\lambda)$ is a zero, then $\rho(\lambda)$, that is, the reflectance is one, which means, the total energy incident on the object is reflected and recorded by sensing systems. The classical example of this type of object is snow (white object). If $\xi(\lambda)$ is one, then (λ) is a zero indicating that whatever the energy incident on the object, is completely absorbed by that object. Black body such as lamp smoke is an example of this type of object. Therefore it can be seen that the reflectance varies from 0 (black body) to 1 (white body). When we divide the incident energy on both sides of the balance equation, we get the proportions of energy reflected, absorbed and transmitted which vary for different features of the earth depending on the material type. These differences provide a clue to differentiate between features of an image. Secondly, from the wavelength dependency of the energy components, it is evident that even within a given feature type, the proportion of reflected, absorbed, and transmitted energies may vary at different wavelengths. Thus two features which are indistinguishable in one spectral range, may exhibit a marked contrast in another wavelength band. Because many remote sensing systems operate in the wavelength regions in which reflected energy predominates, the reflectance properties of terrestrial features are very important.

3.6.1 Spectral Reflectance Curves

A basic assumption made in remote sensing is that a specific target has an individual and characteristic manner of interacting with incident radiation. The manner of interaction is described by the spectral response of the target. The spectral reflectance curves describe the spectral response of a target in a particular wavelength region of electromagnetic spectrum, which, in turn depends upon certain factors, namely, orientation of the sun (solar azimuth), the height of the Sun in the sky (solar elevation angle), the direction in which the sensor is pointing relative to nadir (the look angle) (Fig. 3.14), and nature of the target, that is, state of health of vegetation.

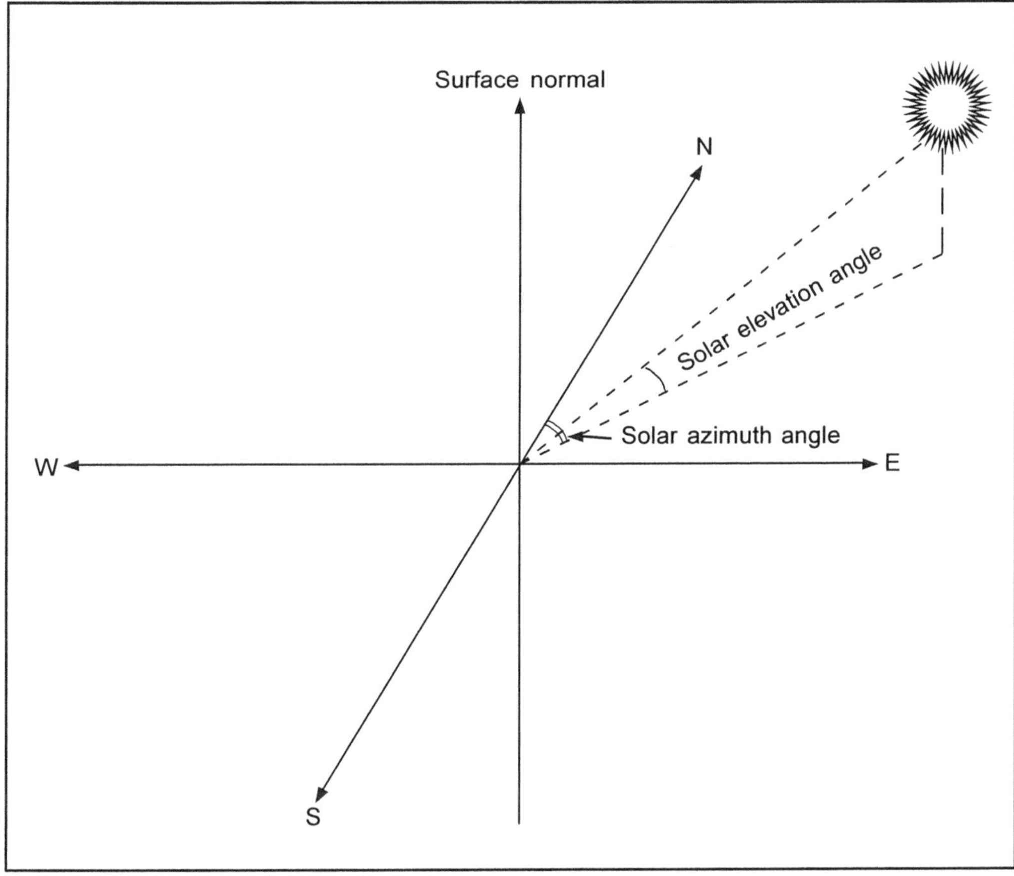

Fig 3.14 Solar elevation and azimuth angles. Elevation is measured upwards from the horizontal plane. Azimuth is measured clockwise from north. The zenith angle is measured from the surface angle, and equals 90 minus elevation angle, in degrees.

Every object on the surface of the earth has its unique spectral reflectance. Fig. 3.15 shows the average spectral reflectance curves for three typical earth's features : vegetation, soil and water. The spectral reflectance curves for vigorous vegetation manifests the "Peak-and-valley" configuration. The valleys in the visible portion of the spectrum are indicative of pigments in plant leaves. Dips in reflectance (Fig. 3.15) that can be seen at wavelengths of 0.65 μm, 1.4 μm and 1.9 μm, are attributable to absorption of water by leaves. The soil curve shows a more regular variation of reflectance. Factors that evidently affect soil reflectance are moisture content, soil texture, surface roughness, and presence of organic matter. The term spectral signature can also be used for spectral reflectance curves. Spectral signature is a set of characteristics by which a material or an object may be identified on any satellite image or photograph within the given range of wavelengths. Sometimes, spectral signatures are used to denote the spectral response of a target.

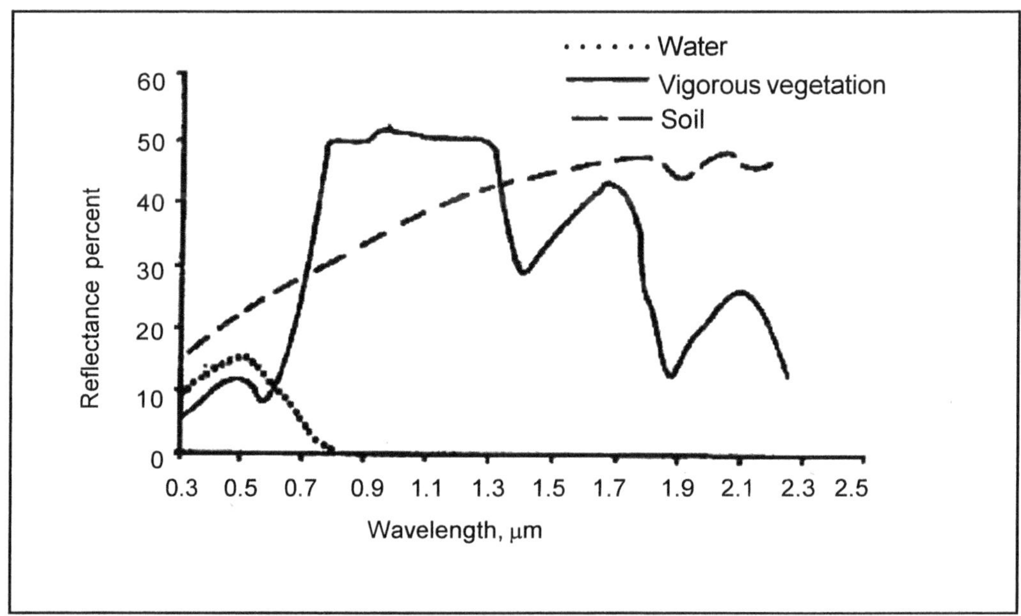

Fig. 3.15 Spectral Reflectance Curves or Spectral Signatures of Typical Features of earth's Surface.

The characteristic spectral reflectance curve Fig. 3.15 for water shows that from about 0.5μm, a reduction in reflectance with increasing wavelength, so that in the near infrared range, the reflectance of deep, clear water is virtually a zero (Mather, 1987). However, the spectral reflectance of water is significantly affected by the presence of dissolved and suspended organic and inorganic material and by the depth of the water body. Fig. 3.16 shows the spectral reflectance curves for visible and near-infrared wavelengths at the surface and at 20 m depth. Suspended solids in water scatter the down welling radiation, the degree of scatter being proportional to

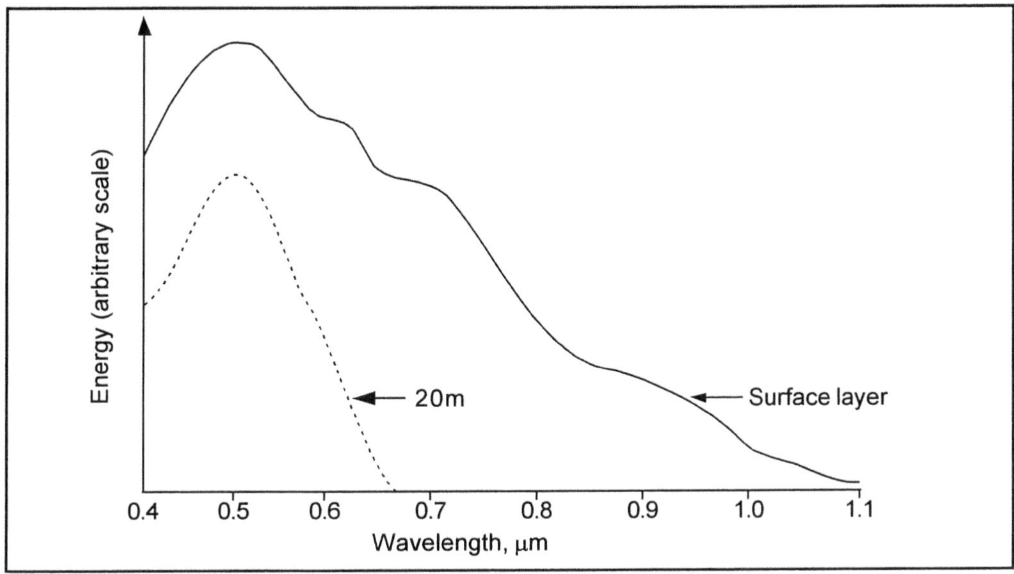

Fig. 3.16 Spectral reflectance curves for water at different depths.

the concentration and the colour of the sediment. Experimental studies in the field and in the laboratory as well as experience with multispectral remote sensing have shown that the specific targets are characterised by an individual spectral response. Indeed the successful development of remote sensing of environment over the past decade bears witness to its validity. In the remaining part of this section, typical and representative spectral reflectance curves for characteristic types of the surface materials are considerd. Imagine a beach on a beautiful tropical island. Fig. 3.17 shows interaction

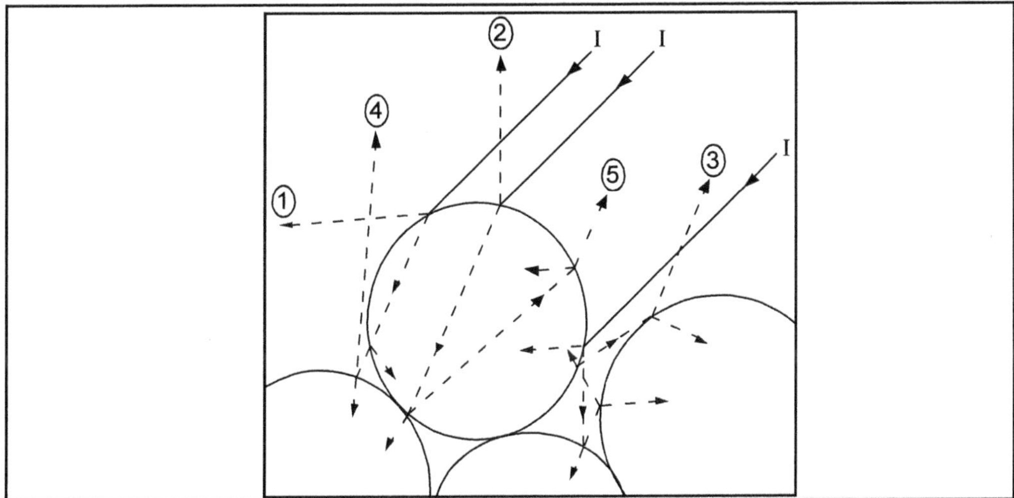

Fig. 3.17 Interaction between electromagnetic radiation and the top layer of particles comprising a mat surface. Solid lines (I) represent incident rays, lines 4 and 5 are volume rays. (After Vincent and Hunt 1968.) (Courtesy of "Applied Optics," Optical Society of America.)

of electromagnetic radiation with the top layer of sand grains on the beach. When an incident ray of electromagnetic radiation strikes an air/grain interface, part of the ray is reflected and part of it is transmitted into the sand grain. The solid lines in the figure represent the incident rays, and dashed lines 1, 2, and 3 represent rays reflected from the surface but have never penetrated a sand grain. The latter are called specular rays by Vincent and Hunt (1968), and surface-scattered rays by Salisbury and Wald (1992); these rays result from first-surface reflection from all grains encountered. For a given reflecting surface, all specular rays reflected in the same direction, such that the angle of reflection (the angle between the reflected rays and the normal, or perpendicular to the reflecting surface) equals the angle of incidence (the angle between the incident rays and the surface normal).

The measure of how much electromagnetic radiation is reflected off a surface is called its reflectance, which is a number between 0 and 1.0. A measure of 1.0 means the 100% of the incident radiation is reflected off the surface, and a measure of 0 means that 0% is reflected. In the case of first-surface reflection, this measure is called the specular reflectance, which will be designated here as $r^s(\lambda)$. The λ in parentheses indicates that specular reflectance is a function of a wavelength. The reason that $r^s(\lambda)$ is a function of a wavelength is that the complex index of refraction of the reflecting surface material is dependent on a wavelength. The term *complex* means that there is a real and imaginary part to the index of refraction. Every material has a complex index of refraction, though for some materials at some wavelengths, only the real part of the complex index of refraction may be nonzero.

For a sand grain with complex index of refraction $N(\lambda) = n(\lambda)[1 - ik(\lambda)]$, the specular reflectance is expressed by Fresnel's equation (Jenkins and White 1957), as follows :

$$r^s(\lambda) = \frac{[n(\lambda)-1]^2 + n^2(\lambda)k^2(\lambda)}{[n(\lambda)+1]^2 + n^2(\lambda)k^2(\lambda)} g(\theta, \phi) \qquad \text{..........(3.12)}$$

where,

$r^s(\lambda)$ = specular reflectance of one reflecting grain ($0 \leq r^s(\lambda) \leq 1$)

$n(\lambda)$ = intrinsic spectral index of refraction of the grain

$k(\lambda)$ = intrinsic spectral index of absorption of the grain

$g(\theta, \phi)$ = generally nonzero function of the angle of incidence (ϕ) and angle of observation (θ) with respect to the macroscopic surface

The final reflectance of a specular ray bouncing off multiple grains of sand is simply the multiplicative product of specular reflectance from all of the encountered air/grain interface. For instance, if the specular reflectance of three grains for a particular wavelength of electromagnetic radiation were 0.9, 0.8 and 0.7, respectively, the final reflectance of a specular ray bouncing off all three grains would be (0.9)(0.8)(0.7) = 0.504. The specular reflectance of the beach surface, $R^s(\lambda)$, is the average of all the individual specular ray reflectance.

Rays of electromagnetic radiation that have been transmitted through some portion of one or more grains are called volume rays. These are shown as dashed lines 4 and 5 in Fig.3.17. The equation for the volume reflectance, $r^s(\lambda)$, of a sand grain is complicated because it depends on both the transmittance of the grain and the interface reflectance of the top of that grain and the underlying grain (s).

The average $r^s(\lambda)$ for all the grains in the beach from which electromagnetic radiation is reflected is defined as the volume reflectance of the beach, $R^s(\lambda)$. The total reflectance of the beach, $R^T(\lambda)$, is the averaged sum of the specular and volume reflectance, as follows :

$$R^T(\lambda) = \frac{[R^S(\lambda) + R^V(\lambda)]}{2} \qquad \text{..........(3.13)}$$

The dependence of $R^s\lambda$ and $R^v(\lambda)$ are markedly different, as demonstrated in Fig.3.18 for the case of a uniform grain size and varying wavelength. Three important observations can be summarised from the above discussion on the beautiful beach island.

(i) The spectral locations of absorption bands depend on chemical composition of the material; for example, quartz and calcite absorption bands in the thermal infrared wavelength region have different spectral locations because Si and O ions in quartz are connected by a "spring" with a different bond strength than that of the " spring" connecting Ca and O ions in calcite.

(ii) The brightness, or magnitude, of the spectral reflectance depends primarily on the size of the reflecting grains.

(iii) Absorption bands appear as reflectance minima in transparent materials (such as quartz and calcite in the visible wavelength region), whereas absorption bands appear as reflectance maxima in opaque materials.

Note that when we use the terms transparent or opaque to explain optical behavior, we must designate both a wavelength region and the material because the complex index of refraction of any material is generally non-constant over a large range of wavelength.

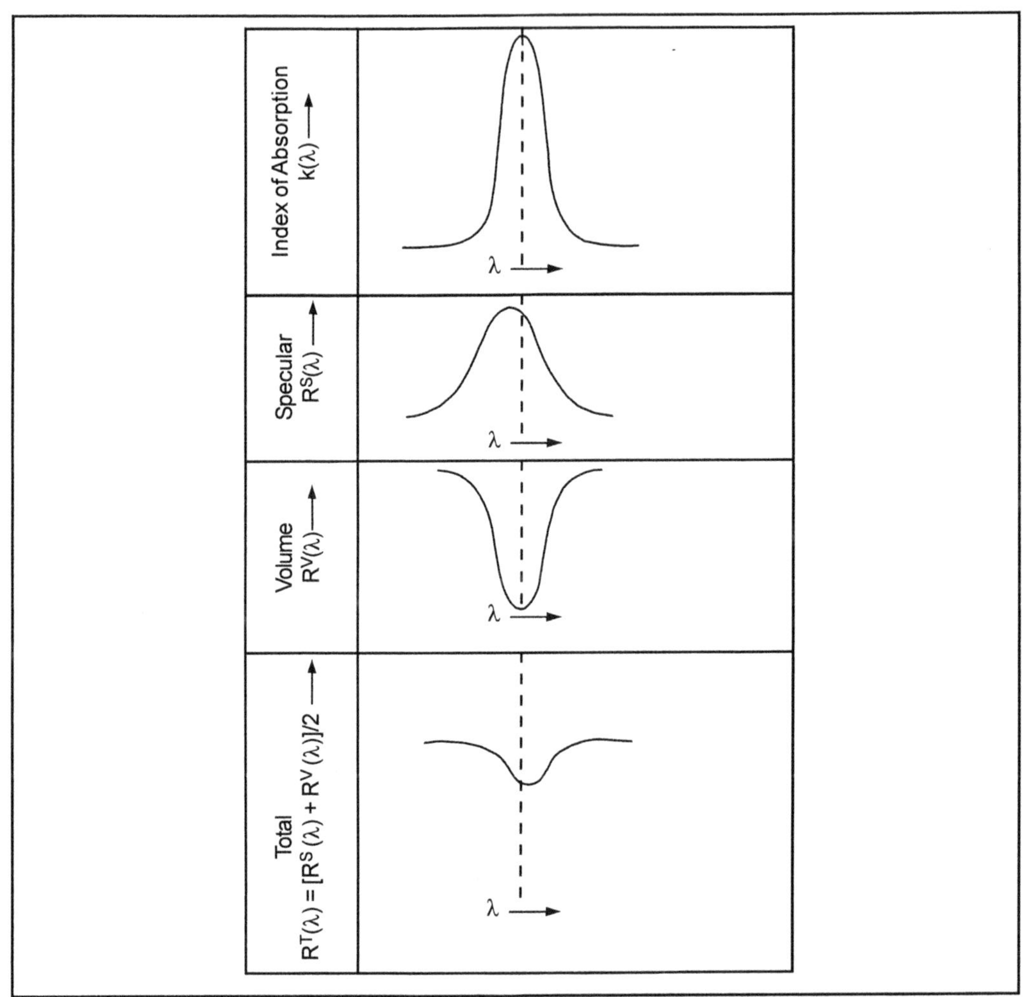

Fig. 3.18 Absorption coefficient, specular reflectance, volume reflectance, and total reflectance vs. wavelength for a spectral feature. (After Vincent and Hunt 1968.) (Courtesy of "Applied Optics," Optical Society of America.)

To consider the effect on reflectance of mixing several minerals together, let us take the simpler case of a particulate medium consisting of several mineral constituents, with air filling the interstices between particles. It is possible for us to estimate the spectral reflectance of a mixed-mineral particulate sample by using a linear combination of the reflectance spectra of its mineral constituents, weighed by the percentage of area on the sample's surface that is covered by each mineral constituent. The following equation demonstrates this estimation for the total spectral reflectance of a mixed particulate sample at wavelength λ (Robert K. Vincent., 1997) (Fig. 3.19).

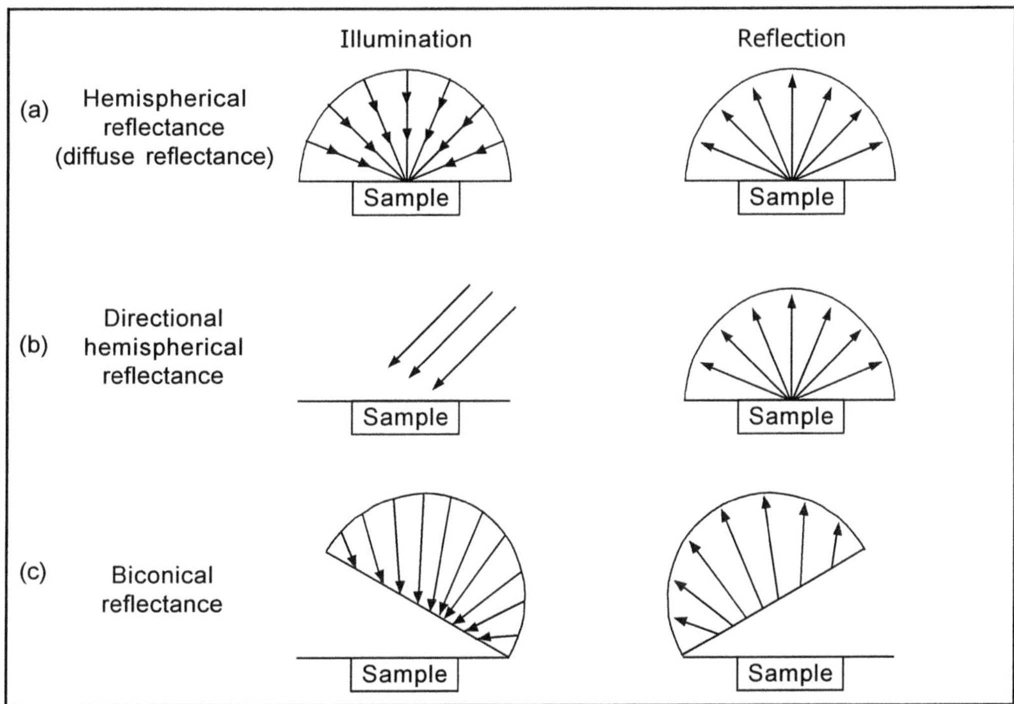

Fig. 3.19 Diagrams of the illumination and reflection geometries of (a) hemispherical reflectance. (b) directional hemispherical reflectance, and (c) biconical reflectance. (Courtesy of R. K. Vincent.)

$$R^T(\lambda) = \sum_{i=1}^{n} f_i R_i^T(\lambda) \qquad \text{..........(3.14)}$$

where

f_1 = fraction or percentage of the ith mineral constituent covering the sample surface, where $\sum_{i=1}^{n} f_i = 1.0$

n = total number of mineral constituents in the particulate sample

$R_i^T(\lambda)$ = spectral reflectance (the total of specular and volume reflectance) at wavelength l for the ith mineral constituent alone.

Thus so far we have talked about volume reflectance and specular reflectance on the basis of whether electromagnetic rays did or did not penetrate one or more grains in a soil or rock surface. Now we need to define some reflectance terms that relate to the manner in which the soil or rock surface is illuminated, as well as, how the reflected energy from its surface is measured. The most fundamental term for reflectance used in this book is defined as spectral hemispherical reflectance or diffuse reflectance.

3.7 Cossine Law

Assume that this is a photographic system (Fig. 3.20). Let

f = Focal length of the lens used

θ = Angle made by the normal with the axis of ray from the object to the image

d_{AS} = Object area

d_{Ai} = Image area

MK = R, distance from object area to lens system

KN = r, distance from image area to lens system.

We know that,

the solid angle subtended by object area = $\dfrac{d_{As}\cos\alpha}{R^2}$

the solid angle subtended by image area = $\dfrac{d_{Ai}\cos\theta}{r^2}$

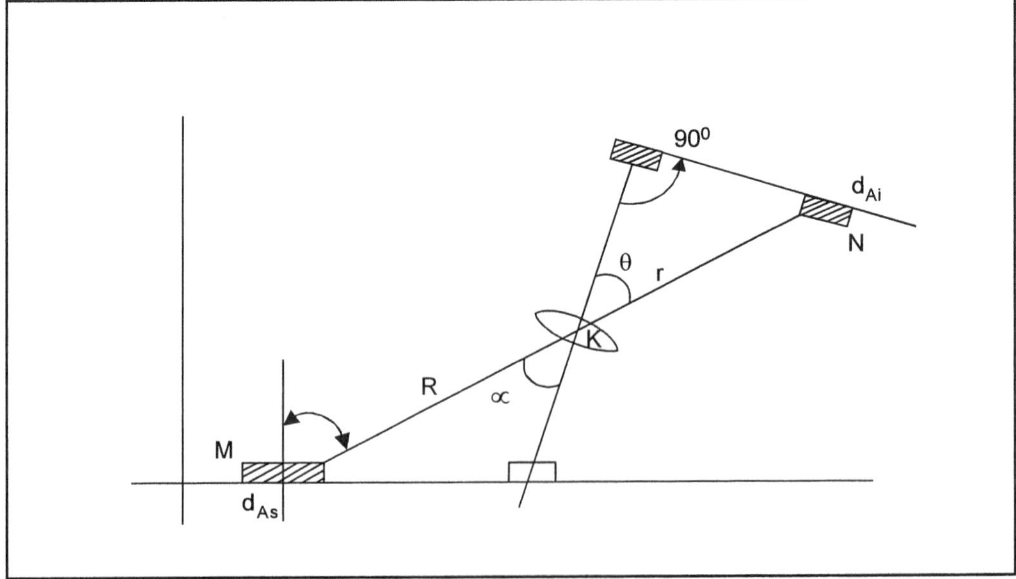

Fig. 3.20 Photographic system that explains the Cossine law.

At the centre of lens the solid angles are equal

That is, $$\frac{d_{As}\cos\alpha}{R^2} = \frac{d_{Ai}\cos\theta}{r^2}$$

$$d_{AS} = \frac{R^2 d_{Ai}\cos\theta}{r^2\cos\alpha} \qquad \text{..........(1)}$$

Now, $$dw = \frac{\text{Normal area}}{R^2}$$

We know that, Normal area = $\pi\,(D/2)^2\cos\alpha$

But we also know that,

$$L\ (\text{Radiance}) = \frac{d^2\theta}{d(A\cos\theta)dw}$$

$$df = Ld_{AS}\cos\alpha\ dw \qquad \text{..........(2)}$$

But, $$dw = \frac{\text{Normal area}}{R^2}$$

$$= \frac{\pi(D/2)^2\cos\alpha}{R^2} = \frac{\pi D^2\cos\alpha}{4R^2}$$

From (2) $$d^2f = L\,d_{AS}\,\text{Cos}\,\alpha\,\frac{\pi D^2\cos\theta}{4R^2}$$

$$= \frac{\pi D^2 L\cos\theta\cos\alpha\,d_{As}}{4R^2}$$

Finally, $$dE = \frac{\pi L\cos^4\gamma}{4N^2}$$

This relation is called Cossine Law. As "α" decreases the intensity of energy increases, and vice versa. This common effect in photographic images is called vignetting.

4

Microwave Remote Sensing

4.1 Introduction

The region of the spectrum composed of electromagnetic radiation with wavelength between 1mm and 1m is called the microwave band. The microwave band is a valuable region for remote sensing in view of two distinctive features, (i) Microwaves are capable of penetrating the atmosphere under almost all conditions. Depending on the wave lengths involved, microwave energy can 'see through' haze, light rain, snow, clouds, and smoke, (ii) Microwave reflections or emissions from earth materials bear no direct relationship to their counterparts in the visible or thermal portions of the spectrum. The surfaces that appear rough in the visible may be smooth in microwave. Remote sensing techniques in the microwave region of electromagnetic spectrum can be classified into two categories (Reeves, 1979): active microwave remote sensing, and passive microwave remote sensing. Active systems provide their own illumination, where as passive systems record the energy of thermal origin emitted from materials. Active microwave sensing systems are of two types and they are imaging sensors and non-imaging sensors. Most imaging sensor or imaging radars used for remote sensing are Side Looking Airborne Radar (SLAR). The radar is an acronym derived from Radio Detector and Ranging. These imaging radars are divided into two categories. The first category is real aperture, and the second one is synthetic aperture systems. In the real aperture system, resolution is determined by the actual beam width and antenna size. The synthetic aperture system utilises signal processing techniques to achieve narrow beam width in the long track direction which provides better resolution. The customary nomenclature used is SLAR for the real aperture system and SAR for synthetic aperture system. Non-imaging remote sensing radars are either scatterometers or altimeters. Any calibrated radar that measures the scattering properties of a surface is called scatterometer designed for back scatter measurements. Passive microwave sensors called radiometers, measure the emissive properties of the earth's surface.

A radar altimeter sends out pulses of microwave signals and record the signal scattered back from the earth surface. The height of the surface can be measured from the time delay of the return signals. A wind scatterometer can be used to measure wind speed and direction over the ocean surface. It sends out pulses of microwaves along several directions and records the magnitude of the signals that are back scattered from the ocean surface. The magnitude of the backscattered signals is related to the ocean surface roughness, which, in turn, is dependent on the sea surface wind conditions, so that the wind speed and direction can be derived. Imaging radars are side looking rather than nadir looking instruments and the geometry is complicated by foreshortening to the extent that the top of a mountain appearing closer to the sensor than the foot of the mountain, and shadow caused by the far side of a mountain or hill is being invisible to the side looking radar sensor. A microwave radiometer is a passive device which records the natural microwave emission from the earth. It can be used to measure the total water content of the atmosphere within its field of view. Application potential of radar remote sensing for various disciplines like soil moisture, agriculture, geology, hydrology, and oceanography, has been demonstrated through various ground based, aircraft and space craft experiments. This chapter provides the principles of radar remote sensing.

4.2 The Radar Principle

Radio Detection and Ranging (RADAR) is an active microwave sensing system which transmits electromagnetic radiation of wave length in the centimeter range as a source of illumination (self-illumination) to detect remote targets. The microwave portion of the spectrum includes wavelength within the approximate range of 1m. In active microwave remote sensing, the radar antenna transmits short burst (pulses) of energy to the target and echoes from these targets carry informations about the position (range) and quality of the illuminated objects. Therefore, the imaging radar consists of the operational units, namely, transmission of microwave pulses directed by reception of echoes as mean terrain information by the same antenna, signal correlation to generate radar raw data and data processing and production of final digital/optical image products. The radar equation relates the influence of the system and terrain parameters to the power received by the antenna (Reeves 1979) as shown below :

$$P_r = \frac{P_t G_t G_r}{(4\pi)^3 R^4} \lambda^2 \sigma \qquad \text{..........(4.1)}$$

where, P_t = transmitted power

p_r = received power

G_r = gain of receiver antenna

G_t = gain of transmitted antenna

R = distance between target and sensor

σ = scattering cross section

λ = wavelength

Most radars use either the same or identical transmitting and receiving antennas. Therefore,

$$G = G_t = G_r$$

Then the received power is given by

$$P_r = \frac{P_t G^2}{(4\pi)^3 R^4} \lambda^2 \sigma \qquad \text{..........(4.2)}$$

This radar equation is modified in view of the nature of remotely sensed areas on the ground in which each resolution cell contains many separate scatterers. Therefore, the power received from a resolution cell is a combined power obtained by adding the powers from these scatterers. Then the above equation can be modified as

$$P_r = \frac{P_t G^2}{(4\pi)^3 R^4} \sum_{i=1}^{N} \frac{P_{ti} G_i^2 \sigma_i}{R_i^4} \quad \text{where N = number of scatterers} \qquad \text{...........(4.3)}$$

The summation of powers of all scatterers can be replaced by an integer using an average value of scattering coefficient per unit area rather than the actual scattering cross section associated with each individual element. The above equation can be converted to

$$P_r = \frac{P_t G^2 \lambda^2 \sigma_0 d_A}{(4\pi)^3 R^4} \qquad \text{..........(4.4)}$$

Where σ_0 scattering coefficient, d_A = resolution of ground area

This form of radar equation is generally used in remote sensing and the quantity measured is back scattering coefficient (σ_0) which describes the properties of target on the ground. The back scattering coefficient, according to Elachi (1988), is defined as the ratio of the energy received by the sensor over the energy that the sensor would have received if the surface had scattered the energy incident on it in isotropic fashion. This is expressed in decibels (dB). Back scattering coefficient describes the terrain contributing factor to the radar image tone, and the radar cross sections per unit area (resolution cell). It is a result of the sensor-target interaction (Fig. 4.1). The backscattering coefficient can be a positive number focusing energy in the back direction or a negative number away from the back direction.

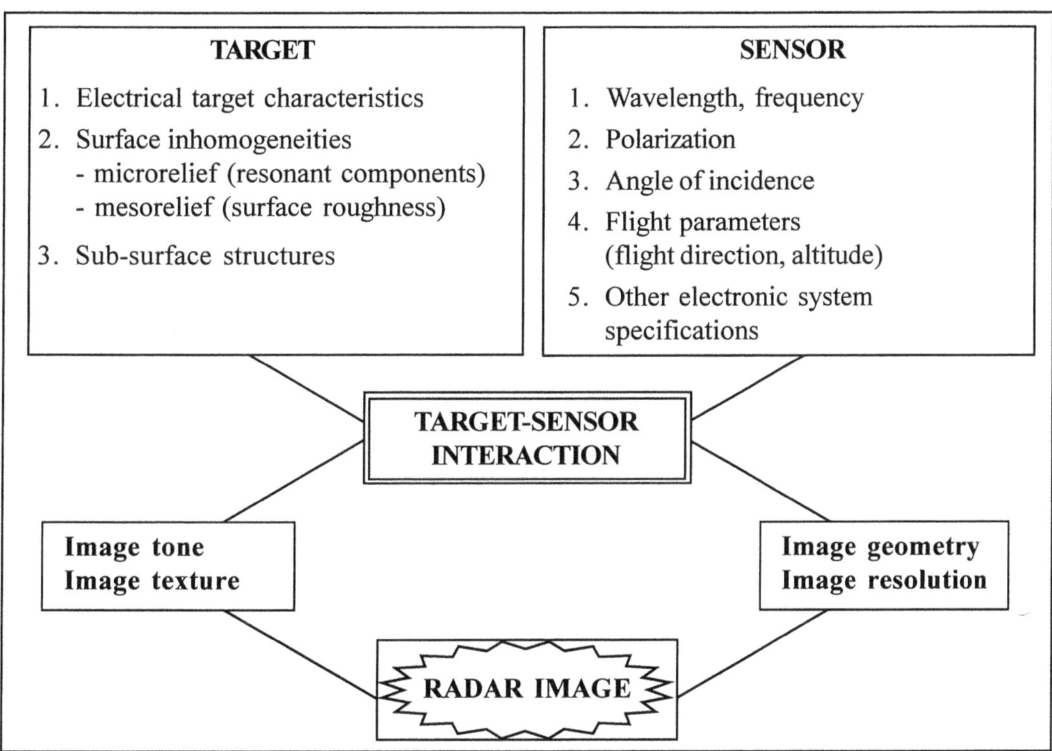

Fig. 4.1 Theoretical Scheme of Sensor - Target Interaction.

4.3 Factors affecting Microwave Measurements

Factors affecting the microwave signatures of an object are governed by the system parameters like frequency, polarisation, incident angle and scatter mechanism, geometric characteristics of radar imagery, transmission characteristics of radar signal, and image characteristics of radar signals. They are also governed by physical and electrical properties of the target. The electromagnetic property of materials is expressed by the complex relative permitivity (dielectric constant)

$$\varepsilon_s = \varepsilon' - j\varepsilon'' \qquad \text{..........(4.5)}$$

The imaginary part ε'' is related to the electrical conductivity, e, by

$$\varepsilon'' = \frac{e}{\omega\varepsilon_0} \qquad \text{..........(4.6)}$$

where ε_0 is the vacuum permitivity (8.854×10^{-12} farad/meter) and ω is the angular frequency. For nonmagnetic media, the index of refraction, n, is related to the permitivity by $n^2 = \varepsilon$. For a conducting medium, the amplitude of a wave propagation in it is attenuated exponentially with distance. The penetration depth, is defined as the depth below the surface at which the magnitude of the power of the transmitted wave is equal to the 37% (or 1/e) of the power of the transmitted wave at a point just beneath the surface.

4.3.1 Surface Roughness

Surface roughness is a function of the incident angle and wavelength. Rayleigh's criterion of surface roughness is given by

$$h < \frac{\lambda}{8 \operatorname{Cos}\theta} \qquad \text{..........(4.7)}$$

where h is the height variations above a plane and θ is the incident angle (Fig.4.2). Surfaces that are relatively smooth ($h = \lambda/8 \operatorname{Cos} \theta$) tend to reflect electromagnetic waves in accordance with Fresenel reflection coefficient (specular reflection). Therefore strong backscatter is observed only in nadir direction. Rough surfaces tend to reradiate uniformity in all the directions (diffuse scattering), so they give relatively strong radar returns in all the directions.

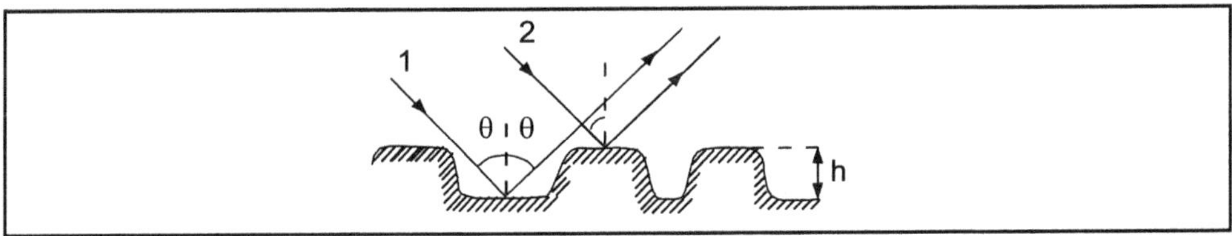

Fig.4.2 A sketch of surface roughness.

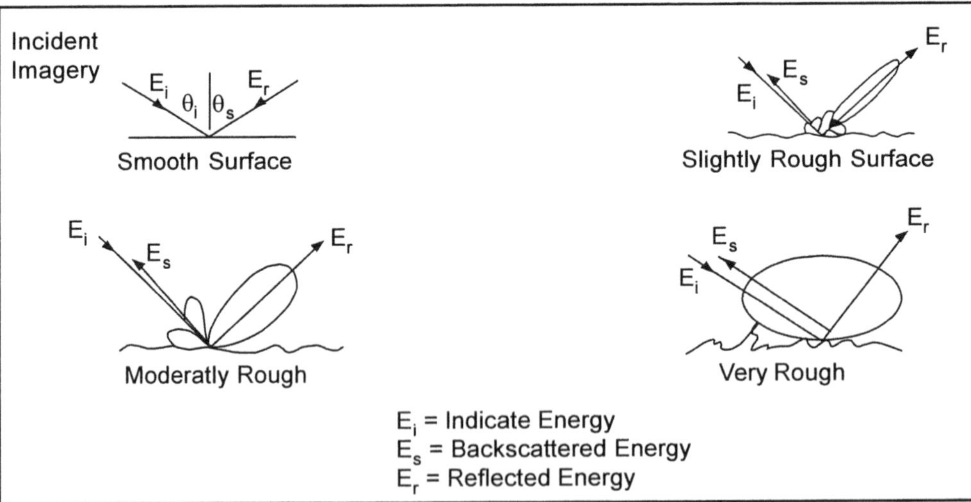

Fig.4.3 Effects of surface roughness on scattering of EM energy.

Scattering as a function of roughness is illustrated in Fig. 4.3. For a smooth surface where the surface roughness scale is much shorter than wavelength, incident energy is reflected off specularly as illustrated in Fig. 4.3. As the roughness scale approaches the same dimension as the wavelength, the scattered energy is dispersed, and when the roughness scale exceeds the wavelength of the incident energy, scattering is nearly uniform over the hemisphere.

More exact classification of surface roughness considering surface slopes is defined by Fung (1976) :

- Slightly rough surface : the height variations are small compared with the wavelength, and the surface slopes are small compared with unity.
- Smooth undulating surface : the height variations are comparable to or larger than the wavelength, and the surface is locally flat relative to the wavelength (Kirchhoff tangent plane approximation).
- Two scale composite rough surface : a large scale roughness satisfying the tangent plane approximation superimposed by a small scale roughness (for example, sea surface with large gravity and small capillary waves).

4.3.2 Radar Scattering Mechanism

Radar backscattering coefficient variation from a terrain may be the result of surface scattering, volume scattering, or both. Surface scattering is caused normally at the air-ground interface, whereas volume scattering is caused by the dielectric discontinuities in a volume. The relative importance of surface and volume scattering is governed by the surface statistics of target boundary, the inhomogeneity of the

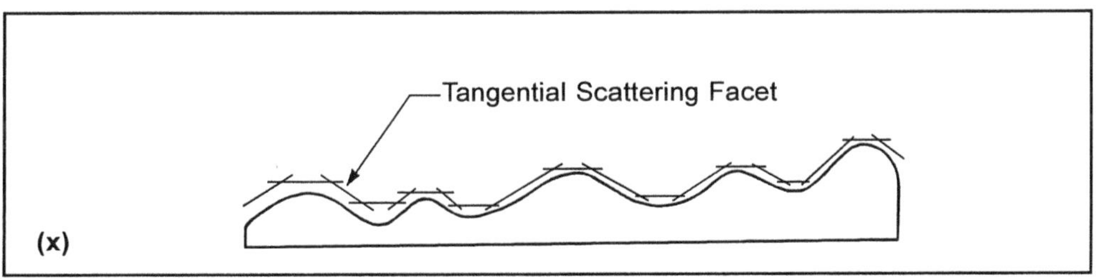

medium underneath the surface, and the penetration depth of the medium. The surface scattering mechanism is an important component of radar scattering process. In general, surface scattering occurs at the air-ground interface. For a perfectly smooth surface, the incident wave will excite the atomic oscillators in the dielectric medium at a relative phase such that the reradiated field consists of two plane waves, namely, reflected wave and refracted wave or transmitted wave (Fig. 4.4). For rough surface, energy is scattered in all directions depending upon the roughness of the surface as well as dielectric properties of the surface. The roughness can be statistically characterised by its standard deviation relative to the mean flat surface. The surface correlation length is the separation after which two points are statistically independent, that is, the length after which auto-correlation function is less than 1/e.

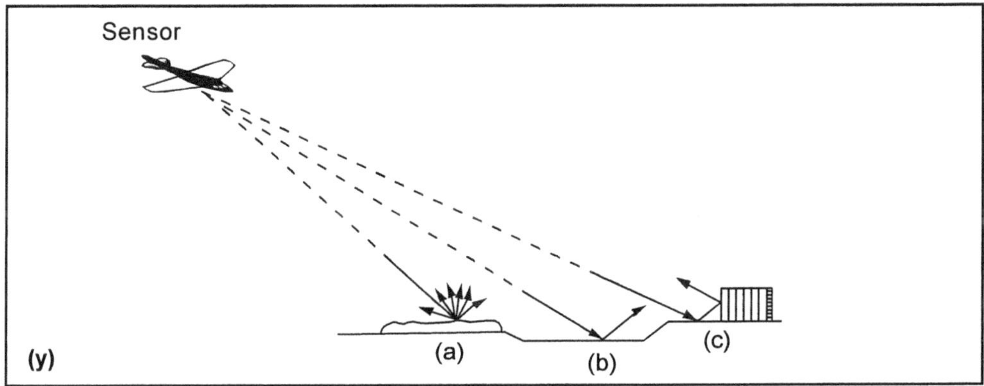

Fig. 4.4(x) Example of surface scattering (a) diffuse scattering
(b) specular reflection
(c) corner reflection.
(y) Representing a rough surface as a collection of facets.

Various models which are applicable for different levels of roughness, distinguish the backscattering mechanism from the terrain. The description of models likel point scattering model, facet model, and Bragg model, are beyond the scope of this handbook.

4.4 Radar Wavebands

Radar wavelength bands are described by a code 'L' and/ or 'C' band. Table 4.1 shows one commonly accepted delimitation of the radar wavelengths. Although a radar signal does not detect color information or temperature information, it detects surface roughness and electrical conductivity information in soil moisture conditions. Hence, the wavelength, depression angle, and the polarisation of the signal are important properties.

Table 4.1 Radar wavebands nomenclature

Band designation	Frequency (MHz)	Wavelength (cm)
P	300 - 1000	30 - 100
L	1000 - 2000	15 - 30
S	2000 - 4000	7.5 - 15
C	4000 - 8000	3.75 - 7.5
X	8000 - 12000	2.5 - 3.75
Ku	12000 - 18000	1.667 - 2.5
K	18000 - 27000	1.111 - 1.667
Ka	27000 - 40000	0.75 - 1.111

4.5 Side Looking Airborne Radar (SLAR) systems

A radar image displays the backscattering characteristics of the earth's surface in the form of a strip map (Fig. 4.5). There are two categories of side-looking airborne radar (SLAR) systems, namely, real aperture and synthetic aperture. The latter is the focus of this review, but the real aperture SLAR systems may be briefly considered so as to understand why synthetic aperture radar (SAR) systems have been developed.

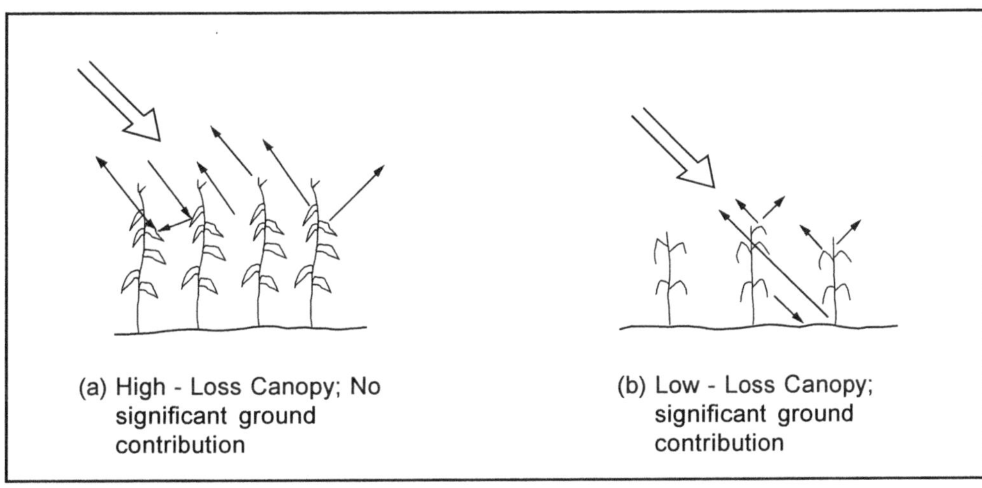

Fig. 4.5 (A) Volume scattering from (a) high loss vegetation canopies and (b) low loss vegetation canopies.

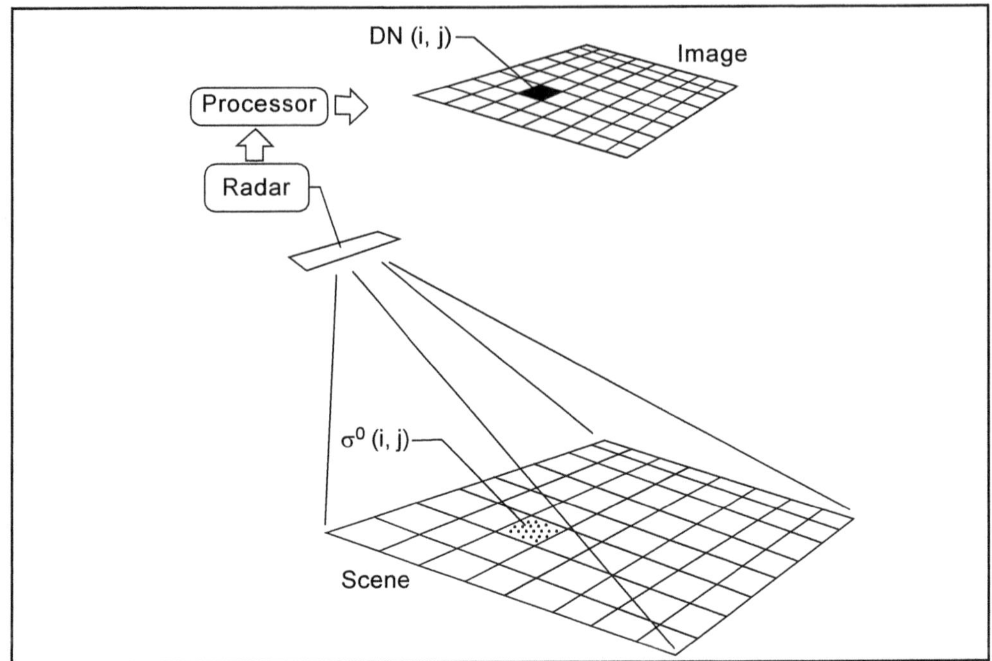

Fig. 4.5 Correspondence of the radar backscatter coefficient at locations (i, j) pixel in the scene, to the (i, j) digital number in the processed image (Anon, 1987).

The basic operating principle of a SLAR system is illustrated in Fig. 4.6. Short pulses or bursts of microwave radiation are transmitted from the antenna, each pulse

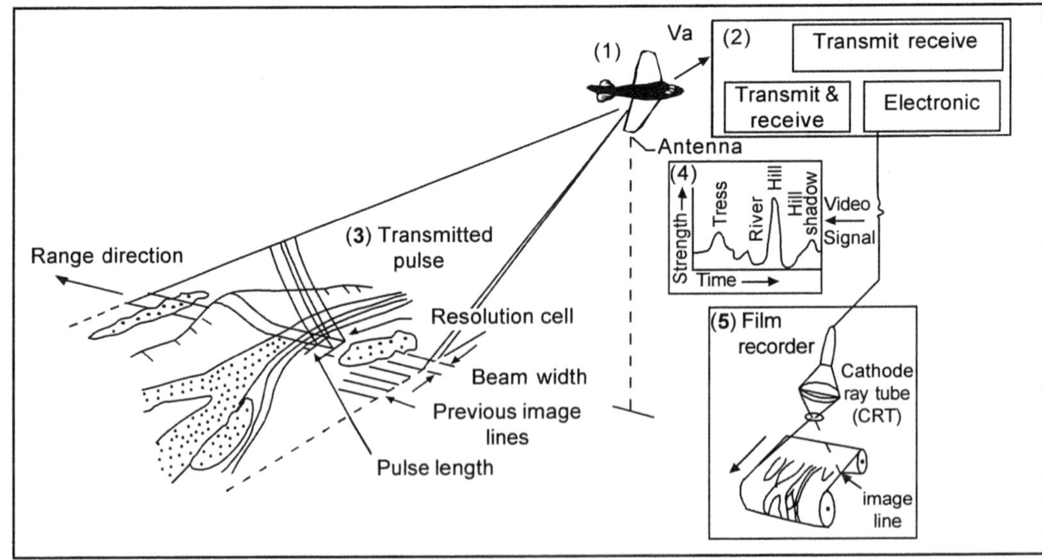

Fig. 4.6 Basic operating principle of side-looking airborne radar (SLAR) (Lillesand & Kiefer, 1979).

being transmitted for only a very brief period of time usually in the order of microseconds (Lillesand and Kiefer, 1979). This pulse moves out radially from the antenna and results in a beam being formed which is vertically wide but horizontally narrow. The time taken by a pulse to move away from the antenna, strike the ground and its backscatter or 'echo' return is measured electronically. From this time measurement, it is possible to determine the distance between the antenna and the object in the slant range. Since the energy propagates in air at approximately the velocity of light 'c', the slant range, to any given object is given by

$$\overline{SR} = \frac{Ct}{2}$$

where, $\overline{SR}$ = slant range (direct distance between transmitter and object)

C = speed of light (3×10^8 m/sec)

t = time between pulse transmission and echoreception.

These echoes are then recorded to produce an amplitude/ time video signal (Fig. 4.7). An image product is generated in a film recorder by using the signal to control the intensity of the beam on a single line cathode ray tube (CRT), and recording this

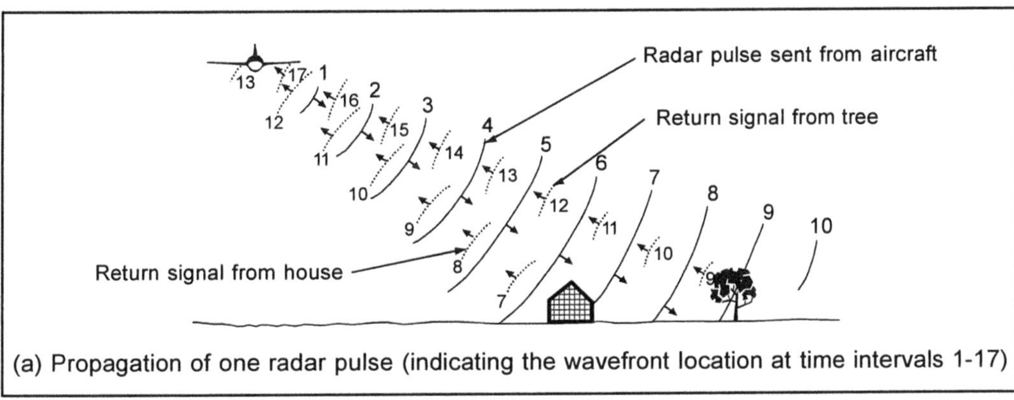

(a) Propagation of one radar pulse (indicating the wavefront location at time intervals 1-17)

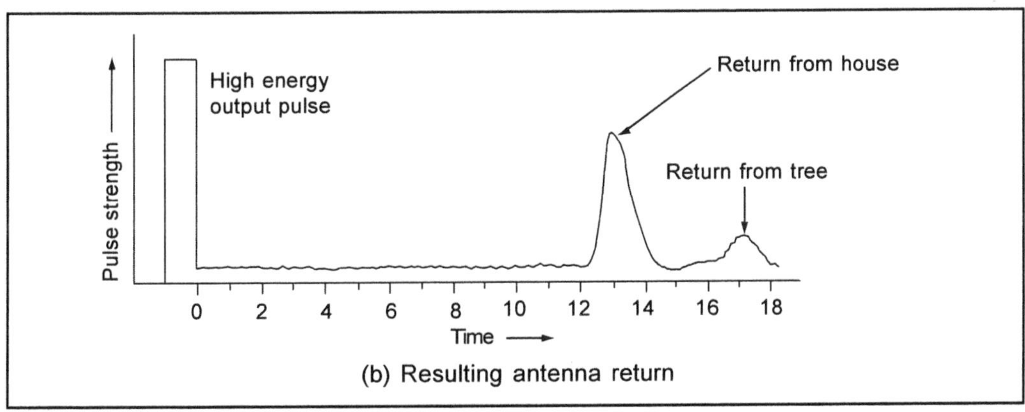

(b) Resulting antenna return

Fig. 4.7 Operating principle of SLR. (Lillesand & Kiefer, 1999)

line on to the film via a lens, to form an image. The film is advanced at a rate proportional to the aircraft's motion. In this way the combined response of many radar pulses is used to generate an image in which each line is the tonal representation of the strength of the signals returned to the radar antenna from a single pulse (Lillesand and Kiefer, 2000). The ground resolution cell size of SLAR system mainly depends on pulse length and antenna beam width. The pulse length is defined as the length of time that the antenna emits its energy. Pulse length determines the spatial resolution in the direction of propagation (Fig. 4.8). This direction is called range resolution. The other resolution: azimuthal resolution, which is based on the width of the antenna beam, determines the resolution cell size in the flight direction. Therefore the resolution in the radar system is measured in two directions, along the track (azimuthal) and across the track (range resolution). The effective resolution is the minimum separated distance that can be determined between two targets with echoes of similar strength. The resolution in the range direction is given by

$$R_r = \frac{C\,t}{2} \qquad \text{..........(4.8)}$$

where C is the velocity of electromagnetic radiation and t is pulse width. The slant range resolution when converted to the ground range can be written as

$$R_g = \frac{C\,t}{2\cos\theta_d} \qquad \text{..........(4.9)}$$

where θ is the angle of incidence.

The resolution in the range direction is thus a function of the pulse length (τ). In the azimuthal direction, however, the resolution (R) is determined by the angular beam width of the antenna, and the slant range distance which can be expressed as :

$$R_a = \beta R = \frac{\lambda}{D} \,.\, R \qquad \text{..........(4.10)}$$

where
β = angular beam width of the antenna
λ = wavelength
D = antenna length
R_a = slant range distance
R = resolution

Hence the resolution in the azimuthal direction is a function of the length of the antenna, and the range R_a therefore varies across the imaged swath. This is because the radar beam 'fans out' with increasing distance from the aircraft. This results in a deterioration of the azimuthal resolution with increasing range, and so objects which are separable close to the flight line are not distinguished further. At spacecraft altitude, the azimuthal resolution becomes too coarse.

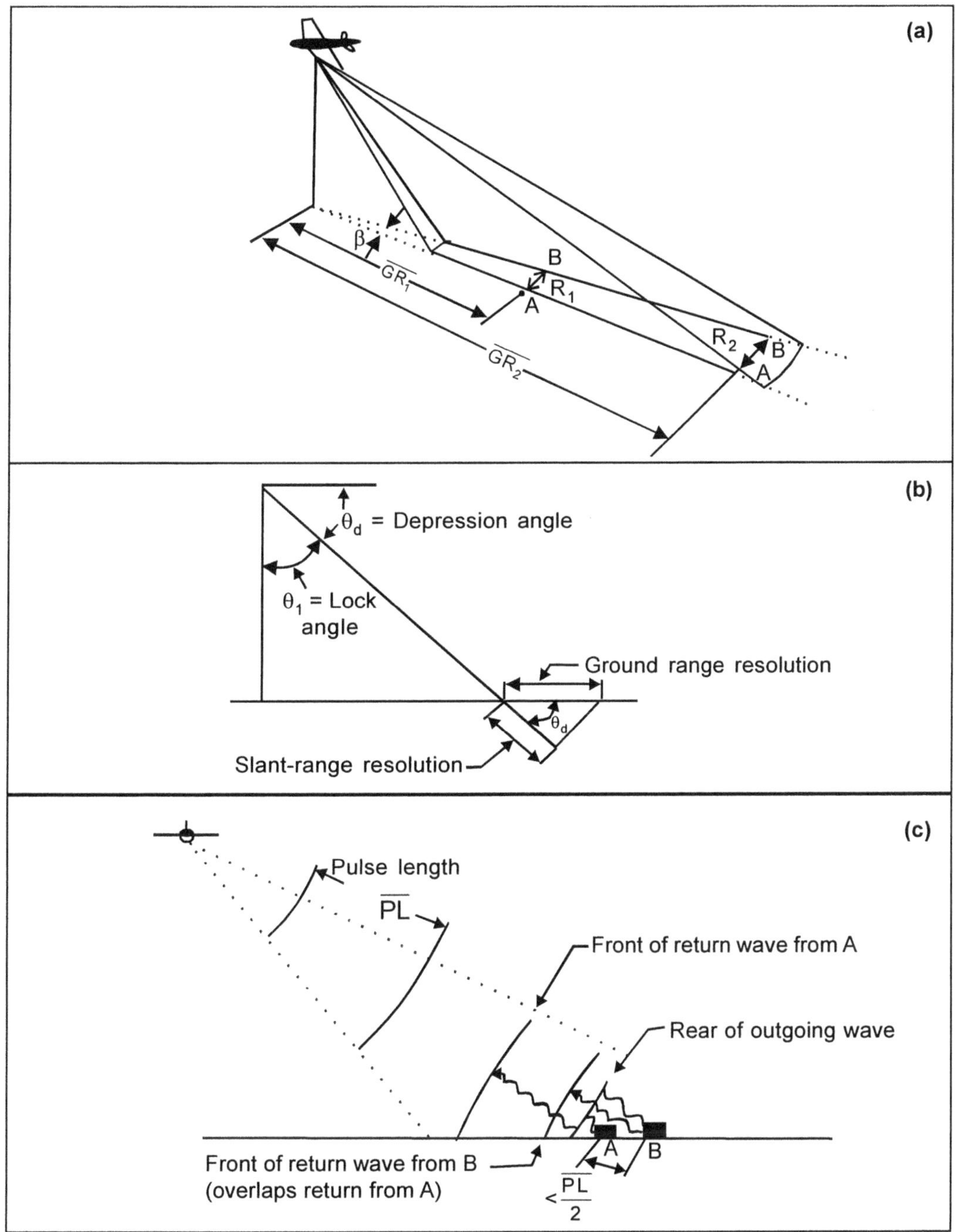

Fig. 4.8 Graphical representation of resolution of radar remote sensing data
(a) Resolution (R) on antenna beamwidth (β) and ground range (GR)
(b) Relationship between slant-range resolution and ground-range resolution
(c) Dependence of range resolution on pulse length

In order to obtain a fine azimuthal resolution, it is necessary to use as long an antenna as possible, which creates logistic problems, and/or use as short a wavelength as possible, which will, however, result in greater atmospheric attenuation and dispersion (Lillesand and Kiefer, 2000). To obtain a fine range resolution shorter pulses would need to be transmitted, and these would require a high peak power. The radar systems in which beam width is controlled by the physical antenna length, are called brute-force, real aperture or non-coherent radars.

4.6 Synthetic Aperture Radar (SAR)

In synthetic aperture radar (SAR) imaging, microwave pulses are transmitted by an antenna towards the earth surface. The microwave energy scattered back to the spacecraft is measured. The SAR makes use of the radar principle to form an image by utilising the time delay of the back scattered signals (Fig. 4.9).

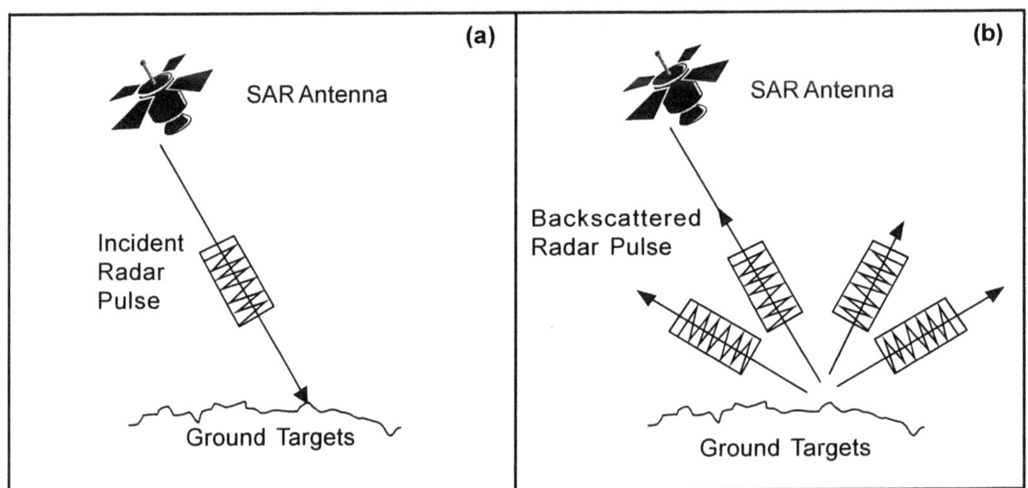

Fig.4.9 (a) A radar pulse is transmitted from the antenna to the ground.
(b) The radar pulse is scattered by the ground targets back to the antenna.

In real aperture radar imaging, the ground resolution is limited by the size of the microwave beam sent out from the antenna. Finer details on the ground can be resolved by using a narrower beam. The beam width is inversely proportional to the size of the antenna, that is, the longer the antenna, the narrower the beam.

It is not feasible for a spacecraft to carry a very long antenna that is required for high resolution imaging of the earth surface. To overcome this limitation, SAR capitalises on the motion of the space craft to emulate a large antenna (about 4 km for the ERS SAR) in place of the small antenna (10 m on the ERS satellite) it actually carries on board. Fig. 4.10 shows the imaging geometry for a typical strip-mapping synthetic aperture radar imaging system.

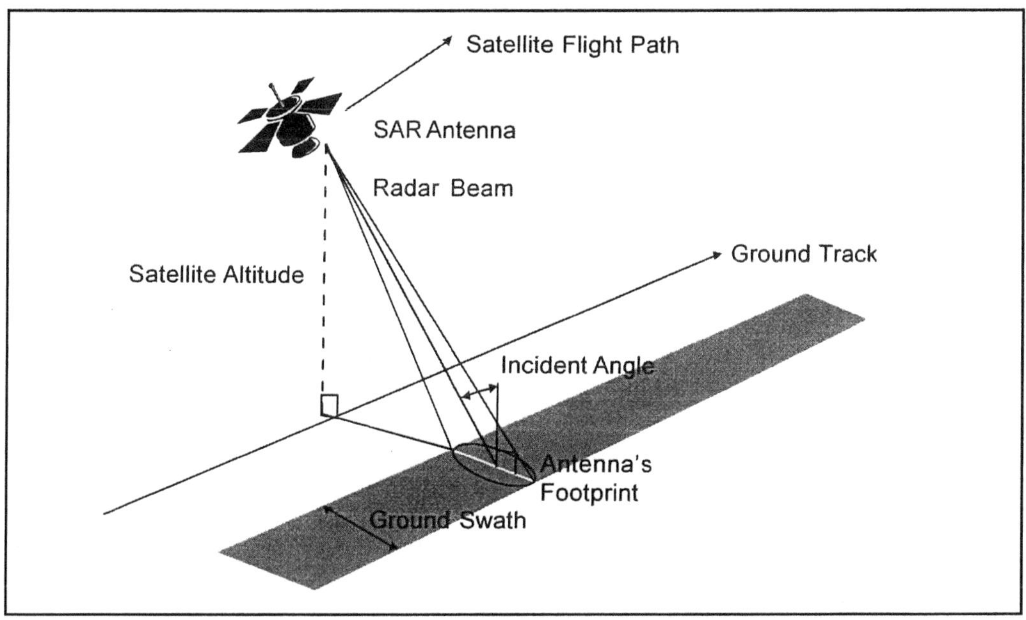

Fig.4.10 Imaging geometry for a typical strip-mapping synthetic aperture radar imaging system. The antenna's footprint sweeps out a strip parallel to the direction of the satellite's ground track.

With a SAR system there is the advantage that the azimuth or along track resolution is improved by making it independent of the range. This is because with a SAR system a physically short antenna is made to behave as if it were much longer. This aperture synthesis is achieved by recording not only the strength of a returned signal from an object on the ground, but also the signal's frequency. With this extra information the beam width can be effectively narrowed when the Doppler shift is used (Lillesand and Kiefer, 1979).

Fig. 4.11 shows the concept of SAR. An antenna with a along track beam width of β_h radians illuminates an object at slant range R and along track location X_0. At this range, the antenna beam has an along track width, L, of

$$L = \beta_h \cdot R \qquad \text{..........(4.11)}$$

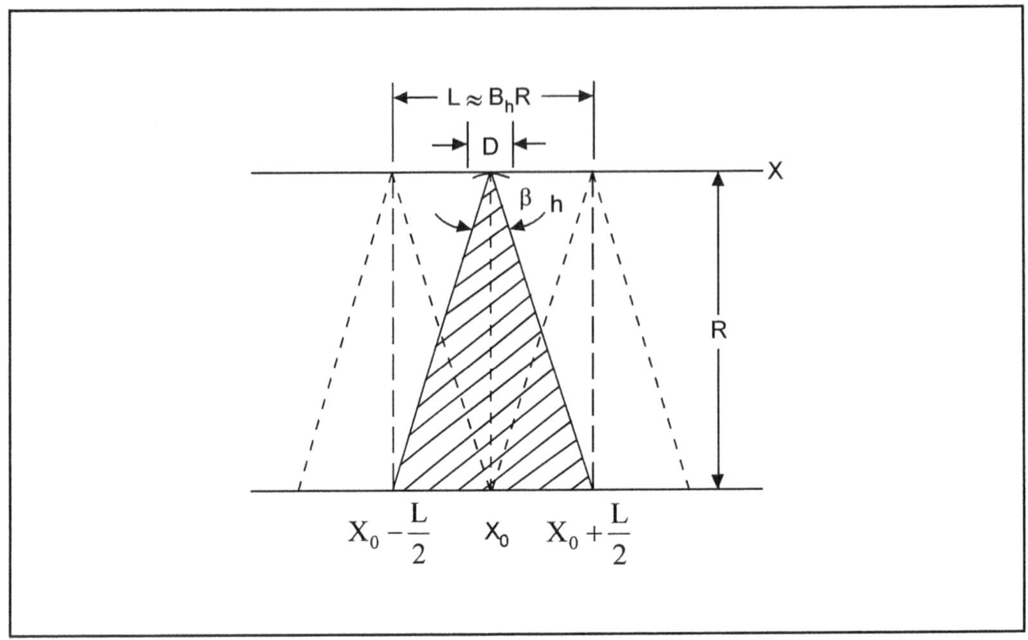

Fig. 4.11 Visibility of a target at X_0 over distance L (Reeves, 1979)

The radar acquires visibility of the object when it reaches the coordinate location of $(X_0 - L/2)$. Therefore, it is possible for a single moving antenna to successively occupy the element positions of X to X_0 in an array of length L. Under these conditions, it is possible in principle to combine observations from the moving antenna to synthesise an array of length $L \sim R\frac{\lambda}{D}$. Because the SAR system is a coherent system and incorporates two-way signal propagation, it can be shown that an array of the form described can provide an effective angular resolution given below :

$$\beta'' \sim \lambda/2L \qquad \text{..........(4.12)}$$

The along track resolution at range R then becomes

$$\gamma_a = R\beta'' \qquad \text{..........(4.13)}$$

Recalling that $L \sim Rb_h \sim R\frac{\lambda}{D}$, this reduces to $g_x \sim D/2$. This calls for the use of a smaller antenna if a fine resolution is to be achieved, unlike the real aperture SLAR. In the range direction the resolution can be expressed as

$$R_r = \frac{C}{2 \propto \sin\theta} \qquad \text{..........(4.14)}$$

where $\propto$ = pulse band width ~ $1/\tau$

This means that short pulses, which require a high peak power, need not be used. The pulse band width, $\propto$, can be made quite large by using chirp techniques without excessive peak power requirements. The Radar system uses either a slant range or ground range presentation for the across track coordinates. The raw image suffers from geometrical and radiometric errors which are discussed in the following sections.

4.7 Interaction between Microwaves and Earth's surface

When microwaves strike a surface, the proportion of energy scattered back to the sensor depends on many factors like physical factors, such as, the dielectric constant of the surface materials; and on the moisture content geometric factors, such as, surface roughness, slopes, orientation of the objects relative to the radar beam direction, the types of landcover (soil, vegetation or man-made objects), microwave frequency, polarisation, and incident angle of microwave remote sensing systems.

4.7.1 Speckle Noise

Unlike optical images, radar images are formed by coherent interaction of the transmitted microwave with the targets. Hence, it suffers from the effects of speckle noise which arises from coherent summation of the signals scattered from ground scatterers distributed randomly within each pixel. A radar image appears more noisy than an optical image. The speckle noise is sometimes suppressed by applying a speckle removal filter on the digital image before display and further analysis.

Because of wave interference, random fluctuations from an extended target appear as speckles on radar imagery. The random variation in the brightness of individual pixels can be large if the pixels are observed once, whereas if they are observed many times, speckles can be reduced. It is for this reason that speckles are more severe in synthetic aperture radar images as compared to real aperture radar images. Fading of a radar signal which causes speckles, complicates image interpretation. It is often necessary to reduce the effect of speckles by the knowledge of spatial gray level distribution.

4.7.2 Backscattered Radar Intensity

A single radar image is usually displayed as a gray scale image, such as the one shown in Fig. 5.15. The intensity of each pixel represents the proportion of microwave backscattered from that area on the ground which depends on a variety of factors, such as, type, size, shape and orientations of the scatterers in the target area, moisture content of the target area, frequency and polarisation of the radar pulses and the incident angles of the radar beam. The pixel intensity values are often converted to a physical quantity called the backscattering coefficient or normalised radar cross-section measured in decibel (dB) units with values ranging from +5 dB for very bright objects to 40 dB for very dark surfaces.

4.8 Interpreting SAR Images

Interpreting a radar image is not a straight forward task. It very often requires some familiarity with the ground conditions of the areas imaged. As a useful rule of thumb, the higher the backscattered intensity, the rougher is the surface being imaged. Flat surfaces, such as, paved roads, runways, or quiet water normally appear as dark areas in a radar image since most of the incident radar pulses are specularly reflected away. Calm sea surfaces appear dark in SAR images. However, rough sea surfaces may appear bright especially when the incident angle is small. The presence of oil films smoothen out the sea surface. Under certain conditions when the sea surface is sufficiently rough, oil films can be detected as dark patches against a bright background.

Trees and other vegetation are usually moderately rough on the wavelength scale. Hence, they appear as moderately bright features in the image. The tropical rain forests have a characteristic backscatter coefficient between –6 and –7 dB, which is spatially homogeneous and remains stable in time. For this reason, the tropical rainforests have been used as calibration targets in performing radiometric calibration of SAR images. Very bright targets may appear in the image due to the corner-reflector or double bounce effect where the radar pulse bounces off the horizontal ground (or the sea) towards the target, and then reflected from one vertical surface of the target back to the sensor as in the case of cargo containers. Built-up areas and many man-made features usually appear as bright patches in a radar image due to the corner reflector effect.

The brightness of areas covered by bare soil may vary from dark to very bright depending on its roughness and moisture content. Typically, rough soil appears bright in the image. For a similar soil roughness, the surface with a higher moisture content will appear brighter. The interpretation of radar images depends on the geometrical properties and other characteristics of imaging systems. The geometrical characteristics that affect the microwave remote sensing data are discussed in the following paragraphs.

4.9 Geometrical Characteristics

Large scale variations in the surface (for example, terrain slopes), affect the backscattering properties. In a radar image, broadly four geometrical characteristics are observed, namely, (i) slope foreshortening, (ii) aspect, (iii) radar shadow, and (iv) layover.

4.9.1 Slope Foreshortening

For similar terrain slopes, or for the same slope recorded at different depression angles, there is a variation in the slope length on the radar image. Slopes are often made to appear shorter than they really are, that is, they are foreshortened. Foreshortening is at a maximum for a slope when the incident angle, θ, is 90^0, and a minimum when θ is at the grazing angle, where the slopes are recorded in true proportion to their length. Fig. 4.12 shows this effect. A slope is recorded at its true length if it slopes away from the radar antenna at an angle at which it is truncated in its slant range presentation.

For slope 1 in Fig. 4.12, which faces the radar and has a gradient equal to the depression angle, both the top and the base of the slope are intercepted simultaneously by the radar beam wavefront, and so are recorded as a single point on the image. Foreshortening thus occurs when a slope is less steep than when perpendicular to the wavefront, with the base of the slope intercepting the wavefront first (Lillesand and Kiefer, 1979), as shown for slope 2 in Fig. 4.12.

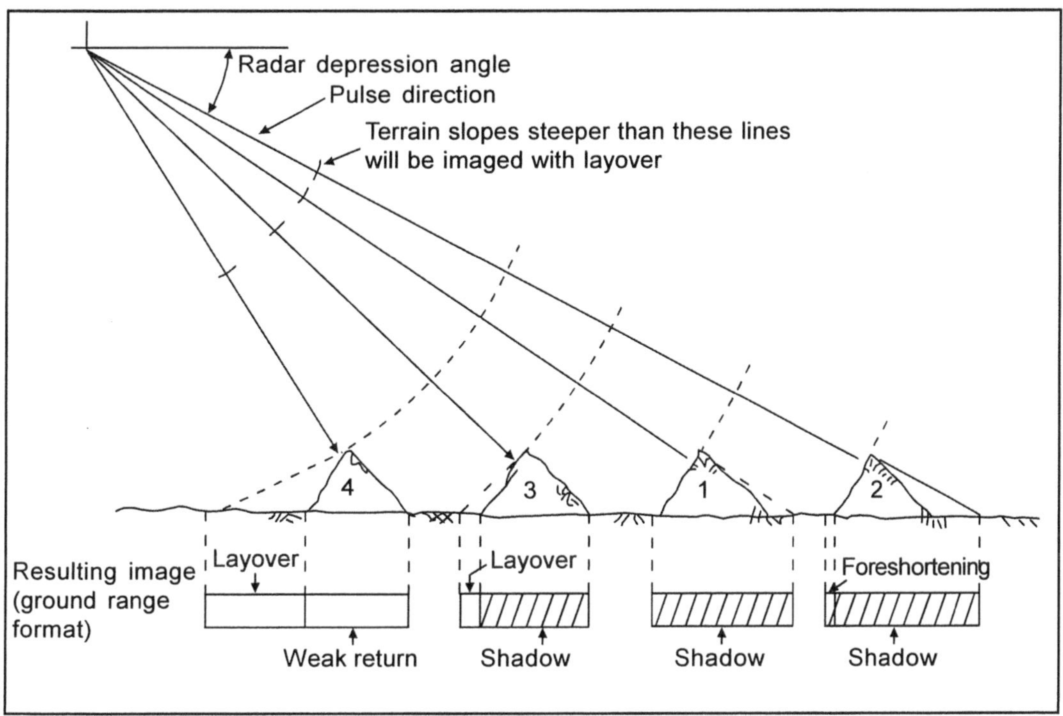

Fig. 4.12 The influence of terrain slope on radar imagery (from Lillesand and Kiefer, 1979).

4.9.2 Layover

If, however, the terrain slope is steeper than a line perpendicular to the incident wavefront, so that its top intercepts the radar beam wavefront before the base of the slope, then layover occurs. This is illustrated in Fig. 4.12 for slopes 3 and 4. It was shown above that foreshortening occurs when a slope facing the radar is not as steep as the wavefront. With layover, however, a slope facing the radar appears to be steeper than it really is, as its slope is steeper than that of the wavefront. From the above discussion of layover and foreshortening, and the fact that the depression angle varies across the image swath, it can be seen that similar terrain slopes at different positions from the flight line will be recorded differently. Layover is most likely to occur in the near range where the depression angle is low.

Layover and foreshortening are important factors, and should be considered while interpreting radar imagery. This is not only because of the distortion they introduce into the image but also because the amount of energy received per unit area varies with the angle at which the energy is received and so with the slope. The angle at which the energy is received at each surface is critical to a distribution of the energy back scatter from that surface. For flat terrain the angle at which the energy arrives varies across the swath, and further complexity is introduced by variations in the surface slope.

4.9.3 Aspect

Aspect is important since, if a slope is orientated away from the flight line, then the wavefront's journey from the base of the slope to its top is further increased, and hence the amount of energy per unit area is reduced. This can greatly affect the backscatter recorded and so the interpretability of the imagery. Slope orientation relative to the radar look direction is therefore important.

4.9.4 Radar Shadow

The other important characteristic of radar imagery is that of radar shadow, which occurs whenever the terrain backslope is steeper than the depression angle. This is because the slope facing the radar beam from illuminating the backslope. The length of the radar shadow is determined by the 'wavefront angle', as are layover and foreshortening. Areas of radar shadow are generally more common in the far range, because in the near range few backslopes are steep enough to be obscured from the radar beam.

5

Remote Sensing Platforms and Sensors

5.1 Introduction

Remote sensing of the surface of the earth has a long history, dating from the use of cameras carried by balloons and pigeons in the eighteenth and nineteenth centuries. The term 'remote sensing' is used to refer to the aircraft mounted systems developed for military purposes during the early part of the 20th century. Air borne camera systems are still a very important source of remotely sensed data (Lillesand kiefer, 1994). Although photographic imaging systems have many uses, this chapter is concerned with image data collected by satellite sensing systems which ultimately generate digital image products.

Space borne sensors are currently used to assist in scientific and socioeconomic activities like weather prediction, crop monitoring, mineral exploration, waste land mapping, cyclone warning, water resources management, and pollution detection. All this has happened in a short period of time. The quality of analysis of remote sensing data and the varied types of applications to which the science of remote sensing is being put to use are increasing enormously as new and improved spacecraft are being placed into the earth's orbit. The primary objectives, characteristics and sensor capabilities of the plethora of remote sensing satellites circling this planet, are discussed in this Chapter. An attempt is made to classify the satellites into three types, namely, earth resources satellites, meteorological satellites, and satellites carrying microwave sensors. This classification is not rigid. For instance, most of the meteorological satellites are also capable of sensing the resources of the earth. Before turning to the individual satellite's description and the corresponding sensors and capabilities, a brief overview of satellite system parameters is presented in the following paragraphs.

5.2 Satellite System Parameters

A brief overview of the most important satellite system parameters which describe the functions and operations of the remote sensing systems are presented in this section. Broadly, the system parameters are of two types : instrumental and viewing. The principal instrumental parameters, namely, wavelength or frequency, polarisation, and sensitivity or radiometric resolution are determined by the design of the transmitter, receiver, antenna, detectors, and data handling system. The principal viewing parameters are determined by both the instrument design and the orbital parameters of the satellite. Revisit internal, resolution, swath width, illumination and/ or observation angle and mission lifetime are the important viewing parameters of any satellite sensing system.

5.2.1 Instrumental Parameters

All of the remote sensing systems make use of information carried by electromagnetic radiation, of which there is an infinite range of possible frequencies or wavelength. In practice, however, the transparency or otherwise of the earth's atmosphere limits the possible wavelength ranges to about 0.4–15 μm (the visible and infrared regions) and 1mm to 1m (the microwave region, corresponding to a frequency range of 300–0.3 GHz). The visible (VIS) and infrared (IR) region is conveniently subdivided into the VIS/near-infrared (NIR) region on one hand, (approx. 0.4–2 μm) the thermal infrared or TIR region (approx. 0.4–2 mμ) on the other hand. Sensors designed to detect atmospheric constituents utilise spectral bands between the atmospheric 'windows' (Robert Massom, 1992).

Naturally occurring radiation from the earth's surface is found in all these ranges of wavelength. In the VIS/NIR band the sunlight is predominantly reflected, and the most important parameters of the target material is thus its reflectance (Chapter 3). In the TIR band, on the other hand, the main source of radiation is, as the name suggests, the blackbody thermal mechanism by which all objects above absolute zero emit radiation. For an object at a typical terrestrial temperature, most of this radiation is emitted at wavelengths around 10 μm. Detected radiation essentially contains information on two parameters, namely the temperature of the target material and its effectiveness in emitting radiation in this waveband, called the emissivity. Emissivity is a unique characteristic of the target material and its state or condition (Wolfe and Zissis. 1989).

Small but significant and measurable amounts of blackbody thermal radiation can also be detected in the microwave band, and this kind of remote sensing is known as passive microwave radiometry. Again, the detected signal is governed by both the target temperature, and its type and condition. The other main use of the microwave region is for active remote sensing, by which radiation is emitted by the remote sensing instrument and detected after its reflection from the target material. Active remote sensing in the microwave region is called radar, and the main observable parameters are the range to the target (from the time delay of the returned signal) and the reflectivity of the material which in turn is determined by many of its physical properties. This concept was discussed in detail in Chapter 4.

The reflective and emissive properties of a material are different for different polarisations, that is, orientations of the electrical field vector in the electromagnetic radiation, where H is horizontal and V is vertical, so that further information on the physical properties of the target material may be obtained by observing different polarisations. In practice, this has so far found any significant application only in passive and active microwave remote sensing, not in the VIS and IR bands (Robert Massom, 1989).

The sensitivity of a remote sensing system measures the response produced by the radiation of a given intensity and wavelength. Other things being equal, it should be as large as possible, but because the output data are usually digitised, they can only span a finite range of values so that a high sensitivity (low value of the minimum detectable signal) implies a low value for the maximum signal that can be detected. This then requires some kind of optimisation, and what is optimal for one kind of target material may not be optimal for another. This may often cause saturation of the detecting system. The spectral resolution and radiometric resolution which are the measures of sensitivity of satellite sensing system are discussed in detail in the following sections.

5.2.2 Viewing Parameters

Viewing parameters, namely, revisit interval, swath width, illumination and observation angle, mission life time, altitude and resolution of remote sensing system are also termed as orbital characteristics of satellites. Satellites used for remote sensing are generally of two types, geostationary and near earth, polar orbiting and sun-synchronous, circular and near polar. Geostationary are stationary with respect to the earth and are at an altitude of about 36000 kms above a point on the equator, that is, geostationary satellites maintain a fixed location with respect to the earth's surface. Conversely, a satellite in a low polar orbit traces out a curving path over the earth's surface, as a consequence of the satellite's orbital motion and precision and of the earth's rotation about its axis. This path (the subsatellite track) wraps itself round the earth from east to west like a ball of string, oscillating between the equal north and south latitudes in a pattern set by the inclination of the orbit. The path may close up on itself if the orbital parameters (inclination, height and eccentricity) are suitably chosen, in which case the satellite will revisit a given location at regular intervals. This interval may in general be any integral number of days, though other constraints on the orbital parameters may limit the choice. The point where the satellite, travelling northwards, passes directly over the equator is called the ascending node. The descending node describes the southward crossing (Stewart, 1985 and Rees, 1990). Owing to their immense distance from the earth, high resolution imaging is difficult. INSAT, GOES and METEOSAT are some of the satellites under this category.

Near the earth, polar orbiting and sun-synchronous satellites are locted at a much lower altitude, generally a few hundred to a few thousand kilometers. In the case of this type of satellites, the time at which the satellite revisits a given location is the same on each occasion. This is very useful for visible and infrared observations, since the level of solar illumination can be chosen. Biological and environmental studies, for example, require specific timings for the collection of data. Fig. 5.1 illustrates the principles of a sun-synchronous orbit. The angle between the orbital plane and the earth's equatorial plane is termed the inclination of the orbit. Changes in the orbit are due largely to precision caused mainly by the slightly non-spherical shape of the earth. Landsat, IRS and SPOT are some of the satellites which have this kind of orbit.

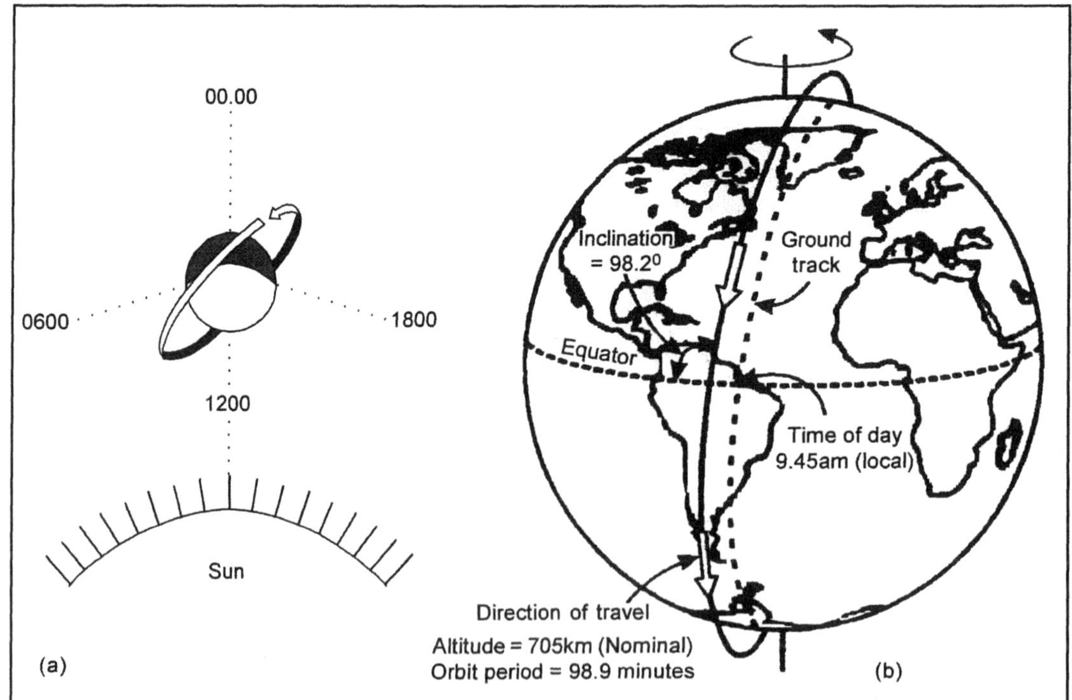

Fig. 5.1 (a) Illustrating the principles of a Sun-synchronous orbit.
(b) Example of Sun-synchronous orbit.

Some earth observing platforms are not in near polar, sun-synchronous orbits. The space shuttle has an equatorial orbit. Thus, the orbit selected for a particular satellite determines not just the time taken to complete one orbit (temporal resolution) but also the nature of the relationships between the satellite and solar illumination direction. Therefore illumination and observation angle are very important parameter on which the precision of satellite data depends.

Many satellite systems observe at the nadir (directly below the satellite viewing angle = 0^0), and this is especially valuable for imaging systems since distortion is reduced if the image and object planes are parallel. Some systems, however, observe at non-zero viewing angles. This is usually to exploit the dependence on incidence angle of the emissivity or reflectivity of the surface, for which a physical model may exist. Observation away from nadir can thus assist in the discrimination of one target material from another, or the measurement of some physical property of the target material.

The resolution of remote sensing system is influenced by the swath width. The swath width is defined as the width of the strip, parallel to the satellite's track, from which radiation is received. The maximum swath width attainable by a sensor is effectively limited by its data bandwidth, the maximum value for which is about 100 megabits per second for present transmission systems. High resolution systems are thus limited as to the swath width that they can cover. Typically, radar and visible scanners achieving a 15 to 20 m resolution are confined to about a 100 to 200 km swath. In contrast, systems with a 1km spatial resolution, such as, the NOAA-Advanced Very High Resolution Radiometer (AVHRR), can scan about 3,000 km across the subsatellite track. MSS of landsat with spatial resolution 80 m are confined to about 185 kms and of LISS I of IRS confined to about 148 km.

It is clear that if one wishes to obtain spatially complete coverage of a given area which is wider than the swath width of a sensor, the satellite should be in an orbit such that spatially-adjacent suborbital tracks are closer together than the swath width. The spacing between adjacent suborbital tracks decreases towards the earth's poles, so this is less of a problem for polar remote sensing than for the study of equatorial regions. Inter-orbital spacing is strongly dependent on the revisit interval, in the sense that small spacings require long revisit intervals, so that the choice of orbital parameters for a satellite mission usually represents a compromise between conflicting requirements.

Satellite remote sensing missions are of much greater duration than airborne operations, and the continuity of data thus provided is one of the substantial advantages offered by satellites. The factors which limit the operational lifetime of a satellite remote sensing mission are the robustness of the equipment carried by the satellite, and decay of the orbit through atmospheric friction. It is not yet routinely possible to 'service' satellites in orbit, so once a piece of equipment fails the corresponding data are forever more degraded or unavailable. The space environment is fairly harsh, and lifetime of a few years are typical. For a satellite in an orbit at a height of 500 km or more, the decay of the orbit itself is negligibly slow, but at a height of say 200 km, the orbit lifetime can be as short as one month, and this becomes the dominant factor. Very low altitude orbits are essentially used for high-resolution military reconnaissance satellites.

5.3 Sensor Parameters

Sensors are devices used for making observations. These consist of mechanisms, usually sophisticated lenses with filter coatings to focus the area observed on a plane in which the detectors are placed. These detectors are sensitive to a particular region in which the sensor is designed to operate and produce outputs which are representative of the observed area.

The major characteristics of an imaging remote sensing instrument operating in the visible and infrared spectral bands are described in terms of its spatial, spectral and radiometric resolution. These three types of resolutions vary from sensor to sensor. Each sensor has its own capability of detecting the energy reflected from the earth's surface features. The details of all these characteristics of various sensing systems are described in the next section and the concepts of resolution are discussed in this section.

5.3.1 Spatial Resolution

Spatial resolution is a complex concept which can, for the purpose of remote sensing of polar regions, be defined as the smallest object that can be detected and distinguished from a point. The most frequently used measure, based upon the geometric properties of an imaging system, is the instantaneous field of view (IFOV) of a sensor. The IFOV is the area on the surface that is theoretically viewed by the instrument from a given altitude at a given time.

The spatial resolution is usually determined by instrumental parameters and by the height of the satellite above the ground. With the exception of active microwave systems, the resolution of a system cannot be better than approximately Hl/D (the diffraction limit), where H is the height, l is the wavelength and D is the diameter of the objective lens, objective mirror or antenna. This limit is typically of the order of 10 to 100 m for VIS and IR systems operating from satellites in low orbits, and typically 1 to 10 km when the satellite is geostationary. For passive microwave observations, the resolution limit is much coarser (of the order of tens of km) because of the larger wavelength measured.

It was stated that the best achievable spatial resolution is of the order of Hl/D (except for some types of radar system), although some non-radar systems may not reach this resolution because of other instrumental effects. Two important examples are sensors in which the incoming radiation is focused on to an image array of discrete detecting elements, and photographic systems. The detecting element or film imposes its own maximum resolution, again proportional to the height H and, if this is poorer than the diffraction-limited resolution, it will dominate.

The spatial resolution achievable by radar systems is very dependent on the way the data from the system are processed. Such systems are often pulsed, and one important factor is the length of the emitted pulse. Synthetic aperture radars (SARs) also integrate the return signal for a period of time while the radar is carried forward on its platform, and the integration time also influences the resolution. It is not possible to give here a statement of the general principles determining radar spatial resolution, and the interested reader is referred to treatments given by Ulaby, Moore and Fung (1981 and 1982), Elachi (1987) and Rees (1990).

Spatial resolution of an imaging system can be measured in a number of different ways. It is the size of the smallest object that can be discriminated by the sensor. The greater the sensor's resolution, the greater the data volume and smaller the area covered. In fact, area coverage and resolution are interdependent and these two factors determine the scale of an imagery. Alternatively, spatial resolution can be said to be the length of the size of the area on the ground represented by a pixel on an image. The basis for the definition of spatial resolution can depend on four criteria, namely, : (i) Geometrical properties of the imaging system, (ii) the ability to distinguish between point targets, (iii) the ability to measure the periodicity of repetitive targets, and (iv) the ability to measure the spectral properties of small targets (Mather, 1999).

Spatial resolution of any satellite sensor applies to the image produced by the system, whereas resolving power of any photograph applies to an imaging system or a component of the system. As mentioned earlier, the most commonly used measure for spatial resolution of any sensor, based on the geometric properties of the imaging system, is the Instantaneous Field of View (IFOV) of a sensor. IFOV is defined as the area on the ground that is viewed by an instrument from a given altitude at any given instant of time. Fig. 5.2 illustrates the relationship between the swath width and the IFOV. The IFOV can be measured in one of the two ways, (i) by measuring angle "α" and (ii) by measuring the distance XY on the ground.

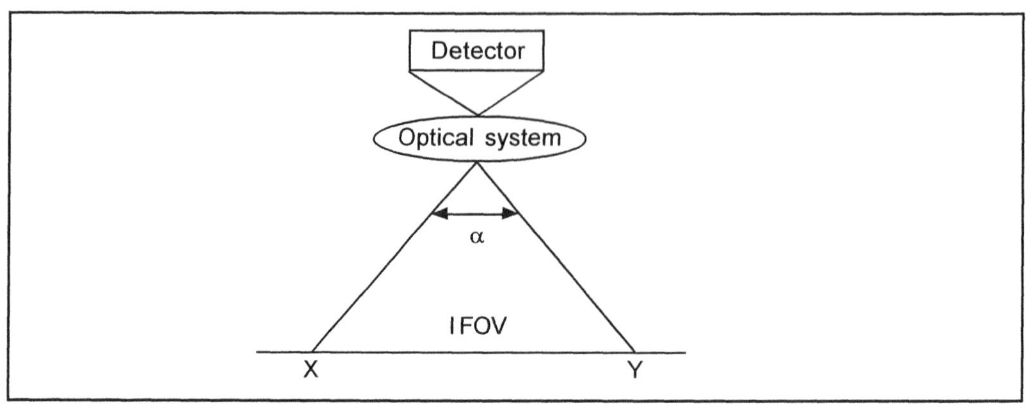

Fig. 5.2 Definition of IFOV.

Eventhough the IFOV is measured by the angle "α" and the cross section XY, it depends upon a number of factors. No satellite has a perfectly stable orbit; its height will vary. If Landsat 1 to 3 have a nominal altitude of 930 km, their actual altitude varies between 880 km. and 940 km. Because of this change in altitude the spatial resolution varies from 79 m to 81 m, but it has been specified as 80 m × 80 m resolution. This measurement of IFOV may not be useful. In order to explain this, let us consider how the radiance from a point source on the ground does not produce a single bright point on the image, but produces an image which has the intensity of a diffused circular region owing to the optical properties. Hence this IFOV can be determined by taking a Gaussian distribution of intensity of circular region over a point source with respect to the amplitude of the energy. This type of distribution function is called Point Spread Function (PSF).(Fig. 5.3.)

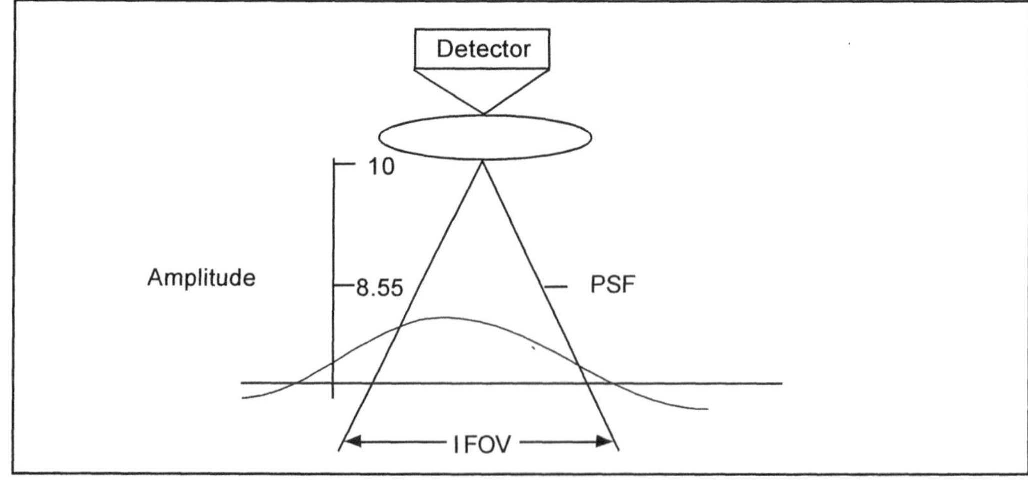

Fig. 5.3 Distribution curve of Point Spread Function.

The spatial resolution of TM in Landsat 4 and 5 is based on PSF and it is measured as 30 in., whereas the spatial resolution of MSS is measured as 90 m. rather than 79 in. This type of measurement can reveal that the presence of relatively bright or dark areas within IFOV of the sensor will increase or decrease the amplitude of PSF thereby making the observed radiance either higher or lower than the surrounding areas. That is why the highest contrast features such as rivers and canals, which are less than 79 in. width on the ground are more detectable on Landsat imagery as also in the case of IRS imagery. Spatial resolution plays an important role in resolving various earth's surface features from the interpretation of satellite imagery. Plate 2 illustrates the effect of spatial resolution in discriminating various features from the IRS satellite image. It has three different images at different spatial resolutions covered by the same terrain. In the satellite image of LISS which has spatial resolution of 72.5 m × 72.5 m indicates that an area of 72.5 × 72.5 sq.m can be represented by a pixel on the image. Similarly, the LISS II image in which 36.25 × 36.25 sq. m area can be shown as a pixel whereas LISS III provides an image with spatial resolution of 23.5 × 23.5 sq.m. This clearly indicates that LISS III image provides very fine details than LISS II image. PAN image of IRS IC/ID can provide much finer details as it has spatial resolution of 5.8 × 5.8 m. Recently, USA has launched a satellite : IKONOS, which provides the spatial resolution of 1m × 1m. Plate 2 shows the image acquired over Sally farm, Hissar by IRS LISS 1, II and III during the same season. Cropped area seen clearly in LISS 1 individual fields are celarly discernible in LISS III sensor. The effect of spatial resolution on crop classification accuracy was re-evaluated using IRS LISS III data acquired on two consecutive dates. The separability between wheat and sugarcane increased significantly on LISS III data, as compared to LISS II data. There was smoothening of signal from crop fields and fewer number of peaks then in LISS III data.

For greater clarity of spatial resolution of any remote sensing imagery, it is more appropriate to introduce the term pixel (picture element). One pixel on the digital satellite imagery represents an IFOV on the ground. Each IFOV is filled by Digital Number (DN). Each digital number indicates some inherent property of the terrain element (chapter 7).

5.3.2 Spectral Resolution

It is the width of the spectral band and the number of spectral bands in which the image is taken. Narrow band widths in certain regions of the electromagnetic spectrum allow us to discriminate between the various features more easily. Consequently, we need to have more number of spectral bands, each having a narrow bandwidth, and these bands should together cover the entire spectral range of interest.

The digital images collected by satellite sensors except microwave sensing systems like Seasat, SIR B Radarsat, have been multi-band or multispectral, individual images separately recorded in discrete spectral bands. Multispectral imaging refers to viewing a given area in several narrow bands to obtain better identification and classification of objects. Multistage imaging refers to the observations of the same area from different positions of the platforms (stereoscopic data). Multistage imaging refers to the observations made over the same area on different dates to monitor the objects like crop growth. This is also called temporal resolution. The term spectral resolution refers to the width of the spectral bands. Spectral resolution can be explained by considering two points, (i) the position of the spectrum, width and number of spectral bands will determine the degree to which individual targets can be determined on the multispectral image, and (ii) the use of multispectral imagery can lead to a higher degree of discriminating power than any single band taken on its own.

Various surface features can be identified by taking these multispectral bands. For example, differences in reflectance of various rock types are separable only if the recorded device is capable of detecting the spectral reflectance of the target in a narrow spectral wave band. In the following Fig. 5.4, it can be seen how healthy vegetation can be discriminated from unhealthy vegetation. Most of the difference occurs in infrared region. Unfortunately, it is not possible to increase the spectral resolution of a sensor simply to suit user's needs, because of signal-to-noise (S/N) ratio which is defined as the ratio of total radiant energy to the noise equivalent power.

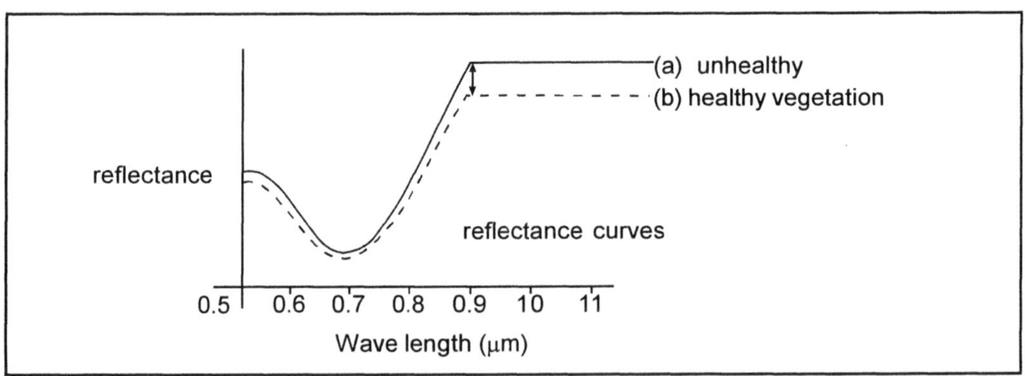

Fig. 5.4 Concept of Spectral Resolution.

Multispectral scanning systems are designed to sense energy over a small IFOV and in narrow wavelength bands. Since IFOV are very small, the amount of energy incident on each of a system's detectors is very low. Hence the detectors must be very sensitive to output a signal, and stronger than the level of noise. The noise is an extraneous and unwanted response which shows a signal from the detector as it would appear in the absence of system noise. The variations in the signals depend on the terrain conditions covered by that time. The large changes in the signal peaks indicate the cover types, subclasses and so on (Fig. 5.5). From Fig. 5.5, it can be noticed that the random noise is added to the signal. The Fig. 5.5(a) shows signal along one scan line without noise, (b) signal with low noise component and (c) with high noise component. Mathematically-S/N ratio can be given as (Lillesand and Kiefer 2000)

$$\left(\frac{S}{N}\right) \alpha \ D_\lambda \beta \left[\frac{H}{V}\right]^{1/2} D_\lambda L_\lambda$$

Where D_λ = Detectivity (measure of detector performance quality)
β = IFOV (Instantaneous Field of View)
H/V = Height/Velocity of sensor platform
L_λ = Spectral radiance of ground feature

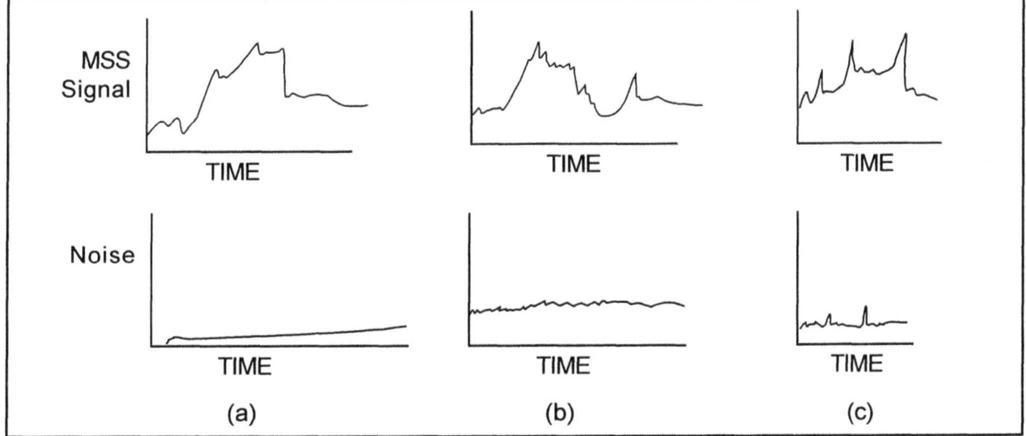

Fig. 5.5 Vibrations in the signal strengths.

A compromise must be sought between the twin requirements of narrow bandwidth (high spectral resolution) and a low S/N ratio. This can be achieved by an arrangement which can provide for a longer look at each scan line elements which is seen in the push broom scanner (eg. LISS of IRS).

5.3.3 Radiometric Resolution

It is the capability to differentiate the spectral reflectance / emittance between various targets. This depends on the number of quantisation levels within the spectral band. In other words, the number of bits of digital data in the spectral band or the number of gray level values, will decide the sensitivity of the sensor. It is the smallest difference in exposure that can be detected in a given film analysis. It is also the ability of a given sensing system to discriminate between density levels. In general, the radiometric resolution is inversely proportional to contrast, so that higher contrast film is able to resolve smaller differences in exposure. Low contrast films have greater radiometric range while highest contrast films have smaller exposure range and lower radiometric range. It is commonly expressed as the number of binary digits (bits) required to store the maximum level value. Thus the number of bits required for 2, 4, 8, 16, 64 and 256 levels is 1, 2, 4, 6 and 8 respectively (Mather, 1987). Fig. 5.6 shows digital images composed of different gray levels. Fig. 5.6(a) shows the image composed of only two levels whereas Fig. 5.6(b) is composed of 16 levels, and an image with 256 levels in which 1 indicates white and 0 indicates black. An image with 16 levels increases the capability for the extraction of information about different features.

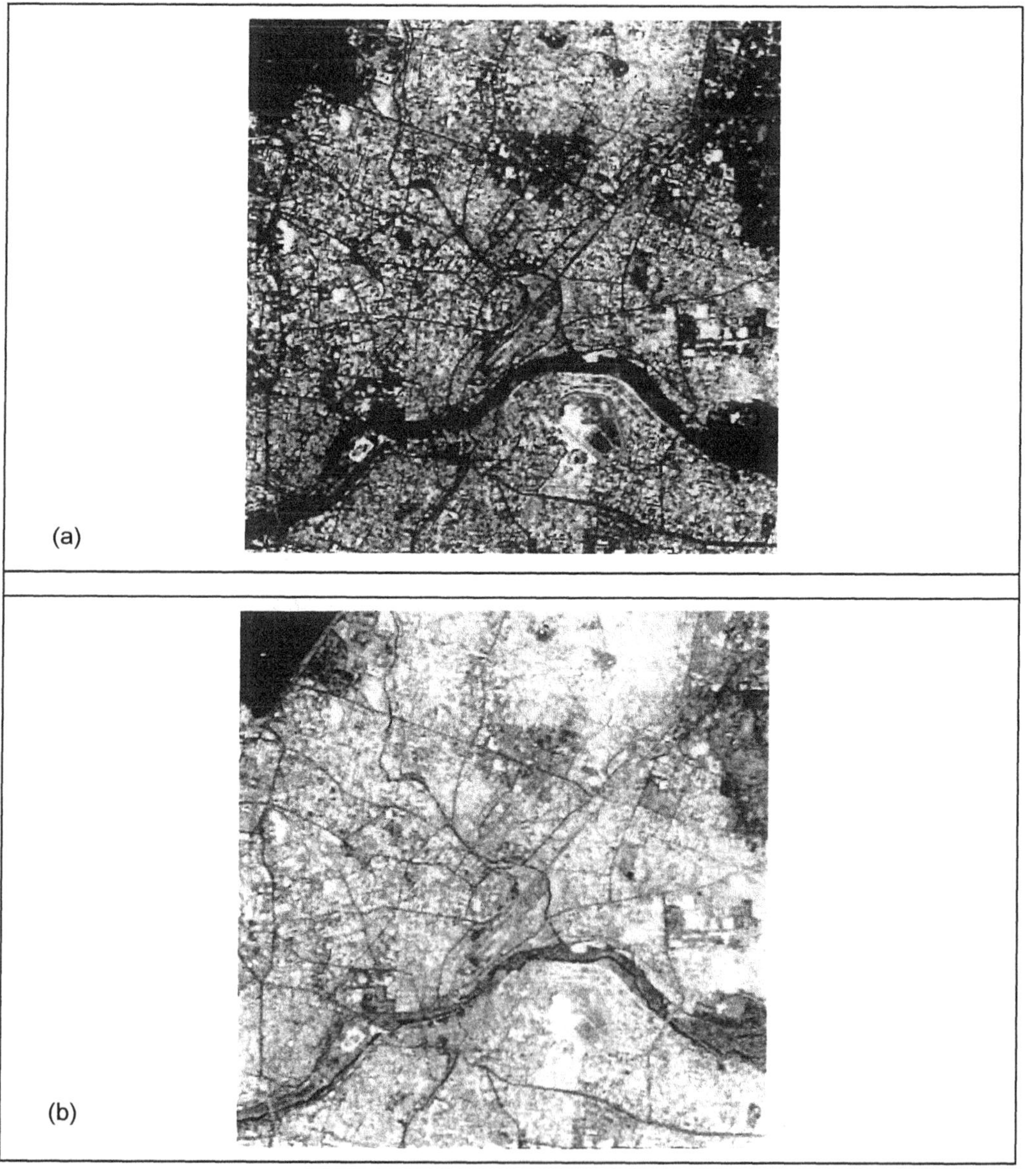

Fig. 5.6 Digital images of different levels of gray values.

5.4 Imaging Sensor Systems

Various components of sensor systems operating in the visible, infrared, thermal and microwave regions of the electromagnetic spectrum are described in this section. Although analogue photographic imagery has many advantages, this book is mainly concerned with image data collected by scanning systems that ultimately generate digital image products. It is apparent that the useful wavebands are mostly in the visible and the infrared for passive remote sensing detectors and in the radar and microwave region for active type of sensors. Accordingly the imaging sensor systems in remote sensing are classified as shown in Fig. 5.7.

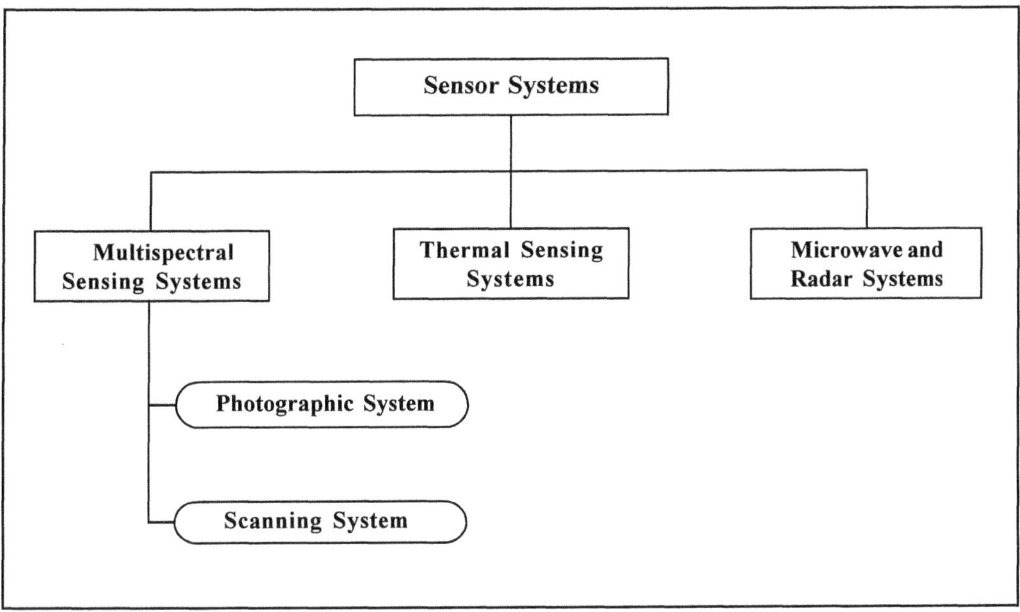

Fig. 5.7 Classification of imaging sensor systems.

Broadly, all the imaging sensor systems are classified based on technical components of the system and the capability of the detection by which the energy reflected by the terrain features is recorded. The classification scheme is (a) Multispectral imaging sensor systems (b) Thermal remote sensing systems, and (c) Microwave radar sensing systems. The multispectral or multiband imaging systems may use conventional type cameras or a combination of them, along with filters for the various bands in the visible part in the scanning system of multiband imaging. This way electromagnetic energy can be recorded by scanning the ground bit by bit.

In some instances, both photographic and scanning systems like Return Beam Vidicon (RBV) sensor of Landsat which is almost similar to an ordinary TV camera, are used.

The thermal system uses radiometers, photometers, spectrometers and/ or thermistors to detect the temperature changes where microwave sensing systems use the antenna arrays for collecting and detecting the energy from the terrain elements.

5.4.1 Multispectral Imaging Sensor Systems

In the case of multiband photographic system, different parts of the spectrum are sensed with different film-filter combinations. Multiband digital camera images and video images are also typically exposed on to the camera's CCD or CMOS sensor (s) through different filters. Electro-optical sensors, such as, the thematic mapper of Landsat, typically sense in atleast several bands of electromagnetic spectrum.

The photographic system suffers from one major defect of considerable distortion at the edges. This is due to a large lens opening. From lens theory, we know that distortions can be minimised and resolution considerably improved by using a narrow beam of light. This can be achieved by a system called scanning system.

A multispectral scanner (MSS) operates on the same principle of selective sensing in multiple spectral bands, but such instruments can sense in many more bands and over a great range of the electromagnetic spectrum. Because of the advancement in utilising electronic detectors, MSS can extend the range of sensing from 0.3 μm to 14 μm. Further MSS can sense in very narrow bands.

Multispectral scanner images are acquired by means of two basic process : across-track and along-track scanning. Multispectral scanner systems build up two-dimensional images of the terrain for a swath beneath the platform. Across-track systems are also called whisk broom scanner systems. This type of scanning system scans the terrain along scanlines that are right angles to the direction of the spaceborne/airborne platform. Fig. 5.8 illustrates the operation across-track system. In this type of scanning system, scanner repeatedly measures the energy from one side of the aircraft to the other. Data are collected within an arc below the aircraft typically of 90^0 to 120^0. Successive scanlines are covered as the aircraft moves forward, yielding a series of contiguous or narrow strips of observation comprising a two-dimensional image of rows (scan lines) and columns. At any instant, the scanner 'sees' the energy within the systems IFOV. This explains the spatial resolution of the sensing systems.

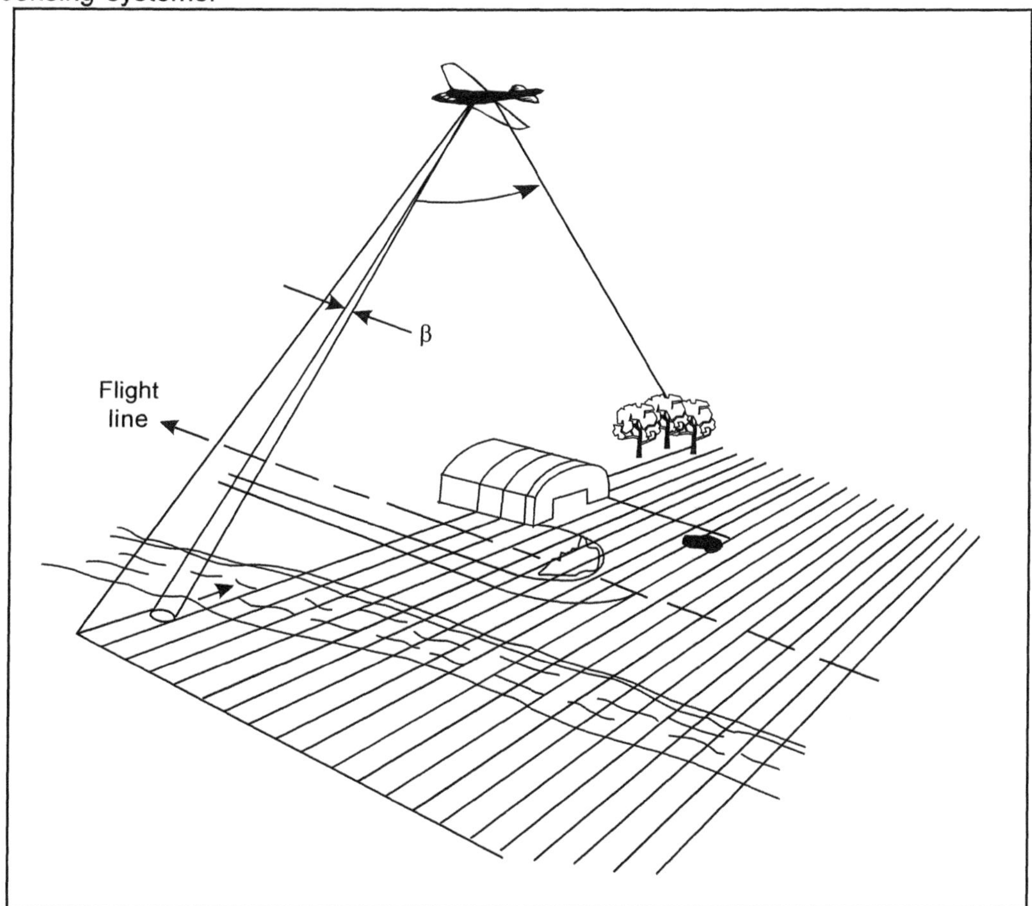

Fig. 5.8 Across-track or whiskbroom scanner system operation.

The second type of multispectral scanning system is along-track scanning system or pushbroom systems. This type of scanners record multiband image data along a swath beneath an aircraft. As the aircraft/spacecraft advances in the forward direction, the scanner scans the earth with respect to the designed swath to build a two-dimensional image by recording successive scanlines that are oriented at right angles to the direction of the aircraft/spacecraft. The system's operation is shown in Fig. 5.9. In along track systems, a linear array of detectors is used instead of a rotating mirror. This is the marked difference between along track and across track scanners.

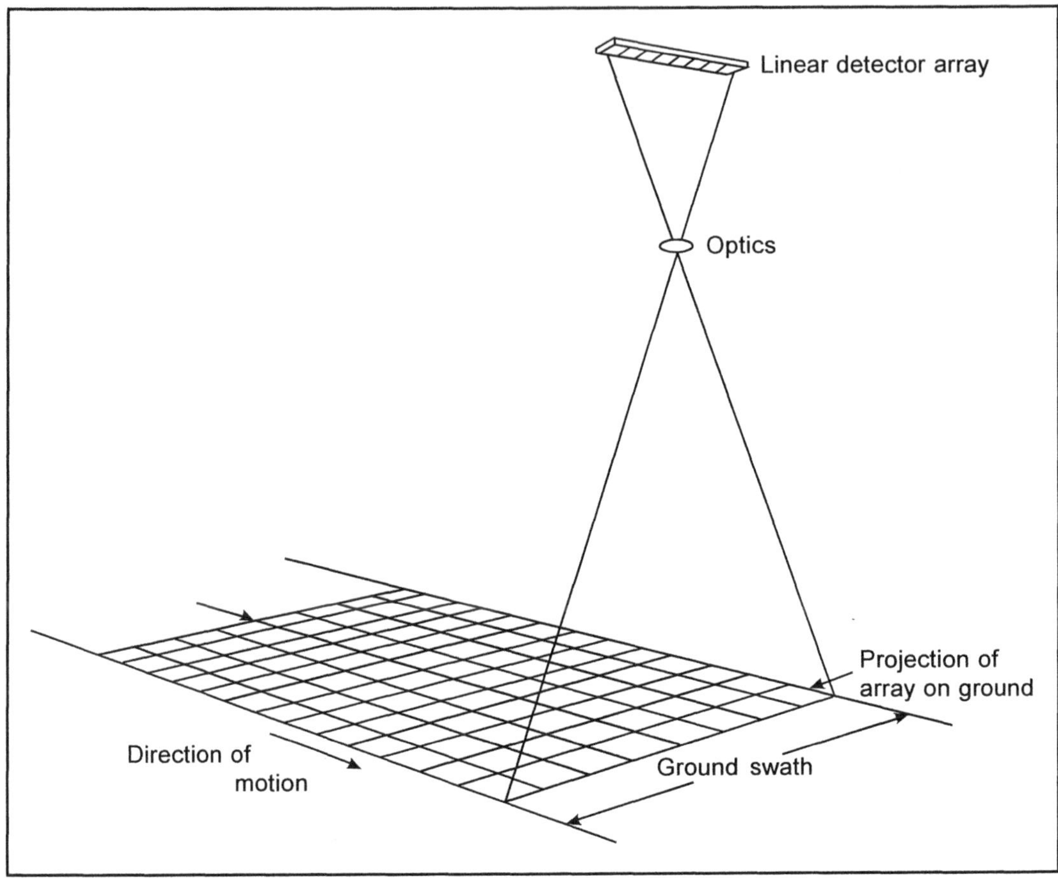

Fig. 5.9 Along-track or pushbroom scanner system operation.

5.4.2 Thermal Sensing Systems

Thermal scanner is one of the most important thermal sensing systems, a particular kind of across track multispectral scanner which senses in the thermal portion of the electromagnetic spectrum by means of inbuilt detectors. These systems are restricted to operating in either 3 to 5 μm or 8 to 14 μm range of wavelengths. The operation and the efficiency of this type of scanning systems are based on the characteristics of the detectors. Quantum or photon detectors are typically used to detect the thermal radiation. These detectors operate on the principle of direct interaction between photons of radiation incident on them and the energy levels of electrical charge carriers within the detector material. The spectral sensitivity range and the operating temperatures of three photon detectors which are common in use are as follows :

Type	Abbreviation	Useful spectral range (μm)
Mercury-doped germanium	Ge:Hg	3 - 14
Indium antimonide	In Sb	3 - 5
Mercury Cadmium telluride	Hg cd Te (MCT)	8 - 14

Fig. 5.10 illustrates schematically the basic operation of a thermal scanner system. A thermal scanner image is a pictorial representation of the detector response on a line-by-line basis. The usual convention when looking at the earth's surface is to have higher radiant temperature areas displayed as lighter toned image areas. Geometrical characteristics of both along track and across track scanner imageries, and radiometric calibrations of these sensing systems should be considered in the design of the thermal scanning systems. The geometrical characteristic of across-track scanner imagery, such as like spatial resolution and ground coverage, tangential scale distortions, resolution cell size variations and one-dimensional relief

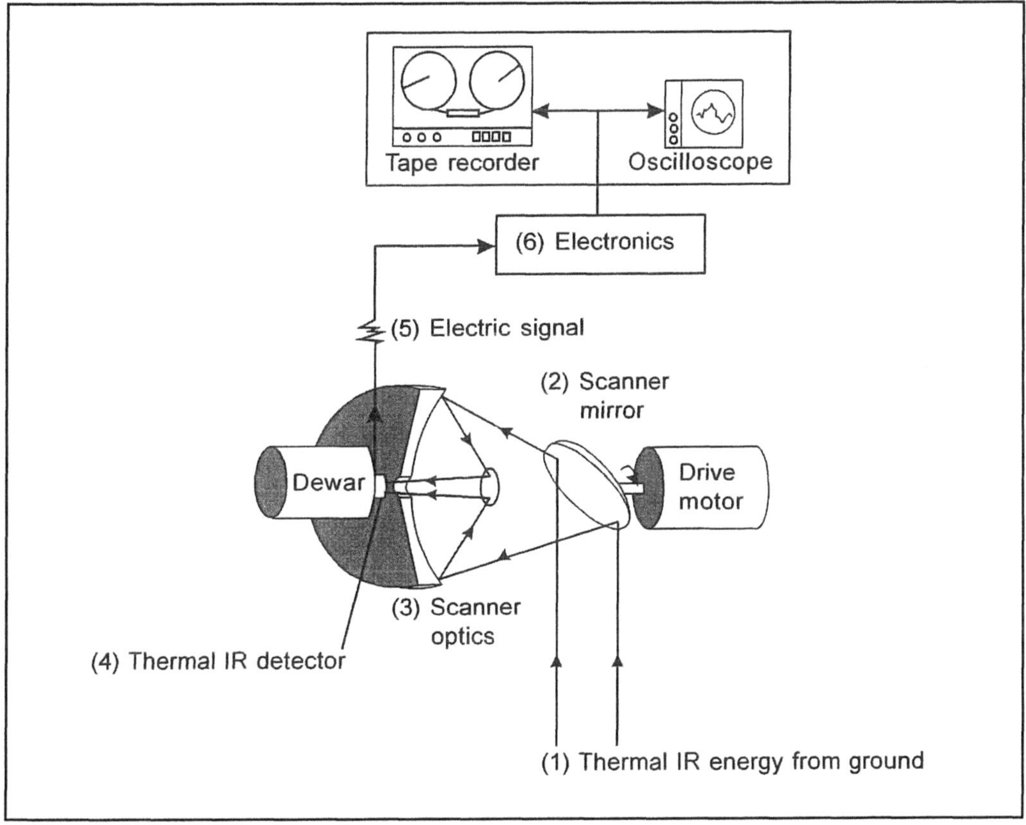

Fig. 5.10 Across-track thermal scanner schematic.

displacement, are some of the geometrical parameters to be considered in the design of thermal and multispectral scanners. Radiometric calibration of thermal scanners can be performed through a number of approaches and each has its own degree of accuracy. Methods used are internal blackbody source referencing and air-to-ground calibration. The description and other principles of these methods are described by Lillesand and Kiefer (2000).

5.4.3 Microwave Imaging Systems

The fundamental principle of microwave sensing and the conceptual design of radar have been discussed in chapter 4, where it is stated that the microwave region of the electromagnetic spectrum includes radiation with wavelengths longer than 1mm-Imaging. Microwave instruments do not, however, rely on the detection of solar or terrestrial emissions. In the following sections of this chapter, the properties of the operational synthetic aperture radar (SAR) systems and Radarsat systems are presented along with other sensing systems.

5.5 Earth Resources Satellites

There are three distinct groups of earth resources satellites. The first group of satellites record visible and near visible wavelengths. The five satellites of Landsat series which are the first generation earth resources satellites are a classic example of this group. The four IRS satellites and the more improved SPOT series of these satellites may be considered the second generation earth resources satellites of the same group. Group two satellites carry sensors that record thermal infrared wavelengths and include the Heat Capacity Mapping Mission satellites, namely, Explorer series. Group three satellites are deployed with sensors that record microwavelengths. The seasat series and the ERS are examples of this group.

5.5.1 Landsat Satellite Programme

National Aeronautics and Space Administration (NASA) of USA with the co-operation of the U.S. Department of Interior planned the launching of a series of Earth Resources Technology Satellites (ERTS). ERTS-1 was launched by a Thor-Delta rocket on July 23, 1972 and it operated until January 6,1978. It represented the first unmanned satellite designed to acquire data about the earth resources on a systematic, repetitive, medium resolution, multispectral basis. Subsequently, NASA renamed the ERTS programme as "Landsat" programme to distinguish it from the series of meteorological and oceanographic satellites that the USA launched later. ERTS-1 was retrospectively named Landsat-1. Five Landsat satellites have been launched so far and this experimental programme has evolved into an operational global resource monitoring programme. Three different types of sensors have been flown in various combinations on the five missions. These are Return Beam Vidicon (RBV) camera system, the Multispectral Scanner (MSS) system and the Thematic

Mapper (TM). Table 5.1 summarizes the characteristics of Landsat-1, through 5, and the sensors used on these satellites and orbital characteristics. Landsat images have found a large number of applications, such as, agriculture, botany, cartography, civil engineering, environmental monitoring, forestry, geography, geology, land resources analysis, landuse planning, oceanography, and water quality analysis.

5.5.2 SPOT Satellite Programme

France, Sweden and Belgium joined together and pooled up their resources to develop the System Pourl' Observation dela Terre (SPOT), an earth observation satellite programme. The first satellite of the series, SPOT-1 was launched from Kourou Launch Range in French Guiana on February 21,1986 aboard an Ariance Launch vehicle (AIV). This is the first earth resource satellite system to include a linear array sensor employing the pushbroom scanning technique. This enables side-to-side oft-nadir viewing capabilities and affords a full scene stereoscopic imaging from two different viewing points of the same area. The high resolution data obtained from SPOT sensors, namely, Thematic Mapper (TM) and High Resolution Visible (HRV), have been extensively used for urban planning, urban growth assessment, transportation planning, besides the conventional applications related to natural resources. The characteristics of SPOT satellite and HRV sensor are given in Table 5.2.

5.5.3 Indian Remote Sensing Satellite (IRS)

The IRS mission envisages the planning and implementation of a satellite based remote sensing system for evaluating the natural resources. The principal components of the mission are: a three axis stabilised polar sunsynchronous satellite with multispectral sensors, a ground based data reception, recording and processing systems for the multispectral data, ground systems for the in-orbit satellite control including the tracking network with the associated supporting systems, and hardware and software elements for the generation of user oriented data products, data analysis and archival. The principal aim of the IRS mission is to use the satellite data in conjunction with supplementary/complementary information from other sources for survey and management of natural resources in important areas, such as, agriculture, geology and hydrology in association with the user agencies. IRS series of satellites are IRS IA, IRS IB, IRS IC, IRS ID and IRS P4 apart from other satellites which were launched by the Government of India. The orbital and sensor characteristics of IRS IA and IB are the same and IRS IC and IRS ID have almost similar characteristics. IRS P4 is an oceanographic satellite, and this will be discussed in the next section. IRS has application potential in a wide range of disciplines such as management of agricultural resources, inventory of forest resources, geological mapping, estimation of water resources, study of coastal hydrodynamics, and water quality surveying.

The sensor payload system consists of two pushbroom cameras (LISS-II) of 36.25 m resolution and one camera (LISS-I) of 72.5 m resolution employing linear Charge Coupled Device (CCD) arrays as detectors. Each camera system images in

Table 5.1 Characteristics of Landsat Satellites and Their Sensors

Satellite Capabilities :

Particulars	Landsat - 1 to 3	Landsat - 4 & 5
Altitude	919 Km	705 Km
Orbit	Near-Polar Sun-Synchronous	Near-Polar Sun-Synchronous
Inclination	99.09 Degree	98.2 Degrees
Period	103 minutes	99 minutes
Equatorial crossing time	0930 Hours	0945 Hours
Repeat Cycle	18 Days	16 Days
Swath Width	185 Km	185 Km
Data rate	15.06 Mbps	84.9 Mbps

Sensor Capabilities :

Sensor	Mission	Channel	Spectral Spatial Resolution (Microns)	Spatial Resolution	Radiometric Resolution
RBV	Landsat 1 to 3	1	0.475-0.575	80 m	6 bits (127 levels)
		2	0.580-0.680	80 m	
		3	0.690-0.830	80 m	
		4	0.505-0.750	80 m	
MSS	Landsat 1 to 5	1	0.5-0.6	79/82 m*	6 bits (127 levels)
		2	0.6-0.7	79/82 m*	
		3	0.7-0.8	79/82 m*	
		4	0.8-1.1	79/82 m*	
		5	10.4-12.6	240 m	
TM	Landsat 4 & 5	1	0.45-0.52	30 m	8 bits (255 levels)
		2	0.52-0.60	30 m	
		3	0.63-0.69	30 m	
		4	0.76-0.90	30 m	
		5	1.55-1.75	30 m	
		6	2.08-2.35	30 m	
		7	10.4-12.5	120 m	

* The Spatial Resolution is 79 m for Landsat-1, 2 & 3. It is 82 m for Landsat 4 & 5.

Table 5.2 Characteristics of SPOT Satellite and HRV Sensor

SPOT Satellite		
Orbit	:	Near-polar Sun-synchronous
Altitude	:	832 km
Inclination	:	98.7 Degrees
Equatorial Crossing Time	:	10.30 Hours
Repeat Cycle	:	26 Days
HRV Sensor		
Channel		**Waveband (Microns) Multispectral**
1		0.50-0.59
2		0.61-0.68
3		0.79-0.89
Panchromatic		
1		0.51-0.73
Spatial resolution	:	20-m (Multispectral) (at nadir) 10 m (panchromatic)
Radiometric resolution	:	8 bits (Multispectral) 6 bits (Panchromatic)
Swath Width	:	117 Km (60 km per HRV, 3 Km overlap)
Angular field of view	:	4.13 Degrees.
Off-nadir viewing	:	$\pm$ 27^0 in 45 steps of 0.6^0 (= $\pm$ Km from nadir)

four spectral bands in the visible and near IR region. The camera system consists of collecting optics, imaging detectors, inflight calibration equipment, and processing devices. The orbital characteristics of the IRS-1A, 1B satellites and the sensor capabilities are given in Table 5.3. As IRS-1D satellite is the latest satellite of the series and hence the system overview of IRS - 1D is provided.

The IRS-1D is a three-axes body stabilized satellite, similar to IRS-1C. Since IRS-1C and 1D are similar in orbital characteristics and sensor capabilities, the details of IRS-1D are discussed as it is a very recent satellite. It will have an operational life of three years in a near polar sunsynchronous orbit at a mean altitude of 780 Km. The payload consists of three sensors, namely, Panchromatic camera (PAN), linear imaging and self-scanning sensor (LISS-III) and wide Field sensor (WiFs). The satellite is equipped with an On-Board Tape Recorder (OBTR) capable of recording limited amount of specified sensor data. Operation of each of the sensors can be programmed. The payload operation sequence for the whole day can be loaded daily on to the on-board command memory when the satellite is within the visibility range. The ground segment consists of a Telemetry Tracking and Command (TTC) segment comprising a TTC network, and an Image segment comprising data acquisition, data processing and product generation system along with data dissemination centre. The over view of IRS-1D mission is to provide optimum satellite operation and a mission control centre for mission management, spacecraft operations and scheduling. The three sensors on board IRS-1D and IRS-1C are described in the following paragraph.

The panchromatic camera provides data with a spatial resolution of 5.2-5.8 m (at nadir) and a ground swath between 63 Km –70 Km (at nadir). It operates in the 0.50 – 0.75 microns spectral band. This camera can be steered upto ± 26 deg. storable upto ±398 Km across the track from nadir, which in turn increases the revisit capability to 3 days for most part of the cycle and 7 days in some extreme cases. Table 5.4 shows the specifications and resolution of PAN. Fig. 5.11 shows the PAN image showing parts of Mumbai city and Fig. 5.12 shows the IRS-1D PAN camera by which this image is obtained. The LISS-III sensor provides multispectral data collected in four bands of the visible, near infra-red (V,NIR) and short wave infra-red (SWIR) regions. While the spatial resolution and swath in the case of visible (two bands) and NIR (one band) regions are between 21.2 m to 23.5 m and 127 Km-141 Km. respectively, they are between 63.6 m to 70.5 m and 133 Km to 148 Km for the data collected in SWIR region. (Table 5.5). Plate 3 shows IRS-1D LISS III FCC image (band 2, 3, 4) and corresponding black and white images of band 2, band 3, and band 4 of path 108, row 56 showing Culcutta and surrounding areas whereas Fig. 5.13 shows the LISS III camera.

The Wide Field Sensor (WiFS) sensor collects data in two spectral bands and has a ground swath between 728 Km to 812 Km with a spatial resolution of 169 m to 188 m. (Table 5.6). The camera by which the WiFS image is formed is shown in Fig. 5.14.

Table 5.3 Particulars of Indian Remote Sensing Satellites (IRS Series)

Characteristics of Satellite		
Orbit	:	Near-polar, Sun-synchronous
Altitude	:	904 Km
Inclination	:	99.03 Degrees
Equatorial Crossing Time	:	10.00 Hours
Repeat Cycle	:	22 days
Eccentricity	:	0.002
Period	:	103 minutes

Sensor Capabilities

Linear Image Scanning System : LISS

No. of LISS Cameras	LRC (One)*	MRC (two)**
No. of Spectral Bands	4	4
IFOV (Microrad)	80	40
Geometric Resolution	72.5	36.25
Swath Width	148 Km	74 Km
Radiometric Resolution	7 bits	7 bits
Band-to-Band	0.5	0.5

* Low Resolution Camera
** Medium Resolution Camera

Table 5.4 Details of PAN of IRS ID

S.No.	Parameter	Specification
1.	Spatial resolution (m)(at Nadir)	5.2 to 5.8
2.	A. Swath (Km) at Nadir B. Swath Steering Range (Deg) C. Step size (Deg) D. Repeatability (Deg)	63 to 70 ± 26 ± 0.09 ± 0.1
3.	Spectral band (micron)	0.50 – 0.75
4.	Camera Square Wave Response (SWR) (at Nyquist frequency)	> 0.20
5.	Quantisation (Bits)	6
6.	Signal to Noise Ratio (SNR) (at saturation radiance)	> 64
7.	Saturation radiance (Gain 1) (mw/cm^2-str-micron)	35 – 40
8.	Integration time (ms)	0.873
9.	Data rate (Mbps)	84.903

Fig. 5.11 PAN image showing parts of Mumbai city

Fig. 5.12 IRS-1D PAN Camera

Table 5.5 Details of LISS III of IRS-1D

Sl.No.	Parameter	Specification	
1.	Spatial resolution (m)	B2,B3,B4 B5	21.2 to 23.5 63.6 to 70.5
2.	Swath (Km)	B2,B3,B4 B5	127 to 141 133 to 148
3.	Spectral band (microns)	B2 B3 B4 B5	0.52 - .059 0.62 – 0.68 0.77 – 0.86 1.55 – 1.70
4.	Camera Square Wave Response (SWR)	B2 B3 B4 B5	>40 >40 >35 >30
5.	Quantisation (bits)	7	
6.	Signal to Noise Ratio	>128	
7.	Saturation Radiance (Gain 1) (mw/cm^2-str-micron)	B2 B3 B4 B5	29 ± 1.5 28 ± 1.5 28 ± 1.5 32.5 ± 2.5
8.	Integration time (ms)	B2, 3 and 4 B5	3.6 10.8
9.	Data rate (Mbps)	B2, 3 and 4 B5	35.7904 1.3906

Fig. 5.13 IRS-1D LISS-III Camera

Fig. 5.14 IRS-1D WiFS Camera

Table 5.6 Details of WiFS of IRS ID

Sl.no.	Parameter	Specification	
1.	Spatial resolution (m)	169 to 188	
2.	Swath (Km)	728 to 812	
3.	Spectral band (micron)	B3	0.62 – 0.68
		B4	0.77 - 0.86
4.	Square Wave Response	B3	> 0.34
		B4	> 0.20
5.	Quantisation (bits)	7	
6.	Signal to Noise Ratio (at saturation)	>128	
7.	Saturation radiance (mw/cm^2-str-micron)	B3	28 ± 1.5
		B4	31 ± 1.5
8.	Integration time (ms)	28.8	
9.	Data rate (Mbps)	2.0616	

5.5.4 AEM Satellites

The (Heat capacity Mapping Mission) HCMM satellite is the first of a small and relatively inexpensive series of NASA's Applications Explorer Mission (AEM) satellites. Launched in April 1978, it lasted till September 1980. Table 5.7 summarises the details of AEM satellite and HCM sensor characteristics. The orbits of the satellite are arranged to ensure that images of each scene are obtained during the periods of maximum and minimum surface temperature for the determination of thermal inertia. The data from HCMM are intended primarily for conversion to thermal inertia maps for geological mapping. However, the images have found wider applications, such as, vegetation mapping, vegetation stress detection, microclimatology, soil moisture mapping, snowmelt prediction, and monitoring industrial thermal pollution.

5.6 Meteorological Satellites

Meteorological satellites designed specifically to assist in weather prediction and monitoring, generally incorporate sensors that have very coarse spatial resolution compared to land-oriented systems. These satellites, however, afford a high frequency global coverage. USA has launched a multiple series of meteorological satellites with a wide range of orbit and sensing system designs. The first of these series is called the NOAA, an acronym for National Oceanic and Atmospheric Administration. These satellites are in near-polar, sunsynchronous orbits similar to those of 'Landsat and IRS'. In contrast, another series of satellites which are of essentially meteorological type, called Geostationary Operational Environmental Satellite (GOES) series and Meteosat operated by European Space Agency, are geostationary, remaining in a constant relative position over the equator.

5.6.1 NOAA Satellites

Several generations of satellites in the NOAA series have been placed in orbit. The satellites NOAA-6 through NOAA-10 contained Advanced Very High Resolution Radiometer (AVHRR). The even-numbered missions have daylight (7.30 A.M.) north-to-south equatorial crossing and the odd-numbered missions have night time (2.30 A.M.) north-to-south equatorial crossing. The basic characteristics of these missions and the AVHRR instrument are listed in Table 5.8. Apart from routine climatological analyses, the AVHRR data have been used extensively in studies of vegetation dynamics, flood monitoring, regional soil moisture analysis, dust and sandstorm monitoring, forest wild fire mapping, sea surface temperature mapping, and various geological applications, including observation of volcanic eruptions, and mapping of regional drainage and physiographic features.

Table 5.7 Details of AEM Satellite and HCMM Sensor

Characteristics of Satellite		
Orbit	:	Near-polar Sun-synchronous
Altitude	:	620 Km
Inclination	:	97.6 Degrees
Equatorial CrossingTime	:	14.00 hrs (Ascending) 02.00 hrs (Decending)
Repeat Cycle	:	16 days

HCMM Sensor Capabilities	
Channel	**Waveband (Microns)**
1	0.5-1.1
2	10.5-12.5
Spatial Resolution	500 m (channel -1) 600 m (channel-2)
Swath width	716 Km

Table 5.8 Particulars of NOAA Satellite and AVHRR Sensor

NOAA Satellite		
Orbit	:	Near-polar, Sun-synchronous
Altitude	:	833-870 Km
Inclination	:	98.7 Degrees
Equatorial crossingTime	:	0730 and 1930 Hours
		1400 and 0200 Hours
Repeat Cycle	:	12 Hours
Period	:	102 Minutes
AVHRR Sensor Capabilities		
Channel		**Waveband (Microns)**
1		0.58 - 0.68
2		0.725 - 1.10
3		3.55 - 3.93
4		10.3 - 11.3
5		11.5 - 12.5
Spatial resolution	:	1.1 Km (at nadir)
Radiometric resolution	:	10 bits (1024 levels)
Swath Width	:	3000 Km.

5.6.2 GOES Satellites

The GOES programme is a cooperative venture between NOAA and NASA. The Geostationary Operational Environmental Satellites (GOES) are part of a global network of meteorological satellites spaced about 70^0 longitude apart around the world. The GOES images are distributed in near real-time for use in local weather forecasting. They have also been used in certain large area analyses such as regional snow cover mapping. The details of GOES satellites and the sensors they are carrying are shown in Table 5.9.

5.6.3 NIMBUS Satellites

This is one of the ocean monitoring satellites launched in October 1978. This satellite carries the Coastal Zone Colour Scanner (CZCS) designed specifically to measure ocean parameters. The details of the six bands in which the CZCS operates and the characteristics of NIMBUS-7 satellite are presented in Table 5.10 The CZCS has been used to measure sea surface temperatures, detection of chlorophyll and suspended solids of near-shore and coastal waters.

5.6.4 Meteosat Series

Meteosat-1 of this series launched in November 1977 failed in November 1979. Meteosat-2 was launched June 1981, and like GOES its sensors recorded images every half an hour in three wavebands (Table 5.11). Data from Meteosat sensors have been used for many meteorological applications like studies of synoptic climatology, sea surface temperature, land surface temperature, windvector determinations and monitoring all types of natural disasters.

5.6.5 Megha-Tropiques

Megha-Tropiques is a satellite mission to study the water cycle in the tropical atmosphere in the context of climate change[1] A collaborative effort between Indian Space Research Organisation (ISRO) and French Centre National d'Etudes Spatiales (CNES), Megha-Tropiques was successfully deployed into orbit by a PSLV rocket in October 2011.

Megha-Tropiques was initially scrapped in 2003, but later revived in 2004 after India increased its contribution and overall costs were lowered.[2][3] With the progress made by GEWEX (Global Energy and Water Cycle Experiment), Megha-Tropiques is designed to understand tropical meteorological and climatic processes, by obtaining reliable statistics on the water and energy budget of the tropical atmosphere. Megha-Tropiques complements other data in the current regional monsoon projects such as MAHASRI and the completed GAME project. Megha-Tropiques also seeks to describe the evolution of major tropical weather systems. The focus will be the repetitive measurement of the tropics.

Specifications of Megha-Tropiques

MEGHA-TROPIQUES	
Operator	ISRO / CNES
Major contractors	ISRO
Mission type	Meteorological Research Satellite
Satellite of	Earth
Launch date	12 October 2011, Sriharikota Andhra Pradesh, India
Launch vehicle	PSLV-CA
Mission duration	3 years (minimal)
COSPAR ID	2011-058A
Homepage	Official website
Mass	1,000 kg (2,205 lb)
Orbital elements	
Eccentricity	near circular
Inclination	19.99ointented:20o
Apoapsis	865 km (537 mi) intented:867
Periapsis	864 km (537 mi) intented:865
Orbital period	101.93 minutes

Table 5.9 Particulars of GOES Satellites and their Sensors

GOES Satellite		
Orbit	:	Geostationary
Altitude	:	33,367-48,390 Km
Inclination	:	1.9 - 0.2 Degrees
Repeat Cycle	:	Twice per hour
Period	:	1430 - 1436 Minutes
Visible Infrared Spin Scan Radiometer (VISSR)		
Channel	**Waveband (Microns)**	**Ground Resolution**
1	0.55-0.70	14 Km
2	10.5-12.6	8 Km
VISSR Atmospheric Sounder (VAS)		
1	0.55-0.70	14 Km
2	4.496-14.81	16 Km

Table 5.10 Characteristics of NIMBUS Satellites and CZCS Sensor

NIMBUS Satellites		
Orbit	:	Near-polar, Sun-synchronous
Altitude	:	955 Km
Inclination	:	99.3 Degrees
Repeat Cycle	:	6 Days)
Period	:	104 Minutes
Overpass	:	1200 Hours

CZCS Sensor Capabilities	
Channel	**Waveband (Microns)**
1	0.433 - 0.453
2	0.510 - 0.530
3	0.540 - 0.560
4	0.660 - 0.680
5	0.700 - 0.800
6	10.50 - 12.50
Spatial resolution	825 m
Swath Width	1566 Km.

Table 5.11 Details of Meteosat and Sensor Wavebands

Meteosat Satellites		
Orbit	:	Geostationary
Altitude	:	35900 Km
Position	:	Equator, 0° Longitude
Sensor Wavebands		
Channel		**Wavelength (Microns)**
1		0.4 - 1.1
2		5.7 - 7.1
3		10.5 - 12.5
Spatial Resolution	:	2.4 Km (channel - 1)
	:	5 KM (channel-2&3)
Radiometric Resolution	:	8 bits (256 levels)

5.7 Satellites Carrying Microwave Sensors

Microwave imaging is gaining increasing importance with a better understanding of the relation between image tone and the earth's surface characteristics. The clear advantage of microwave sensor is its capacity to penetrate cloud cover. Satellites that carry microwave sensors are seasat with Synthetic Aperture Radar (SAR), European Remote Sensing Satellite (ERS)-1 and Radarsat.

5.7.1 Seasat

It is an experimental satellite designed by NASA to establish the utility of microwave sensors for remote sensing of oceans. Images of the land are also obtained giving environmental scientists their first synoptic view of the earth in microwavelengths. The satellite carries five sensors, two of which are of potential interest to environmental scientists, these being a radiometer and synthetic aperture radar. The characteristics of the satellite and these two sensors are given in Table 5.12. The data obtained from seasat are used for ocean studies, mapping sea ice, urban patterns landcover, geology and hydrological features.

5.7.2 European Remote Sensing Satellite-1

ERS-1 launched in May 1991 has a synthetic aperture radar capable of penetrating clouds (Table 5.13). The objectives of this mission are study of coastal zones, dynamics of ice cover, and monitoring of global weather and landuse.

5.7.3 Radarsat

Radarsat - 1 launched on November 28, 1995 is the first satellite of Canadian Space Agency. The orbit for Radarsat is sunsynchronous at an altitude of 798 km and inclination of 98.6^0. The orbit period is 10.7 min. and the repeat cycle is 24 days. Radarsat is a right -looking sensor facing east during the ascending orbit and west during the descending orbit. The Radarsat-SAR is a C-band (5.6 cm) system with HH polarisation. The system can be operated in a variety of beam selection modes providing various swath widths, resolutions and look angles. Fig. 5.15 shows the radarsat image of the Zhao Qing, China . The primary applications of Radarsat are land cover mapping, agricultural and forestry monitoring, disaster monitoring, landslide identification, flood monitoring, and snow distribution mapping.

5.7.4 RISAT-1

Radar Imaging Satellite 1, or **RISAT-1**, is an Indian remote sensing satellite which was built and is operated by the Indian Space Research Organisation (ISRO). The second RISAT satellite to be launched, it uses C-band Synthetic Aperture Radar for earth observation irrespective of the light and weather conditions of the area being imaged.

The launch of RISAT-1 came several years after that of RISAT-2, which carried an Israeli-built x-band radar. The RISAT-2 mission was prioritised over RISAT-1 following the 2008 Mumbai attacks, resulting in RISAT-1 being delayed by several years.

Specifications of RISAT-1

RISAT-1	
Operator	ISRO
Mission type	Radar imaging
Launch date	26 April 2012, 00:17 UTC
Carrier rocket	PSLV-C19
Launch site	Satish Dhawan Space Centre FLP
Mission duration	5 years (planned)
COSPAR ID	2012-017A
Homepage	ISRO - RISAT-1
Mass	1,858 kg (4,100 lb)
Orbital elements	
Regime	Sun-synchronous
Inclination	97°
Altitude	536 km[1]
Orbits per day	14
Instruments	
Main instruments	C-band SAR
Imaging resolution	1m - 50m[2]

5.7.5 RISAT-2

RISAT-2, or Radar Imaging Satellite 2 is an Indian radar reconnaissance satellite that is part of India's RISAT programme. It was built by the Israel Aerospace Industries and successfully launched aboard a PSLV-CA rocket at 01:15 GMT on April 20, 2009 from the Second Launch Pad at the Satish Dhawan Space Centre. The specifications of RISAT is shown in Table.

RISAT-2's main sensor is an X-band synthetic aperture radar from Israel Aerospace Industries (IAI).[2] It is designed to monitor India's borders and as part of anti-infiltration and anti-terrorist operations.[3] The satellite has a mass of 300 kilograms (660 lb).

Specifications of RISAT-2

Operator	Indian Air Force ISRO
Major contractors	ISRO IAI
Mission type	Radar imaging
Launch date	20 April 2009 01:15 GMT
Carrier rocket	PSLV-CA C12
Launch site	SDSC Second Launch Pad (SLP)
COSPAR ID	[1] 2009-019A
Mass	300 kilograms (660 lb)
	Orbital elements
Regime	Sun-Synchronous Circular orbit
Inclination	41°
Apoapsis	550.0 kilometres (341.8 mi)
Periapsis	550.0 kilometres (341.8 mi)
Orbital period	90 minutes

5.8 Ocean Series

5.8.1 Oceansat-1

The successful launch of IRS-P4 satellite at Sriharikota, India on May 26,1999, was a fully indigenous effort which will help India to know more about its, oceans, their wealth and the costal ecosystems. This satellite has onboard two main sensors, namely, the Ocean Color Monitor (OCM) and Multi Frequency Scanning Microwave Radiometer (MSMR). The potential scientific investigations, application areas, and calibration and validation of OCM and MCMR are listed below as outlined by ISRO (1999). The satellite and sensor characteristics of IRSP4 are given in Table 5.14 and 5.15.

Ocean colour Monitor (OCM)

Scientific Investigations

Atmospheric correction algorithms
Aerosol characterization
Ocean primary productivity
Models for fish stock assessment
Inverse modelling techniques for retrievel of oceanic parameters
Space-time variability of plankton growth due to eco-physiological parameters
Models for quantitative estimation of sediments
Optical properties of water
Regional and global climate and environmental changes
Bio-geochemical cycles

Table 5.12 Seasat Satellite and Its Sensors

Seasat Satellite		
Orbit	:	Near-polar Circular
Altitude	:	790 Km
Inclination	:	108 Degrees
Repeat Cycle	:	152 Days
Period	:	101 Minutes
Synthetic Aperture Radar (SAR) Sensor		
Frequency	:	1274.8 GHz
Wavelength	:	23.5 Cm (L-Band)
Polaraization	:	H H (Horizontal & Horizontal)
Depression Angle	:	20^0
Spatial Resolution	:	25 m
Swath width	:	100 Km
Radiometer Sensor		
Visible band	:	0.47-0.94 microns
Spatial Resolution	:	2 Km.
Thermal infrared band	:	10.5 - 12.5 microns
Spatial resolution	:	4 Km.

Table 5.13 SAR Imaging Mode Characteristics

ERS - 1 Satellite		
Orbit	:	Sun - Synchronous
Altitude	:	785 Km
Repeat cycle	:	3 days
Equator crossing time	:	10.30 A.M.
ERS-1 SAR		
Frequency	:	C - band (5.3 0Hz + 0.2 MHz)
Swath width	:	100 Km
Incident angle	:	23 Degrees mid swath (350 in roll-tilt)
Spatial resolution	:	30 m x 30 m
Radiometric resolution	:	2.5 dB for 0o = -18 db
Chirp band width	:	15,55 = 0.1 MHz
Chirp length	:	37.1+0.05 US
Peak power	:	4.8 Kw
Antenna size	:	10 mxl. m
A/D sampling rate	:	18.96 M samples
Quantisalion	:	51, SQ, OGRC (Nominal 61, 6Q OBRC (experimental)

Table 5.14 Specifications of IRS-P4 OCM

1.	Ground resolution (m) Real-Time Mode (Full resolution Mode)		:	360 (across-track) 236 (along-track)
2.	Swath (km)		:	1420
3.	Instrument Spectral Bands	Wavelength Range (nm)	SNR at Ocean Radiance's Radiance*	Expected SNR
	C-1	402-422	9.1	307.6
	C-2	433-453	8.4	440.7
	C-3	480-500	6.6	415.0
	C-4	500-520	5.6	408.8
	C-5	545-565	4.6	412.2
	C-6	660-680	2.5	345.6
	C-7	745-785	1.6	393.7
	C-8	845-885	1.1	253.6
	*In mw/(cm^2-Sr-mm)			
4.	MTF (at Nyquist)	> 0.2		
5.	Quantisation	12 bits		
6.	Data Rate (Mbps)	20.8		
7.	Transmission Frequency	X-band		
8.	Along track steering (To avoid sunlight)	$\pm 20^0$		
	Source : ISRO, 1999			

Source : Canadian Space Agency/Agence spatial canadenna 1996 Data received by Canada centre for remote sensing processed and distributed by RADARSAT international

Fig. 5.15 Radarsat image.

Table : 5.15 Specifications of IRS – P4 MSMR

1.	Frequencies (GHz)	6.6 10.65 18.0 21.0
2.	Polarisation	V & H for all frequencies
3.	Spatial Resolution (km x km)	6.6 (V & H) : 105 × 68 10.65 (V & H) : 66 × 43 18.0 (V & H) : 40 × 26 21.0 (V & H) : 34 × 22
4.	Swath	1360 km for all frequencies
5.	Nominal Incidence Angle (deg.)	49.7
6.	Radiometric Sensitivity	Better than 1^{O} K
7.	Dynamic Temp. K	Range 10 – 330
8.	Data Rate (kbps)	6.4

Modeling Carbon fluxes
Synergistic use of data from multiple missions
Development of new techniques for image analysis and classification

Application Demonstration

Synergistic use of SST-Chlorophyll in fishery forecast
Identification and monitoring of phytoplankton bloom
Coastal processes, Sediment dynamics and Coastal pollution
Vegetation dynamics / agriculture applications
Snow-ice studies

Calibration and Validation

Characterization of sensor performance
Algorithm development
Geophysical parameter validation
Inter-comparison with similar ocean colour sensor from different missions.

Multifrequency Scanning Microwave Radiometer

Application Demonstration

Sea state predictions
Oceanic process studies
Weather and climate studies
Monsoon and atmospheric predictions
Cyclone genesis and development
Air-sea interaction studies
Polar ice studies
Circulation and wave studies
Mixed layer depth studies
Rainfall estimation
Latent heat flux estimation
Soil moisture estimation

Calibration and Validation

Characterisation of sensors
Algorithm development
Geophysical parameter validation
Data assimilation in numerical models
Inter-comparison with similar sensors from different missions.

5.8.2 Oceansat-2

Oceansat-2 is an Indian satellite designed to provide service continuity for operational users of the Ocean Colour Monitor (OCM)instrument on Oceansat-1. It will also enhance the potential of applications in other areas. The main objectives of OceanSat-2 are to study surface winds and ocean surface strata, observation of chlorophyll concentrations, monitoring of phytoplankton blooms, study of atmospheric aerosols and suspended sediments in the water.

Oceansat-2 is ISRO's second in the series of Indian Remote Sensing satellites dedicated to ocean research, and will provide continuity to the applications of Oceansat-1 (launched in 1999). Oceansat-2 will carry three payloads including an Ocean Colour Monitor (OCM-2), similar to the device carried on Oceansat-1. Data from all instruments will be made available to the global scientific community after the post-launch sensor characterization, which is expected to be completed within 6 months of the launch. Oceansat-2 was launched from SDSC (SHAR) on 23 September 2009 using PSLV-C14.

Specifications of Oceansat-2

Operator	Indian Space Research Organisation
Bus	IRS
Mission type	Oceanography
Launch date	23 September 2009
Carrier rocket	PSLV-C14
Launch site	Satish Dhawan Space Centre
COSPAR ID	OCEANS2
Mass	960 kilograms (2,100 lb)
Orbital elements	
Regime	Sun-Synchronous Circular orbit
Inclination	98.280^{o}
Apoapsis	720 kilometres (450 mi)
Periapsis	720 kilometres (450 mi)
Orbital period	99.31 minutes

5.9 IKONOS Satellite Series

IKONOS satellite is one of the high resolution satellites launched on September 24th, 1999 by USA. This is a very recent satellite with a resolution higher than that of other existing satellites. The revisit time will be more frequent for higher latitudes and less frequent for latitude closer to the equator. This satellite is designed to occupy a 681 Km sunsynchronous orbit having an equatorial crossing time of 10:30 AM. The ground track of the systems repeats every 11 days, but the revisit time for imaging is less than 11 days, based on latitude and the tilt of the system. The system collects the data at angles upto 45^0 from vertical in both along-track and across-track directions. This provides an opportunity for frequently covering an area, and also enables the use of the data for stereoscopic or 3-D study of the terrain.

IKONOS employs linear array technology and collects data in a Panchromatic band (0.45-0.90 μm) with spatial resolution of 1m. It also covers the terrain in Multispectral bands at a nominal spatial resolution. These spectral bands include 0.45 to 0.52 μm (blue), 0.52 to 0.60 μm (green), 0.63 to 0.69 μm (red) and 0.76 to 0.90 μm (near infrared). The Panchromatic and multispectral bands can be combined to produce pan-sharpened multispectral imagery, with an effective resolution of 1m data collection over 2048 gray levels (11 bits). The ground processing software has the capability to rapidly process on mosaic imagery so as to create seamless image products with a consistent fixed ground sample distance. The IKONOS data has been extensively used for urban growth assessment studies, municipal planning, utility management, facility management and so on. It is very much useful in application areas like the creation of Cadastral Information System (CIS) where large scale mapping is necessary. Table 5.16 shows the orbital characteristics of IKONOS. Fig. 5.16 is an IKONOS image of a part of Beijing, China with 1m resolution.

Table 5.16 Orbital Information of IKONOS Satellite

Altitude	681 km.
Inclination	98.1 degrees
Speed	7 km/sec.
Descending nodel	10:30 a.m.
Orbit time	98 minutes
Orbit type	Sun-synchronous
Swath	11 km at nadir.
Image area	11 km × 11 km.

Fig. 5.16 IKONOS Image

5.10 Communication Satellites

5.10.1 GSAT

GSAT-3 or EDUSAT is a communications satellite which was launched on 20 September 2004 by the Indian Space Research Organisation. EDUSAT is the first Indian satellite built exclusively to serve the educational sector. It is mainly intended to meet the demand for an interactive satellite-based distance education system for the country.

EDUSAT carries five Ku band transponders providing spot beams, one Ku band transponder providing a national beam and six extended C band transponders providing national coverage beams. EDUSAT was successfully launched into a Geosynchronous Transfer Orbit on the first operational launch of the Geosynchronous Satellite Launch Vehicle, which flew from the First Launch Pad at the Satish Dhawan Space Centre in Sriharikota. EDUSAT was initially placed into a transfer orbit with a perigee of 180 kilometres (110 mi) and an apogee of 35,985 kilometres (22,360 mi) and a period of 10.5 hours, inclined at 19.2 degrees to the equator.

Specifications of EDUSAT

EDUSAT	
Operator	ISRO
Major contractors	ISRO Satellite Centre, Space Applications Centre
Mission type	Communication satellite
Launch date	20 September 2004 10:31:00 UTC]
Carrier rocket	GSLV F01
Launch site	Satish Dhawan FLP
COSPAR ID	2004-036A
Homepage	http://www.isro.org/Edusat/
Mass	1,950 kilograms (4,300 lb)
Power	2040 W from solar panels
Batteries	24 Ah Ni-Mh
Orbital elements	
Regime	Geosynchronous
Inclination	0°
Apoapsis	35,802 kilometres (22,246 mi)
Periapsis	35,770 kilometres (22,230 mi)
Orbital period	1436.06 minutes
Longitude	74° East

Launch sequence. EDUSAT or GSAT-3 is a communications satellite which was launched on 20 September 2004 by the Indian Space Research Organisation. EDUSAT is the first Indian satellite built exclusively to serve the educational sector. It is mainly intended to meet the demand for an interactive satellite-based distance education system for the country.

GSAT-4, also known as HealthSat, was an experimental communication and navigation satellite launched in April 2010 by the Indian Space Research Organisation on the maiden flight of the Geosynchronous Satellite Launch Vehicle Mk.II rocket. It failed to reach orbit after the rocket's third stage malfunctioned. The third stage was the first Indian-built cryogenic-fuelled upper stage, and was making its first flight. The ISRO suspects that the failure was caused by the third stage not igniting.

Specifications of GSAT-4

Operator	ISRO
Major contractors	ISRO Satellite Centre, Space Applications Centre
Mission type	Communication
Launch date	15 April 2010 10:57 GMT
Carrier rocket	GSLV Mk.II (s/n D3)
Launch site	Satish Dhawan Space Centre SLP
Mission duration	7 years (planned) Failed to orbit
Mass	2,220 kg
Power	2,760 watts from solar panels
Batteries	36 A·h NiMH
Orbital elements	
Regime	Geosynchronous (planned)
Longitude	82°E (planned)
Instruments	
Main instruments	GAGAN
Transponders	
Transponders	Ka-band bent pipe and regenerative transponder

GSAT-5P, or GSAT-5 Prime, was an Indian communications satellite which was lost in a launch failure in December 2010. Part of the Indian National Satellite System, it was intended to operate in geosynchronous orbit as a replacement for

INSAT-3E Satellite GSAT-5P was a 2,310-kilogram (5,100 lb) spacecraft, which was built by the Indian Space Research Organisation based around the I-2K satellite bus. ISRO also launched the satellite, and was to have been responsible for its operation. It was equipped with 36 transponders operating in the G/H band of the NATO-defined spectrum, or the C band of the older IEEE spectrum. Twelve of the transponders operated on extended frequencies within the band.[2] GSAT-5P was expected to operate for at least 12 years, and would have been placed at a longitude of 55 degrees east.

Specifications of GSAT-5P

GSAT-5P	
Operator	ISRO
Major contractors	ISRO
Bus	I-2K
Mission type	Communication
Launch date	25 December 2010 10:34 UTC
Carrier rocket	GSLV Mk.I F06
Launch site	SDSC SLP
Mission duration	12-14 years (planned) Failed to orbit
Mass	2,310 kilograms (5,100 lb)
Orbital elements	
Regime	Geosynchronous (planned)
Longitude	55° East (planned)
Transponders	
Transponders	36 G/H band (IEEE C band)

GSAT-8 is communication satellite. It was constructed by the Indian Space Research Organisation, as part of INSAT system. GSAT-8 was launched on May 21st, 2011 from Kourou, French Guiana. The rocket, an Ariane 5 was the carrier, marketed by the European Arianespace. Prior to launch, the spacecraft was transported from India to Cayenne – Rochambeau Airport in French Guiana by an Antonov An-124 cargo aircraft. The success of the launch is said to have made up for the previous loss of two satellites on the indigenous GSLV rocket.

Specifications of GSAT-8

Operator	ISRO
Major contractors	Indian Space Research Organisation
Mission type	Communication
Launch date	21 May 2011 03:58 hrs (IST)
Carrier rocket	Ariane-5 VA-202
Launch site	Kourou, French Guiana
Mission duration	More Than 12 Years
COSPAR ID	2011-022A
Power	6,242 watts from Lithium cell
Batteries	100 Ah Lithium-ion battery
Orbital elements	
Regime	Geosynchronous
Longitude	55°E
Instruments	
Main instruments	GAGAN
Transponders	
Transponders	24 Ku band transponders

GSAT-12 is communication satellite developed by the Indian Space Research Organisation, as part of INSAT system. GSAT-12 had been launched on July 15th , 2011. The satellite had been injected by PSLV-C17(XL) into sub-Geosynchronous Transfer Orbit of 284 km perigee and an apogee of 21000 km and later be transferred to Geostationary orbit of 36000km by a series of maneuvers using on board Liquid engine motors. The launch had taken place from Satish Dhawan Space Centre in Sriharikota.

Specifications of GSAT-12

GSAT-12	
Operator	ISRO
Mission type	Communications Satellite
Satellite of	Earth
Carrier rocket	PSLV-C17
Launch site	Satish Dhawan Space Centre SLP
Mission duration	8 years
COSPAR ID	2011-034A

Contd.....

Mass	1,412 kilograms (3,110 lb)
Power	Solar
	Orbital elements
Regime	Geostationary Orbit
Inclination	0
Apoapsis	35,786 km
Periapsis	35,786 km
Longitude	83 E
	Instruments
Spectral band	C-band
	Transponders
Transponders	12

5.10.2 HAMSAT

HAMSAT also known as HAMSAT INDIA and VU2SAT is a microsatellite weighing 42.5 kilograms (93.7 lb), providing amateur radio satellite communications services for Indian and international amateur radio operators. This satellite carries the in-orbit designation of VO-52, and is an OSCAR series satellite. It was launched by PSLV-C6 on May 5, 2005. The main payload was an Indian Remote Sensing satellite, CARTOSAT-1 weighing 1,560 kilograms (3,440 lb). HAMSAT was placed into a polar sun synchronous orbit. It carries two transponders, one built by William Leijenaar (Call Sign: PE1RAH), a Dutch Radio Amateur and graduate engineering student from the Higher Technical Institute at Venlo and the other by enthusiasts at ISRO (Indian Space Research Organisation).

The satellite joins many previous satellites in amateur radio service; mostly launched by HAMSAT.

5.10.3 INSAT

It was launched in March 2007 by the European Ariane launch vehicle. Configured with payloads identical to that of INSAT-4A, INSAT-4B carries 12 Ku band and 12 C-band transponders to provide EIRP of 52 dBW and 39 dBW respectively. Two Tx/Rx dual grid offset fed shaped beam reflectors of 2.2 m diameter for Ku band and 2 m diameter for C-band are used. INSAT-4B augments the high power transponder capacity over India in Ku band and over a wider region in C-band. It is co-located with INSAT-3A at 93.5 degree E longitude.

The national space agency Indian Space Research Organisation (ISRO) has allotted nearly seven Ku band transponders to Sun Direct; a DTH service provider from South India, and the other five to Doordarshan's DD Direct Plus. 12 transponders in the C band are for TV, radio and telecommunication purposes.

INSAT-4CR was launched on 2 September 2007 by GSLV-F04. It is a replacement satellite of INSAT-4C which was lost when GSLV-F02 failed and had to be destroyed on its course. It carries 12 Ku band 36 MHz bandwidth transponders employing 140 W TWTAs to provide an Effective Isotropic Radiated Power of 51.5 dBW at Edge of Coverage with footprint covering Indian mainland. It also incorporates a Ku band Beacon as an aid to tracking the satellite.

On 8 September 2007 ISRO reported the satellite had reached a near geosynchronous orbit, and would be stabilized in its intended orbital position of 74 degrees E longitude by 15 September.[12] The satellite is designed for a mission life in of ten years. There were reports that the mission life of the satellite had decreased by five years as the thrusters had to burn this much fuel to restore the satellite to its correct orbit. However, the ISRO later refuted this claim dismissing it as false.

5.11 Cartosat Series

5.11.1 CARTOSAT-1

CARTOSAT-1 was successfully launched by the ninth flight of ISRO's Polar Satellite Launch Vehicle, PSLV-C6, from Satish Dhawan Space Centre (SDSC) SHAR, Sriharikota, India on May 05,2005.The CARTOSAT-1 carries two panchromatic cameras that take black-and-white stereoscopic pictures in the visible region of the electromagnetic spectrum. The imageries have a spatial resolution of 2.5 meter and cover a swath of 30 km. Imageries from both the cameras were received at the National Remote Sensing Agency's Data Reception Station at Shadnagar, near Hyderabad. Preliminary analysis confirms the excellent performance of the cameras. Initial Imageries received from the satellite are over Punjab and Gujarat regions of India on May 8, 2005.

Further calibration of data, validation of the ground processing systems for decoding the Signals and other initial operations are continuing. Fine-tuning the orbit to the required 618 km circular with an inclination of 97.89 deg with respect to the equator is also being carried out. During the intial phase, Cartosat has produced good data over various areas in India as well as over some regions of the world. The specifications of cartosat-1 are shown in Table.

Operator	ISRO
Mission type	Cartography
Satellite of	**Earth**
Launch date	May 5, 2005, 04:44:00 UTC
Carrier rocket	PSLV
Launch site	SLP,
Mission duration	5 years
COSPAR ID	2005-017A

Contd.....

Mass	1560.0 kg
Power	46 W from solar panels
Orbital elements	
Regime	Sun-Synchronous Circular orbit
Eccentricity	1.4287754E-4
Inclination	97.9°
Apoapsis	620.0 kilometres (385.3 mi)
Periapsis	622.0 kilometres (386.5 mi)
Orbital period	97.1 minutes

Cartosat-1 is a stereoscopic Earth observation satellite in a sun-synchronous orbit. The satellite was built, launched and maintained by the Indian Space Research Organisation (ISRO). Weighing around 1560 kg at launch, its applications will mainly be towards cartography in India. It was launched by the PSLV-C6 on 5 May 2005 from the newly built second launch pad at Sriharikota. Images from the satellite will be available from GeoEye for worldwide distribution. Cartosat-1 carries two state-of-the-art panchromatic (PAN) cameras that take black and white stereoscopic pictures of the earth in the visible region of the electromagnetic spectrum. The swath covered by these high resolution PAN cameras is 30 km and their spatial resolution is 2.5 metres. The cameras are mounted on the satellite in such a way that near simultaneous imaging of the same area from two different angles is possible. This facilitates the generation of accurate three-dimensional maps. The cameras are steerable across the direction of the satellite's movement to facilitate the imaging of an area more frequently. The images taken by Cartosat-1 cameras are compressed, encrypted, formatted and transmitted to the ground stations. The images are reconstructed from the data received at the ground stations.

Cartosat-1 also carries a solid state recorder with a capacity of 120 Giga Bits to store the images taken by its cameras. The stored images can be transmitted when the satellite comes within the visibility zone of a ground station.

5.11.2 CARTOSAT-2

Cartosat-2 is an Earth observation satellite in a sun-synchronous orbit. The satellite was built, launched and maintained by the Indian Space Research Organisation. Weighing around 680 kg at launch, its applications will mainly be towards cartography in India. It was launched by the PSLV on January 10, 2007. The specifications of cartosat-2 are shown in Table.

Cartosat-2 carries a state-of-the-art panchromatic (PAN) camera that take black and white pictures of the earth in the visible region of the electromagnetic spectrum. The swath covered by these high resolution PAN cameras is 9.6 km and their spatial resolution is less than 1 metre. The satellite can be steered up to 45 degrees along as well as across the track.

Cartosat-2 is an advanced remote sensing satellite capable of providing scene-specific spot imagery. The data from the satellite will be used for detailed mapping and other cartographic applications at cadastral level, urban and rural infrastructure development and management, as well as applications in Land Information System (LIS) and Geographical Information System (GIS).

The first imagery, received on January 12, 2007, covered a length of 240 km from Paonta Sahib in Shivalik region to Delhi. Another set of imagery of about 50 km length covered Radha Nagari to Sagoan in Goa. Analysis of the first imagery received at National Remote Sensing Agency's data reception station at Shadnagar, in Hyderabad, confirmed excellent performance of the on-board camera.

Cartosat-2 can produce images of up to 80 cm in resolution (black and white only), compared to the 80 cm offered by Ikonos which is multispectral. In the past, India has been buying images from Ikonos at about $20 per square kilometre of imagery (where regional pricing was valid) otherwise the price is $7.70 per square kilometer. With Cartosat-2 offering better resolution at twenty times lower cost per sq m of imagery, buying images from Ikonos is likely to decline in future. Currently, India buys images worth about Rupees 2 crore a year from Ikonos.

Specifications of Cartosat-2

Cartosat-2	
Operator	ISRO
Mission type	5 years Earth observation
Satellite of	**Earth**
Orbits	14
Launch date	2007-01-10, 03:45 GMT
Carrier rocket	PSLV-C7
Launch site	SDSC - First Launch Pad[1]
COSPAR ID	2007-001B
Homepage	PSLV C7
Mass	680 kg
Power	Solar array generating 900 W
Batteries	Two 18 Ah Ni-Cd batteries
Orbital elements	
Regime	SSO
Inclination	97.91°
Apoapsis	630 kilometres (390 mi)
Periapsis	630 kilometres (390 mi)
Orbital period	~90 minutes
Swath width	9.6 km

Contd....

Instruments	
Main instruments	One panchromatic camera
Spatial resolution	Less than 1 metre
Spectral band	0.5 µm - 0.85 µm
Solid state recorder	64 GB

5.11.3 CARTOSAT-2A

Cartosat-2A is an Earth observation satellite in a sun-synchronous orbit. The satellite is the thirteenth satellite in the Indian Remote Sensing (IRS) satellite series to be built, launched and maintained by the Indian Space Research Organisation. It was launched by the Polar Satellite Launch Vehicle - C9 on April 28, 2008 along with the 87 kg Indian Mini Satellite (IMS-1) and eight nano research satellites belonging to research facilities in Canada, Denmark, Germany, Japan and the Netherlands. This satellite is a Ministry of Defence mission for the Government of India. The designation of the satellite could be presumed to be its similarity with its civilian namesake launched on January 10, 2007.[citation needed] It is a dedicated satellite for the Indian Armed Forces which is in the process of establishing an Aerospace Command.

The satellite carries a panchromatic (PAN) camera capable of taking black-and-white pictures in the visible region of electromagnetic spectrum. The highly agile Cartosat-2A can be steered up to 45 degrees along as well as across the direction of its movement to facilitate imaging of any area more frequently. The satellite's health is continuously monitored from the Spacecraft Control Centre at Bangalore with the help of ISTRAC network of stations at Bangalore, Lucknow, Mauritius, Bearslake in Russia, Biak in Indonesia and Svalbard in Norway.

Specifications of Cartosat-2A

Cartosat-2A	
Operator	ISRO
Mission type	Earth observation
Satellite of	**Earth**
Launch date	28 April 2008, 03:53 GMT
Carrier rocket	PSLV-C9
Launch site	SLP, Satish Dhawan Space Centre
Mission duration	5 years
COSPAR ID	2008-021A
Mass	690 kg
Power	900 watts
Batteries	Two 18 Ah Ni-Cd batteries

Contd....

Orbital elements	
Regime	LEO
Apoapsis	630 kilometres (390 mi)
Periapsis	630 kilometres (390 mi)
Orbital period	97.4 minutes
Repeat interval	4 days
Swath width	About 9.6 kilometre
Instruments	
Main instruments	One panchromatic camera
Spatial resolution	Less than 1 metre
Spectral band	0.5 – 0.85 micrometre
Data rate	105 Mbit/s after compression
Solid state recorder	64 GB

5.11.4 CARTOSAT-2B

CARTOSAT 2B is an Earth observation satellite in a sun-synchronous orbit. The satellite is the seventeenth satellite in the Indian Remote Sensing (IRS) satellite series to be built by the Indian Space Research Organisation. It was launched along with the 116 kg Algerian satellite, one nano satellite each from Canada and Switzerland, and StudSat, a pico satellite, on 12 July 2010 in a Polar Satellite Launch Vehicle (PSLV) rocket from the spaceport at Sriharikota.

The satellite carries a panchromatic (PAN) camera capable of taking black-and-white pictures in the visible region of electromagnetic spectrum. The highly agile CARTOSAT-2B can be steered up to 26 degrees along as well as across the direction of its movement to facilitate imaging of any area more frequently.

Specifications of Cartosat-2B

Cartosat-2B	
Operator	ISRO
Mission type	Earth observation
Launch date	12 July 2010
Carrier rocket	PSLV-CA (s/n C15)
Launch site	Satish Dhawan FLP
Mission duration	5 years
Mass	690 kg
Power	930 watts
Batteries	Two 18 Ah Ni-Cd batteries

Contd....

Orbital elements	
Regime	LEO[1]
Apoapsis	630 kilometres (390 mi)
Periapsis	630 kilometres (390 mi)
Orbital period	97.4 minutes[1]
Repeat interval	4 days
Swath width	About 9.6 kilometres
Instruments	
Main instruments	One panchromatic camera
Spatial resolution	Less than 1 metre
Spectral band	0.5 – 0.85 micrometre
Data rate	105 Mbit/s after compression
Solid state recorder	64 GB

5.12 Resource SAT Series

5.12.1 Resourcesat – 1

Resourcesat-1 is conceptualized and designed to provide continuity in operational remote sensing with its superior capabilities. The main objective of Resourcesat–1 is not only to provide continued remote sensing data for integrated land and water management and agricultural and it's related applications, but also to provide additional capabilities for applications. Apart from making data available in real time to the Ground Stations in its visibility area Resourcesat – 1 with its ability to record data any where in the world with its advanced On Board Solid State Recorder, has entered into new dimensions of meeting the requirements of Resource Managers globally.

Deployed Configuration of Resourcesat – 1 (Fig 5.17)

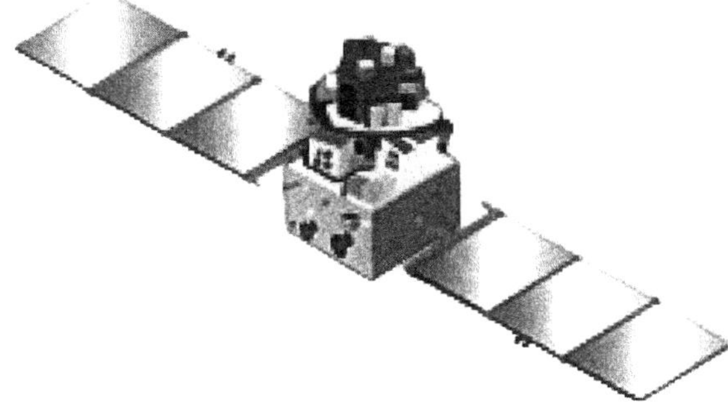

Fig. 5.17 Deployed configuration of resourcesat-1 Space Craft.

5.12.2 RESOURCESAT-2

Resourcesat-2 is a follow on mission to Resourcesat-1 and and the eighteenth Remote Sensing satellite built by ISRO.RESOURCESAT-2 is intended to continue the remote sensing data services to global users provided by RESOURCESAT-1, and to provide data with enhanced multispectral and spatial coverage as well. Compared to Resourcesat-1, LISS-4 multispectral swath has been enhanced from 23 km to 70 km based on user needs. Suitable changes including miniaturization in payload electronics have been incorporated in Resourcesat-2. Resourcesat-2 along with Youthsat and X-Sat (Singapore) was launched on PSLV-C16 on 20 April 2011.

The satellite contains 3 multispectral cameras onboard.

1. Advanced Wide-Field Sensor (AWiFS) with 56 meter spatial resolution,
2. the Linear Imaging Self-Scanning Sensor (LISS-III) with 23.5 meter spatial resolution and
3. LISS-IV Camera with 5.8 meter spatial resolution.

Specifications of RESOURCESAT-2

Resourcesat-2	
Operator	Indian Space Research Organisation
Mission type	Remote-sensing
Launch date	April 20th , 2011
Carrier rocket	PSLV-C16
Launch site	Satish Dhawan Space Centre
Mission duration	5 years
COSPAR ID	2011-015A
Mass	1,206 kg
Orbital elements	
Regime	Polar Sun-Synchronous Circular orbit
Inclination	98.72^{o}
Apoapsis	817 kilometres (508 mi)
Periapsis	817 kilometres (508 mi)
Orbital period	102 minutes

5.13 Quick Bird

Quick Bird was launched on the 18th October 2001. This is currently the highest resolution commercial satellite data available. After an on-orbit calibration and commissioning period of about 90 days, Quick Bird is now acquiring 61 centimeter (2 foot) resolution panchromatic (black and white) and 2.44 meter (8 foot) multispectral (colour) imagery. At 61 centimeter resolution, buildings, roads, bridges and other detailed infrastructure become visible. The imagery will be used for a wide range of applications, focusing on the assessment and management of land, infrastructure, and natural resources. Quick Bird collects an industry-leading 16.5 kilometer (10.3 mile) swath of imagery that enables greater collection of large areas.

Rush Priority

The image acquisition window is 1-14 days, and the time from order to first acquisition is 48 hours. There is no clould cover threshold. This service involves a price uplift of 100%. Multispectral band information is given in Table. 5.17.

Table. 5.17 Multispectral band information for quick bird

	Band Width	Spatial Resolution
Band 1	0.45 – 0.52 mm (blue)	2.44 – 2.88 meters
Band 2	0.52 – 0.60 mm (green)	2.44 – 2.88 meters
Band 3	0.63 – 0.69 mm (red)	2.44 – 2.88 meters
Band 4	0.76 – 0.90 mm (near infra-red)	2.44 – 2.88 meters

Quick Bird pan sharpened 0.61 metre image of Tampa FL (Fig. 5.18a).

Fig. 5.18(a) Quick Bird pan sharpened 0.61 m

Quick Bird pan sharpened 0.61 metre image of DC Airport (Fig. 5.18b).

Fig. 5.18(b) Quick bird pan sharpened.

Quick Bird 2.44 metre multispectral image of McMurdo Sound (Fig. 5.18c).

Fig. 5.18(c) Quick bird 2.44 metre multispectral.

Quick bird 62 centimeter data over the port of Abu Dhabi (orientated ESE up). This is a merge of the black and white 62 cm data with the lower spatial resolution colour data. Can be output at 1: 2500 scale or better (Fig. 5.18d)

Fig. 5.18(d) Quick bird 62 centimetre data of Abu Dhabi.

5.14 CHANDRAYAAN-1

Chandrayaan-1, lit: Moon vehicle pronunciation (help•info)) was India's first unmanned lunar probe. It was launched by the Indian Space Research Organisation in October 2008, and operated until August 2009. The mission included a lunar orbiter and an impactor. India launched the spacecraft with a modified version of the PSLV, PSLV C11 on 22 October 2008 from Satish Dhawan Space Centre, Sriharikota, Nellore District, Andhra Pradesh, about 80 km north of Chennai, at 06:22 IST (00:52 UTC). Former prime minister Atal Bihari Vajpayee announced the project on course in his Independence Day speech on 15 August 2003. The mission was a major boost to India's space program, as India researched and developed its own technology in order to explore the Moon. The vehicle was successfully inserted into lunar orbit on 8 November 2008.

On 14 November 2008, the Moon Impact Probe separated from the Chandrayaan orbiter at 20:06 and struck the south pole in a controlled manner, making India the fourth country to place its flag on the Moon. The probe impacted near Shackleton Crater at 20:31 ejecting underground soil that could be analysed for the presence of lunar water ice.

The estimated cost for the project was 3.86 billion Indian rupees (US$90 million).

The remote sensing lunar satellite had a mass of 1,380 kilograms (3,042 lb) at launch and 675 kilograms (1,488 lb) in lunar orbit. It carried high resolution remote sensing equipment for visible, near infrared, and soft and hard X-ray frequencies. Over a two-year period, it was intended to survey the lunar surface to produce a complete map of its chemical characteristics and three-dimensional topography. The polar regions are of special interest as they might contain ice. The lunar mission carries five ISRO payloads and six payloads from other space agencies including NASA, ESA, and the Bulgarian Aerospace Agency, which were carried free of cost.

After suffering from several technical issues including failure of the star sensors and poor thermal shielding, Chandrayaan stopped sending radio signals at 1:30 AM IST on 29 August 2009 shortly after which, the ISRO officially declared the mission over. The main culprit is said to be the failure of onboard DC-DC Converter manufactured by MDI Power U.S.A. The converters failed to meet the radiation specifications for the intended mission time. Chandrayaan operated for 312 days as opposed to the intended two years but the mission achieved 95 percent of its planned objectives. Among its many achievements was the discovery of the widespread presence of water molecules in lunar soil.

Specifications of Chandrayaan-1

Chandrayaan-1	
Operator	Indian Space Research Organisation
Mission type	Orbiter
Satellite of	Moon
Orbital insertion date	8 November 2008
Orbits	3400 orbits around the Moon.[1]
Launch date	22 October 2008 00:52 UTC
Launch vehicle	PSLV-XL[2]
Launch site	SDSC, Sriharikota
Mission duration	Intended: 2 years Achieved: 312 days
COSPAR ID	2008-052A
Homepage	Chandrayaan-1
Mass	1,380 kg (3,042 lb)
Orbital elements	
Eccentricity	near circular
Inclination	Polar
Apoapsis	initial 7,500 km (4,660 mi), final 100 km (62 mi), final (wef 19 May 2009) 200 km (124 mi)
Periapsis	initial 500 km (311 mi), final 100 km (62 mi), final (wef 19 May 2009) 200 km (124 mi)

6

Visual Image Interpretation

6.1 Introduction

Remote sensing is the process of sensing and measuring objects from a distance without directly coming physically into contact with them. According to the physics of remote sensing, different surface objects return different amounts of energy in different wavelengths of the electromagnetic spectrum. Detection and measurements of these energies, called, spectral signatures enable identification of surface objects both from the air borne and from the spaceborne platforms. Remote sensing through airborne and satellite based sensors covers surveying and monitoring which are essential for the planning and management of national natural resources. This technology, integrated with traditional techniques, is emerging as an efficient, speedy, cost effective and important tool for the national development efforts.

The study of landuse, land cover or land form from the air has been a focus of interest since the early days of aerial photography. It has been gaining momentum again with the availability of new remote sensing techniques using aircraft and spacecraft as platforms with the capacity for operating outside the visible part of the electromagnetic spectrum through microwaves (radar) and thermal radiation. Thus photo interpretation is now complemented by the interpretation of other kinds of imagery which are different from photography. The limitation of 'photo-interpretation' has now been changed to broad spectrum of 'image interpretation. In order to derive the maximum benefits through the use of these new technique's, it is important to ensure that planners and managers and other potential users appreciate properly the value of this new tool through proper application of the principles and methodology of image interpretation either by visual interpretation or by digital analysis. This data can be collected by the remote sensing devices including passive and active systems and employ different bands in the visible, near infrared, middle infrared and far infrared as well as microwave regions. In the passive remote sensing, the reflected or emitted electromagnetic energy is measured by sensors operating in different selected spectral bands where the original source of energy is the sun. But in active remote sensing method the earth surface is illuminated by an artificial source of energy. The emitted and reflected energy detected by the sensors onboard are transmitted to the earth station. The data then it is processed (after various corrections) and made ready for the users.

The analysis of pictorial data can be performed using visual image interpretation techniques. Visual image interpretation has been applied in many fields including agriculture, archaeology conservation, engineering, forestry, geology, geography, meteorology, military intelligence, natural resource management, oceanography, soil science, and, urban and regional planning (Wolf, 1983). Competence has been achieved in the use of this technique, more particularly in areas like land use planning, wasteland mapping, land evaluation, water quality monitoring, topographic mapping, and numerous other areas.

6.2 Types of Pictoral Data Products

The remote sensing data products are available to the users in the form of (a) photographic products such as paper prints, film negatives, diapositives of black and white, and false colour composite (FCC) on a variety of scales and (b) digital form as computer compatible tape (CCTs) after necessary corrections (NRSA, 1995). Broadly, satellite data products can be classified into different types based on satellite and sensor, level of preprocessing and the media. Data products acquired for the specific period can be generated if the data pertaining to the period of interest is available in archives. Depending upon the corrections applied and on the level of processing, data products can be classified as : raw data, partially corrected products, standard products, geocoded products, and precision products. The raw data is radiometrically and geometrically uncorrected data with ancillary information (stereo products for photogrammetric studies). Standard products are radiometrically and geometrically corrected for systematic errors. Geocoded products are systematically and geometrically corrected products. The systematic corrections are based on the standard survey of India toposheet and rotation of pixels to align to true north and resampled to standard square pixel. Precision products are radiometric and geometric corrections refined with the use of ground control points to achieve greater locational accuracy.

Data products can be broadly classified into two types depending upon the output media, as photographic and digital. Fig. 6.1 shows the types of products based on media. Photographic products can either be in black and white, or colour. Further they could be either film or paper products, and in films it is possible to have either positive film or negative film. The sizes of photographic products can vary depending

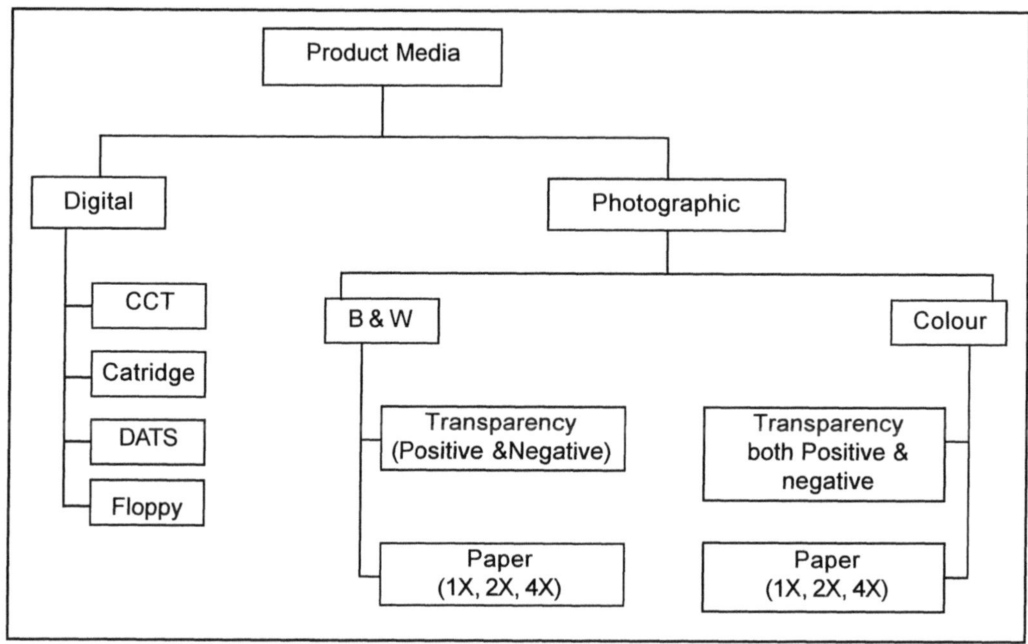

Fig. 6.1 Types of product media (NRSA, 1999).

on the enlargement needed, and this is specified as 1X, 2X, 4X and so on. The size of film recorders is generally 240 mm and this is the basic master output from which further products are generated. When we say colour photographic products, it generally means false colour composites (plate 3). FCCs are generated by combining the data contained in 3 different spectral bands into one image by assigning blue, green and red colours to the data in three spectral bands respectively during the exposure of a colour negative. The choice of band combinations can be determined depending upon the application on hand.

Different types of photographic products supplied by National Remote Sensing Agency (NRSA) data centre, Govt., of India (NDC) are : Standard B/W and FCC, films. Standard products are available in colour, and black and white in the form of 240 mm films, either as negatives or positives. Fig. 6.2 shows the various photographic products of different sizes and different media of printing.

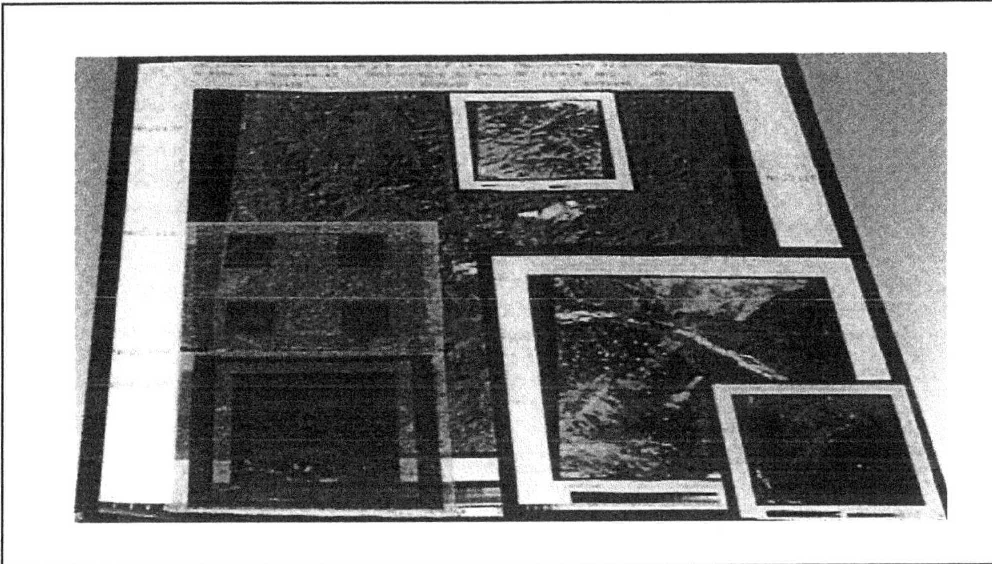

Fig. 6.2 Photographic products.

Paper prints both B and W and FCC are supplied in various scales. They are 1 X (contact prints), 2X (two times enlarged) and 4X (four times enlarged) and 5X (5 times enlarged). Depending upon the enlargement the scale of the product varies (IRS handbook, 1998). The photographic products contain certain details annotated on the margins. These are useful for identifying the scene, sensor, date of pass, processing level, band combination, and so on. A list of details generally annotated on the images are given in Table 6.1 and Fig. 6.3 shows the annotation on photographic products.

Table 6.1 Details generally annotated on the satellite imagery

Product type	Data and time of acquisition of data
Sensor used	Sub-scene
Band numbers	Gain settings
Path and row number	Latitude/longitude coordinates
Sun elevation and azimuth	Resampling method
Map projection	Enhancement
Orbit cycle and day	Orbit number
Uncorrected centre coordinates	Geographic marks
Name of data receiving station	Product generation agency
Product generation date	Satellite identification

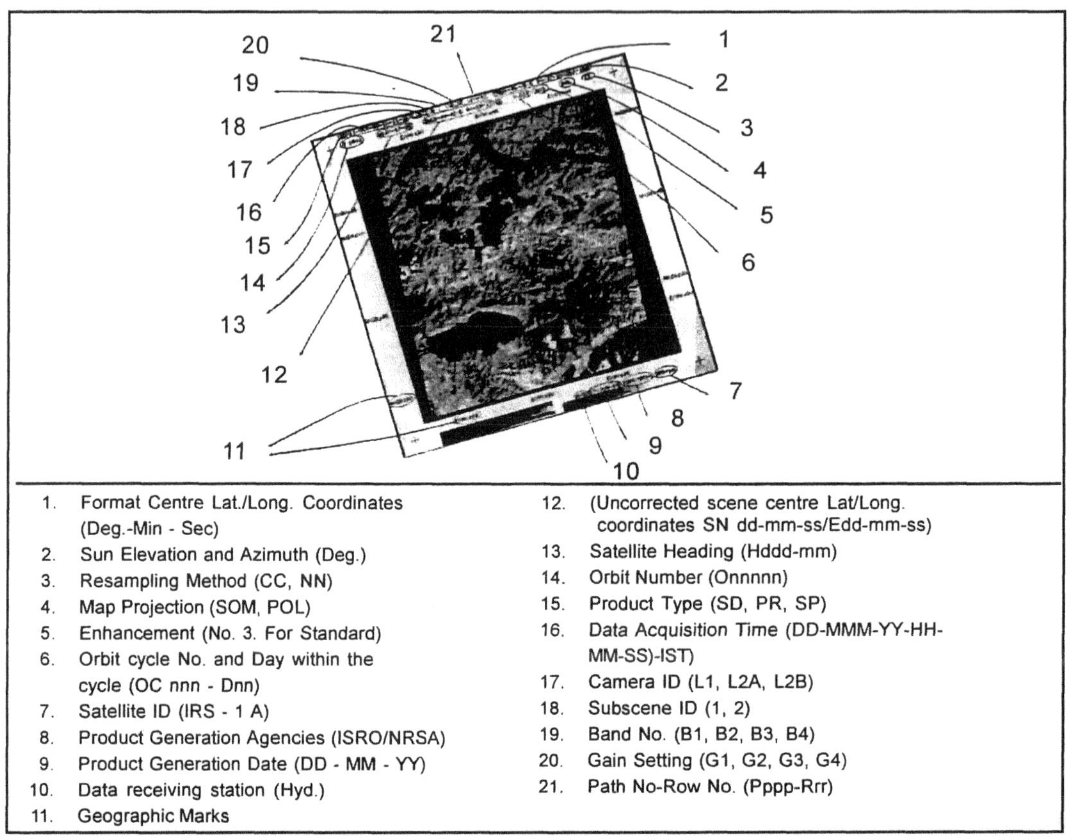

Fig. 6.3 Annotation of Photographic products.

Basically, the visual interpretation of the remote sensing data is based on the False Colour Composites (FCCs). Even after the digital techniques, the results are visually interpreted. Hence, this technique is very important and requires the specific training for the users. Scientists, analysts and other users may interpret the same scene for different purposes. In fact it is one of the rare sources of information which can generate multiple themes, such as, water resources, soil, landuse, and urban sprawl. The visual interpretation has become more important to create spatial database for GIS. A case study on the application of visual image interpretation techniques is presented in subsequent chapters.

6.3 Image Interpretation Strategy

Visual Image interpretation is a process of identifying what we see on the images and communicate the information obtained from these images to others for evaluating this significance. This process, however, is not restricted to making decisions concerning what objects appear in images, but it also usually includes a determination of their relative locations and extents. This requires the application of at least some of the elementary photogrammetric measurement and mapping techniques. This chapter concentrates on the use of standard data products like aerial photographs, satellite single band imageries, false colour composites for performing image interpretations to extract thematic information for subsequent input to a GIS. Aerial photographs and satellite images contain a detailed record of the earth surface features and the interpreter or image analyst examines these pictorial data products to extract such features. In addition to this cursory examination, the interpreter may also study other supplementary data products, such as, maps, existing records, and reports of field work. Success in visual image interpretation will vary with the training and experience on the pictorial data analysis, a study of the nature of the terrain, quality of the data, and visual equipment used. If the interpreter has an artistic and photographic sense, the information derived from the interpretation process may be more authentic and reliable.

6.3.1 Levels of Interpretation Keys

The image interpretation process can involve various levels of complexity from a simple direct recognition of objects in the scene to the inference of site conditions. An example of this can be a national highway or a major river on the satellite imagery, more particularly on a false colour composite. If the interpreter has some experience, the interpretation of these linear features, road and river, may be straight forward. On the other hand, the interpreter has to adopt some other image characteristics in order to infer the appearance of the objects on the image. For example, interpretation of a IRS. ICLISS III false colour composite imagery for the identification of 18 km pipeline from Patancheruvu to Amberpet of Hyderabad city, which carries the industrial effluents, is an indirect approach. In this case the actual pipeline can not be seen, but there are often changes at the ground surface caused by the buried pipeline which are visible on FCC. The appearance of a light -toned linear streak across the image is associated with the highway. This leads the identification and mapping of pipeline from FCC. This is supported by the field work, that is, the ground truth analysis. The process of interpretation should also consider the dates to identify the ground-cover types. For example, crop development stages or knowledge of crop-development stages (crop calendar) for an area would determine if a particular crop is likely to be visible on an imagery of a particular date.

Keys that provide useful reference of refresher materials and valuable training aids for novice interpreters are called image interpretation keys. These image interpretation keys are very much useful for the interpretation of complex imageries or photographs. These keys provide a method of organising the information in a consistent manner and provide guidance about the correct identification of features or conditions on the images. Ideally, it consists of two basic parts : (i) a collection of annotated or captioned images (stereopairs) illustrative of the features or conditions to be identified, and (ii) a graphic or word description that sets forth in some systematic fashion the image recognition characteristics of those features or conditions.

There are two types of keys : selective key and elimination key. A selective key is also called reference key which contains numerous example images with supporting text. The interpreter selects one example image that most nearly resembles the feature or condition found on the image under study. An elimination key is arranged so that the interpretation proceeds step by step from the general to the specific, and leads to the elimination of all features or conditions except the one being identified. Elimination keys are also called dichotomous keys where the interpreter makes a series of choices between two alternatives and progressively eliminates all but one possible answer. Fig. 6.4. shows a dichotomous key prepared for the identification of fruit and nut crops. This is a classical example of this type of key. This key gives more positive answers than selective keys. But the elimination key sometimes gives more erroneous answers if the interpreter is forced to make uncertain choice

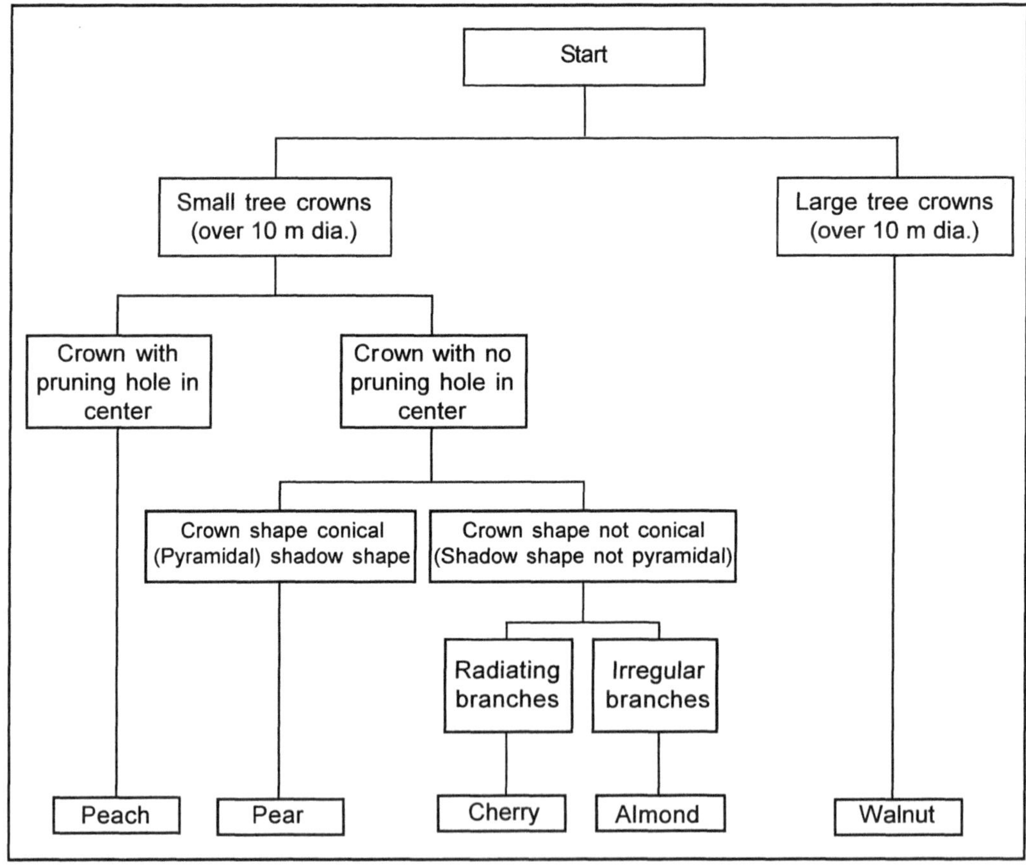

Fig. 6.4 Dichotomous airphoto interpretation key to fruit and not crops in the Sacramento Valley, CA, designed for use with 1:6000 scale panchromatic aerial photographs. (Adapted from American Society of Photogrammetry, 1983. Copyright © 1975, American Society of photogrammetry. Reproduced with permission.)

between any two unfamiliar image characteristics. The elimination key is used and successfully employed for agricultural studies, and forestry applications. An important point is that, these keys are normally developed and used on a region-by-region and season-by-season basis, that is these keys are dependent upon location and season. Generally, keys are more easily constructed and reliably used for the cultural feature identification and extraction, of objects like houses, bridges, roads, water tanks, vegetation and land form. Therefore temporal aspects are also important for image interpretation because of changes of the earth surface features. For example, vegetation growth and soil moisture vary during the year. For crop identification, more positive results can be drawn from the interpretation of multiperiod satellite imageries. Similarly, soil moisture conditions vary during the day or two different seasons, the timing of image acquisitions for soil studies being very critical.

6.4 Process of Image Interpretation

Image interpretation or analysis is defined as the "act of examining images for the purpose of identifying objects and judging their significance". Interpreters study remotely sensed data and attempt through logical process to detect, identify, classify, measure, and evaluate the significances of physical and cultural objects, their patterns and spatial relationship. As previously mentioned, the image interpretation is a complex process of physical and psychological activities occurring in a sequence of time. The sequence begins with the detection and identification of images and later by their measurements. The image interpretation has various aspects which have overlapping functions. These aspects are detection,recognition and identification, analysis, classification, and idealisation. The detection is a process of 'picking out' an object or element from photo or image through interpretation techniques. It may be detection of point or line or a location, such as, agricultural field and a small settlement. Recognition and Identification is a process of classification or trying to distinguish an object by its characteristics or patterns which are familiar on the image. Sometimes it is also termed as photo reading of water features, streams, tanks, sands, and so on.

Analysis is a process of resolving or separating a set of objects or features having a similar set of characters. In analysis, lines of separation are drawn between a group of objects and the degree of reliability of these lines can also be indicated. Classification is a process of identification and grouping of objects or features resolved by analysis. This is the most important aspect and the conceptual view of digital image classification which is discussed in Chapter 7. Idealisation is a process of drawing ideal or standard representation from what is actually identified and interpreted from the image or map, such as a set of symbols or colours to be adopted in waste land maps, and geomorphic landforms.

There is no single "right" way to approach the interpretation process (Lillesand and Kiefer, 2000). The interpretation process depends upon the type of data product, interpretation equipment and a particular interpretation task undertaken. For example, in an application of land use/land cover mapping, the image analyst need to commit various discrete objects like dwelling units, recreational areas, and so on. In some other applications, the image analyst might survey large areas identifying analogous conditions like the sources of pollution, entering a stream and areas of forest. In some cases, interpreting process involves delineation of discrete areal units throughout the images like mapping of various forest covers requiring the delineation of boundaries. The interpreter should take care to draw the boundary not as a discrete edge, but as a fuzzy edge.

Before an interpreter undertakes the task of performing visual interpretation, two important issues should be addressed. The first is the definition of classification system or criteria to be used to separate the various categories of features occurring in the images. For example, in the process of mapping of land use the analyst must firmly fix the level of classification system and the specific characteristics which determine whether the area is residential, industrial, or commercial. Similarly in the case of forest resources, mapping under various categories to be delineated in a particular species, should be defined before processing the satellite data.

The second important issue is the selection of minimum mapping unit (MMU) to be applied on the image interpretation. MMU refers to the smallest size areal entity to be mapped as a discrete area. Selection of the MMU depends upon the application and the extent of detail mapped by an interpreter. The minimum mapping unit for the forest mapping under various categories are illustrated in Fig. 6.5. In some applications like landuse/land cover mapping, the interpreter might choose to delineate photomorphic regions. The delineation of such regions often serves as a stratification tool in the interpretation process.

Fig. 6.5 Influence of minimum mapping unit size on interpretation detail. (a) Forest types mapped using a small MMU : O, oak; M, maple; W, white pine; J, jack pine; R, red pine; S, spruce. (b) Forest types mapped using a large MMU: D, deciduous; E, evergreen.

6.5 Interpretation of aerial photo

In case of Aerial photo the characteristics of the image of earth's surface relates to tone, colour, shadow, texture, shape and size and other associations between features and their surroundings. In addition certain ground characteristics such as land form, vegetation, land use, drainage, erosion and lineaments help in interpreting the image.

6.6 General procedure for photo interpretation

The following procedure can be adopted for both aerial photo and remote sensed imagery:

- Preliminary examination
- Detailed examination
- Interpretation
- Compilation

6.6.1 Preliminary stage

In this stage it is necessary for the interpreter to become familiar with the site and its surrounding area. At this examination, easily identifiable images are correlated with ground features such as roads, rivers etc. The principle of working from "whole to part" is also followed here from overall viewing of terrain using aerial photos. This is achieved by making a photo-mosaic of the aerial photos (by joining successive and overlapping photos and removing common area).

6.6.2 Detailed examination

The next stage, the detailed examination is to examine overlapping photos under a stereoscope to see the third dimension. viz: topography of the terrain and its variation. Overlays are then prepared by tracing from the photos or tracing paper showing features, which have been identified.

6.6.3 Interpretation stage

Many times the stages of detailed examination and interpretation can be combined and can be carried out side by side. Some of the general and common features to be noted from aerial photos, which are common to most of the project are as follows:

- Topography – hills, valleys, flood plain, lakes, steep and gentle slopes, depression, quarries, embankments, costal features etc
- Drainage and drainage patterns–springs, seepage, major rivers, streams, channels, marshy lands, ponds, lakes etc.
- General geology viz: soil boundaries, soil types
- Exiting hazards like abandoned mining sites, trenches, buried foundation etc.
- Site history such as earlier use of presently abandoned buildings, waste land, removal of vegetation etc.

6.6.4 Compilation Stage

Consists of making maps from aerial photos and showing features of significance as identified by interpretation. Depending upon the topography and tilt of the photos, any existing photogrammetric methods such as rectification may be employed to get a final map with symbols depicting the features of the terrain and other information of the terrain such as land use etc.

6.7 Three dimensional interpretation Method

The basic requirement is a set of overlapping photographs (as per specification) and a stereoscope to obtain a three dimensional view of the terrain.

6.7.1 Stereoscopic depth perception

Stereoscopic vision is binocular vision when the left and right eye are focussed on a certain point, the optical axes of the two eyes converge on that point intersecting at an angle called the parallactic angle. The nearer the object the greater the parallactic angle and vice versa.

In the Fig. 6.6 L and R are the left and right eyes. The distances between the two eyes are called the eye base. (b_e)

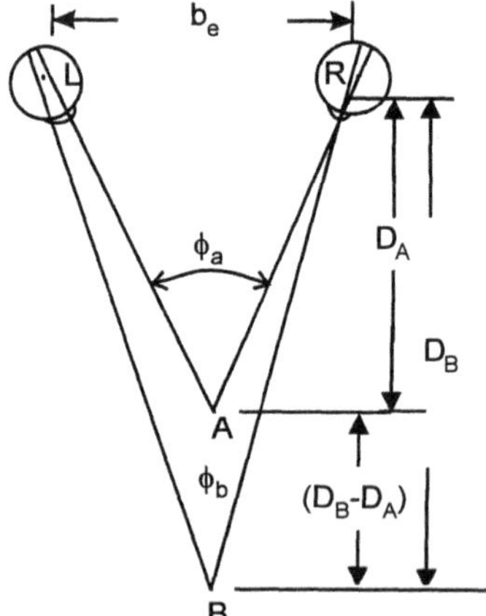

Fig. 6.6 Stereoscopic depth perception

When the eyes are focussed at point A, the optical axes converge forming parallactic angle ϕ_a. Similarly when sighting at B the optical axes converge at B forming parallactic angle ϕ_b. The brain associates the distance D_A and D_B with corresponding parallactic angle ϕ_a and ϕ_b. The depth between the object A and B is D_B - D_A and is perceived as the difference between these two parallactic angles. Thus the basic concept of creating three dimensional or stereoscopic impression of objects is by viewing identical images of objects placed at different distances. The same principle is used for creation of three dimensional effect by viewing aerial photographs taken from two different exposure stations. The two overlapping photos are laid on a table and viewed in such a way that left eye see only left photo and right eye sees only right photo. The brain judges the heights (3-D model) of each overlapping identical objects, by associating the depth with their corresponding parallactic angles. The effect is the viewing of a virtual three demensional model, which are associated with the continuous changing parallactic angles. The three dimensional model thus formed is called a stereoscopic model and the overlapping pair of photos is called a stereo pair.

6.7.2 Stereo scope

It is difficult to view stereo photographs stereoscopically without the aid of optical devices. The difficulties can be overcome by an instrument called stereoscope. There are number of stereoscopes used for viewing photographs. The most commonly used are:

- Pocket stereoscope
- Mirror stereoscope

6.7.2.1 Pocket stereoscope

Most commonly used. Simple construction. Consists of two simple convex lenses mounted on a frame. The spacing between the lenses can be varied to accommodate various eye bases. For stereo viewing the photographs are placed so that the corresponding images are slightly less than the eye base apart, two inches. (Fig. 6.7)

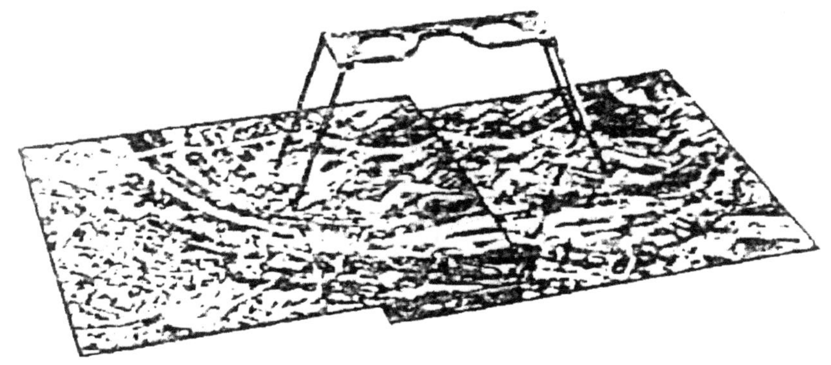

Fig. 6.7 Pocket Stereoscope

6.7.2.2 Mirror stereoscope

Mirror Stereoscope has two large wing mirrors and two smaller eye piece mirrors. The light rays from the photo points a_1,a_2 are reflected from the mirror surfaces, and according to the principle of reflection are received at the eyes form the parallactic angle ϕ_a. Similarly for point b_1,b_2 also forming parallactic angle ϕ_b. The brain automatically associates the depths of the points A and B, with the respective parallactic angles ϕ_a & ϕ_b. This happens for the infinite number of points reflected from the left and right photo, which ultimately generates the 3-D stereoscope viewing of the overlapping area.

Fig. 6.8 Shows the photograph of a mirror stereoscope.

Fig. 6.9 Shows the Principle of working of mirror sterescope.

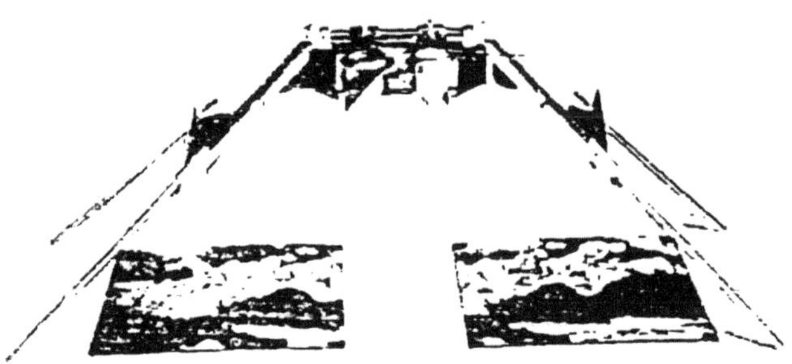

Fig. 6.8 Mirror Stereoscope

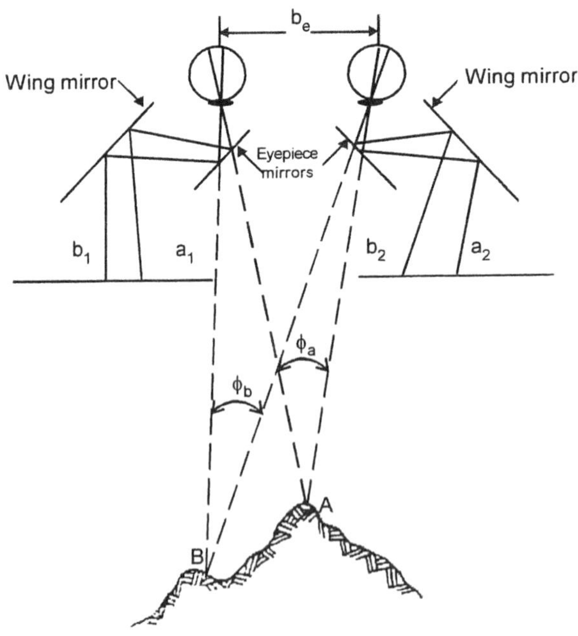

Fig. 6.9 Principle of working of mirror stereoscope

6.8 Basic elements of image interpretation

A systematic study of aerial photographs and satellite imageries usually involves several characteristics of features shown on an image. The following characteristics (elements) are called fundamental picture elements. These elements aid visual interpretation process of aerial photos and/or satellite imagery.

(i) *Tone*

Ground objects of different colour reflect the incident radiation differently depending upon the incident wave length, physical and chemical constituents of the objects. The imagery as recorded in remote sensing is in different shades or tones. For example, ploughed and cultivated lands record differently from fallow fields. Tone is expressed qualitatively as light, medium and dark. In SLAR imagery, for example, the shadows cast by non-return of the microwaves appear darker than those parts where greater reflection takes place. These parts appear of lighter tone. Similarly in thermal imagery objects at higher temperature are recorded of lighter tone compared to objects at lower temperature, which appear of medium to darker tone. Similarly top soil appears as of dark tone compared to soil containing quartz sand. The coniferous trees appear in lighter tone compared to broad leave tree clumps.

Tone, therefore, refers to the colour or reflective brightness. Tone along with texture and shadow (as described below) help in interpretation and hence is a very important key. Differences in moisture content of the soil or rock result in differences in tone. In a black and white photograph dark tone indicates dark bodies, namely, greater moisture contents and grey or white tone reflect the dry soil. The aerial photos with good contrast bring out tonal differences and hence help in better interpretation. Tonal contrast can be enhanced by use of high contrast film, high contrast paper or by specialized image processing techniques such as 'Dodging' or 'Digital Enhancement'. Some times Infrared film can give better contrast but it can also reduce resolution and loss of detail in shadows.

(ii) *Texture*

Texture is an expression of roughness or smoothness as exhibited by the imagery. It is the rate of change of tonal values. Mathematically it is given as dD/dx where D is the Density and 'x' the distance measured from one arbitrary starting point, and can be measured numerically by the use of micro-densitometer. Changes of density 'D' from point 'A' of the imagery to point 'B' as measured by the micro-densitometer divided by the distance gives the texture values numerically. Texture is dependent upon (a) photographic tone (b) shape, (c) size, (d) pattern and scale of the imagery. Any slight variation of these can change the texture.

Texture can qualitatively be expressed as course, medium and fine. The texture is a combination of several image characteristics such as tone, shadow, size, shape and pattern etc., and is produced by a mixture of features too small to be seen individually because the texture by definition is the frequency of tonal changes. As an example, leaves of a tree are too small to be seen on an aerial photo collectively along with shadow they give what is called texture, which in turn helps to differentiate between shrubs and trees. Texture sometimes can be very important factor in determining the slope stability. In the case of a humid ground, the blockage of water or bad drainage a characteristic texture results. Even spring and seepage of water from the base of clay give a kind of 'turbulant' texture. So is the case with mud flows. The term texture is also, sometimes, used to denote drainage density and the degree of dissection of land surface.

(iii) *Association*

The relation of a particular feature to its surroundings is an important key to interpretation. Sometimes a single feature by itself may not be distinctive enough to permit its identification. For example, sink holes appears as dark spots on an imagery where the surface or immediate subsurface soil consists of lime stones, Thus the appearance of sink holes is always associated with surface lime stone formation. An example is that of kettle holes which appear as depressions on photos due to terminal moraine and glacial terrain. An another example is that of dark–toned features associated with a flood plain of a river, which can be interpreted as infilled oxbow lakes.

(iv) *Shape*

Some ground features have typical shapes due to the structure or topography. For example air fields and football stadium easily can be interpreted because of their finite ground shapes and geometry whereas volcanic covers, sand, river terraces, cliffs, gullies can be identified because of their characteristics shape controlled by geology and topography.

(v) *Size*

The size of an image also helps for its identification whether it is relative or absolute. Sometimes the measurements of height (as by using parallax bar) also gives clues to the nature of the object. For example, measurement of height of different clumps of trees gives an idea of the different species, similarly the measurement of dip and strike of rock formation help in identifying sedimentary formation. Similarly the measurements of width of roads help in discriminating roads of different categories i.e. national, state, local etc. Size of course, is dependent upon the scale of imagery.

(vi) *Shadows*

Shadows cast by objects are sometimes important clues to their identification and interpretation. For example, shadow of a suspension bridge can easily be discriminated from that of cantilever bridge. Similarly circular shadows are indicative of coniferous trees. Tall buildings and chimneys, and towers etc., can easily be identified for their characteristic shadows. Shadows on the other hand can sometimes render interpretation difficult i.e. dark slope shadows covering important detail.

(vii) *Site factor or Topographic Location*

Relative elevation or specific location of objects can be helpful to identify certain features. For example, sudden appearance or disappearance of vegetation is a good clue to the underlying soil type or drainage conditions.

(viii) *Pattern*

Pattern is the orderly spatial arrangement of geological topographic or vegetation features. This spatial arrangement may be two-dimensional (plan view) or 3-dimensional (space).

Geological pattern may be linear or curved. Linear pattern are formed of a very large number of continuous or discontinuous short ticks which when viewed by eye appear to be continuous lines. Examples of linear geological pattern are faults, fractures, joints, dykes, bedding planes, anticlines etc., Examples of curved features are plunging anticlines and folds.

Lineaments or lineations may be short, medium or long running for several hundred kilometers. These are very important expressions of the lithologic characters of the underlying rocks and the attitude of the rock bodies, spacing of planes of bedding and other structural weaknesses and the control extended by them over the surface features.

Vegetation pattern may be of the 'Block' type or 'Alignment' type.

The 'Alignment' type may be further subdivided into the Linear, Parallel and Curved type.

Alignments are due to narrow rockbands or faults. Since faults retain moisture, vegetation is aligned along the fault lines.

Example of topographic pattern are the typical drainage patterns (controlled and uncontrolled type). The uncontrolled type are those, which are purely governed by topography, i.e., the slops whereas the controlled type are those, which are governed by the underlying geological formations.

The well-known drainage patterns are:

(i) Dendritic, (ii) Trellis, (iii) Annualsr, (iv) Radial, (v) Rectangular, (vi) Parallel Type, (vii) Braided, (viii) Anastomotic, (ix) Asymmetrical, (x) Collinear.

Drainage patterns take an important place amongst the various interpretation elements, used as criteria for identification on geological and geomorphological phenomena. Factors to be considered are amongst other, the density which is measured for the erodibility of the rocks; the amount of geological control on the drainage pattern and the integration and homogeneity of the pattern.

Main drainage patterns are given in Fig. 6.10

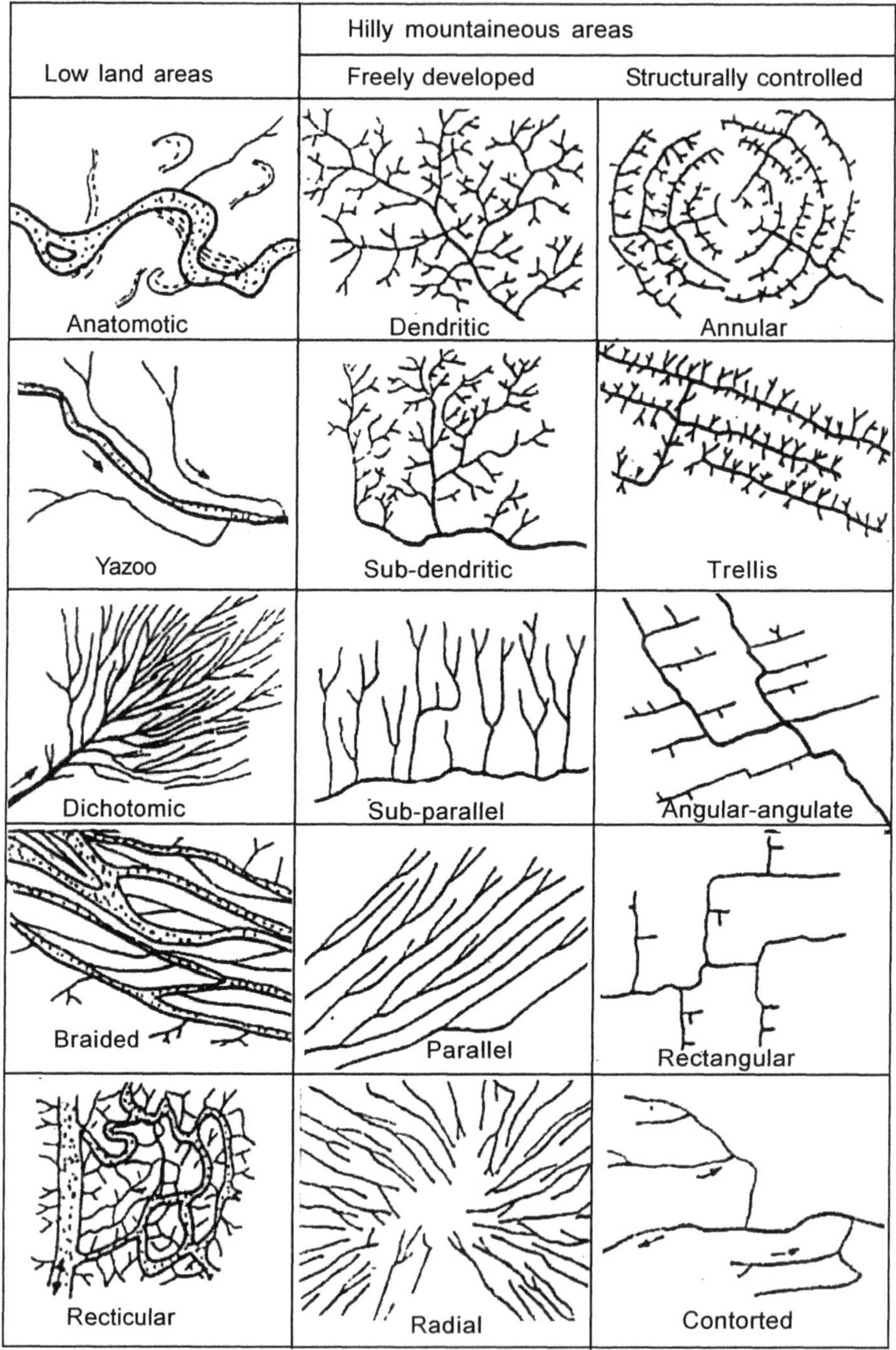

Fig. 6.10 Main drainage patterns

6.9 Application of Aerial Photo Interpretation

Aerial photos have been used for several applications. Using the principles of visual photo interpretation, information about the earth's surface features, can be obtained. Some of the areas where aerial photos have been extensively used are:

(i) Topographical mapping,
(ii) Geology,
(iii) Soil mapping,
(iv) Forestry,
(v) Terrain evaluation,
(vi) Land use/Land cover mapping,
(vii) Agriculture,
(viii) Water resources, and
(ix) Environmental studies/flood damage studies.

(i) Topographical Mapping

The basis of topographical mapping using aerial photograph is the common overlap between two successive photos in the forward and lateral direction. The forward direction is the direction of the aircraft flight and the minimum forward overlap required is 60 percent. The lateral overlap, at right angles to the forward direction can vary from 25 to 30 percent.

The overlapping photos are placed in the so-called stereo projectors in such a manner that the model of the ground is recreated and the observer sees the three-dimensional view of the ground. For this it is necessary to create geometrical conditions between the two photos in the stereo projector in such a way that the inclination between the two photos in the stereo projector in such a way that the inclination between the two successive photos is exactly the same as at the time of taking pictures by the aerial camera. It is possible to draw not only the planimetry but also the contours at desired scale and interval by suitable selection of ground control points. These ground control points appear on the aerial photos. The biggest advantage of aerial mapping lies in cost and efficiency. Whereas a very large number of such control points are required for mapping by traditional ground survey methods, such as plane tabling, a much less density of control points is required by aerial methods. The time required by photo mapping is also much less compared to ground methods. At present, the accuracy achieved is so much that the aerial mapping has almost become a universal technique of mapping.

(ii) *Geology*

The geological mapping involves both surface and subsurface mapping for mineral and other geophysical exploration. An aerial photo gives an excellent birds eye view, a so called 'synoptic' view of the terrain from a given height. Higher the elevation from where the photo has been taken greater is the synoptic view. In case of geological surface features it is some time very difficult to recognize it from the ground. For example, a fault of width of a few metres may not be recognized while standing on the ground but if seen from a height the fault will appear very clearly on the photo. This is due to what is called the lineament effect. Human eye can some how recognize features if they extend over a long distance even though the width of the feature is beyond the resolution of the eye. Using the principles of photo interpretation and by the help of so-called 'key' it is possible to do geological mapping for the following:

(i) The photographic tone of the rock body to that of adjacent rocks,
(ii) resistance to erosion,
(iii) boundaries of rock types,
(iv) topographic expression,
(v) boundaries of outcrops,
(vi) the joints and their pattern,
(vii) faults and their pattern,
(viii) drainage pattern
(ix) the vegetation cover, and
(x) the lineament pattern.

(iii) *Soil Mapping*

It is well known that soils are derived from rocks and that soils are in general, surface features and hence cannot be directly mapped the same way as in forestry, geology or topographic mappings. A photo will serve to prepare a Base Map on which ground information can be inserted and a correlation between the type of soil and corresponding tone and texture on the aerial photo be identified. Using this the experienced soil scientist can use aerial photos to his benefit. In addition viewing the photos under a stereoscope gives him additional 3-D view. Other factors which help in soil type mapping are the soil moisture content, soil texture, surface roughness, the presence of iron oxide and the organic matter content.

The unpredictability and indirectness of relationship between individual landscape element and soils has led to an emphasis on the use of so called 'Physiographic Systems' in photo interpretation. Such a physiographic system is defined as specific set of physiographic conditions which can be inferred by the use of 'Converging evidence' produced by different individual landscape elements on the photos. These physiographic systems may be related more specifically to conditions created by stratigraphy, block faulting, ground water and surface water hydrology, geomorphology, erosion and weathering processes, wind action or sedimentology, but more generally they are related to combinations of these conditions and processes. An example of this is the physiographic system of Bhabar and Tarai below the foothills of the Himalayas, which is not only a hydrological system but has also aspects related to sedimentology and soils. An another well known example is the physiographic system produced by block faulting processes north of Dehradun which has both geological and sedimentological aspects, but which is also influenced by resulting erosion, weathering and seepage processes. The third example is the system produced by gently dipping Gondwana sandstones overlying Gondwana shales. In all these situations these physiographic systems are expressed on the aerial photos in not just one but in combinations of almost all the individual photo interpretation elements.

A model of the process and conditions governing certain physiographic system and their resulting soil patterns can generally be presented in the forms of a detailed cross-section. A model like this is established during detailed studies in sample areas. After it is fully understood, its influence on soil patterns studied and its different expressions on the photographs established, the model can be applied to the other parts of the survey area for which the sample is representative. The validity of this extrapolation of knowledge is then checked by some rapid traverses in the rest of the area. These traverses may take the form of a field check and a "predicted soil map" based on the physiographic model drawn up in the sample area. If the predicted soil patterns along these traverses prove to be correct we may conclude that the model is applicable. If not, either the model as a working hypothesis was not correct or the sample area was not truly representative. In the latter case we may have to select new sample areas and change our first interpretation.

(iv) *Forestry*

The main application of photo interpretation to Forestry involves:

(i) Preparation of a Base Map, (ii) Identification of tree species, (iii) Quantitative measurements about the density of trees in a given area, height of trees and crown shapes and volumes.

The principles of photo interpretation as explained earlier such as shape, size, pattern, shadow, tone and texture are applied for (i), (ii) mentioned above.

Individual tree species have their characteristic shape and size. With some ground knowledge and experience it is possible to identify these on the air photos. Shadows of trees help to know their shape in profile. Changes of tone and texture also help in identification of species. Tress also give a peculiar pattern depending upon underground moisture condition. Trees along a straight line may be due to underground water channel. Trees also occur in clumps. Knowledge about tree height can easily be obtained using a simple parallax bar with viewing under stereoscope. Physiographic conditions also help in identification of tree types. For example, Sal trees generally grow on low hills up to 1000 meters and Pine trees up to 2000 meters or more of terrain heights above sea level.

(v) *Terrain Evaluation*

The study of terrain is essential and a prerequisite for proper planning and utilization of land resources. The purpose could be a short term military requirement for certain localized zones, a long term peace time need like development and exploitation of mineral resources, availability of construction material, exploration for ground water etc. The aim of terrain study is to gather maximum and systematic information on various aspects of the ground so that proper evaluation of this information can be done to meet the requirements of different users. A system of classification of terrain is, therefore, necessary where a given area can be subdivided into basic units. The question becomes more important for an inaccessible area where the only available tool is the aerial photo and the concept of 'known to the unknown' will hold good. In fact the aerial photos, when seen under a stereoscope, bring the 'Ground' to the laboratory.

The two important units in terrain classification are Pattern are Facet. A Facet is the fundamental unit in the classification. It is desirable as a piece of ground possessing uniform physical properties for all practical purposes. It is classified on the basis of surface configurations, nature of surficial deposits, surface and subsurface water region, land use and its associations with other terrain units. A landscape pattern is an area or areas of regularly occurring pattern of topography and surficial deposits including soils. The landscape pattern is recognized on the basis of geologic set up, climate and topography. The following information is recorded for each terrain unit :

Terrain Unit	:	Type of Information Card
Pattern	:	Pattern Card, Pattern Description, Climate
Facet	:	Facet Card, Ground Description, Surficial deposit and soils, Land use, Vegetation, Soil Properties and Classification, Engineering Resources and Water Supply.

The detailed method of Terrain study is as follows:

Area of study

1. **Library Work**

   Collection of basic material and background information

2. **Laboratory Exercise**

   Scanning of air photos, checking of quality, scale etc., study of maps and delineation of major units, preliminary list of landscape Patterns and Facets

3. **Ground Reconnaissance**

   Initial Field Traverses

4. **Study of Aerial Photos**

   Delineation of facets and planning of field work

5. **Detailed Field Work**

   (a) Checking of photo-characteristics, recording terrain data, collecting soils samples, field tests for moisture content and soil strengths (b) Final list of Pattern and Facets

6. **Laboratory Tests**

   Soil tests, rock identification etc.

7. **Completion of Reports**

   Confirmatory recce, if required finalization of report, typing of Cards and preparation of maps

   Main Report, Appendices (pattern, Facets and information cards, Facet maps).

   The process of photo-interpretation can be divided into steps like observation and measurements, logical reasoning and influence. The equipment required is simple and constitutes pocket stereoscope for quick scanning in the field. Mirror stereoscope for detailed work, parallax bar for height measurement and sketch master to transfer the interpreted data to the base map. In the case of terrain studies presenting in progress the details are directly transferred to existing toposheets.

   The recognition elements used for photo interpretation aids to describe different terrain units are :

(i) Position in landscape and association with other terrain units.

(ii) Morphology, surface configuration and micro relief.

(iii) Drainage pattern, texture and density, internal and external drainage.

(iv) Erosional features and gully sections.

(v) Land use and cultural features.

(vi) Vegetation.

(vii) Tonal variation and special photo textures and pattern if any.

(vi) *Land use/Land Cover*

The term land use refers to the human activity associated with a specific piece of land whereas the term land cover related to the type of feature present on the surface of the earth. The united States Geological Survey has devised a classification scheme at different levels for Land use/Land Cover system for use with aerial photos. Examples of Level I and Level II are :

Level I	Level II
1. Urban or Built up Land	1.1. Residential
	1.2. Commercial and Services
	1.3. Industrial
	1.4. Transportation, Communication and Utilities
	1.5. Industrial and Commercial Complexes
	1.6. Mixed Urban
	1.7. Other Urban

Similarly under Urban or built up land levels I, II, III are :

Level I	Level II	Level III
Urban or Built up	1.1 Residential	1.1.1. Single Family
land	1.2. Commercial	1.1.2 Multi Family
	1.3. Industrial	1.1.3. Group Quarters
	1.4. Lakes	1.1.4 Residential Hotels
	1.5. Parks	
	1.6. Orchards	
	1.7. Agricultural	

All these land use/land cover types can easily be delineated on aerial photos seen under stereoscopic vision. The resolution or the smallest distance that can be identified depends upon the scale. The larger the scale of photos greater is the clarity and resolution. In fact for land use/ land cover work photos on large scale such as 1: 10,000 to 1: 5000 are recommended. Color photos enhance the features much more and hence give much more information. Shadows play an important role in identifying multistoried houses and tall structures. Ground verification follows the photo interpretation map.

(vii) ***Agricultural***

The three major areas in which photo interpretation can help in the discipline of agriculture are :

(i) Crop condition assessment, (ii) Crop type classification, (iii) Crop yield estimation. Crop type is characterized by characteristic pattern and texture on an aerial photo whereas crop condition i.e. healthy or diseased are identified by virtue of tonal and grey level changes. The crop can be damaged due to several causes such as insects, moisture excesses, iron deficiency, nitrogen deficiency, soil salinity and air pollution. The crop yields can easily be determined by determining the area under cultivation for a particular variety.

The photo interpretation steps involved in agricultural studies relate to determination of drainage pattern and analysis, erosion pattern and analysis, photo tones, textures and vegetative feature and their pattern.

(viii) ***Water Resources***

The application of photo interpretation techniques of water resources involve two types (i) mapping surface water bodies, (ii) subsurface or ground water potential. The first part is simple as surface water bodies like streams, rivers, lakes etc., can be easily identified on the aerial photos. Water bodies appear darker in tone compared to rest of the features because of the simple fact that most of the sunlight that enters a clear water body is absorbed within about two meters of the surface. The degree of absorption depends upon the incident wave length. This infrared light in the region of 1 to 2 micrometer wave lengths is absorbed within a length of a meter for the surface.

The ground water location on the other hand is not so simple as this is a sub-surface feature. The indicators of ground water are topography and vegetations namely presence of springs and wells and other seepages. Present photo interpretation techniques can not be applied for estimation or determination of ground water depth. The other features of the interpretation techniques are the delineation of watershed and its assessment for reservoir selection and snow cover mapping, which help in determination water available from snow melt. The underground or buried stream can also be mapped for vegetation patterns on the surface along their streams and which also help in locating possible ground water zones.

(ix) *Environmental Studies/Flood Damage Studies*

Under thus heading, the areas of study are (i) Water pollution, (ii) Deforestation and denudation, (iii) Industrial pollution. Water pollution is caused by organic waste from domestic sewage and industrial wastes, production of excessive algae and water weeds and sediments brought down by rivers etc. Some times pollution can often be seen on aerial photos and delineated due to different tones and textures. The pure water appears with light grey and polluted water with darker tone. Deposits of oil on water also show darker tone. Use of aerial photos for flood damage can also be done by taking aerial photos before and after floods. Areas prone to successive flooding can be marked and preventive measures can be adopted.

6.10 Interpretation of Satellite Imagery

Since the satellite imagery is available both in graphic forms (on contact prints or film negatives and positives) and in Digital form (on CCT"s) there are two possibilities for interpretation. For the graphical form the same technique is used as for aerial photographs. However, due to very small scale of imagery (1:1 million or so) special emphasis has to be laid on additional help for the imagery on different bands or channels. For example, water features appear quite dark on Band 7 imagery compared to other bands. Use should also be made of additional imagery obtained at different dates. This helps in forestry and vegetation studies as the colour and leaves of trees change with time. These also are helpful in other areas such as natural hazards, flooding, landslide, earthquakes and other feature which change with time. The satellite imagery of the same area available every 18 days for landsat and for similar time for other satellite. The repetitive imagery and imagery in different spectral bands are two main advantages of satellite imagery photo over interpretation. The main drawback is the small scale of imagery, which render identification of small features impossible. For example, an image of a 1 mm in size of the satellite imagery of 1:1m scale of 1 km size of the object. Thus individual building or feature of 80 m in dimensions are extremely difficult to identify. But there is one advantage in this case. It is that of linear image. A road of 30 or 10 m in width is too narrow to be seen as such but the fact that it appears as a linear feature can help in its identification. In fact mapping of lineaments using the satellite imagery has greatly improved the geological interpretation because the satellite imagery, because of the large area it covers, gives a good synoptic view of the terrain, so necessary for geological mapping.

6.11 Key Elements of Visual Image Interpretation

Various terrain characteristics are very important to soil scientists, geologists, geomorphologists, geographers, real estate developers, irrigation engineers,

(ix) *Environmental Studies/Flood Damage Studies*

Under thus heading, the areas of study are (i) Water pollution, (ii) Deforestation and denudation, (iii) Industrial pollution. Water pollution is caused by organic waste from domestic sewage and industrial wastes, production of excessive algae and water weeds and sediments brought down by rivers etc. Some times pollution can often be seen on aerial photos and delineated due to different tones and textures. The pure water appears with light grey and polluted water with darker tone. Deposits of oil on water also show darker tone. Use of aerial photos for flood damage can also be done by taking aerial photos before and after floods. Areas prone to successive flooding can be marked and preventive measures can be adopted.

6.10 Interpretation of Satellite Imagery

Since the satellite imagery is available both in graphic forms (on contact prints or film negatives and positives) and in Digital form (on CCT"s) there are two possibilities for interpretation. For the graphical form the same technique is used as for aerial photographs. However, due to very small scale of imagery (1:1 million or so) special emphasis has to be laid on additional help for the imagery on different bands or channels. For example, water features appear quite dark on Band 7 imagery compared to other bands. Use should also be made of additional imagery obtained at different dates. This helps in forestry and vegetation studies as the colour and leaves of trees change with time. These also are helpful in other areas such as natural hazards, flooding, landslide, earthquakes and other feature which change with time. The satellite imagery of the same area available every 18 days for landsat and for similar time for other satellite. The repetitive imagery and imagery in different spectral bands are two main advantages of satellite imagery photo over interpretation. The main drawback is the small scale of imagery, which render identification of small features impossible. For example, an image of a 1 mm in size of the satellite imagery of 1:1m scale of 1 km size of the object. Thus individual building or feature of 80 m in dimensions are extremely difficult to identify. But there is one advantage in this case. It is that of linear image. A road of 30 or 10 m in width is too narrow to be seen as such but the fact that it appears as a linear feature can help in its identification. In fact mapping of lineaments using the satellite imagery has greatly improved the geological interpretation because the satellite imagery, because of the large area it covers, gives a good synoptic view of the terrain, so necessary for geological mapping.

6.11 Key Elements of Visual Image Interpretation

Various terrain characteristics are very important to soil scientists, geologists, geomorphologists, geographers, real estate developers, irrigation engineers,

6.11.1 Visual interpretation of topographic features based on reflection characterstics of images is given in table 6.2 below.

Table 6.2 Topographic features based on reflection characterstics.

Factor	Reflection Characteristics	Tonal Range
1. *Topographic*		
(a) Flat surface	Specular reflection if surface is smooth: no return	Dark
(b) Sloping surface facing antenna/ camera	Relatively high return due to orientation effects	Medium to light tones
(c) Sloping surface facing away from antenna or camera	Relatively low return due to orientation effects	Medium to dark tones
(d) High relief	No return from shadow areas	Dark tones
2. *Geologic*		
(a) Rough surface	Diffused Reflection, medium to high return	Medium to light tones
(b) Smooth surface	Specular reflection if surface is flat: no return reflection influenced by topographic factors	Dark tone, lighter tones produced by orientation effects
(c) Natural corner reflectors produced by in bed rock by weathering	Maximum reflection, high return	Very light tones
3. *Vegetation*		
(a) Trees	Diffused reflection	Light tones
(b) Woods and Forests	High returns	Light tones
(c) Bush	Higher returns with increasing densing of occurrence	Light tones in number in sub humid areas medium to dark tone in arid area environments
(d) Natural grass	Diffuse reflections, medium to low returns, dry sparse vegetation produces less scatter than bush rich-vegetation	Medium tones in humid to sub-humid areas. Dark tones in arid environment
Broad leaf crops with high moisture content	Diffuse reflection high returns	Light tones
Small leaf crops	Diffuse reflection medium return	Medium tones

6.11.2 Drainage Pattern and Texture

The drainage pattern and texture seen on aerial and space images are indicators of landform and bedrock type and suggest soil characteristics and site drainage conditions. The drainage pattern, which is a surface expression and can be discerned from the airphoto or satellite image, is a clue/key to infer something about the subsurface phenomena. Fig. 6.13 shows six of the most common drainage patterns that can be observed on different types of terrains : dendritic drainage pattern, rectangular drainage pattern, trellis drainage pattern, radial drainage pattern, centripetal drainage pattern and deranged drainage pattern.

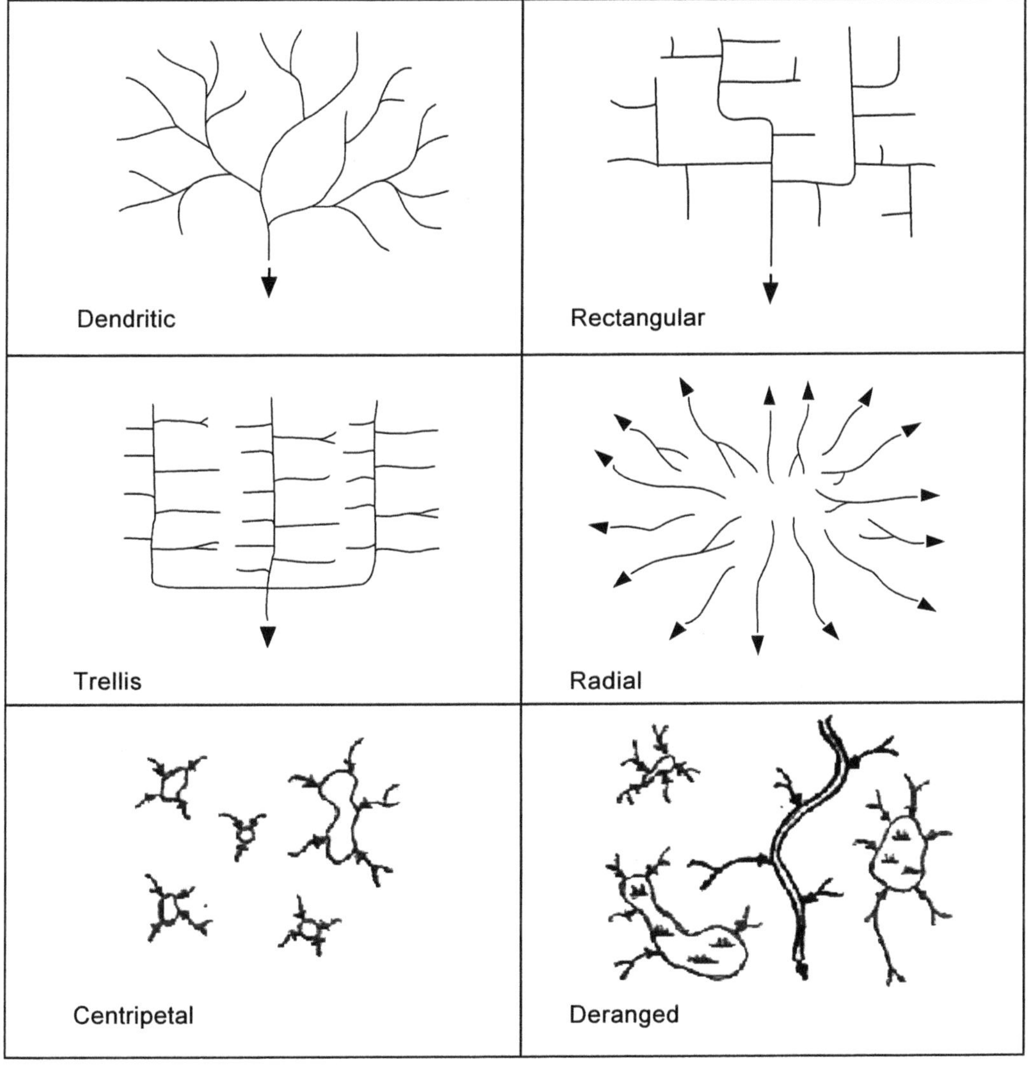

Fig. 6.13 Drainage patterns.

Dendritic drainage pattern is a well integrated pattern formed by a main stream with its tributaries branching and rebranching freely in all directions. This type of drainage pattern commonly occurs on relatively homogeneous materials, such as, horizontally bedded sedimentary rocks and granite.

Rectangular drainage patterns are basically dendritic patterns and are modified by structural bed rock control such that the tributaries meet at right angles to its main stream. This type can be found on flat laying massive sand stone formations with a well-developed joint system.

Trellis drainage pattern consists of a number of streams having one dominant direction with subtributaries as right angles to it. It can be found in areas of folded sedimentary rocks. **Radial drainage patterns** are formed from a central area and are radiated outward from this central area. All the sub-streams radiate away from a single point. These can be found on an area full of volcanoes and domes. **Centripetal drainage pattern** is the reverse of the radial drainage pattern. It can be found in the areas of limestone, sinkholes, volcanic craters and other depressions. **Deranged drainage pattern** is a dis-ordered pattern, irregularly developed and directed short streams, ponds, wetland areas, and glacial till areas.

These six drainage patterns are called destructional or erosional drainage patterns. The depositional constructional drainage patterns are remnants (products) of origin of landforms such as alluvial fans and glacial outwash plains.

Drainage texture is a combination of drainage or integration of different kinds of drainage patterns. Texture can be termed as "coarse textured and fine drainage patterns" (Fig. 6.14). Coarse textured patterns develop where the soils and rocks have good internal drainage with little surface runoff. Fine textured patterns develop where the soils and rocks have poor internal drainage and high surface run-off. Also fine textured drainage patterns develop on soft, easily eroded rocks, such as, shale, whereas coarse textured patterns develop on hard, massive rocks, such as, granite.

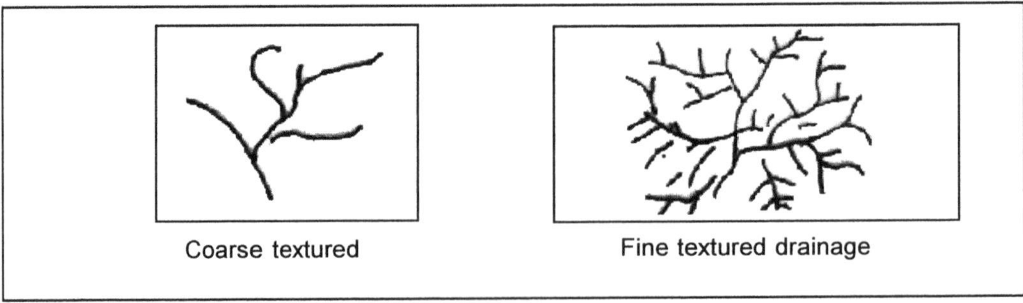

Fig. 6.14 Coarse and Fine drainage texture.

A measure of texture of the drainage pattern is texture ratio (T). It is defined as the ratio of the highest number of stream order of a basin to perimeter of the basin (Vente Chow, 1979), and is expressed as

$$T = N/P$$

where, N is highest number of stream order of a basin and
P is perimeter of the basin.

The high texture ratio indicates high runoff and low texture ratio indicates low runoff. For example, the texture ratio of watershed shown in Fig. 6.15 lies between 0.9 to 3.5, and indicates high runoff.

Drainage Density (D_d)

Drainage density (D_d) is a measure of the texture of the drainage basin and is defined as the ratio of the total stream length cumulated for all orders in the basin to the basin area, that is,

$$D_d = \frac{\Sigma L}{A}$$

where, D_d = Drainage density in km/sq.km
L = Total length of all streams with in a basin in kms
A = Total area of the basin in sq.km.

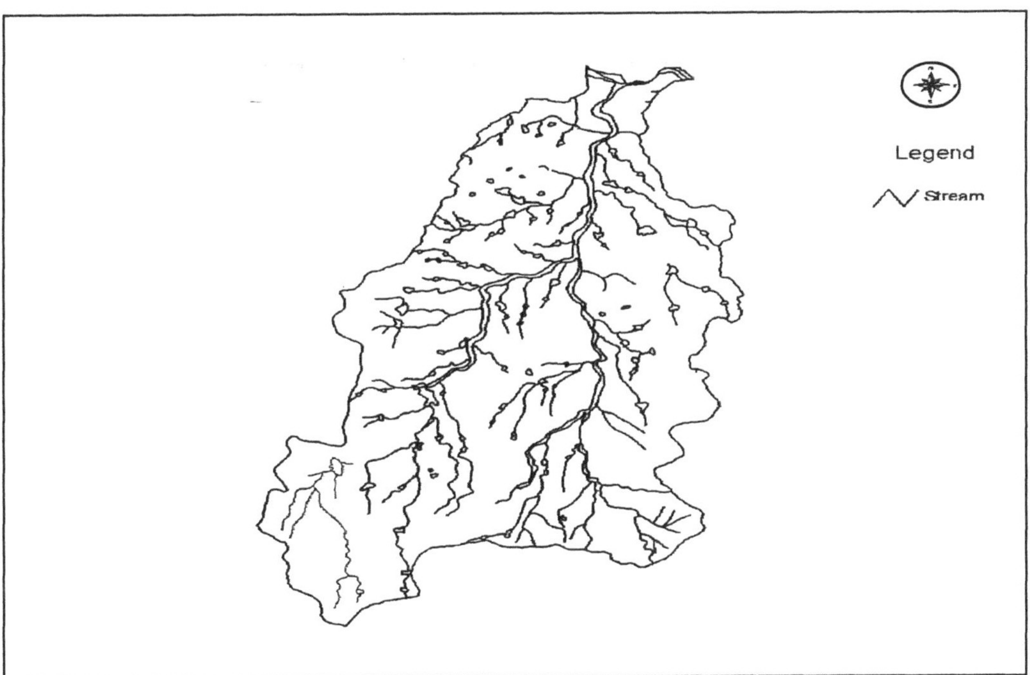

Fig. 6.15 Drainage map of Bheemgal watershed.

The difference in drainage density are commonly attributed to difference of rainfall or relief, infiltration capacity of soil or terrain, and initial resistivity of the terrain to the erosion. The low drainage density favoured generally in the regions of highly resistant or highly permeable subsoil material under dense vegetative cover at low relief. On the other hand, in regions of weak impermeable, subsurface material, sparse vegetation and mountainous relief favour high drainage density.

Horton (1932) classified steep impervious areas as those areas that have drainage density (D_d) ranging between 0.90 to 1.29 km/sq. km, and the areas having drainage density values less than 0.93 km/sq. km as permeable high infiltration rates. Long Bein (1941) suggested a range of drainage density from 0.55 to 2.09 km/sq.km in steep impervious areas and in humid regions with stream density of 1.03 km/sq.km.

Smith (1950) and Strahler (1957) described D_d value less than 5.00 as coarse, between 5.0 to 13.7 as medium, between 13.7 to 155.3 as ultrafine. So, high drainage density indicates low infiltration, high runoff, high relief and impermeable stratum, while low drainage density indicates high infiltration, low runoff, low relief and permeable stratum. For example, D_d of Racharla drainage basin of Prakasam district, Andhra pradesh, India lies between 1.1 to 2.4 km/sq.km. which in turn indicates low infiltration, high runoff, high relief and impermeable stratum.

6.11.3 Erosion

Gullies are the smallest drainage features that can be seen in aerial photographs or imageries and may be as small as a metre wide and 100 m long. Gullies result from the erosion of unconsolidated material by runoff and develop where rainfall cannot adequately percolate into the ground. These small gullies are some times called small rivulets. These initial rivulets enlarge and take on a particular shape. This shape is a characteristic feature of the material type in which they are formed. These gullies are generally named as V -shaped, U - shaped and gently rounded cross sections (Fig. 6.16).

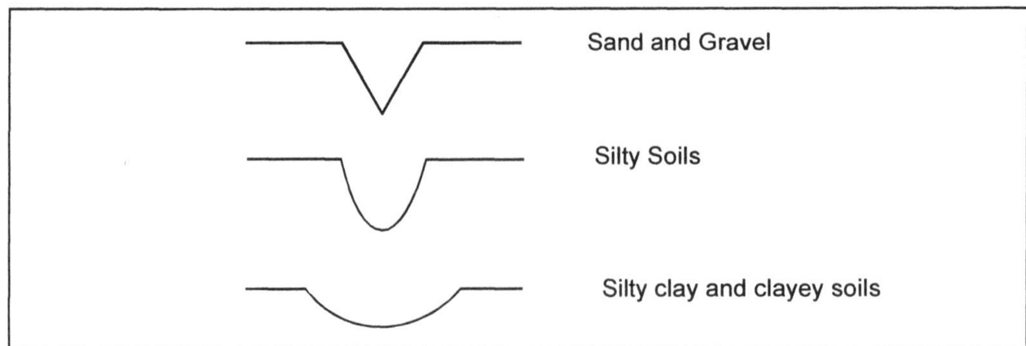

Fig. 6.16 Gulley sections.

6.11.4 Image Tone

The term image tone refers to the brightness at any point in a photograph or space image. The absolute value of the image tone depends not only on certain terrain characteristics but also on photographic factors, such as, film, film filter combination, exposure and photographic processing, image acquisition factors, spectral bonds used for both multispectral and hyperspectral scanning, combination of bonds used for false colour composite, and data processing methods. Image tone also depends on meteorological and climatological factors, such as, atmospheric haze, sun angle and cloud shades. Because of these factors which are not related to terrain, image interpretation should be based on relative tones instead of absolute tonal values. The relative tone values are important because they often form distinct image patterns that may be of great significance in image interpretation.

6.11.5 Vegetation and Land Use

Vegetation and land use can be considered one of the most essential key elements as the variations of vegetation and land use from one area to the other act as an indicator of terrain condition. For example, coconut trees are generally located on deltaic regions, whereas truck farming activities are often based on highly organic soils, such as, peat deposits. In almost all applications of remote sensing and GIS, a knowledge of a land use / land cover is important, and without this, the planning and management activities are biased. The land use / land cover is considered an essential element for modeling and understanding the earth as a system. The use of panchromatic, medium-scale aerial photographs to map land use has been an accepted practice since the 1940's. More recently, small-scale aerial photographs and satellite images have been utilized for land use/land cover mapping (Lillesand Kiefer, 2000).

The term land cover relates to the type of feature present on the surface of the earth. Corn fields, lakes, maple trees, and concrete highways are all examples of land cover types. The term land use relates to the human activity or economic function associated with a specific piece of land. An example, of this is a tract of land on the fringe of an urban area that may be used for a single-family housing. In chapter 16, a detailed procedure of preparing a land use/land cover map using visual interpretation of IRS satellite image, the conversions of this map from vector to raster digital database for the GIS analysis and the final GIS output showing various land use/land cover patterns used for urban and municipal planning of Hyderabad city, is presented.

6.12 Concept of Converging Evidence

Terrain information can be derived from the visual image interpretation of aerial or space image through the identification, evaluation and analysis of all the above key elements. This image interpretation process is like the work of a detective trying to put all the pieces of evidence together to solve a mystery (Lillesand and Kifer, 2000).

Hence the information derived through the analysis of the above key terrain elements can be converged and by combining all the evidences of image element identification, the inferences can be drawn. The inferences will be useful for GIS data input, manipulation and analysis. The process of converging the description of all the interpreted results or informations of these key elements is called "convergence evidence" (Fig. 6.17). The interpreter uses the process of

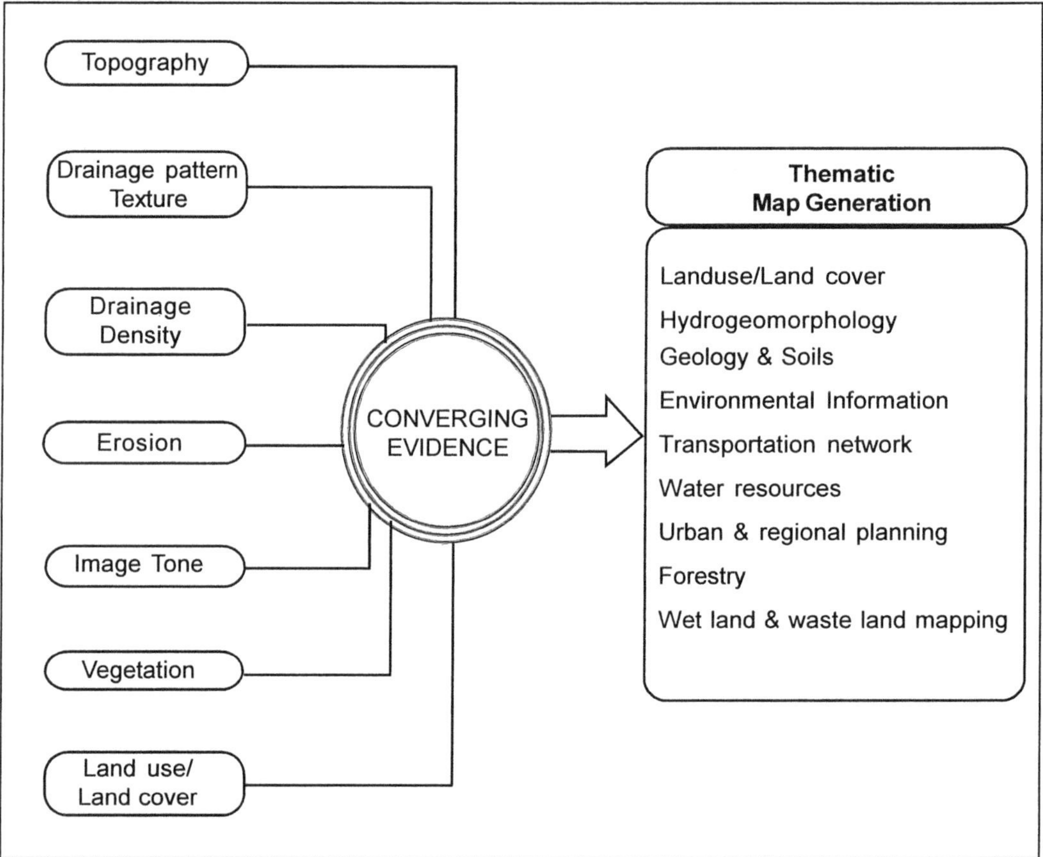

Fig. 6.17 Concept of converging evidence.

convergence of evidence to successfully increase the accuracy and details of the results of visual interpretation. The key elements described above are the most commonly used elements for any application in general. But based on the application, some more key elements derived from these fundamental key elements may be required.

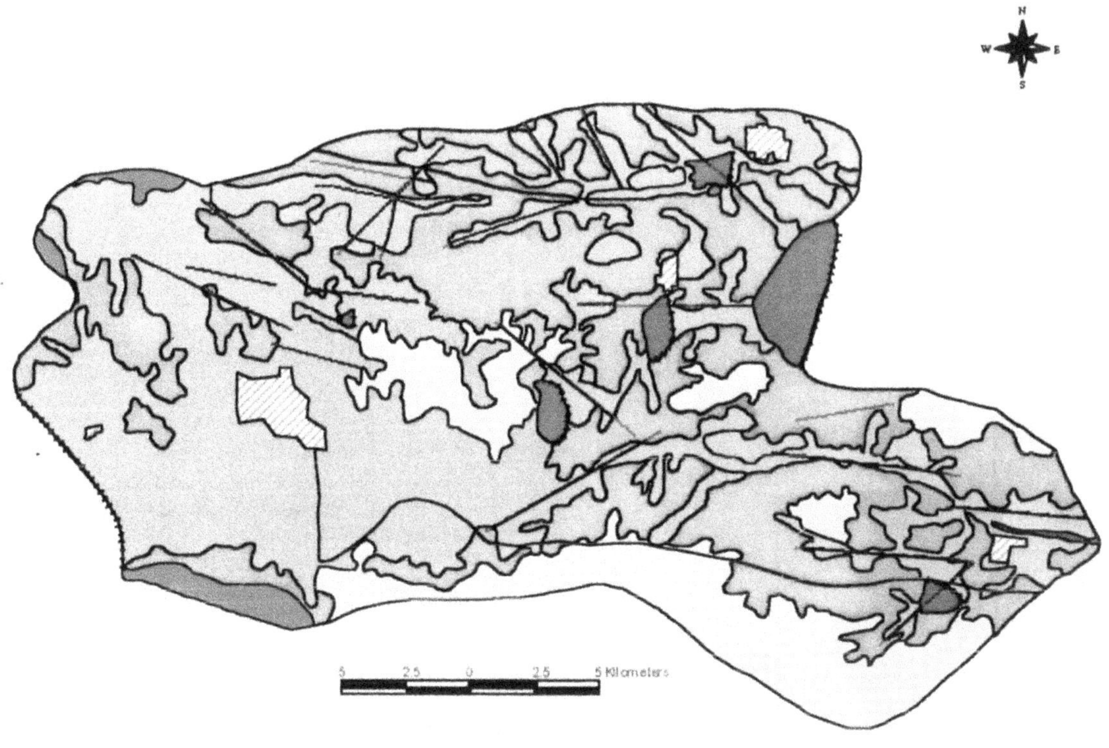

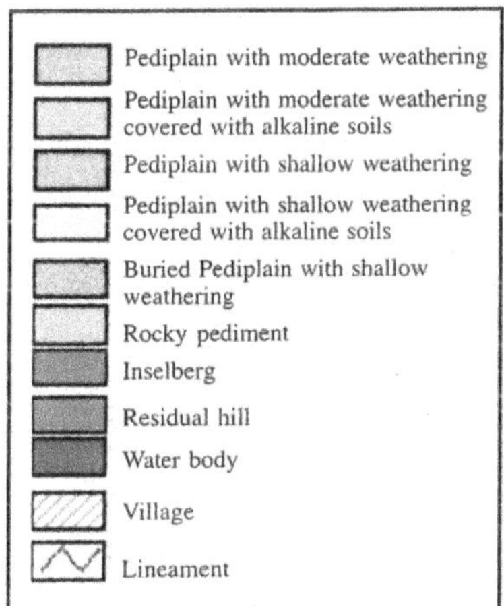

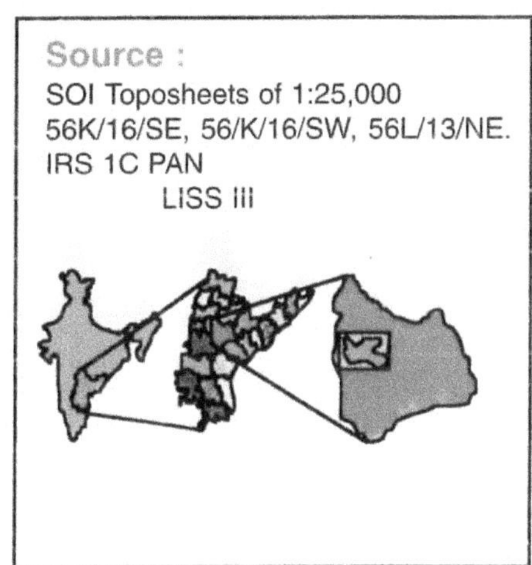

Hydrogeomorphological Map of Shivannagudem Watershed, Andhra Pradesh, India.

Plate No. 1

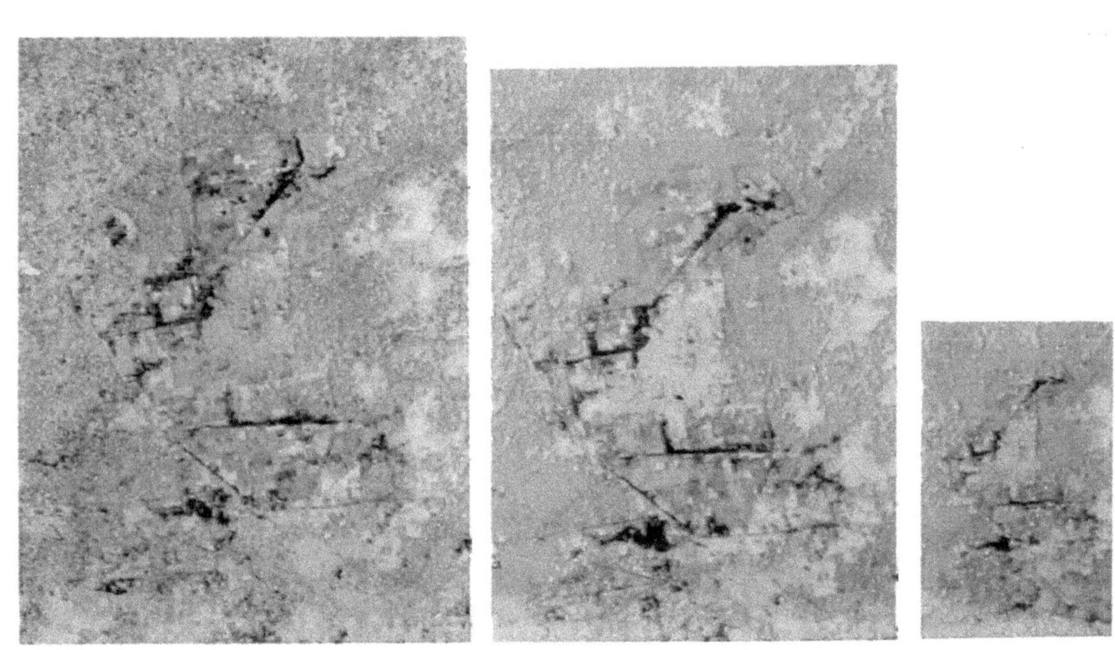

Sally Farms as seen by LISS-I, LISS-II and LISS-III sensors.

Plate No. 2

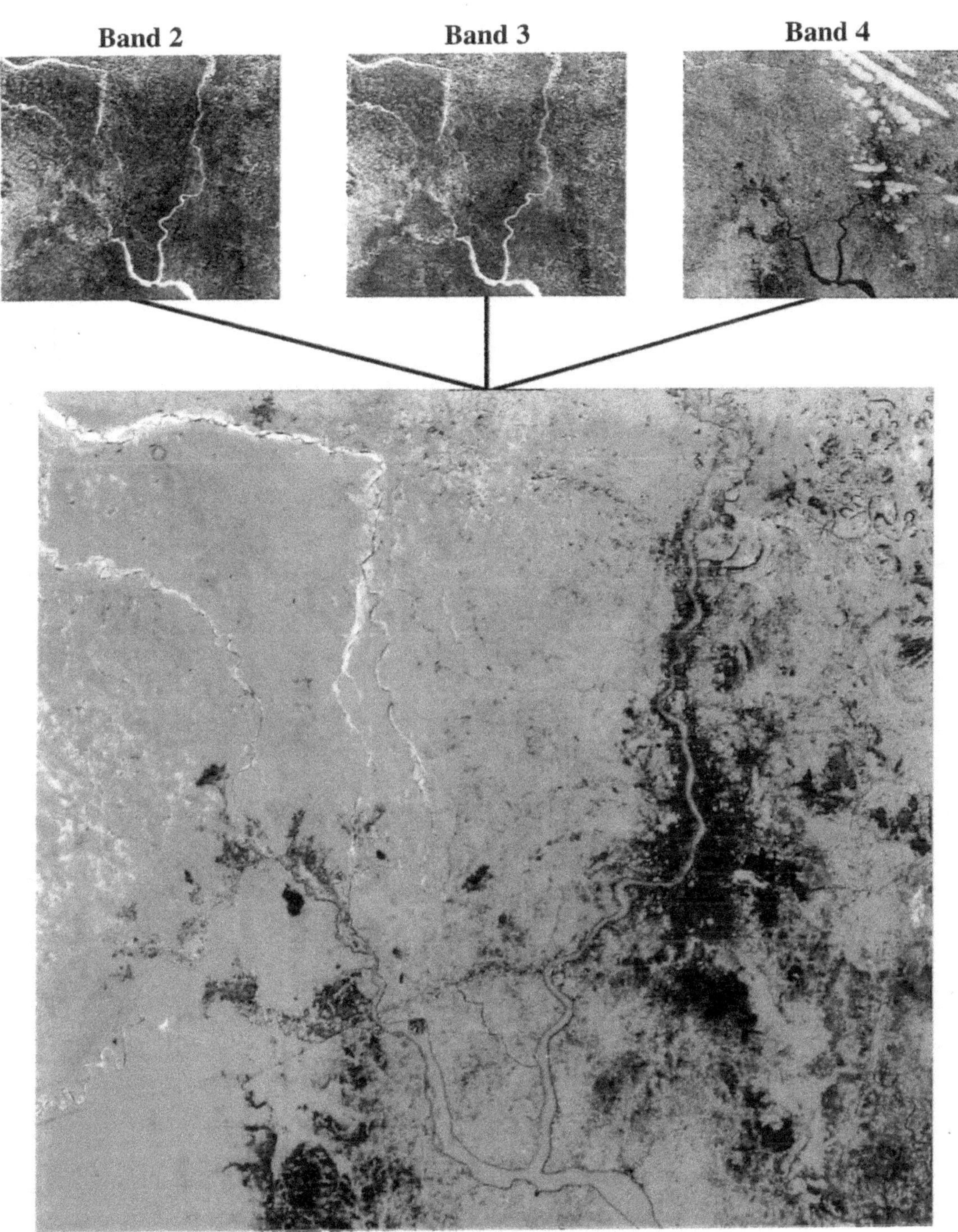

IRS-1D LISS-III FCC image (bands 2, 3, 4) and corresponding Black and White Images of band 2, band 3 and band 4 data of path 108, row 56, showing Calcutta and Surrounding Areas.

Plate No. 3

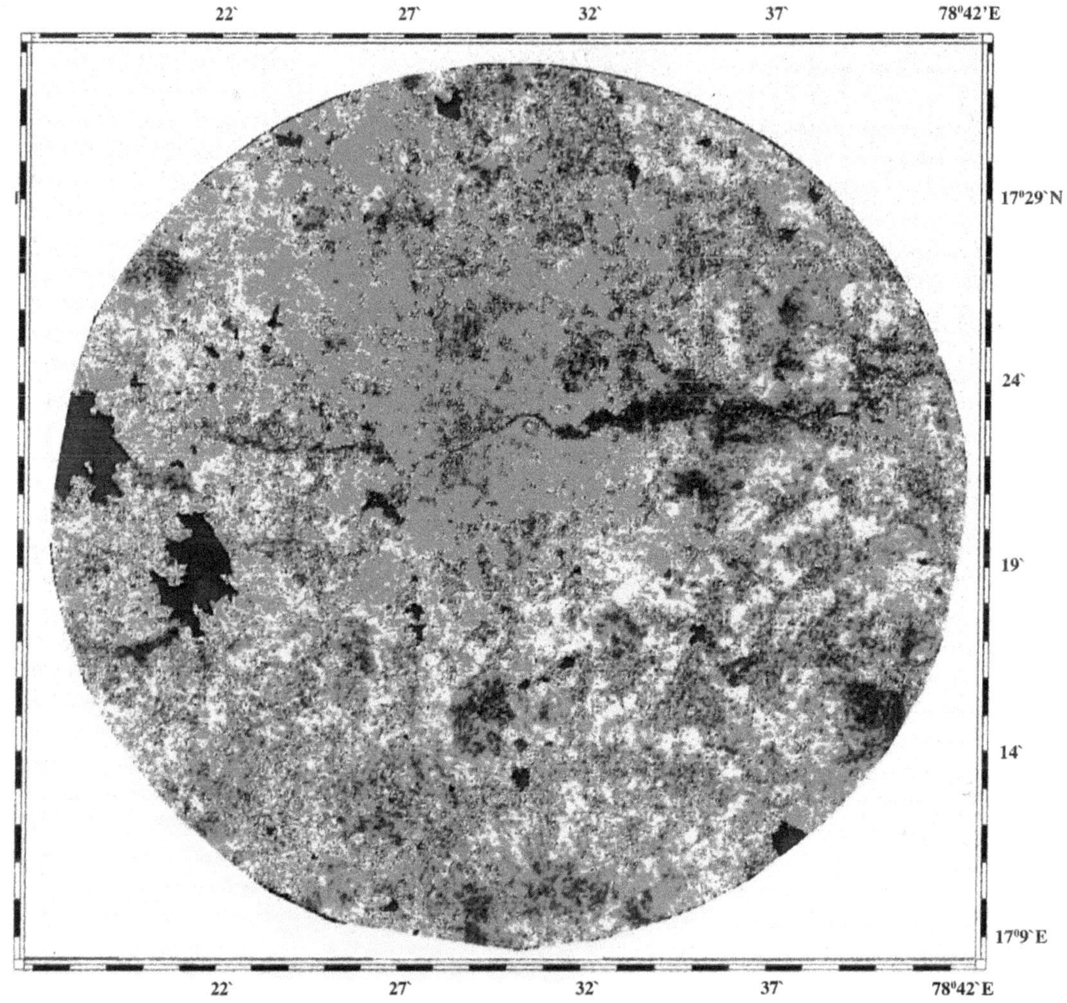

Pseudocolour Image showing NDVI of Hyderabad City used for Environmental Planning under Clean and Green Program.

Plate No. 4

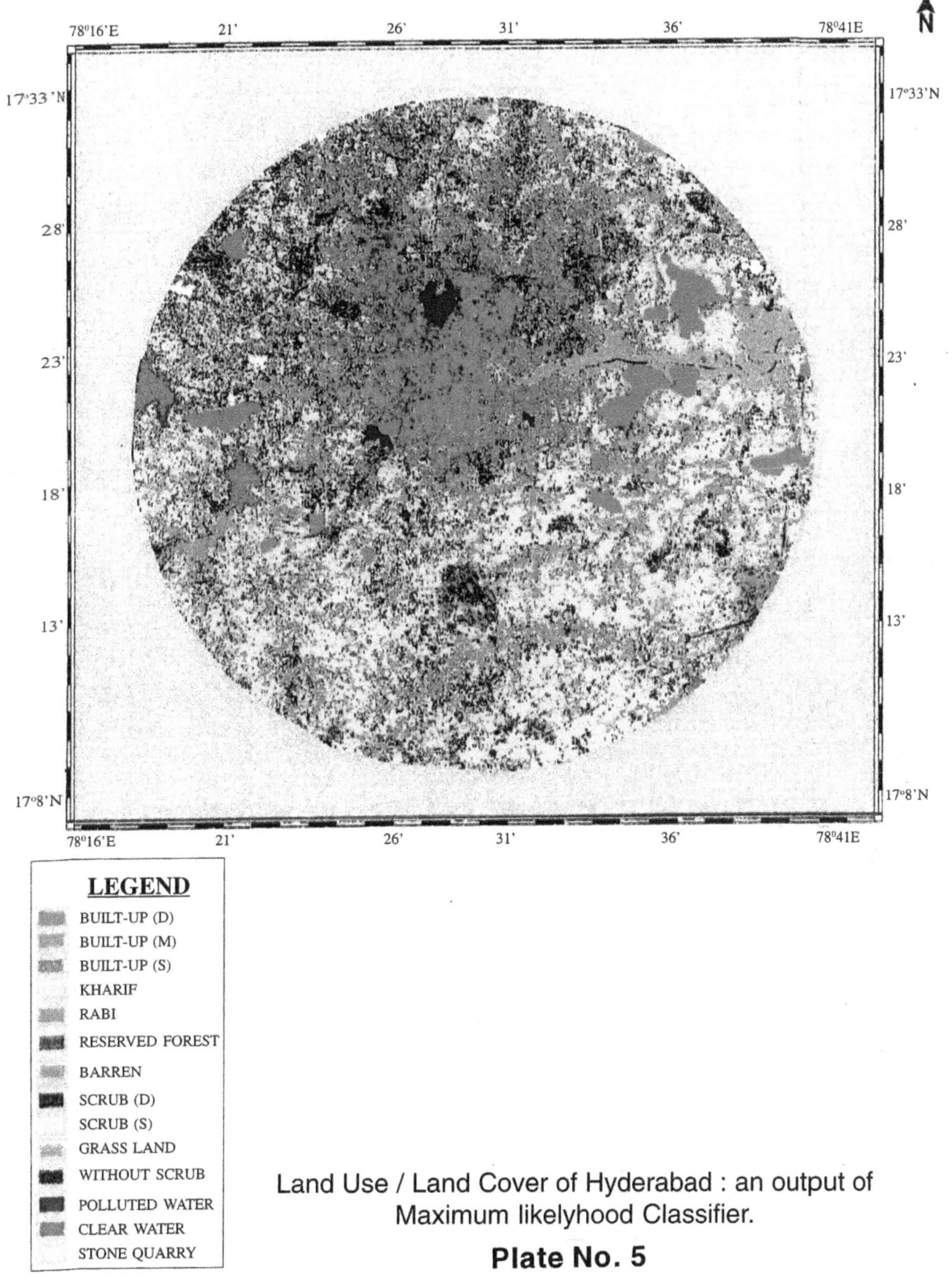

Land Use / Land Cover of Hyderabad : an output of Maximum likelyhood Classifier.

Plate No. 5

Digital Terrain Model (DTM)

Plate No. 6

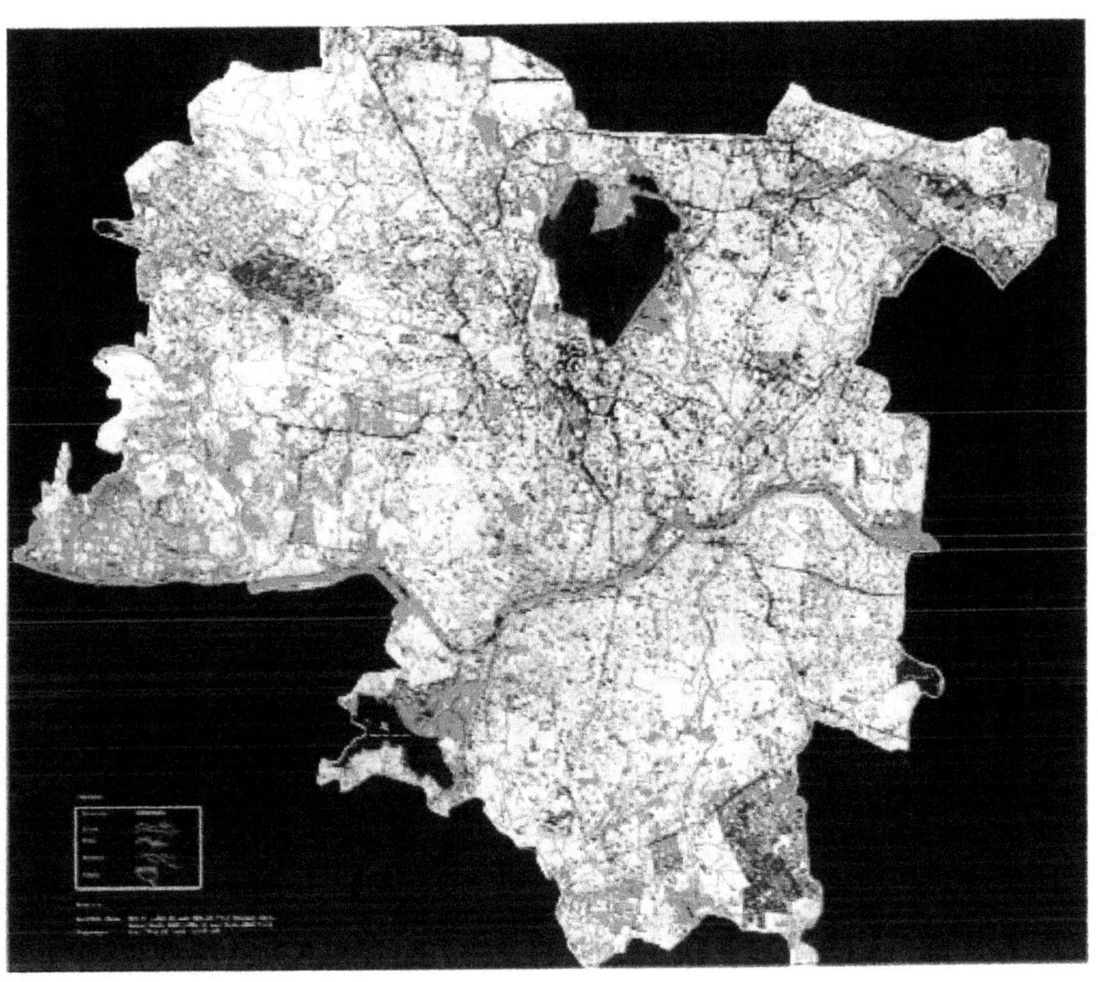

PAN + LISS III Image of IRS-1D of Hyderabad with its
Muncipal Corporation Boundaries.

Plate No. 7

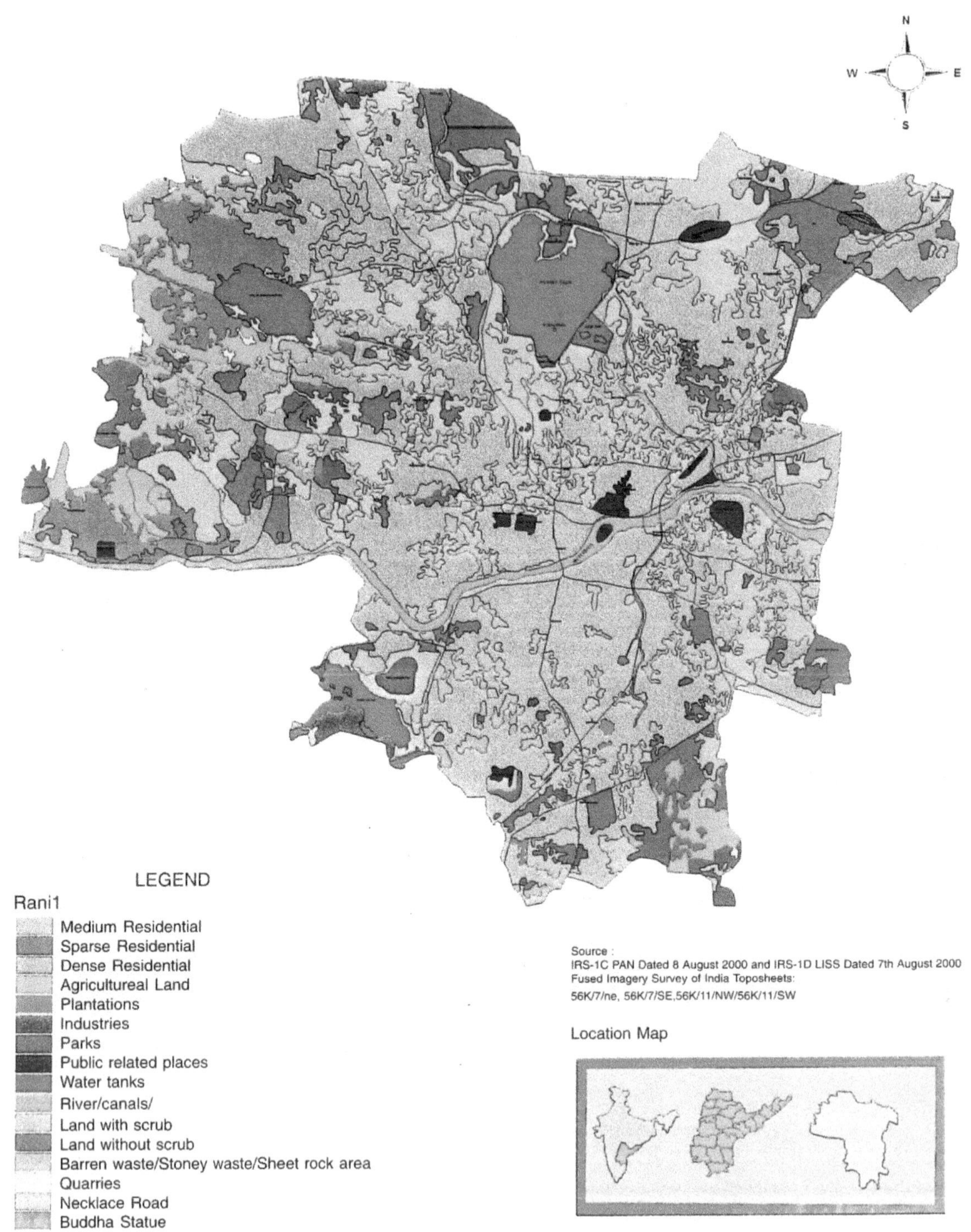

GIS Output : A Land Use/Land Cover Map of Area under Municipal Corporation of Hyderabad (MCH) jurisdiction.

Plate No. 8

7

Digital Image Processing

7.1 Introduction

As seen in the earlier chapters, remote sensing data can be analysed using visual image interpretation techniques if the data are in the hardcopy or pictorial form. It is used extensively to locate specific features and conditions, which are then geocoded for inclusion in GIS. Visual image interpretation techniques have certain disadvantages and may require extensive training and are labour intensive. In this technique, the spectral characteristics are not always fully evaluated because of the limited ability of the eye to discern tonal values and analyse the spectral changes. If the data are in digital mode, the remote sensing data can be analysed using digital image processing techniques and such a database can be used in raster GIS. In applications where spectral patterns are more informative, it is preferable to analyse digital data rather than pictorial data.

7.2 Basic Character of Digital Image

The basic character of digital image data is illustrated in Fig. 7.1. Though the image shown in (a) appears to be a continuous tone photograph, it is actually composed of two-dimensional array of discrete picture elements or pixels. The intensity of each pixel corresponds to the average brightness or radiance measured electronically over the ground area corresponding to each pixel. A total of 200 rows and 200 columns of pixels are shown in Fig. 7.1(a).

Whereas the individual pixels are virtually impossible to discrern in (a) they are readily observable in the enlargements shown in (b) and a 20 row × 20 column enlargement is included in (c).

Part (d) shows the individual digital number (DN) corresponding to the average radiance measured in each pixel shown in (c). These values are simply positive integers that result from quantizing the original electrical signal from the sensor into positive integer values using a process called analog-to-digital (A to D) signal conversion.

(a)

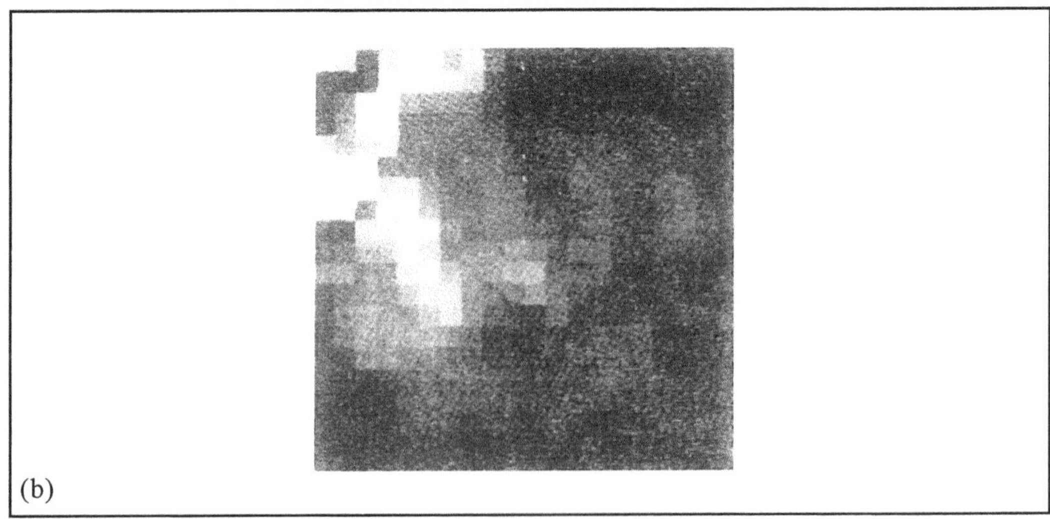

(b)

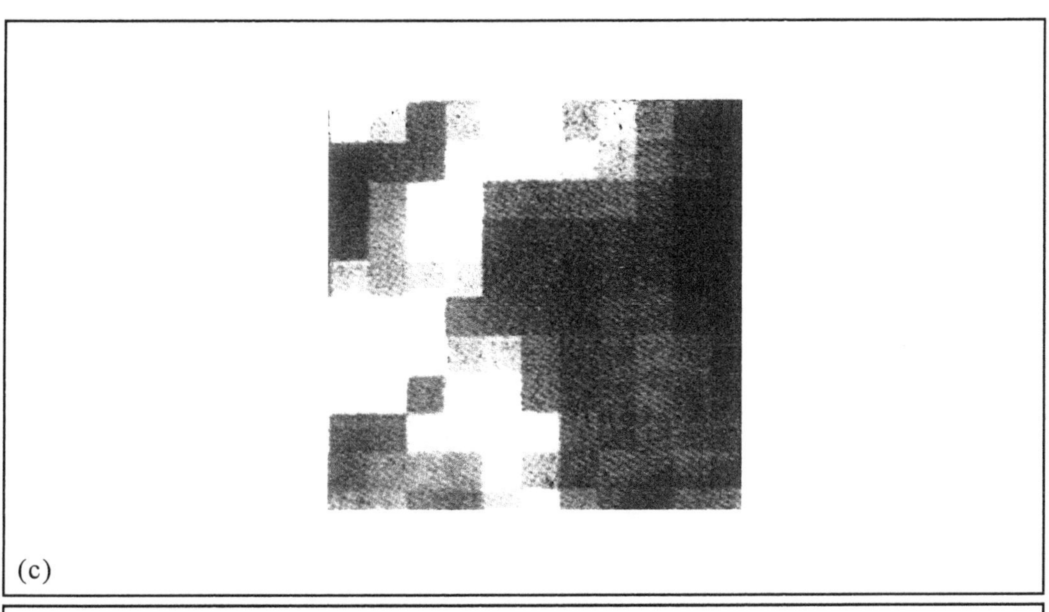

(c)

10	9	20	25	26	34	35	11	20	25	31	32	35	36	28
76	34	11	15	41	42	77	14	30	24	22	11	16	69	70
47	25	13	11	10	55	41	39	42	65	27	79	18	20	35
42	19	15	19	22	52	71	56	34	35	41	50	55	11	38
34	20	16	60	27	85	35	17	25	16	10	31	52	13	40
25	30	29	23	18	49	38	78	19	29	18	26	85	12	41
30	42	20	13	57	23	73	74	60	20	57	63	82	19	71
19	60	45	28	40	11	79	32	48	45	51	72	42	21	36
64	36	55	58	38	13	80	28	64	55	61	44	49	32	37
36	70	31	62	15	15	70	58	76	66	40	13	23	42	72
70	68	22	14	75	19	36	62	47	59	38	71	11	50	34
68	43	27	39	54	60	37	52	36	80	15	53	19	51	72
43	48	41	56	46	16	72	21	70	23	75	33	60	59	73
60	64	10	17	51	18	34	49	68	67	54	37	13	75	79
48	76	18	78	61	82	72	69	43	73	46	12	15	25	80

(d)

Fig. 7.1 Basic character of digital image data

(a) original 200 × 200 digital image

(b) enlargement showing 20 × 20 of pixels

(c) 10 × 10 enlargment

(d) digital numbers corresponding to radiance of each pixel

Fig. 7.2 is graphical representation of the A to D conversion process. The original electrical signal from the sensor is a continuous analog signal shown by the continuous line plotted in the figure. This continuous signal is sampled at a set time interval (ΔT) and recorded numerically at each sample point (a, b, ..., i, k). The sampling rate for a particular signal is determined by the least to twice the highest frequency presents in the original signal in order to adequately represent the variation in the signal.

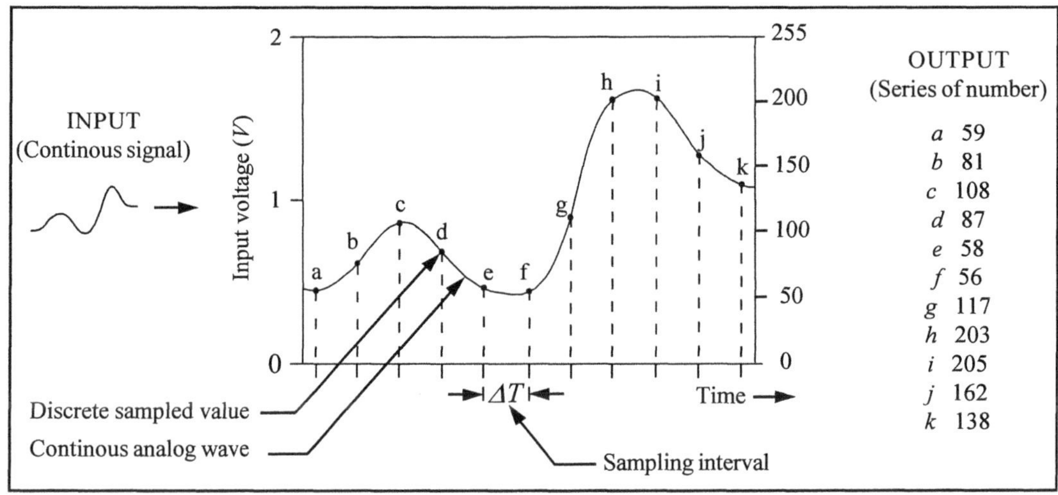

Fig. 7.2 Analog-to-digital conversion process. (Lillesand and Kiefer, 2000)

In Fig. 7.2 we illustrate the incoming sensor signal in terms of an electrical voltage value ranging between 0 and 2 V. The DN output values are integers ranging from 0 to 255. Accordingly, a sampled voltage of 0.46 recorded by the sensor would be recorded as a DN of 59. Typically, the DNs constituting a digital image are recorded over such numerical ranges as 0 to 255, 0 to 511,0 to 1023, or higher. These ranges represent the set of integers that can be recorded using 8–, 9–, and 10 - bit binary computer coding scales, respectively (That is, $2^8 = 256$, $2^9 = 512$, and $2^{10} = 1024$). In such numerical formats, the image data can be readily analyzed with the aid of a computer. This is what we call radiometric resolution of remote sensing data.

A digital image is defined as a matrix of digital numbers (DNs). Each digital number is the output of the process of analog to digital conversion. The image display subsystem carries out the conversion operation, that of taking a digital quantity of an individual pixel value from an image. Fig. 7.3 illustrates the surface of the ground divided into a number of parcels. Each parcel of land can be represented as a pixel (picture element) on the image and each pixel is occupied by a digital number and is called pixel value. This pixel value or digital number shows the radiometric resolution of remote sensing data.

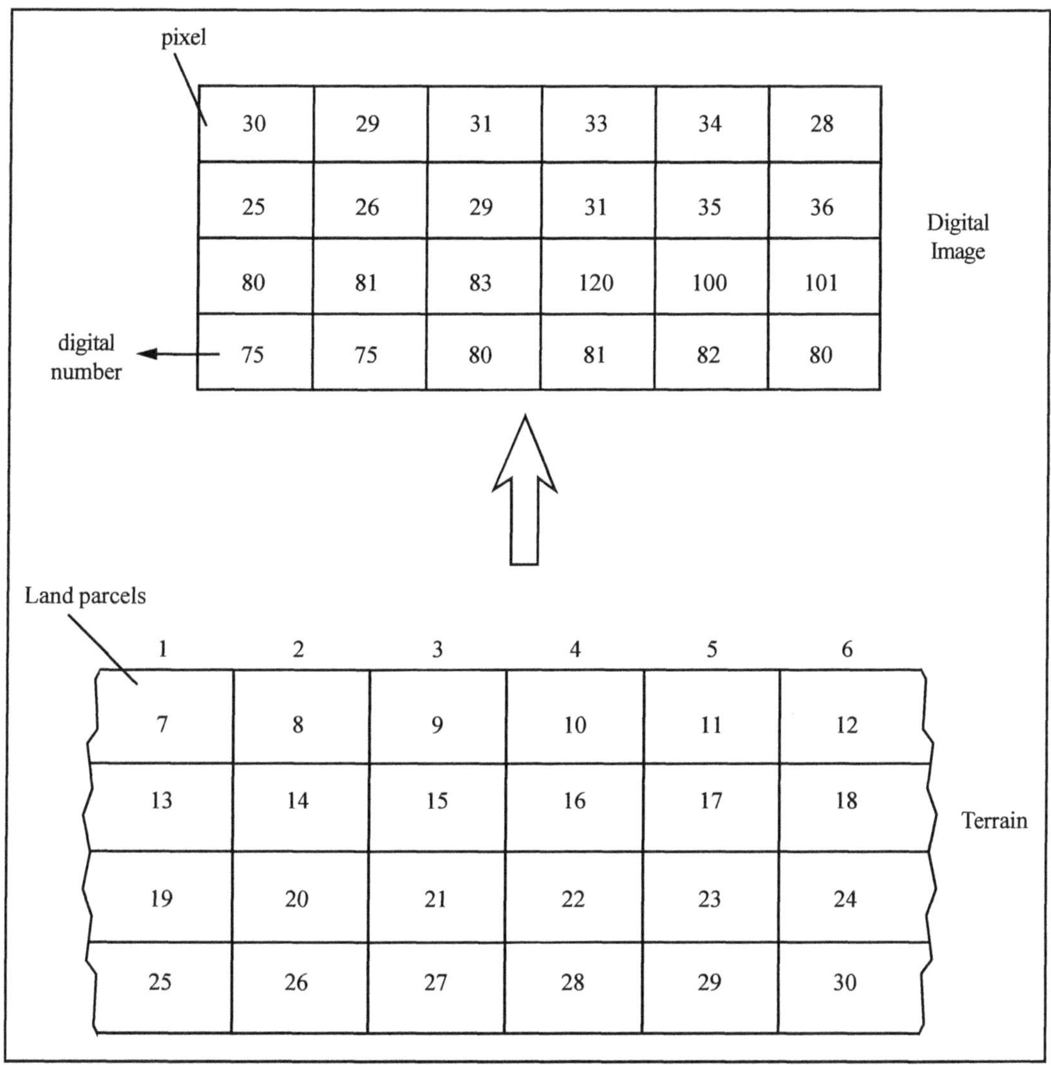

Fig. 7.3 Representation of land parcels with pixel locations and corresponding energy levels/digital numbers.

The use of computer-assisted analysis techniques permits the spectral patterns in remote sensing data to be more fully examined. It also permits the data analysis process to be largely automated, providing cost advantages over visual interpretation techniques. However, just as human beings are limited in their ability to evaluate spectral patterns, computers are limited in their ability to interpret spatial patterns. Therefore visual and numerical techniques are complementary in nature, and consideration must be given to the approach (or combination of approaches) that best fits a particular application.

Satellite remote sensing data in general and digital data in particular have been used as basic inputs for the inventory and mapping of natural resources of the earth surface like agriculture, soils, forestry, and geology. The final product in most of the applications (classified outputs) is likely to complement or supplement the map. Space borne remote sensing data suffer from a variety of radiometric and geometric errors caused by satellite motion, the sensor system, earth's rotation and so on. These distortions would diminish the accuracy of the information extracted and reduce the utility of the data.

In order to update and compile maps with high accuracy, the satellite digital data have to be manipulated using image processing techniques. The central idea behind digital image processing is that, the digital image is fed into a computer, one pixel at a time. The computer is programmed to insert these data into an equation or a series of equations, and then store the results of the computation for each pixel. These results are called look-up-table (LUT) values for a new image that may be manipulated further to extract information of user's interest. Virtually, all the procedures may be grouped into one or more of the following broad types of operations, namely,

(i) Preprocessing	(ii) Image Registration
(iii) Image enhancement	(iv) Image filtering
(v) Image transformation	(vi) Image classification

All the topics of digital image processing of remotely sensing data are divided into these six groups to provide the reader with conceptual road map for studying this chapter.

7.3 Preprocessing

Remotely sensed raw data, received from imaging sensor mounted on satellite platforms generally contain flaws and deficiences. The correction of deficiences and removal of flaws present in the data through some methods are termed as pre-processing methods. This correction model involves the initial processing of raw image data to correct geometric distortions, to calibrate the data radiometrically and to eliminate the noise present in the data. All pre-processing methods are considered under three heads, namely, (i) geometric correction methods, (ii) radiometric correction methods, and (iii) atmospheric correction methods.

7.3.1 Geometric Correction Methods

Remotely sensed images are not maps (Mather, 2000). Frequently information extracted from remotely sensed images is integrated with map data in a geographical information system. The transformation of a remotely sensed image into a map with a scale and projection properties is called geometric correction.

Geometric correction of remotely sensed images is required when the image or product derived from the image such as a vegetation index or a classified image, is to be used in one of the following circumstances (Kardoulas et.al, 1996) :

* to transform an image to match a map projection
* to locate points of interest on map and image
* to bring adjacent images into registration
* to overlay temporal sequences of images of the same area, perhaps acquired by different sensors
* to overlay images and maps within GIS, and
* to integrate remote sensing data with GIS.

To correct sensor data, both internal and external errors must be determined and be either predictable or measurable. Internal errors are due to sensor effects, being systematic or stationary, or, constant for all practical purposes. External errors are due to platform perturbations and scene characteristics, which are variable in nature and can be determined from ground control and tracking data. The sources of image geometric errors are listed in Table 7.1

Table 7.1 Sources of effects of geometric errors of Image

S.No.	Effect	Source of error
1.	Platform	altitude, attitude, scan-skew mirror, scan velocity
2.	Scene effect	earth rotation, map projection
3.	Sensor effect	Mirror sweep
4.	Scene and sensor effect	panorama, perspective

The errors of altitude and attitude of the space craft, and scan skewing are due to the platform instability (Bannari et. al, 1995). Departures of the spacecraft from nominal altitude of the landsat (e.g., Landsat, 920 km) produce scale distortion in the sensor data. For the MSS, this distortion is along-scan only and varies with time, the magnitude of correction being 1.5 kms. Normally, the sensor axis system is maintained with one axis normal to the earth's surface and another parallel to the spacecraft velocity vector. For the MSS, the complete attitude time-history known to pitch is 12 kms, roll is 12 kms and yaw is 2.46 kms. The pitch rate, roll rate and yaw rate are 0.93 km, 0.54 km, and 0.049 km respectively. During the time required for the sensor mirror to complete an active scan, the spacecraft moves along the ground track. Thus the ground swath scanned is not normal to the ground track, but is slightly skewed. This produces cross-scan geometric distortion. The magnitude of correction, for example, of MSS is 0.082 kms.

The mechanical scanner completes one full scan and it takes some time for one full sweep across the track. During this time, the spacecraft moves along the track. Because of this phenomenon, the ground swath is not normal to the ground track, but is slightly skewed. This produces cross scan geometric distortion (Bernstein and Ferneyhough, 1975).

In some situations, spacecraft velocity departs from nominal values, as the ground track covered by a given number of successive mirror sweeps, changes producing along track scale distortion. The magnitude of correction is 1.5 km. If the spacecraft velocity deviates from the nominal, the ground track covered by a fixed number of successive mirror sweeps, changes. This causes a cross scans scale distortion. This kind of distortion may be corrected, on the basis of the following model suggested by Bernstein (1983).

$$X_3 = \left(1 + \frac{dv}{v}\right) X_2$$

where X_3 = actual ground track

X_2 = nominal ground track

$\frac{dv}{v}$ = normalized spacecraft velocity error

v = space craft nominal velocity

The scanning mirror of the scanning system like MSS, nominally moves at a constant angular rate (non-linear sweep). In practice, the mirror rate is not constant. Since the data samples are taken at regular intervals of time, the changing mirror rate produces geometric distortion along scan line. Since the modern linear array scanning system does not have this kind of error, the correction was not dealt in the case of IRS satellite (LISS I, II & III) camera which is free from this defect, the error being insignificant.

As the sensor (MSS) mirror completes successive scans, the earth rotates beneath the sensor. Thus there will be a gradual westward shift of the ground swath scanned, which causes along-scan distortion. The correction is 13.3 km for Landsat MSS. For earth resources use, image data are usually required in a specific map projection. Although map projection does not constitute a geometric error, it does require a geometric transformation of the input data, and this can be accomplished by the same operations that compensate for distortions in the data. The magnitude of correction, for example, for Landsat MSS, is 3.7 kms along scan and along track.

Mirror sweep is an error caused owing to sensor effects. The sensor (MSS) mirror-scanning rate varies non-linearly across a scan, because of imperfections in the electro mechanical driving mechanism. Since data samples are taken at regular intervals of time, the varying scan rate produces along scan distortion. The magnitude of correction is 0.37 km for MSS.

Panorama is a very serious scene and sensor effect. The imaged ground area is proportional to the tangent of the scan angle rather than to the angle itself, and since data samples are taken at regular intervals, this produces along scan distortion. The magnitude of correction is 0.12 km for MSS. For most earth resources applications, the desired landsat images represent the projection of points on the earth, on a plane, tangent to the earth at the nadir, with all projection lines normal to the plane. The sensor data, however, represent perspective projections whose lines meet at a point above the tangent plane. For the MSS, this produces only along scan distortion. The magnitude of correction is 0.08 km.

7.3.2 Radiometric Correction Methods

The primary function of remote sensing data quality evaluation is to monitor the performance of the sensors. The performance of the sensors is continuously monitored by applying radiometric correction models on digital image data sets. The radiance measured by any given system over a given object is influenced by factors, such as, changes in scene illumination, atmospheric conditions, viewing geometry and instrument response characteristics (Lillesand and Kiefer, 2000). One of the most important radiometric data processing activity involved in many quantitative applications of digital image data is conversion of digital numbers to absolute physical values; namely, radiance and reflectance.

The digital numbers have been directly used by some investigators to statistically classify cover types in a single image, identify map terrain features, create continuous images for digital mosaic of several images or ratio spectral bands to eliminate differential illumination effects. The results of such analyses are questionable because the digital numbers do not quantitatively represent any real physical feature. It is therefore necessary to first convert the digital data into physically meaningful values such as radiance and reflectance so that they can be used in further analysis.

Computation of Radiance (L)

Radiance is a measure of the radiant energy given out by an object and picked up by a remote sensor Fig. (7.4). Spectral radiance (L) is defined as the

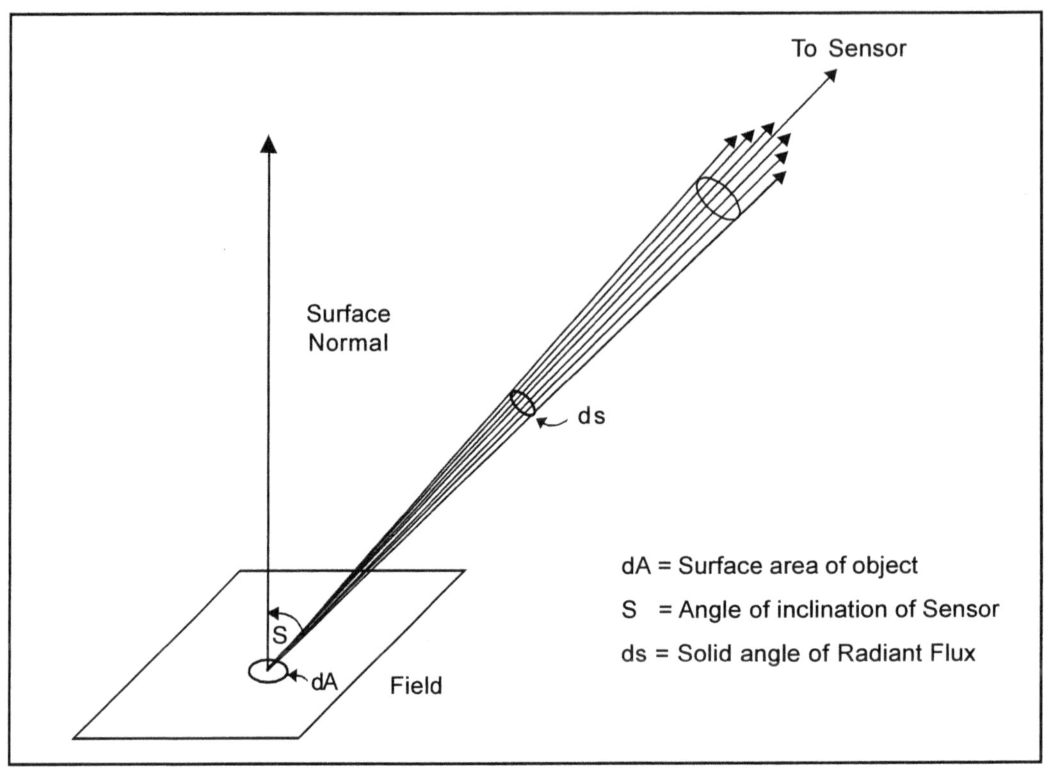

Fig. 7.4 Radiant Flux Received by a sensor.

energy within a wavelength band radiated by a unit area per unit solid angle of measurement. Radiance in a single band is calculated by using the following formula (Robinove, 1982) :

$$\text{Radiance } (L_\lambda) = (D_n/D_{max})(L_{max} - L_{min}) + L_{min} \qquad \text{..........(7.1)}$$

where, D_n = digital value of a pixel from the computer-compatible tape (CCT).

D_{max} = maximum digital number recorded on the CCT = 127 for bands 1, 2, 3 and 4 of IRS.

L_{max} = maximum radiance measured at detector saturation in mW cm^{-2} S_r^{-1}.

L_{min} = minimum radiance measured at detector saturation in mW cm^{-2} S_r^{-1}.

The maximum and minimum values of spectral radiance for LISS-1, LISS-IIA and LISS-IIB sensors of IRS-IA IRS-1B satellites are given in Table 7.2 (IRS hand book, 1989).

Table 7.2 Maximum and minimum values of spectral radiance for LISS I, LISS II sensor of IRS IA, IRS IB

	IRS-IA						IRS-IB					
BAND	LISS I		LISS IIA		LISS IIB		LISS I		LISS IIA		LISS IIB	
	L_{min}	L_{max}	L_{min}	L_{max}	L_{min}	L_{max}	L_{min}	L_{max}	L_{min}	L_{max}	L_{min}	L_{max}
1.	0	15.667	0	16.644	0	14.069	0	16.888	0	14.512	0	15.131
2.	0	24.466	0	22.811	0	22.653	0	17.826	0	30.675	0	24.353
3.	0	16.100	0	20.422	0	18.018	0	16.645	0	14.783	0	15.141
4.	0	15.831	0	16.418	0	16.445	0	16.972	0	13.290	0	14.929

Computation of Reflectance

Unlike radiance which is a measure of radiant energy, reflectance is an energy ratio. It is a function of radiance and is defined by the following formula :

$$\text{Reflectance} = \pi\ (\text{Radiance})\ /E \sin \alpha \qquad \text{..........(7.2)}$$

where, E = irradiance in mW cm^{-2} at the top of atmosphere, and

α = solar elevation angle available in the header file of CCT.

Cosmetic Operations

Two topics are considered in this section. The first is the correction of digital images containing either partially or entirely missing scan lines. The second is the correction of images because of destripping of the imagery. For example, Landsat MSS has six detectors for each band. When it scans, it scans six lines at a time per band. It is expected that all the six detectors should record similar signatures from the same object on the earth surface. But, in practice, it was noticed that the sensitivity of all the detectors is not uniform. This means sometimes detector recorded irradiance (reflected) for the same object may differ. This causes the stripping in remote sensing imagery. The second phenomenon is called line drop. A similar situation may also be common in Landsat TM and IRS system. For modern linear array detectors like IRS LISS-I and II sensor stripping may be along the column instead of line, since the scanning mechanism is different from the Landsat MSS and TM sensor systems. Some of the radiometric corrections like dropouts, line stripping and random noise removal are discussed below.

The line dropout or missing scan line is usually overcome by replacing the zero value by the mean values of the pixels of the previous and the following line. For example, if the10th line is the dropout line, then all the pixels in the 10th line will be replaced by the corresponding mean of the 9th and the 11th line. The stripping of the remote sensing data (MSS) can be corrected with the help of the detector's master calibration curve. This is discussed by Bernstein and Ferneyhough (1975), Bernstein (1983) and Jenson (1983). However, the simplistic method which is widely practised in the digital technique is as follows :

$$Y_k(i, j) = \frac{s}{s_k} \times [X_k(i, j) - M_k] + M \qquad \text{..........(7.3)}$$

where $Y_k(i, j)$ = Output pixel gray value

$X_k(i, j)$ = Input pixel gray value

M = Mean of the full image

M_k = Mean of the k^{-th} detector

s_k = Standard deviation of the k^{-th} detector

s = Standard deviation of the full image

Random Noise Removal

Image noise is any unwanted disturbance in image data that is due to limitations in the sensing and data recording process. The random noise problems in digital data are characterised by nonsystematic variations in gray levels from pixel to pixel called bit errors. Such a noise is often referred to as being 'spiky' in character and it causes images to have a 'salt and pepper' or snowy appearance. Bit errors are handled by recognising that noise values normally change much more abruptly than true image values. Thus, noise can be identified by comparing each pixel in an image with its neighbours. If the difference between a given pixel value and its surrounding values exceeds an analyst specified threshold the pixel is assumed to contain noise. The noisy pixel value can then be replaced by the average of its neighbouring values. Moving window 3 × 3 or 5 × 5 pixel are typically used in such procedures. Fig. 7.5 and Fig. 7.6 illustrates just one of many noise suppression algorithms using such a window.

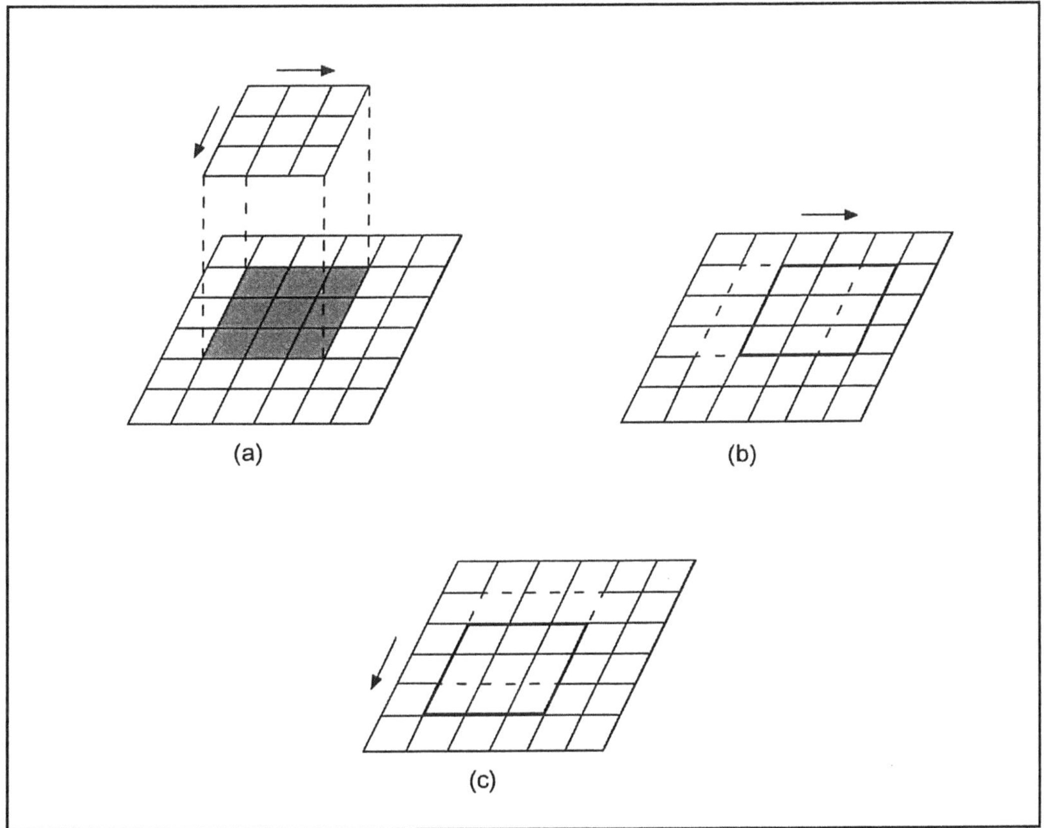

Fig. 7.5 The moving window concept : projection of 3 × 3 pixel window in image being processed; movement of window from line to line.

DN_1	DN_2	DN_3
DN_4	DN_5	DN_6
DN_7	DN_8	DN_9

$AVE_3 = (DN_1 + DN_3 + DN_7 + DN_9)/4$

$AVE_8 = (DN_2 + DN_4 + DN_5 + DN_8)/4$

$DIFF = |AVE_A - AVE_8|$

$THRESH = DIFF \times WEIGHT$

IF : $|DN_5 - AVE_3|$ or $|DN_5 - AVE_8| > THRESH$

THEN : $DN'_5 = AVE_8$ OTHERWISE $DN'_5 = DN_5$

Fig. 7.6 Typical noise correction algorithm employing a 3 × 3 pixel neighbourhood.
Note : "WEIGHT" is an analyst-specified weighting factor. The lower the weight, the greater the number of pixels considered to be noise in an image.

7.3.3 Atmospheric Correction Methods

As discussed in chapter 3, according to Rayleigh scattering, the effect of scattering is inversely proportional to the fourth power of wavelength of energy, that is, scattering is more in the lower wavelength (visible) than in the higher wavelength (infrared band). Further scattering effect increases the signal value (bias). Let us consider a case in which the energy in various wavelengths is interacting with the earth surface features, for example, water body. Assume that the sky is very clear and there is no atmospheric scattering or haze. If the sky is clear with no scattering, then the radiance reflected from the earth surface feature in any of the region of the electromagnetic spectrum should be the same. This is the ideal case. In reality, because of the presence of haze, fog, or atmospheric scattering, there always exists some kind of unwanted signal value called bias.

The bias is the amount of offset for each spectral band. Bias can be determined by regressing the visible band vs. infrared bands. First of all, it is essential to identify some areas like an airport, deep homogenous non-turbid water bodies (Jenson, 1983) and some shadow area in the scene. The brightness values of all these features from each band are then extracted. The gray value of the visible band (for example, IRS-band 1) is plotted against corresponding values at the same pixel location in the infrared band (IRS-band 4). The plot will result in a scatter diagram (Fig. 7.7). Following the regression analysis a regression line is to be fitted. It is assumed that if the data is free from atmospheric scattering, the best fitting line should pass through the origin. However, this rarely happens. The line generally intercepts the X-axis (Band 1). The X-intercept determines the bias caused by the atmospheric effect. The procedure is to be repeated for all the visible bands.

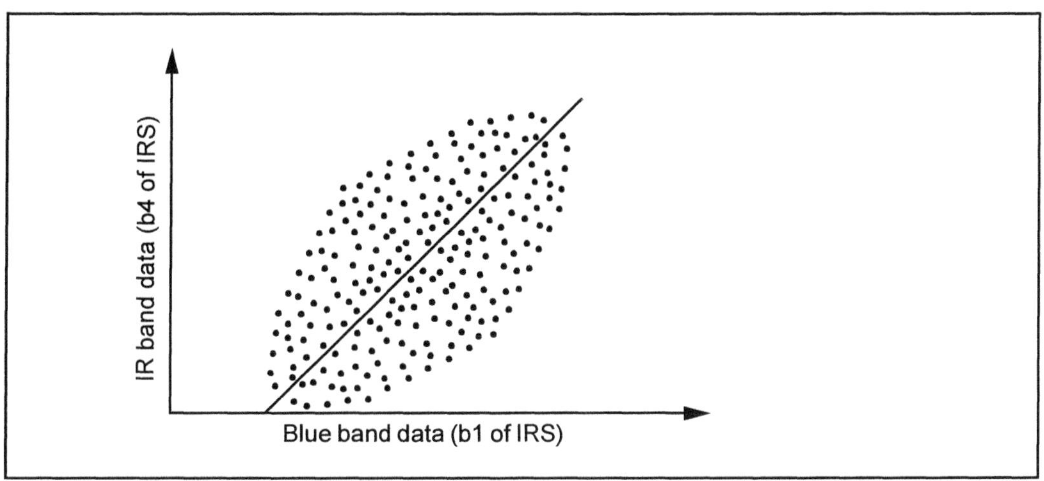

Fig . 7.7 Plot of Band 4 (IR) Vs Band (visible)

7.4 Image Registration

Image registration is the translation and rotation alignment process by which two images/ maps of like geometries and of the same set of objects are positioned coincident with respect to one another so that corresponding elements of the same ground area appear in the same place on the registered images (Haralick, 1973). This is often called image-to-image registration. One more important concept with respect to geometry of satellite image is rectification. Rectification is the process by which the geometry of an image area is made planimetric (Haralick, 1973). It may not remove the distortion caused by topographic replacement in images. This process almost always involves relating GCP pixel coordinates (row and column) with map coordinate counterparts. This is the most precise geometric correction since each pixel can be referenced not only by its row and column in a digital image matrix after rectification is completed, but it is also rigorously referenced in degrees, feet or meters in a standard map projection. Whenever accurate data, direction and distance measurements are required, geometric rectification is required. This is often called as an image-to-map rectification. The fundamental concept of registration process is explained in the following paragraphs (Anji reddy, 1995). To obtain the pixel data of a point on the ground, it is necessary to convert the latitude and longitude of the ground point measured on the survey of India toposheet to its corresponding line number and pixel number on the digital Image. The procedure used for the extraction of digital data for all bands is given below :

7.4.1 Conversion of Geographical Coordinates to Conical Orthomorphic Coordinates

Generally, countries like India, having a large extent in the North-South direction, use Conical Orthomorphic Projection (COP) to convert the geographical coordinates on the ground to the corresponding coordinates on the satellite imagery (Rampal, 1982). For the conversion, the geographical latitudes (ϕ_P) and longitudes (L_P) of the 4 corners of the corresponding satellite imagery (Fig. 7.8) are obtained from the annotation record stored in CCT. The origin O is chosen outside the imagery. The Conical Orthomorphic Coordinates of a point on the map are computed using the following equation (Rampal, 1982).

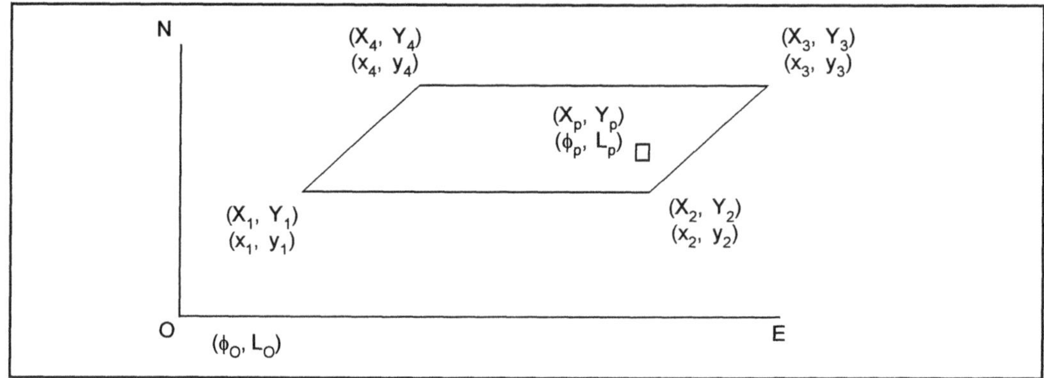

Fig. 7.8 Transformation of geographical coordinates to line number and pixel number on satellite imagery.

$$Y = (P - m') \text{ Sin } \gamma \quad \text{..........(7.4)}$$

$$X = m' + Y \tan(\gamma/2) \quad \text{..........(7.5)}$$

where, $$\gamma = \Delta L \sin \phi_0 \quad \text{..........(7.6)}$$

$$\Delta L = L_P - L_0$$

where (ϕ_P, L_P) are coordinates of the origin of the ground coordinate system and 'ΔL' is the distance computed by measuring the difference between the coordinates of any point P and the origin on the map.

$$m' = m \frac{m^3}{6R_0N_0} + \frac{m^4 \tan \phi_0}{24R_0N_0^2} + \frac{m^5(5+3\tan^2\phi_0)}{120R_0N_{0_0}} \quad \text{..........(7.7)}$$

where $$R_0 = \frac{a(1-e^2)}{(1-e^2\sin^2\phi_0)^{3/2}} \quad \text{..........(7.8)}$$

$$N_0 = \frac{a}{\left(1 - e^2 \sin^2 \phi_0\right)^{1/2}} \qquad \text{..........(7.9)}$$

$m = R_m (\phi_p - \phi_0)$

$P = N_0 C_0 + \phi_0$

and, a = major axis of the reference ellipsoid

b = minor axis of the reference ellipsoid

$e^2 = 1 - b^2/a^2$ ....(7.10)

The Everest ellipsoid is adopted for India and the values of a, b and e for this ellipsoid are (Rampal, 1992),

a = 6377277.6 m

b = 6356075.0 m

e = 1/300.8

R_0 is the radius of curvature of the earth and N_0 is the normal to the surface at ϕ_0.

7.4.2 Transformation of Conical Orthomorphic Coordinates to Digital Imagery Coordinates

The transformation from the conical orthomorphic coordinates of any point to the corresponding line number (x) and pixel number (y) is made by the affine transformations.

$X = A_1x + B_1y + C_1$ ....(7.11)

$Y = A_2x + B_2y + C_2$

where, (X, Y) are the conical Orthomorphic Coordinates and (x, y) are the corresponding line number and pixel number on the IRS imagery. A_1, B_1, C_1 and A_2, B_2, C_2 are constants, whose values are to be determined from a minimum of 3 ground control points. The 3 points are any three corners of the imagery, whose (X, Y) coordinates are computed from the above equations. The line number and the pixel number of the 3 points are known since they are the corner of the scene, and scene of LISS II contains 2048 lines and 2008 pixels per each line (Fig. 7.9). The set of equations used for computing A_1, B_1, C_1 and A_2, B_2, C_2 are :

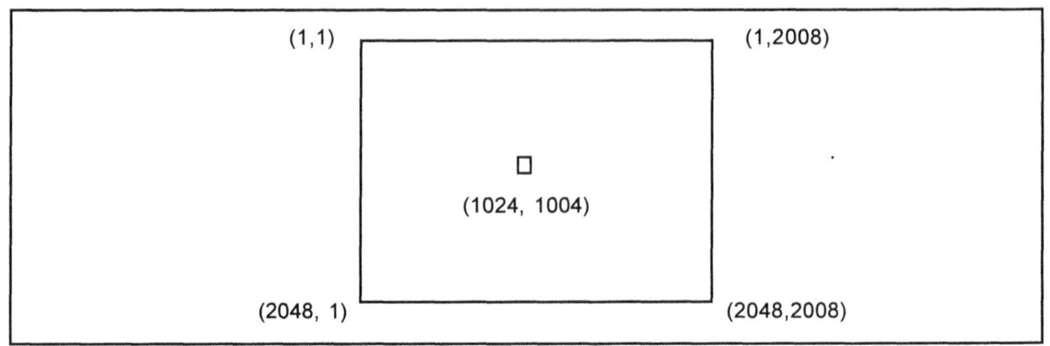

Fig. 7.9 Definition of corner coordinates on the satellite scene in terms of line number and pixel number

For corner point - 1:

$$\left.\begin{aligned} X_1 &= A_1 x_1 + B_1 y_1 + C_1 \\ Y_1 &= A_2 x_1 + B_2 y_1 + C_2 \end{aligned}\right] \qquad \text{..........(7.12)}$$

For corner point - 2 :

$$\left.\begin{aligned} X_2 &= A_1 x_2 + B_1 y_2 + C_1 \\ Y_2 &= A_2 x_2 + B_2 y_2 + C_2 \end{aligned}\right] \qquad \text{..........(7.13)}$$

For corner point - 3

$$\left.\begin{aligned} X_3 &= A_1 x_3 + B_1 y_3 + C_1 \\ Y_3 &= A_2 x_3 + B_2 y_3 + C_2 \end{aligned}\right] \qquad \text{..........(7.14)}$$

These three equations are solved using the Gauss elimination method. The computed values of the constants are checked at the 4th corner of the imagery. The line number and pixel number (x_P, y_P) of any point on the imagery can be found by solving the equations :

$$\left.\begin{aligned} X_P &= A_1 x_p + B_1 y_P + C_1 \\ Y_p &= A_2 x_P + B_2 y_P + C_2 \end{aligned}\right] \qquad \text{..........(7.15)}$$

where, (X_p, Y_P) are the conical orthomorphic coordinates of the point to be determined. After obtaining the line number and pixel number of each point in this manner, the digital data in the four spectral bands can be extracted from CCT or satellite image for any given project area. The above equations are used to specify how to determine the distorted image positions corresponding to correct or undistorted and map positions. During the geometric correction process, we first define an undistorted output matrix of empty map cells and then fill in each cell with the grey level of corresponding pixel or pixels in the distorted image (Lillesand and Kiefer, 2000). After producing transformation function (Eq. 7.15) a process called resampling is used to determine the pixel values to fill into the output matrix as determined on the basis of the pixel values which surround its transformed position in the original input matrix.

7.5 Image Enhancement Techniques

Low sensitivity of the detectors, weak signal of the objects present on the earth surface, similar reflectance of different objects and environmental conditions at the time of recording are the major causes of low contrast of the image. Another problem that complicates photographic display of digital image is that the human eye is poor at discriminating the slight radiometric or spectral differences that may characterise the features. The main aim of digital enhancement is to amplify these slight differences for better clarity of the image scene. This means digital enhancement increases the separability (contrast) between the interested classes or features. The digital image enhancement may be defined as some mathematical operations that are to be applied to digital remote sensing input data to improve the visual appearance of an image for better interpretability or subsequent digital analysis (Lillesand and Keifer, 1979). Since the image quality is a subjective measure varying from person to person, there is no simple rule which may produce a single best result. Normally, two or more operations on the input image may suffice to fulfil the desire of the analyst, although the enhanced product may have a fraction of the total information stored in the original image. This will be realized after seeing the different contrast enhancement techniques in this chapter. There are a number of general categories of enhancement techniques. As in many outer areas of knowledge, the distinction between one type of analysis and another is a matter of personal taste and need of the interpreter.

In remote sensing literature, many digital enhancement algorithms are available. They are contrast stretching enhancement, ratioing, linear combinations, principal component analysis, and spatial filtering. Broadly, the enhancement techniques are categorised as point operations and local operations. Point operations modify the values of each pixel in an image data set independently, whereas local operations modify the values of each pixel in the context of the pixel values surrounding it. Point operations include contrast enhancement and band combinations, but spatial filtering is an example of local operations. In this section, contrast enhancement, linear contrast stretch, histogram equalisation, logarithmic contrast enhancement, and exponential contrast enhancement are considered.

7.5.1 Contrast Enhancement

The sensors mounted on board the aircraft and satellites have to be capable of detecting upwelling radiance levels ranging from low (oceans) to very high (snow or ice). For any particular area that is being imaged it is unlikely that the full dynamic range of the sensor will be used and the corresponding image is dull and lacking in contrast or overbright. In terms of the RGB model, the pixel values are clustered in a narrow range of grey levels. If this narrow range of gray levels could be altered so as to fit the full range of grey levels, then the contrast between the dark and light areas of the image would be improved while maintaining the relative distribution of the gray levels. It is indeed the manipulation of look-up table values. The enhancement operations are normally applied to image data after the appropriate restoration procedures have been performed. The most commonly applied digital enhancement techniques will be considered now.

The sensitivity of remote sensing detectors was designed to record a wide range of terrain brightness from black asphalt and basaltic rocks to white sea ice under a wide range of lighting conditions. In general, few of the scenes have the full brightness range (Sabin, 1986). To produce an image with optimum contrast ratio, it is inevitable to utilize the entire dynamic range (Jenson, 1986). Digital contrast enhancement is thus of prime importance. The objective of contrast stretching is to expand the narrow dynamic range of gray values (digital numbers) typically present in an input image (Fig. 7.10) over a wide range of gray values for the desired output image. A variety of contrast stretching algorithms are available and are broadly categorized as linear contrast stretching and non-linear contrast stretching.

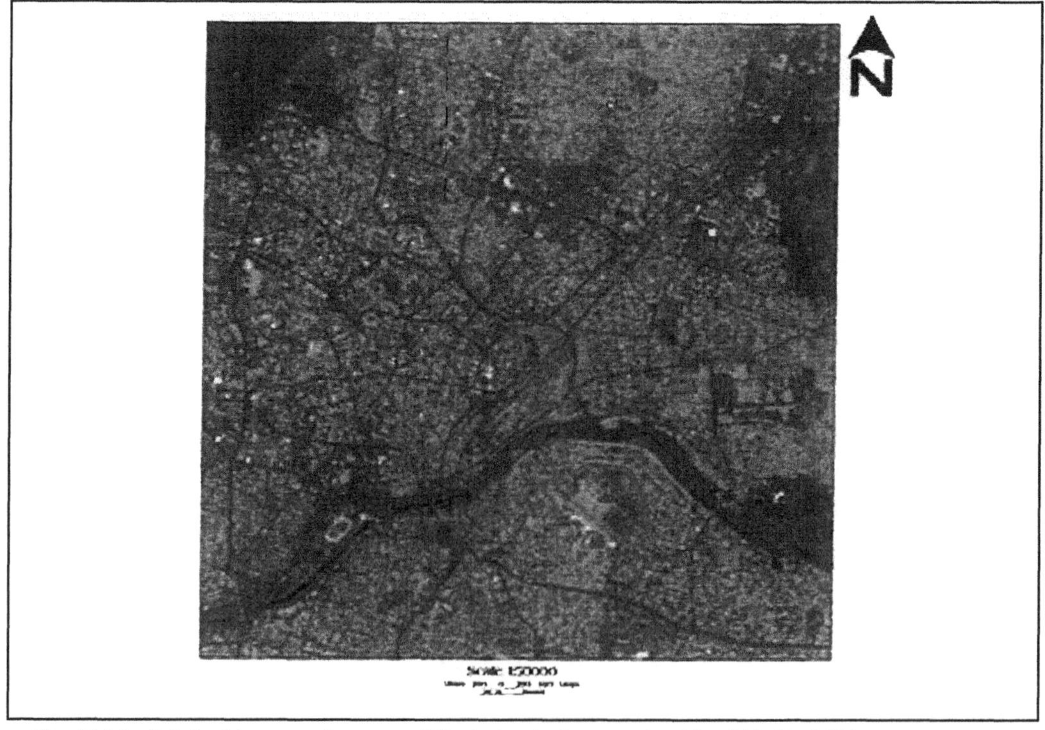

Fig. 7.10 Original image of a part of Hyderbad city acquired by IRS-ID PAN sensor.

In linear contrast stretch, the digital number (DN) value in the lower end of the original histogram is assigned to digital number zero (DN = 0) that is, extremely black, and a value at the higher end is assigned to extremely white (that is DN = 127). The intermediate values are interpolated between 0 and 255 by following a linear relationship, as given below (Fig. 7.11).

$$Y = a + bX \qquad \text{..........(7.16)}$$

Where X and Y are the input gray value of any pixel and output gray value of the same pixel. a and b are intercept and slope respectively.

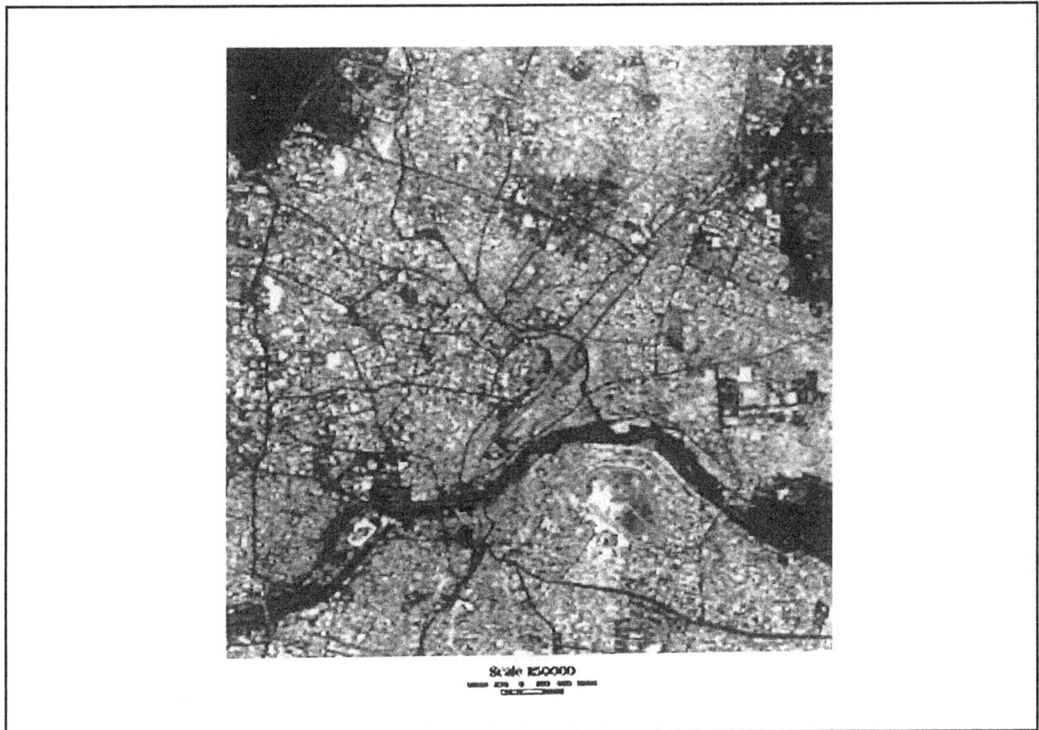

Fig. 7.11 Image after applying the linear contrast enhancement technique.

The linear contrast stretch is widely used to improve the contrast of most of the original brightness values, but there is a loss of contrast of the extreme high and low ends at the tail of the histogram (Sabin, 1986). Another advantage of digital enhancement is that the analyst can enhance any portion of the dynamic range of the histogram to achieve the desired results, since the digital image is flexible. The logarithmic contrast enhancement is very much useful for non-linear contrast enhancement. Here the output pixel grey values (Yij) will be generated from input pixel grey values (Xij) following some logarithmic expressions, as follows :

$$Yij = a \log (Xij) + b \qquad \text{..........(7.17)}$$

The coefficients 'a' and 'b' are determined by taking the maximum and minimum grey values of the input image and the corresponding maximum and minimum values in the output image. This transformation only highlights the features lying in the darker region of the histogram (Hall, 1979; Jenson, 1986). Fig. 7.12 shows the enhanced

Fig. 7.12 Image output of logarithmic contrast Enhancement.

image shown in Fig. 7.10 using logarithmic enhancement algorithm. The advantages and important characteristics of logarithmic contrast enhancement (Hall, 1978) are; (i) It makes the low contrast details more visible by enhancing low contrast edges; (ii) It provides a contrast signal to noise ratio; (iii) It some what matches the responses of the human visual systems; (iv) It usually provides a more equal distribution of gray values and (v) It transforms multiplicative noise into additive noise.

Exponential contrast enhancement is also considered as a non-linear contrast enhancement. The grey values (Xij) in the input image transform to the gray values (Yij) in the output image is as follows:

$$Yij = a\, e^{bxij} + c \qquad \text{..........(7.18)}$$

Where a, b and c are constants, b is arbitrarily chosen between 0.01 and 0.1 to avoid higher value of e. Further, 'a' and `b' scale the dynamic range of the grey values of the output image with 0 and 255. The effect of the exponential transformation on the edges in an image is to compress low contrast edges, while expanding high contrast edges. This generally highlights the feature having higher grey values. The effect is just reverse of logarithmic contrast transformation. This technique generally produces an image with less visible detail than the original and is thus of limited use for image with less visible detail than the original and is thus of a limited use for images enhancement (Hall, 1978).

Histogram equalization is widely used for contrast manipulation in digital image processing because it is very simple for implementation. It needs minimum information from the analyst. In this enhancement, the original histogram has been readjusted to produce a uniform population density of pixels along the horizontal grey value (DN) axis. Gonzalez and Wintz (1977) demonstrated the technique for the modification of the digital number (DN) or grey values from the original image. It involves in two steps: firstly, it computes the histograms of the original image and the cumulative frequency density percentage. Secondly, computation of transformation function based on which the contrast manipulation takes place in the output scene. An improvement of contrast of the image Fig. (7.11) can be seen in the Fig. (7.13). In histogram equalisation method, the shape as well as the extent of the histogram is taken into consideration. Its underlying principle is straight forward. It is assumed that each histogram class in the displayed image must contain an approximately equal number of pixel values, so that the histogram of these displayed values are uniform throughout the classes. If this is done the entropy of the image, a measure of information content, is increased. The method involves

(a) the calculation of the target number of pixel values in each class of equalised histogram. i.e.,

$$\text{Target number of } n_t = \frac{\text{total number of pixel values}}{\text{total number of classes}}$$

(b) the histogram of the input image is converted to cumulative form with the number of pixels in classes 0 to j and is denoted by Cj i.e.,

$$C_j = n_0 + n_1 + n_2 + \ldots.. + n_j$$

n_j = number of pixel values taking greyscale value j.

Then the output level for class j is simply (C_j/n_t) if $j_0 > n_t$; then j_0 or less should be assigned in output as 0.

This procedure can be repeated for all the classes by converting the old look-up-table values to corresponding new look-up table values. Note that the smaller classes in the input image have been amalgamated, reducing the contrast. The number of pixels allocated to each non-empty class varies considerably because discrete input classes can not logically be split into sub-classes. (Fig. 7.13)

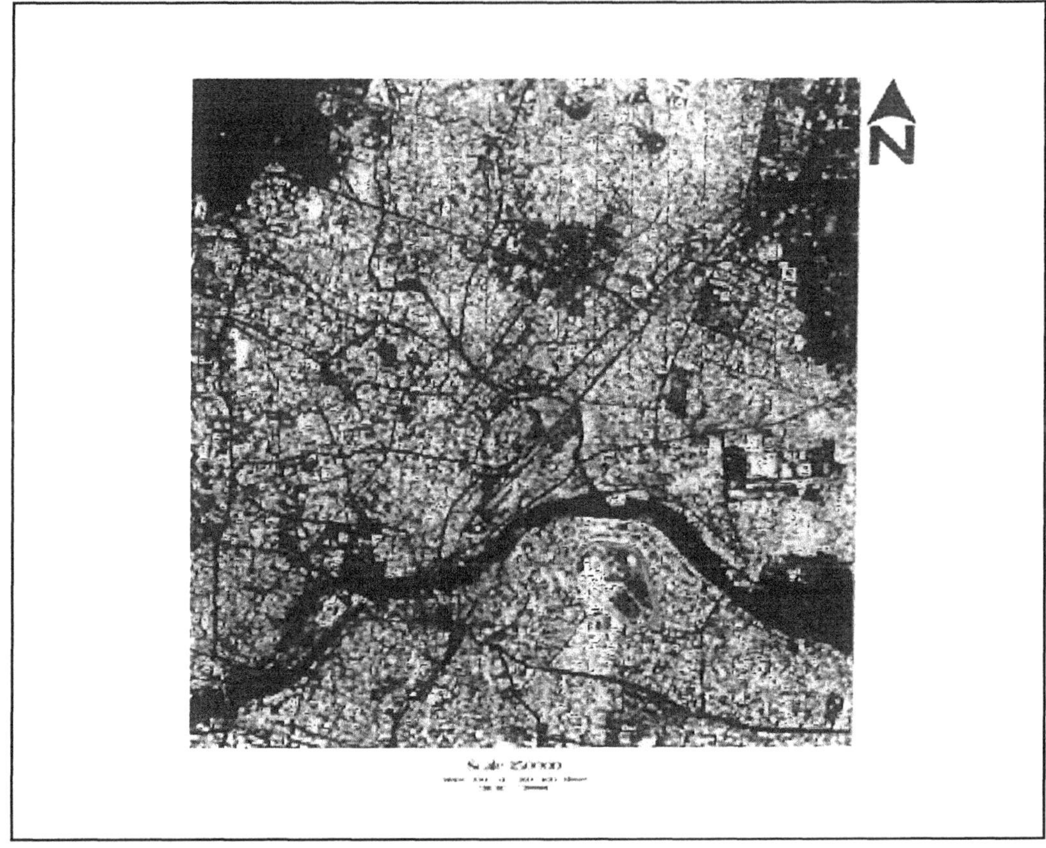

Fig. 7.13 Image output of Histogram Equalisation.

The other method of contrast enhancement based on the histogram of input pixel values and their manipulation is called Gaussion stretch. This is called Gaussion stretch because it involves fitting of Normal Histogram to Gaussion Histogram. i.e, for example,

$$f(x) = Ce^{-\sigma x2}$$

where, $$C = \left(\frac{\sigma}{\pi}\right)^{0.5}$$

σ the standard deviation range of x for which f (x) drops by a factor of $e^{-0.5}$ or 0.607 of its max. value and the Max. Value = $\frac{1}{(2a)^{0.5}}$.

Thus 60.7% of the values of a normally distributed variable lie within one standard deviation of the mean. In this method of enhancement each pixel value of input image can be converted to the LUT value based on the probability of each pixel value with respect to a class following the Gaussian law. The normal distribution curve is shown in Fig. 7.14. In both the cases of contrast enhancement based on histogram analysis of input image values, the range of levels allocated to the output image exceeds the range of levels of pixel values in the input image. This results in the overall brightening of the displayed image. (Gonzalez and wintz, 1983)

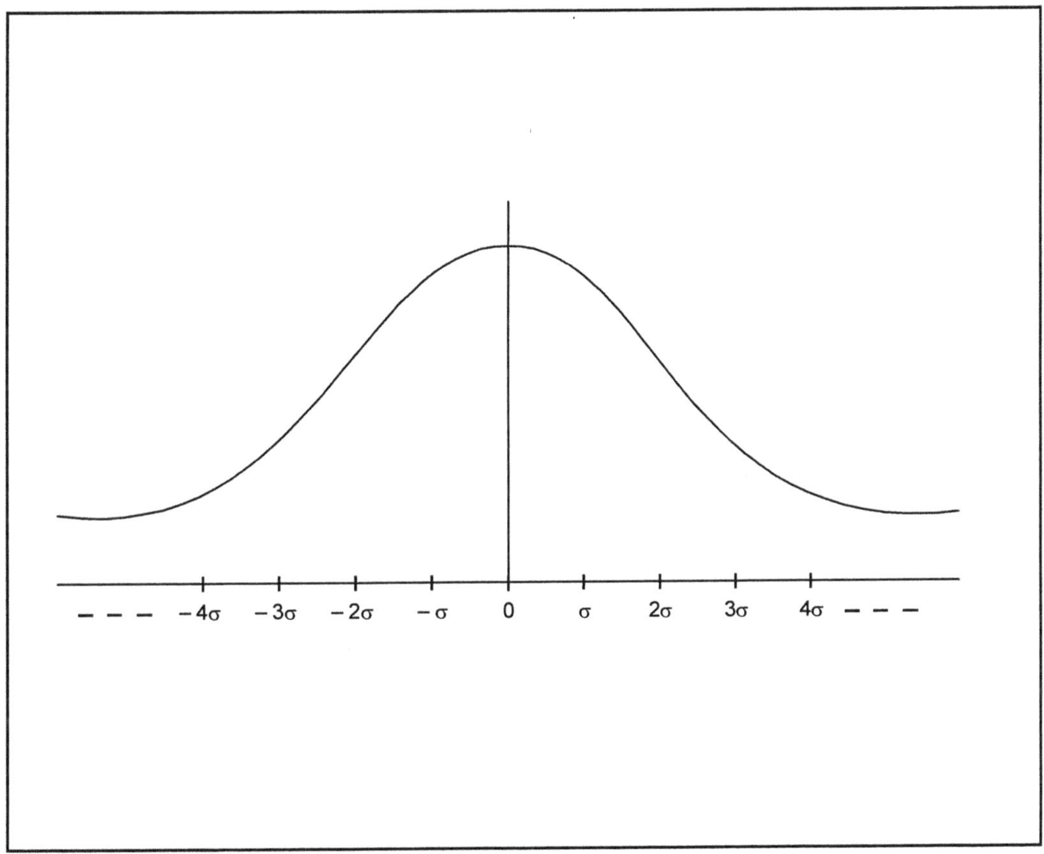

Fig. 7.14 Normal distribution.

7.6 Spatial Filtering Techniques

A characteristic of remotely sensed images is a parameter called spatial frequency, defined as the number of changes in brightness values per unit distance for any particular part of an image. If there are few changes in brightness value over a given area it is termed as a low frequency area. If the brightness values changes dramatically over very short distances, this is called high frequency area. Algorithms which perform image enhancement are called filters because they suppress certain frequencies and pass (emphasise) others. Filters that pass high frequencies while emphasising fine detail and edges called high frequency filters, and filters that pass low frequencies called low frequency filters.

Filtering is performed by using convolution windows. These windows are called mask, template filter or kernel. In the process of filtering, the window is moved over the input image from extreme top left hand corner of the scene. The discrete mathematical function transforming the original input image digital number to a new digital value. First it will move along the line. As soon as the line is complete , it will restart for the next line for covering the entire image (Sabins, 1976 ; Lillisand and Kiefer, 1986).

The mask window may be rectangular (1 x 3, or 1 x 5 pixels) size or square (3 x 3, 5 × 5 or 7 × 7 pixels size). Each pixel of the window is given a weightage. For low pass filters all the weights in the window will be positive and for high pass filter all the values may be negative or zero, but the central pixel will be positive with higher weightage value. In the case of high pass filter the algebraic sum of all the weights in the window will be a zero. Many types of mask windows of different sizes can be designed by changing the size and varying weightage within the window . The simplest form of mathematical function performed in filtering operation is neighbourhood averaging. Another commonly used discrete function is to calculate the sum of the products given by the elements of the mask and the input image pixel digital numbers of the central pixel digital number in the moving window. Some of the most commonly used filtering techniques are explained below.

7.6.1 Low Pass Filters

Fig. 7.15(a) shows the cross section before applying the filter and Fig. 7.15(b) shows the cross section after the application of Low Pass Filter. From plot (Fig. 7.15 (a)) it is inferred that the raw data has a random noise, whereas the other lines can provide some information with greater clarity. This is achieved after the application of low pass filters. There are a number of low pass filters, such as, mean filter, median

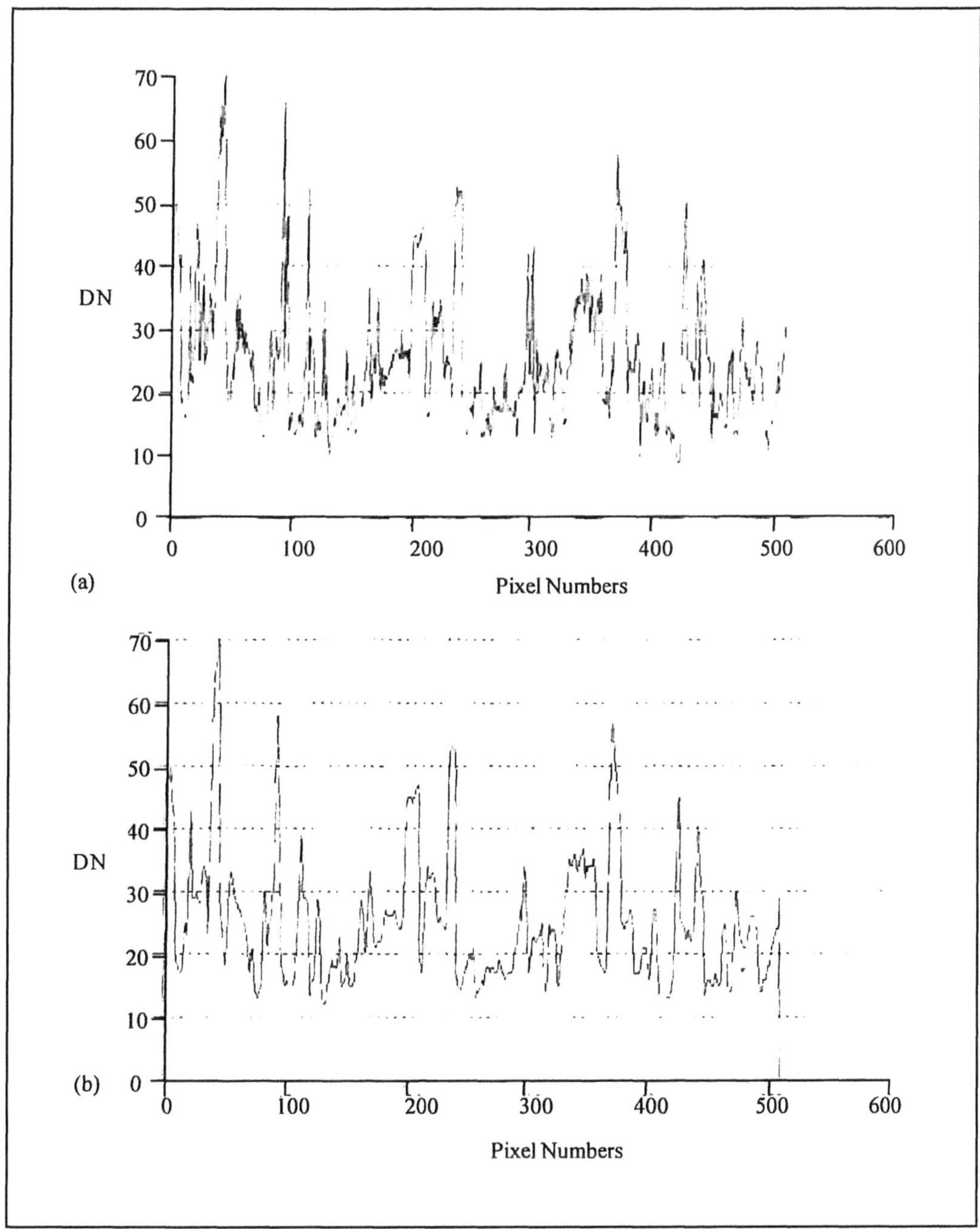

Fig. 7.15 (a) Cross-section from the top left to the bottom right corner of the band 7 image shown in Fig. 7.3 (b) Cross-section between the same points as used in (a) after the application of a smoothing filter.

filter, and adaptive filters, and the principle behind all these filters are discussed in this section. Low pass filtering is used by Crippen (1989), Eliason and Mc Ewen (1990) to remove banding effects on remotely sensed images while Dale et. al (1996) use a low-pass filter to smooth the image in the process of registration with the technique of mean filtering each pixel is sequentially examined and if the pixel digital number (DN) is greater than the average brightness (DN) of its surrounding pixels by some threshold (t), it is replaced by the means of the 3 x 3 pixels window. The window may be of any size, such as 5 x 5, 7 x 7 and so on. The larger the window size, the more will be computational time. The following window demonstrates the comparison of computation time of varying window. The central pixel 'X' has no correlation with the neighbouring pixels and is replaced by the mean value of surrounding pixels.

a_1	a_2	a_3
a_4	X	a_5
a_6	a_7	a_8

$$x = \frac{1}{8}\sum_{a_i=1}^{8} \text{ if } \left(X - \frac{\Sigma a_i}{8}\right) > t \qquad \text{..........(7.19)}$$

Eight raw data values (a_i) centered on point summed an averaged to produce one output value (x). For the output image values from a matrix that has fewer rows and columns than the input image, the unfiltered margin corresponds to the rows and columns that can not be reached. Generally these missing scanlines and columns are filled with zeros in order to keep the input and output images of the same size. The effect of the moving average filter is to reduce the overall variability of the image and lower its contrast. Fig. 7.16(a) shows the output of mean filter. The main problem of mean filter is that the resultant image becomes blurred. So the edges between the features becomes fuzzy (Nag and Kudrat, 1998).

Median Filter

An alternative type of smoothing filter utilises the median of the neighbourhood rather than the mean. The median filter is generally thought to be superior to the moving average filter for two reasons. First, the median of a set of 'n' numbers is always one of the data values present in the set, when n is an odd integer. Secondly, the median is less sensitive to errors or to extreme data values. Let us consider a set of nine pixel values in the digital image, (2, 1, 25, 7, 28, 5, 8, 30, 82) (neighbourhood) of 3 × 3 window, thus the median is the central value when the data are ranked in an

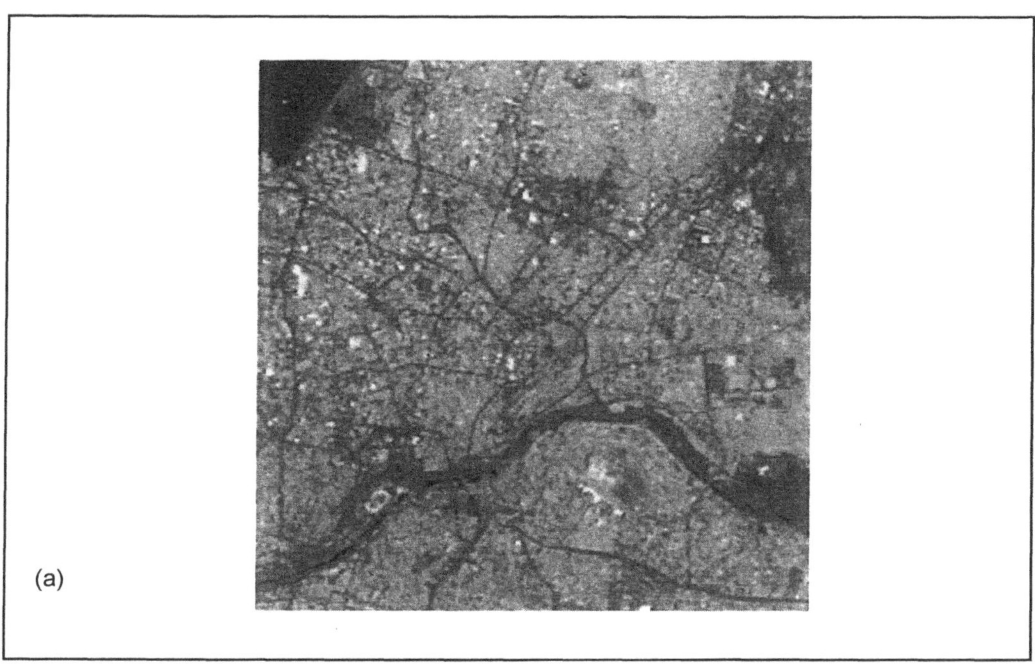

(a)

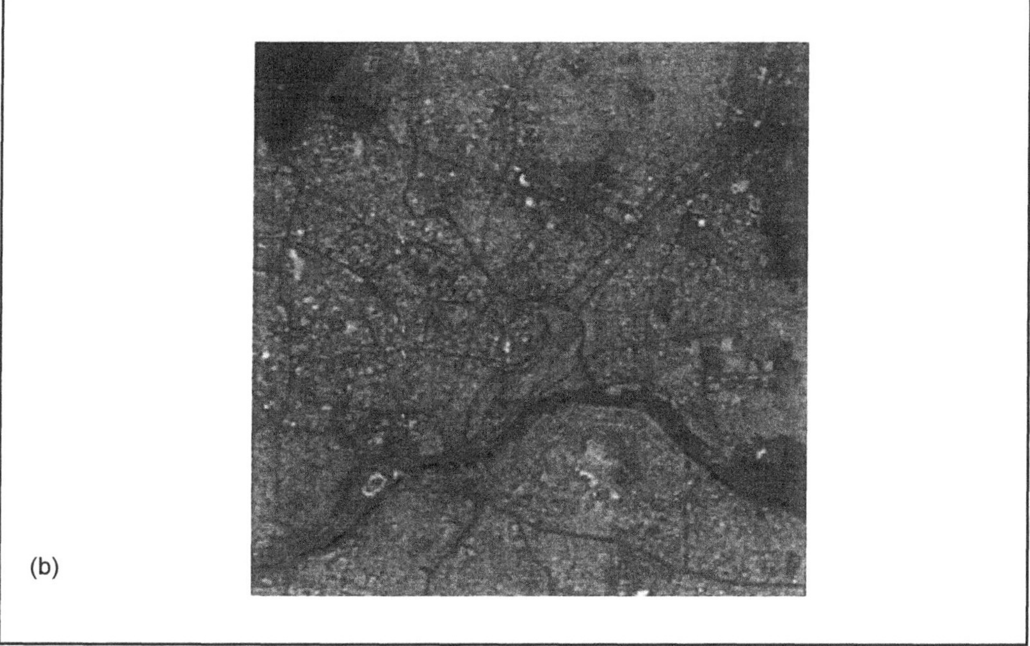

(b)

Fig. 7.16 Output of mean and median filters.
(a) Mean filtered image
(b) Median filtered image

ascending or descending order of magnitude. In this example the ranked values are { 1, 2, 5, 7, 8, 25, 28, 30, 82} giving a median value of 8. The mean 33.10, would be rounded up to the value of 33. The value 33 is not present in the original data, unlike the median value 8. Also the mean value is larger than 8 observed values and may be thought to be unduly influenced by the extreme data values, which might represent noise, are removed by the median filter. The median filter preserves edges better than a mean filter. Fig. 7.16(b) shows the output of a median filter.

The median filtering is an alternative approach to spatial averaging. In this filtering technique, the concept is very much similar to averaging. Here, the abnormal or spatially uncorrelated pixel digital count (x) will be replaced by the median within the 3 x 3 window. This method eliminates isolated values related to noise spikes, while giving rise to lower resolution loss than with averaging. Both the median and mean filter image irrespective of the variability of grey levels. The smoothing methods in which the filter weights are calculated for each window position, and the calculations being based on the mean and variance of the grey levels in the area of the image underlying the window, are called adaptive filters.

7.6.2 High Pass Filters

A simple high pass filter may be implemented by subtracting a low pass filtered image (pixels by pixel) from the original, unprocessed image. The high frequency component image enhances the spatial detail in the image at the expense of the large area brightness information. High pass filtering can be performed by means of image subtraction method or derivative based methods.

It is well known that an image can be considered to be the sum of its low and high frequency components. The low frequency image can be subtracted from the original, unfiltered leaving behind the high-frequency component. The resulting image can be added back to the original. Therefore, it can be effectively doubling the high frequency part. Thomas et. al (1981) developed and used model as,

$$I^* = I - fI' + C \qquad \text{..........(7.20)}$$

where, I^* = filtered pixel value

I = original pixel value

I' = average of the window

f = a proportion vary from 0 to 1

C = a constant.

This method is called image subtraction method. The second method in high pass filtering is a derivative based method and is based on mathematical concept of the derivative. The derivative of a continuous function is the rate of change of that function at a point (Mather, 2000).

7.6.3 Filtering for Edge Enhancement

Edge or boundary between two features is very important for separating the different features. Edge is characterized by high frequencies. In nature, edge between two features or objects may not always be distinct. Thus edge enhancement or edge crispening is required for better interpretation of an image. Edge enhancement filtering technique enhances the edge only. High frequencies signify the large degree of tonal variation within a small area or within a small spatial distance. This refers to those pixels that occur at the transition between two categories. The edge crispening filtering technique is typically employed to exaggerate the edges between contrasting cover type or bring out linear trend of geologic significance (Pratt, 1978; Sabins, 1986). It is also useful in mapping lineaments and drainage pattern. This can be accomplished by a number of filtering techniques. In this section, three of the many filters are discussed. These three filtering techniques are discrete convolution filtering, Laplacian edge enhancement filtering and Robert edge enhancement filtering. Fig. 7.17 shows the output of various high pass filters applied on the Hyderabad image recorded by IRS imaging system.

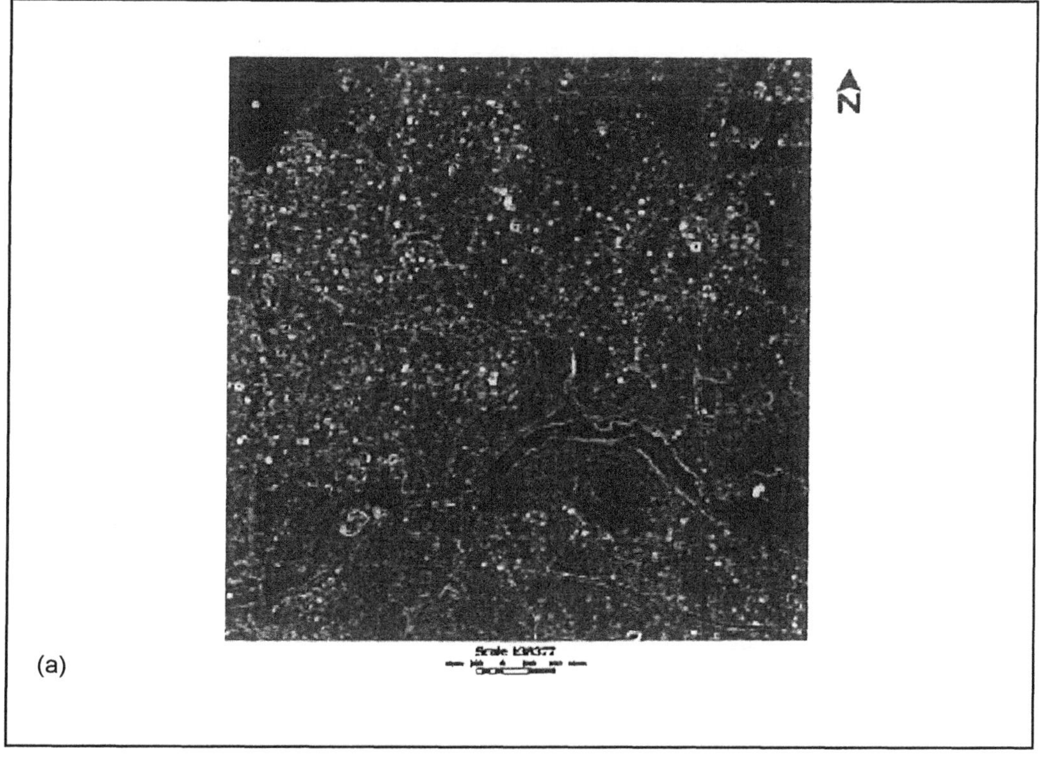

(a)

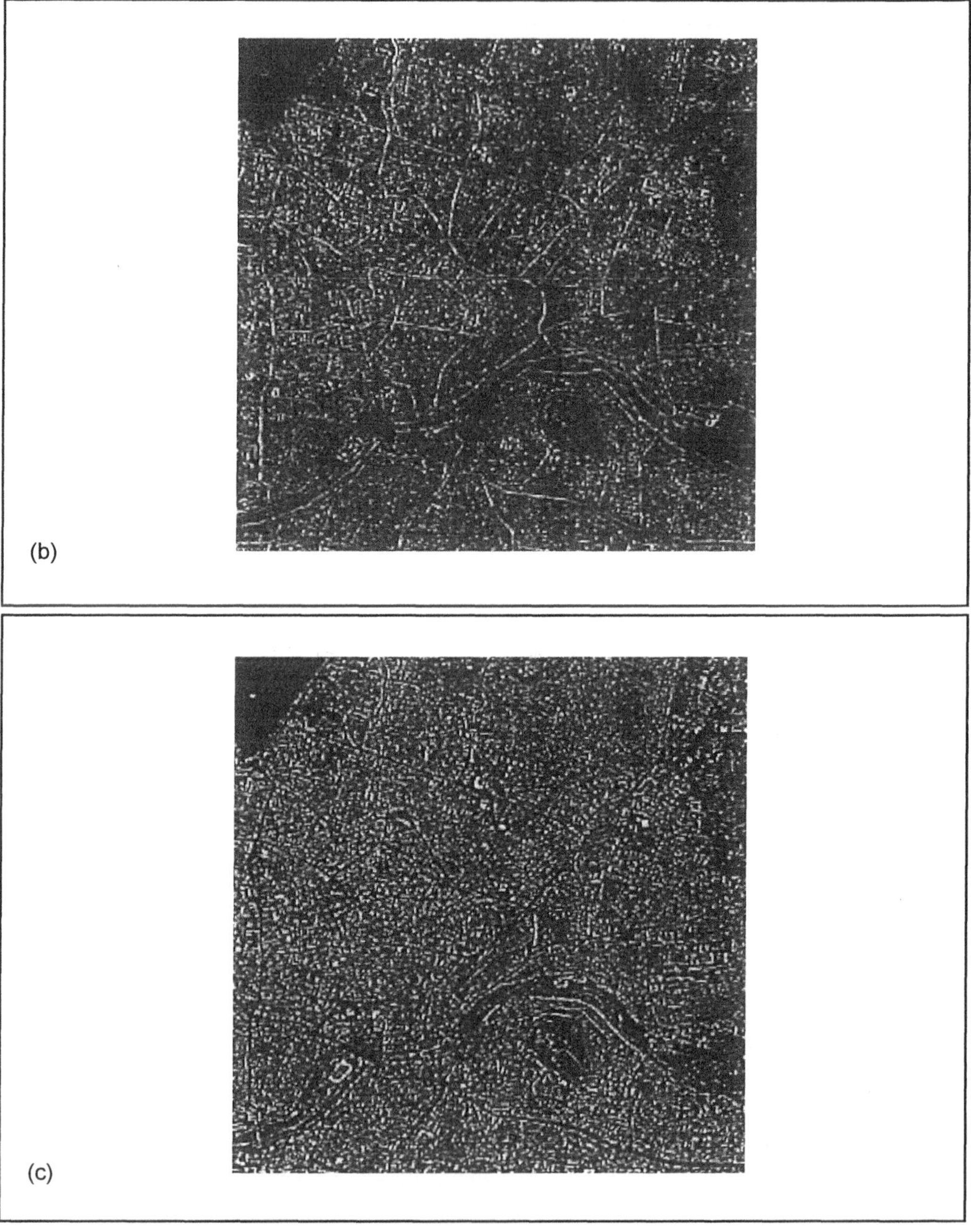

Fig. 7.17 Output of various highpass filters
(a) Discrete Convolution
(b) Laplacian Edge Filter
(c) Robert Edge-Enhancement Filters.

In indiscrete convolution filtering, the input pixel array can be modified by using the following Eq. (7.21) to generate output pixel array:

$$Y(m_{1,} m_2) = åå F_{ij} \times H \qquad ....(7.21)$$

where, F_{ij} = Input pixel array of an image.

$Y(m_l, m_2)$ = output image pixel array of an image

H = Convolution array (high pass filtering as shown below).

$$H = \begin{bmatrix} 0 & -1 & 0 \\ -1 & 5 & -1 \\ 0 & -1 & 0 \end{bmatrix} \text{ or}$$

$$H = \begin{bmatrix} 1 & -2 & 1 \\ -2 & 5 & -2 \\ 1 & -2 & 1 \end{bmatrix} \text{ or}$$

$$H = \begin{bmatrix} -1 & -1 & -1 \\ -1 & -9 & -1 \\ -1 & -1 & -1 \end{bmatrix}$$

Output image will be generated from the input image in steps. First all the digital number of an input image will be multiplied with the corresponding values in the mask (H) and then the algebraic sum of all the multiplied values will replace the central pixel of the window or the mask.

For example, Input image within 3 x 3 window :

F_{ij}			H			out image		
a_1	a_2	a_3	0	−1	0	x	x	x
a_4	a_5	a_6	−1	5	−1	x	y	x
a_7	a_8	a_9	0	−1	0	x	x	x

where, $$Y = 5a_5 - (a_2 + a_4 + a_6 + a_8) \qquad ....(7.22)$$

Then in the output pixel a_5 will be replaced by Y. The important point to note is that the sum of all elements in the mask (H) window is unity, whereas in other high pass filters like Laplacian operation the sum is a zero. Here since the sum of all the coefficient in the mask matrix is equal to unity, the digital number would have been retained at the same value as in the original, while the exchange takes place at the edges (Galbaiti, 1990) only.

The edge enhancement can be performed using Laplacian Edge filtering technique which uses the Laplacian operation on 3 x 3 neighbourhood. It can be explained as follows:

Input image	Filter or Mask	Output image
a_1 a_2 a_3	0 −1 0	a_1 a_2 a_3
a_4 X a_5	−1 4 −1	a_4 Y a_5
a_6 a_7 a_8	0 −1 0	a_6 a_7 a_8

The central pixel (Y) of the output image is $4x - (a_2 + a_4 + a_5 + a_7)$. This is conceptually very much similar to earlier convolution filtering operation. The only operating window is different. The sum of all the elements within the filtering window is a zero . Hence it is a kind of high pass filtering. Basically the Laplacian filtering operation computes the differences between the digital counts of the central pixel and the average of the DN values of four adjacent pixels in the horizontal and vertical location. The above equation can be explained as

$$Y = (X - a_4) + (X - a_5) + (X - a_2) + (X + a_7) \qquad \text{..........(7.23)}$$

So the output image is nothing but the sum of the partial differences in the horizontal and vertical pixels within the operator (Hall, 1978; Gonzalez and Wintz., 1983). In Laplacian filtering operation edges at any direction will be highlighted and it is sometimes called 'omnidirectional' filtering operation. This operation is very much sensitive to noise and this filtering technique is also used for removal of noise.

Robert edge enhancement filtering technique is generally used to highlight the edges between the features. In this operation the central pixel of 3 x 3 sub-image will be replaced by some out put digital number. This operation uses the diagonal derivative to estimate gradient at the point. It highlights any feature extending along the diagonal of the scene. This is very useful for geologic lineament analysis and lithology (Sabins, 1986). All the edge enhancement-filtering techniques are to ascertain the boundaries between agricultural fields, lakes, forests and other terrain regions

7.7 Image Transformations

The term 'transform' is used a little loosely in this chapter, for the arithmetic operators of addition, subtraction, multiplication, and division are included although they are not strictly transformations. All the transformations in image processing of remotely sensed data allow the generation of a new image based on the arithmetic operations, mathematical statistics and fourier transformations. The new image or a composite image is derived by means of two or more band combinations, arithmetics of various band data individually and/or application of mathematics of multiple band data. The resulting image may well have properties that make it more suited to a particular purpose than the original. For example, the division of band 1 and band 2 (B1/B2) and data collected by the NIMBUS satellite sensor (CZCS), may be used to detect the chlorophyll (a) (Prasad et. al 1997). Near infrared and red bands of an image set are widely used as a vegetation index, as an attribute of vegetative cover, with a particular biomass and green leaf area index of the area covered by the image. Table 7.3 shows individual bands, band ratios and band combinations used by different investigators for the determination, estimation and mapping of water quality parameters.

The most widely used technique based on the statistical principles is principal component analysis (PCA) which is a method of re-expressing the information content of multispectral image set of 'm' images in terms of 'm' principal components. These components have two particular properties : a zero correlation between the 'm' principal components, and maximum variance. The analysis of principal components is useful in generating false colour composite image. The third transformation is Fourier transform, which provides for the representation of image data in terms of a coordinate frame work that is based upon spatial frequencies rather than upon distance from an image. This transform is very useful in designing filters for special purposes and in colour-coding the scale components of the image.

In this section, two transformations which are commonly used are presented since they are of a particular relevance to remote sensing community. More specifically these two techniques are Normalised Difference Vegetation Index (NDVI) transformation and Principal Components Analysis (PCA) transformation.

7.7.1 NDVI Transformation

The remote sensing data is used extensively for large area vegetation monitoring. Typically the spectral bands used for this purpose are visible and near IR bands. Various mathematical combinations of these bands have been used for the computation of NDVI, which is an indicator of the presence and condition of green vegetation. These mathematical quantities are referred to as Vegetation Indices. There are three such indices , Simple Vegetation Indices, Rational Vegetation Indices, and Normalised Differential Vegetation Indices.

These indices are computed from the equations

$$\left.\begin{aligned} VI &= NEAR\ IR - RED \\ RVI &= RED/NEAR\ IR \\ NDVI &= (NEAR\ IR - RED)/(NEAR\ IR + RED) \end{aligned}\right] \qquad ....(7.24)$$

For example, of for the four bands of IRS 1C LISS III digital data, the bands corresponding to Red and Near IR are B2 and B3 respectively, the above equation becomes

$$NDVI = (B3 - B2)/(B3 + B2) \text{ of IRS 1C LISS III DATA} \qquad ....(7.25)$$

This formula is applied and the values of NDVI can be computed for the entire image of the project area.

Land surfaces are characterised by a high degree of spatial heterogeneity of surface cover. These types have associated references in emmissivity, thermal properties and reflectance characteristics (Mathews and Rossow, 1987). Satellite studies show that the soil moisture and vegetation parameter like vegetation status, leaf area density and photo synthetic activity are not independent. Naturally, vegetated surfaces can fill moisture from within the soil and may show signs of stress rather than adjacent arable land. Therefore there is a close relationship between meteorologic drought indicators and satellite based indices of vegetation activity (Walsh, 1987).

Photo synthetically active vegetation typically has a reflectance of < 20% in the narrow band visible (0.5 – 0.7μm) but a much higher reflectance upto 60% in the near IR (0.7 – 1.3 mm). It can be referred to from Fig. 7.18 as the generalised spectral

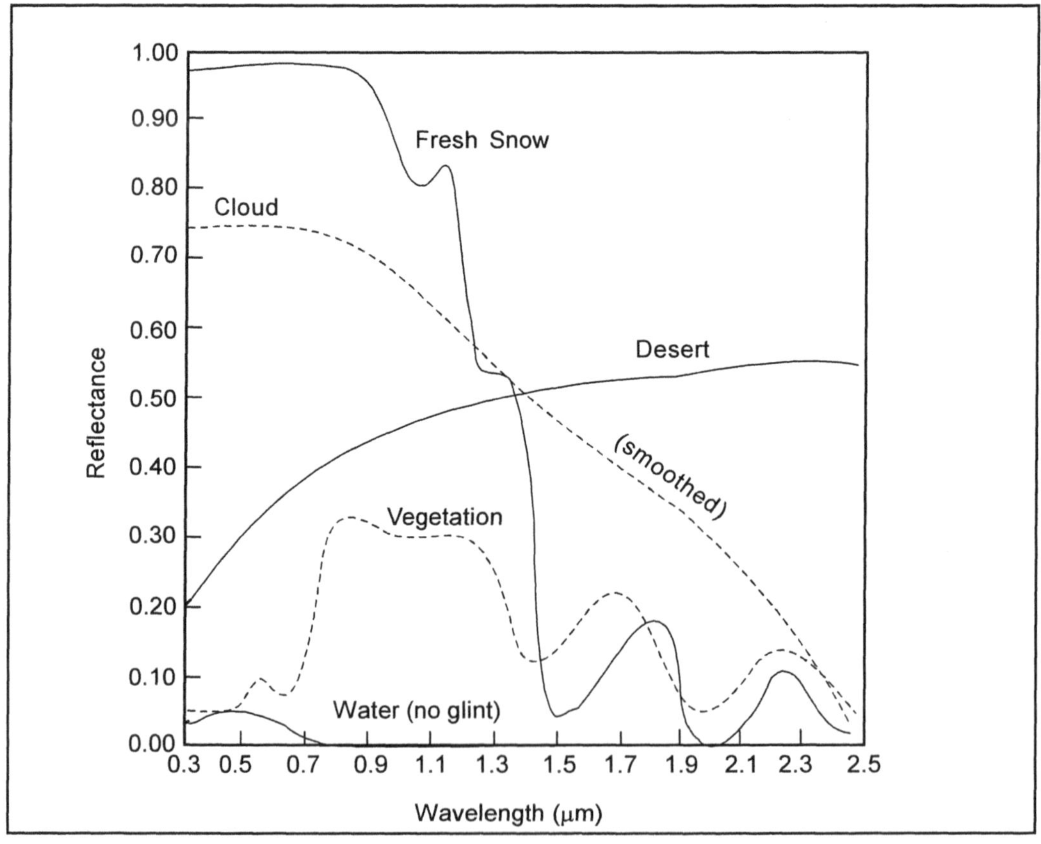

Fig. 7.18 Generalised spectral reflectance curves of five different targets in the 0.3-2.5 μm (visible, near-IR) region. The curve for clouds does not consider particular types. (From Davis et al., 1984.)

reflectance curves of five different targets in the 0.3 – 25 μm (visible and near IR regions). These positions of the electromagnetic spectrum almost correspond with IRS 1C LISS III channels of two and three. An NDVI (Tarpley et. al 1984) can be derived which capitalised on the spectral sensitivities of different land cover types. Thus the NDVI can be quantified by using,

NDVI = (Band3 – Band2)/(Band3 + Band2) of IRS 1C LISS III DATA

This NDVI is bounded ratio that ranges between –1 to +1. Clouds, water and snow have negative NDVI since they are more reflective in visible than near IR wave lengths. Soil and rock have a broadly similar reflectance giving NDVI close to '0'. Only active vegetation has a positive NDVI being typically between about 0.1 and 0.6 values at the higher end of the range indicating increased photosynthetic activity and a greater density of the canopy (Tarpley et. al 1984). This is explained by applying or transforming the digital numbers/spectral reflectance values/pixel values of NDVI using the formula given above for a part of Hyderabad city by adopting the following algorithm:

(i) Computation of NDVI values for the entire study area by conversion of spectral reflectance values into NDVI values.

(ii) Conversion of these NDVI values to a scaled channel values by using density-slicing method that measures apparent reflectance to sensor values.

(iii) Display of image with NDVI and creation of a legend keeping the threshold values and the ranges that are shown in the Plate 4.

Plate (5) is a NDVI image for the Hyderabad city obtained on a clear day, March 22, 1998. Most of the variance in the data occurred in the agricultural land, forest and along the Musi river grasslands. A proportionally small variance exists in the other features like builtup wastelands.

The greenness range is divided into discrete classes by slicing the range of NDVI values into seven ranges by fixing the thresholds. A cursory examination of pseudo colour image of NDVI and classified output of Land use/Land cover (Plate 5) reveals that along the water bodies, builtup lands yield negative values, their reflectance being more visible than near IR wavelengths. All the agricultural lands located around Yenkapalle, Jallapalli, Charlapalli, and some parts of the forest areas yield moderate values of NDVI. Other areas like Musi river grasslands and some forests yield high index values because of the increased photosynthetic activity and greater density of canopy. Areas like Ameerpet, Sanatnagar, Balanagar, and Kukatpally in Hyderabad city have values closer to '0' values (–0.1 to 0.1) because in red and near IR wavelength regions the reflectance is the same giving NDVI close to '0'.

7.7.2 PCA Transformation

Principal components analysis, also referred to as PCA, has been proven to be of a significant value in the analysis of remotely sensed digital data. The application of PCA on raw remote sensing data produces a new image which is more interpretable than the original data. The main advantage of PCA is that it may be used to compress the information content of a number of bands into just two or three transformed principal component images. To perform PCA we apply a transformation to a corrected set of multispectral data. Adjacent bands in multispectral remotely sensed images are generally correlated. Multiband visible / near infrared images of vegetated areas will show negative correlation between the near-infrared and visible bands and positive correlations among visible bands (Fig 7.19). This is due to spectral characteristics

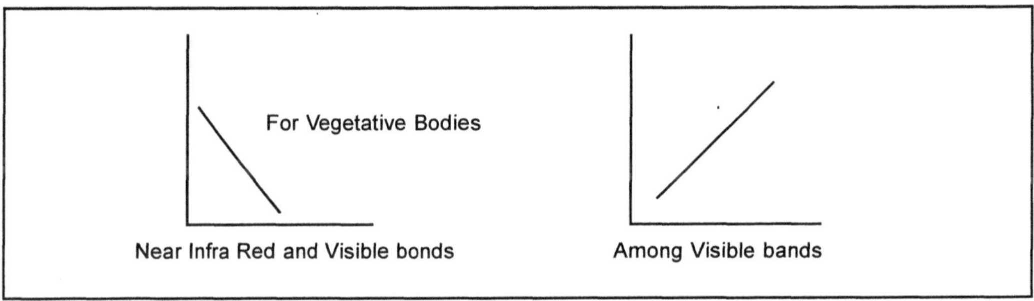

Fig. 7.19 Correlations.

of vegetation and its greenness. The presence of correlation among the bands of multispectral images implies that there is redundancy in the data. Some information is being repeated. It is repetition of information between the bands. It can be illustrated as follows. Let us select two variables (spectral bands) x and y, and assume that these two variables are perfectly correlated as in Fig 7.20.

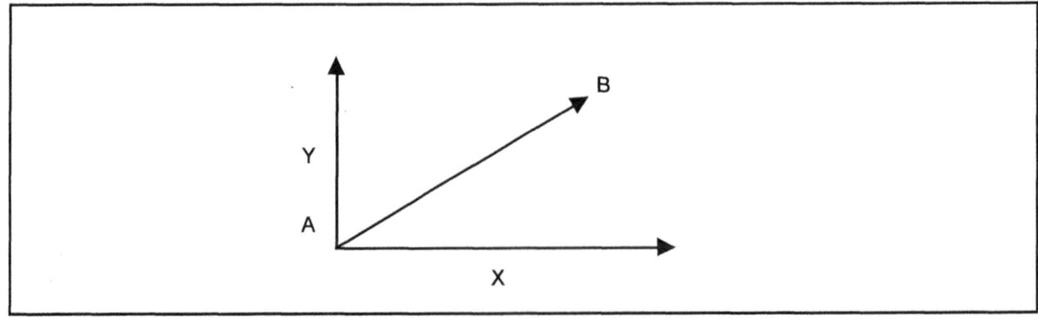

Fig. 7.20 Perfect Correlation.

Since these two variables are correlated, the resultant will be a straight line. Let us name it 'AB'. Instead of analysing two variables 'x' and 'y', if we analyze the line 'AB' we can get more information conveyed by that line which is a dimensional axis. If these two variables are not perfectly correlated there may be a dominant direction of variability. This dominant direction of variability is chosen as a major axis and the second minor axis is perpendicular to this dominant direction. A plot in Fig. 7.21 using the axes AB and CD rather than the conventional axes 'x' and the variability along the CD direction is very small, so it can be neglected. The use of a single axis AB rather than both x and y axes accomplishes the reduction in the size of the dataset since a single coordinate axis AB replaces the separate 'x' and 'y' axes. The information conveyed by the set of coordinates on the axis AB is greater than the information conveyed by the set of measurements on 'x' or 'y' alone. Here, the data means the multispectral data and variability means the variance or scatter about the mean. If we have two variables, there we can have one dimensionality, that is, data sets or bands < number of spectral bands. At this stage, we can know the purpose of PC Analysis, which is to define the number of dimensions that are present in a dataset and to fix the coefficients which specify the positions of that set of uncorelated axes. Therefore the PC transform is to define the dimensionality of the data-set and, identify the principal axes of variability within the data. These properties of PC analysis may be used in evaluating remote sensing data to group the pixels of one category (land cover types) with respect to the new axes system rather than in terms of original spectral bands.

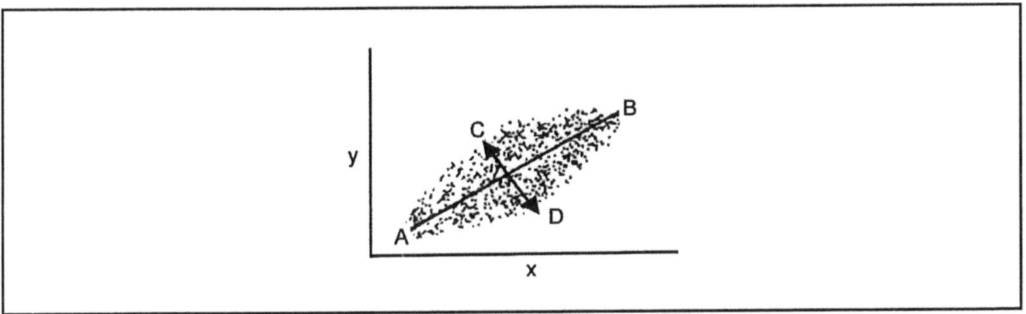

Fig. 7.21 Rotated Coordinate axis used in PCA.

If we use two band data, the principal components can be computed graphically. If the variables or spectral bands are more than two, the geometrical solution is impracticable. Hence we have to follow the algebraic solutions which are beyond the scope of this book. The algebraic method involves the computation of variance-covariance matrix, correlation matrix and a confusion matrix. If the data is 'p' – dimensional, then we have to compute the variance-covariance matrix with 'p' variables or spectral bands. The relationship between the correlation matrix and the variance-covariance matrix leads to form a confusion matrix. In general the terms 'S' and 'R' are used for variance-covariance matrix and the correlation matrix respectively. If there are 'p' bands, each matrix contains 'p' rows and 'p' columns. Each direction in 'p' dimensional space will give the components. This is a feature on the earth surface. These component image data can be computed from eigen values and eigen vectors computed from the matrices S and R, raw image data and corrected data, (Jensen, 1986). Fig. 7.22 shows the principal components of IRS LISS bands 1-4 of a part of Hyderabad city. These principal components image is derived from the matrix of correlations among the LISS bands.

Fig. 7.22 Transformed data resulting from principal component analysis of IRS-ISS data.

7.8 Image Classification

Image classification is a procedure to automatically categorize all pixels in an image of a terrain into land cover classes. Normally, multispectral data are used to perform the classification of the spectral pattern present within the data for each pixel is used as the numerical basis for categorization. This concept is dealt under the broad subject, namely, Pattern Recognition. Spectral pattern recognition refers to the family of classification procedures that utilises this pixel-by-pixel spectral information as the basis for automated land cover classification. Spatial pattern recognition involves the categorization of image pixels on the basis of the spatial relationship with pixels surrounding them. Image classification techniques are grouped into two types, namely supervised and unsupervised. The classification process may also include features, such as, land surface elevation and the soil type that are not derived from the image. A pattern is thus a set of measurements on the chosen features for the individual to be classified. The classification process may therefore be considered a form of pattern recognition, that is, the identification of the pattern associated with each pixel position in an image in terms of the characteristics of the objects or on the earth's surface.

7.8.1 Supervised Classification

A supervised classification algorithm requires a *training sample* for each class, that is, a collection of data points known to have come from the class of interest. The classification is thus based on how "close" a point to be classified is to each training sample. We shall not attempt to define the word "close" other than to say that both geometric and statistical distance measures are used in practical pattern recognition algorithms. The training samples are representative of the known classes of interest to the analyst. Classification methods that relay on use of training patterns are called supervised classification methods. The three basic steps (Fig. 7.23) involved in a typical supervised classification procedure are as follows :

(i) *Training stage* : The analyst identifies representative training areas and develops numerical descriptions of the spectral signatures of each land cover type of interest in the scene.

(ii) *The classification stage :* Each pixel in the image data set is categorised into the land cover class it most closely resembles. If the pixel is insufficiently similar to any training data set it is usually labeled 'Unknown'.

(iii) *The output stage* : The results may be used in a number of different ways. Three typical forms of output products are thematic maps, tables and digital data files which become input data for GIS. The output of image classification becomes input for GIS for spatial analysis of the terrain. Fig. 7.24 depicts the flow of operations to be performed during image classification of remotely sensed data of an area which ultimately leads to create database as an input for GIS. Plate 6 shows the land use/ land cover colour coded image, which is an output of image classification.

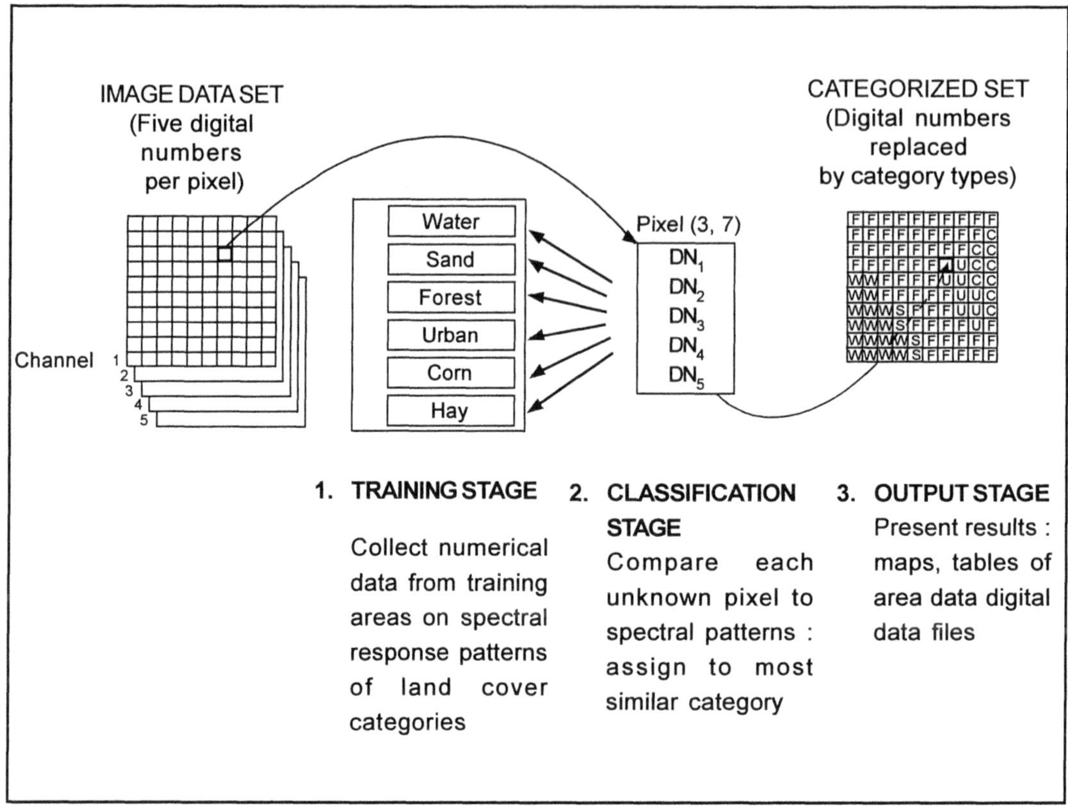

Fig. 7.23 Basic steps supervised classification.

There are a number of powerful supervised classifiers based on the statistics, which are commonly, used for various applications. A few of them are a minimum distance to means method, average distance method, parallelepiped method, maximum likelihood method, modified maximum likelihood method, Baysian's method, decision tree classification, and discriminant functions. The principles and working algorithms of all these supervised classifiers are available in almost all standard books on remote sensing and so details are not provided here. Moreover, in the following section 6.9, a detailed study on the operation/application of maximum likelihood classifier on the study is made of performance analysis of IRS bands for classification of Land use/ Land cover patterns on a part of Hyderabad city. Since all the supervised classification methods use training data samples, it is more appropriate to consider some of the fundamental characteristics of training data.

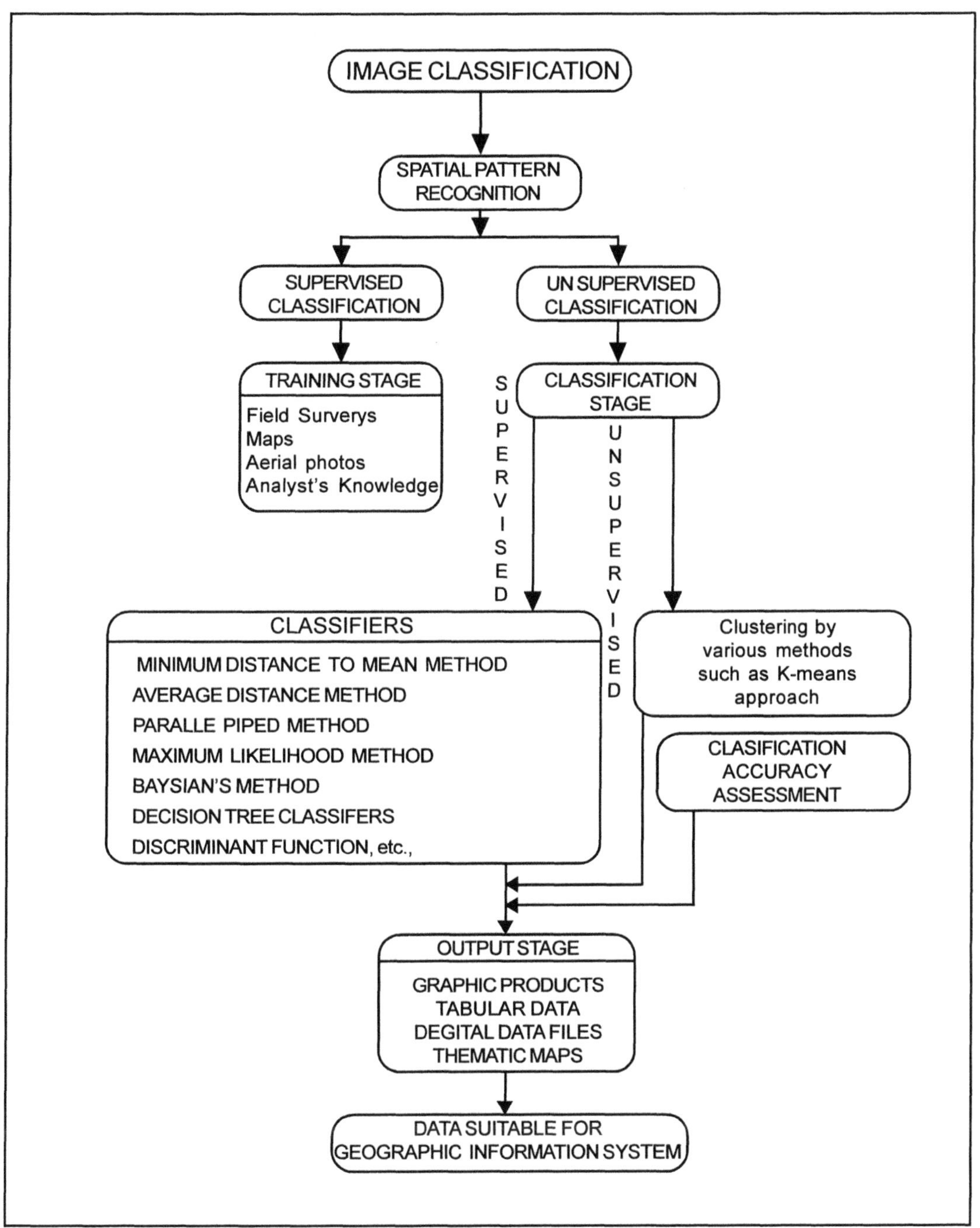

Fig. 7.24 Flow Chart showing Image Classification.

7.8.2 Training Dataset

A training dataset is a set of measurements (points from an image) whose category membership is known by the analyst. This set must be selected based on additional information derived from maps, field surveys, aerial photographs, and analyst's knowledge of usual spectral signatures of different cover classes. Selecting a good set of training points is one of the most critical aspects of the classification procedure. However, an attempt is made to provide some guidelines based on the author's experience. These guidelines are as following :

(i) Select sufficient number of points for each class.

If each measurement vector has N features, then select N+1 points per class and the practical minimum is 10*N per class.

If the class shows a lot of variability (the scatter plot showing considerable spreading or scatter among training points), select a larger number of points, subject to practical limits of time, effort and expense. The more the training points, the better the "extra points" to evaluate the accuracy of the classifier. The more the points, the more accurate the classification will be.

(ii) Select training data sets which are representative of the classes of interest that show both typical average feature values and a typical degree of variability. For each class, select several training areas on the image, instead of just one. Each training area should contain a moderately large number of pixels.

Pick training areas from seemingly heterogeneous or appearing regions. Pick training areas that are widely and spatially dispersed, across the full image.

For each class, select the training areas which are uniformly distributed across the image and with high density.

(iii) Check that selected areas have unimodel *distributions* (histograms). A bimodal histogram suggests that pixels from two different classes may be included in the training sample.

(iv) Select training sets (physically) using a computer-based classification system :

Poorest method **:** Using coordinates of training points or training regions directly.

Better method : using joystick, trackball, lightpen, directly on the image. for example. EASI/PACE :

The program should show the histograms, mean and standard deviations for each region selected, and for each class in total.

The Program should allow to iterate; do classification using one set of training points, then come back and modify training sets and class definitions without starting all over again. There should be options to combine classes from previous classification.

(v) The program should allow one to designate half of the points as training points, and the other half to test the accuracy of the trained classifier. Before it is used, the training set should be evaluated by examining scatterplots and/or histograms for each class. It should show unimodel distributions, hopefully approximating normal distributions. If not unimodel, one may want to select new training sets. After the discriminant functions and the classification rule is derived, accuracy must be tested. Two acceptable techniques which are commonly used are :

   (a) Designate a randomly selected half of the training points as test points, before developing classifier. Use the other half for training. Then classify the half of the data not used for training. Develop contingency table (confusion matrix) to indicate probability of error in each class. This procedure is actually a measure of the consistency of the classifier.

   (b) Randomly select a set of pixel regions from the image of an unknown class. Classify them using the discriminant function and rules developed from the training set. Then verify the correctness of the classification (again with a confusion matrix) by checking the identity of these regions using external information sources like maps and aerial photos.

(vi) Separability of classes : So far, we have looked at an ideal situation where there is no overlap between different classes. In reality the classes are likely to overlap. It can be seen that the less the overlap between classes the lower the chance of misclassifying a given pixel. Classes that have little overlap is said to be highly separable.

7.8.3 Unsupervised Classification

Unsupervised classification algorithms do not compare points to be classified with training data. Rather, unsupervised algorithms examine a large number of unknown data vectors and divide them into classes based on properties inherent to the data themselves. The classes that result stem from differences observed in the data. In particular, use is made of the notion that data vectors within a class should be in some sense mutually close together in the measurement space, whereas data vectors in different classes should be comparatively well separated. If the components of the data vectors represent the responses in different spectral bands, the resulting classes might be referred to as spectral classes, as opposed to information classes, which represent the ground cover types of interest to the analyst.

The two types of classes described above, information classes and spectral classes, may not exactly correspond to each other. For instance, two information classes, corn and soyabeans, may look alike spectrally. We would say that the two classes are not separable spectrally. At certain times of the growing season corn and soyabeans are not spectrally distinct while at other times they are. On the other hand a single information class may be composed of two spectral classes. Differences in planting dates or seed variety might result in the information class " corn" being reflectance differences of tasseled and untasseled corn. To be useful, a class must be of informational value and be separable from other classes in the data.

7.9 Performance analysis of IRS-bands for land use/land cover classification system using Maximum Likelihood Classifier

Remote sensing has become a powerful tool for the regional mapping of natural resources and geological features. Starting with the use of image during the early stages of development of remote sensing in the mid-seventies, enough progress has been achieved in data interpretation with the easy availability of digital data. Digital processing of remote sensing data have gained momentum in the last decade. In India, with the establishment of remote sensing centres all over the country in recent years, attention has been focused on the large scale data processing for natural resources evaluation. One important aspect in remote sensing is the characterisation and classification of spectral measurements taken from satellites into various features of the land surface. Pattern recognition can be carried out if appropriate procedures are adopted for classification. Although several classification methods were developed in the field of statistics, their applicability to the processing of data was limited due to the spatial variation of natural features. In classification studies, it is often desirable to know how well the classes can be separated by observing the values of some feature vector for a set of samples (Toll 1984). In other words, one wants to know how much information the features provide for distinguishing the classes. To answer these questions a measure is needed to quantify the amount of information on the features.

The objective of this study is to improve the classification accuracy of obtaining accurate and cost effective information about the features. This study utilises a multiband data set to determine the effectiveness in improving the classification. The study assesses the utility of the multi-band data for the study of the urban environments, the land covers which are often difficult to examine accurately with remotely sensed data. It also attempts to examine the classification accuracy of a number of land cover classes for different band combinations and the potential of the classification method. The emphasis is on the use of Maximum Likelihood Calcification and the derivation of meaningful confidence level to all land cover classes. (Anji Reddy and S. Srinivasulu, 1994)

7.9.1 Classification Methodology

During the testing of a maximum likelihood classifier for land use classification in the Hyderabad region, the need arises for an account of the application of statistical confidence level assessments in remote sensing. Such an account is necessary in evaluation of the classification which consists of the following components :

(i) The acquisition of data.

(ii) A decision to the level of class separability desired and attainable.

(iii) The selection of training areas which will suit a given computer-based classification software package.

(iv) The effective operation of the package.

(v) The selection of an appropriate threshold for each class to apply likelihood distribution of that class.

(vi) The creation of appropriate output production.

In the present study, Indian Remote Sensing Satellite is used as the data acquisition system and IDRISI software package for the analysis of data. This software uses a Maximum Likelihood Classifier. Each pixel is classified, using IDRISI, into the most likely class type.

7.9.2 The Land Use and Land Cover Classification System

From multi-spectral data, one needs to identify and isolate particular objects. To proceed in a smooth and systematic manner, the data need to be grouped in a suitable framework. The framework should not only be flexible in nomenclature and definitions but also be capable of incorporating new information obtained from the same source or different sources. We have used level-1 classification. The classification categories at level-1 are identified in the study area given below :

(i) Water	(ii) Scrub	(iii) Forest
(iv) Vacant	(v) Commercial	(vi) Residential

Training samples

The validity of any training sample depends on two factors and namely, the size and the representativeness of the sample. Sample size is related to the number of spectral bands whose statistical properties are to be estimated (Forshaw et. al 1983). The user should have a prior knowledge of the data and the study area. The number and statistical characters to be extracted from training samples can be identified by their geographical location using maps and ground truth data. Once they are selected accurately, the results could be estimated after using classification decision rules.

Need of a confidence level

Any computer-derived classification, that will lead ultimately to a ground-cover thematic map, is based on ground truth data gathered by the user from the selected training areas. This applies whether unsupervised or supervised classification is employed and whether parametric or non-parametric techniques are used. The accuracy of the thematic map depends on the ability to extrapolate data successfully from the training areas to the whole mapped area. Unless we have some statistical measure of the efficiency of extrapolation process, the level of confidence in the classification cannot be estimated. Once a confidence level is so quantified, then the user of the classification data can relate it by means of the probability of the correct classification, to actuality over the whole classified area. The classification is thus related to the ground actuality by the confidence level. (Thomas and Allcock 1984).

7.9.3 Data Analysis

A number of procedures for testing the relative performances of spectral variables produced from popular image processing algorithms have been discussed in detail. The original IRS spectral variables, such as, band 1 (0.45-0.5 mm), 2 (0.52-0.59 mm), 3 (0.62 -0.68 mm) and 4 (0.77 - 0.86 mm) are used as standards for comparing the performances of spectral variables. General statistical parameters, such as, mean, standard deviation, variance-covariance matrix and correlation coefficients of different classes like water, scrub, forest, vacant, industrial, and residential, different band data have been analysed supplementary to the test of performance. The sample data are subjected to classification by the application of the Maximum Likelihood Classifier for all four bands. Variance-covariance matrices and correlation coefficient matrices for the three band data (1, 2, 3) reflect the accuracy of the classification in each of the classes (Mead and Szajgin 1982).

IRS spectral variables

The IRS data are transformed into spectral variables and divided into four types which are shown in Table 7.4

(i) The original (02) variables (1-5) consist of unprocessed two-band combinations (02 stands for type 1 variables).

(ii) The original (03) variables (8-11) consist of unprocessed three-band combinations (03 stands for type 2 variables).

(iii) The ratio (R) variables (6-7) consist of linear combinations of band 1, band 2, band 3 and band 4 (R stands for type 3 variables).

(iv) The original (04) variable (12) consists of unprocessed variables with all four bands (04 stands for type 4 variables).

7.9.4 Classification Accuracy Approach

The examination of training samples by using the back classification approach was carried out. Graphs of the six classes of the different combinations were drawn in the accuracy of the variable shown in Fig. 7.25 to Fig. 7.30. Tables 7.5 to 7.8 show the confusion matrices for the six classes of combinations band 1, band 2, band 1/ band 2, band 3/band 4, bands 1, 2 and 3 and bands 1, 2, 3 and 4. Tables 7.9 and 7.10 show the overall accuracies and the detailed analysis.

For many applications redundant or highly correlated bands may be retained in order to minimise the information loss. Analysis of the overall accuracies revealed that the discrimination of various categories of land use classification system was possible if the percentage classification accuracy was known for different variables. The variable for the type 1 for water has been selected keeping the overall accuracy which was higher than the others in the same type. In the same way, the selection of variables table was prepared keeping the overall accuracies of all types of the six classes in order to determine the confidence intervals. These values are shown in Table 7.10.

The overall accuracies of type-3 variables are more than type-1 and type-3 variables. The overall classification accuracy of variables band 1, band 2, and bands 3 are better than any other variable. For water, bands 2 and 4 combinations give 85 percent accuracy which is the best among the two-band variables. The band combination 2, 3 and 4 gives the best performance which is 93 percent. For scrubs, the bands 1 and 2 combination gives 73 percent which is the best among two-band variables and the band combination 1, 3 and 4 gives the best performance of 86 percent among all variables. For vacant land, the 2-3 band combination produced 75 per cent accuracy. Among all the bands the 1, 2 and 3 band combination gives 85 per cent accuracy. For industrial land, the 2-3 band combination gives 74 percent accuracy which is the best among the two-band variable and the band combination 1, 2 and 3 produced 86 percent which is the best among all the variables. For residential land, among the two-band variables the 2-3 band combination gives 76 percent accuracy and the 1, 2, and 3 band combination produced 86 percent accuracy which is the best among all the variables (Anji Reddy, 1994).

From the study of the Maximum Likelihood Classifier in the performance analysis of IRS-bands for land use/land cover classification system it can be deduced that :

(i) The statistical background to the derivation of a confidence level based on the probability distribution function can be taken for a classification exercise.

(ii) This method not only has the advantage as the adaptive feature selected and the sequential classification by the usual layered classifier, it also improves the accuracy of the performance and reduces the amount of computation needed. So, to some extent, we can obtain reasonably good classification results without using excessive computer time.

(iii) This study has concluded that the use of multi-band datasets for the land use / land cover classification system will provide more accurate results than an independent set of sensor data. The best classification occurred when the data contained all the major portions of an electromagnetic spectrum.

(iv) The method of classification accuracy approach reduces the processing time substantially by providing more classes on the pre-classification stage and it may be comparable to that of the other classifications to get more accurate results.

(v) More work is required to validate this and additional tests on other areas having the high spatial complexity needed in order to understand the limits of remote sensor data and the performance of variables, should be checked for the level-II classification system.

7.10 Image Classification and GIS

The early years of the new millennium will see a very considerable increase in the volumes of earth observation data being collected from space platforms and much greater power will be needed if the maximum value is to be obtained from these data (Mather, 2000). An integrated approach to geographical data analysis is now being adopted and a significant effect on the usage of a particular image classification method is performed. The use of non-remotely sensed data in the image classification process is providing the possibility of greater accuracy, while in turn the greater reliability of image based products is improving the capabilities of environmental GIS, particularly with respect to studies of temporal change.

All these factors will present challenges to remote sensing and GIS user communities. Therefore, the focus of research on remote sensing data integration with GIS will move away from specialised algorithm development to the search methods that satisfy user needs and are broader in scope. Some of the existing image processing and GIS software are providing this facility of integration, but still needs to improve. If progress is to be made, then high quality interdisciplinary research is needed involving mathematicians, statisticians, computer professionals, engineers, earth scientists, and geographers.

Table 7.3 Bands, Band Ratios and Band Combinations Used by Different Investigators

Sl. No.	Name of Investigator and Satellite Sensor used	Band No. and Wavelength range (microns)	Water Quality Parameter	Variable used in Model
1.	Ritchie, et. al (1976) Landsat MSS	6 : 0.7 - 0.8	Suspended sediment	B_6
2.	Bartolucci, et. al (1977) Landsat MSS	5 : 0.6 - 0.7	Turbidity	B_5
3.	Khorram (1982) Landsat MSS	4 : 0.5 - 0.6 6 : 0.7 - 0.8	Salinity	B_4, B_6 and B_7
4.	Prasad, et. al (1987) NIMBUS CZCS	1 : 0.43 - 0.45 2 : 0.51 - 0.53 3 : 0.54 - 0.56	Chlorophyll (a)	(B_1/B_2) ad (B_2/B_3)
5.	Aranvachapun and Walling (1988) Landsat MSS	5 : 0.6 - 0.7	Suspended sediments	B_5
6.	Lathrop, et. al (1989) SPOT HRV	1 : 0.5 - 0.59 2 : 0.61 - 0.68 3 : 0.79 - 0.89	Chlorophyll (a) & Suspended sediments	B_1, B_2 and B_3
7.	Prangsma & Roozekrans (1989) NOAA AVHRR	1 : 0.58 - 0.68	Suspended sediment	B_1
8.	-do-	1 : 0.58 - 0.68 2 : 0.72 - 1.1	Floating Blue algae	
9.	Ritchie, et. al (1990)	1 : 10.5 - 12.5	Temperature	B_6
10.	Albanakis (1990) Landsat MSS	4 : 0.5 - 0.6 5 : 0.6 - 0.7 6 : 0.7 - 0.8	Suspended sediment	
11.	Garcia & Robinson (1991) Airborne TM	1 : 0.42 - 0.45 3 : 0.52 - 0.60 5 : 0.67 - 0.69	Chlorophyll (a)	
12.	Anji Reddy (1993) Landsat MSS	1 : 0.5 - 0.6 2 : 0.6 - 0.7 3 : 0.7 - 0.8 4 : 0.8 - 1.1	Suspended sediment	B_1, B_2, B_3 and B_4

Table 7.4 Spectral variables for the test scene

Sample no.	Variable (band combination)	Name of the Variable type
1.	Bands 1 and 2	02
2.	Bands 2 and 3	02
3.	Bands 3 and 4	02
4.	Bands 1 and 3	02
5.	Bands 2 and 4	02
6.	Bands 1/band 2, band 3/band 4 and 2	R
7.	Bands 1/band 3, band 2/band 4 and 2	R
8.	Bands 1, 2 and 3	03
9.	Bands 2, 3, and 4	03
10.	Bands 1, 2 and 4	03
11.	Bands 1, 3 and 4	03
12.	Bands 1, 2, 3 and 4	04

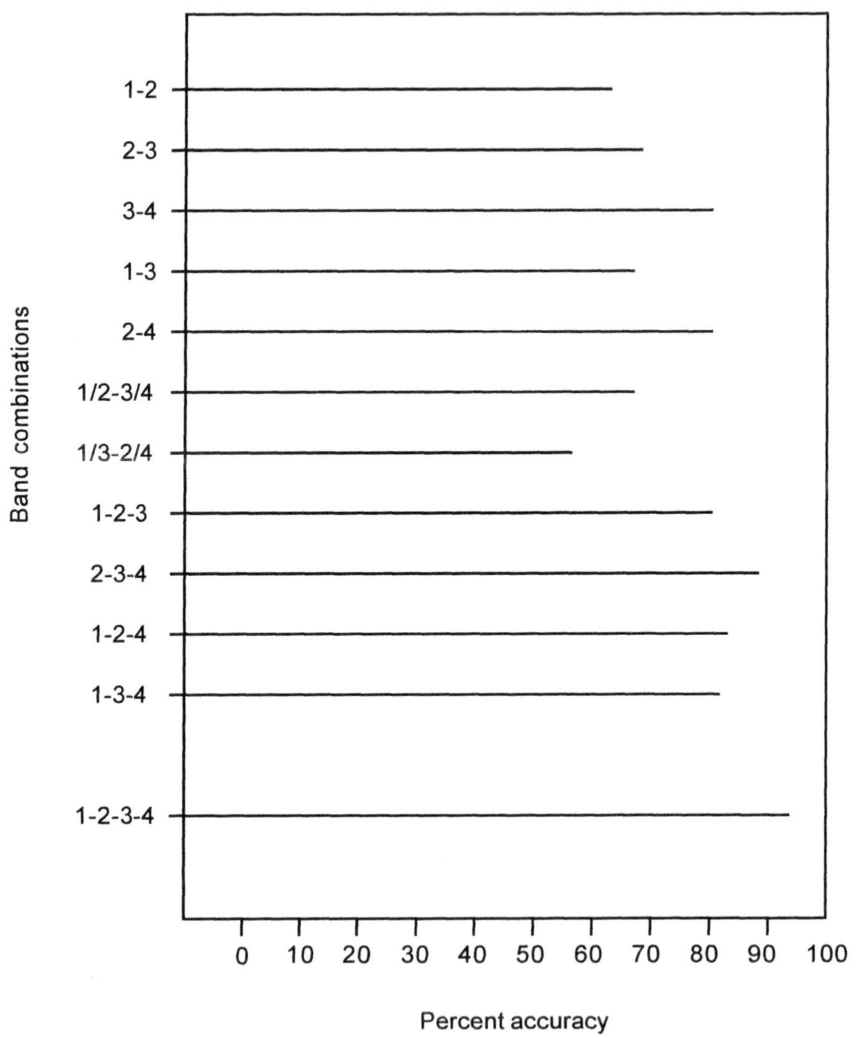

Fig. 7.25 Percentage classification accuracy of water.

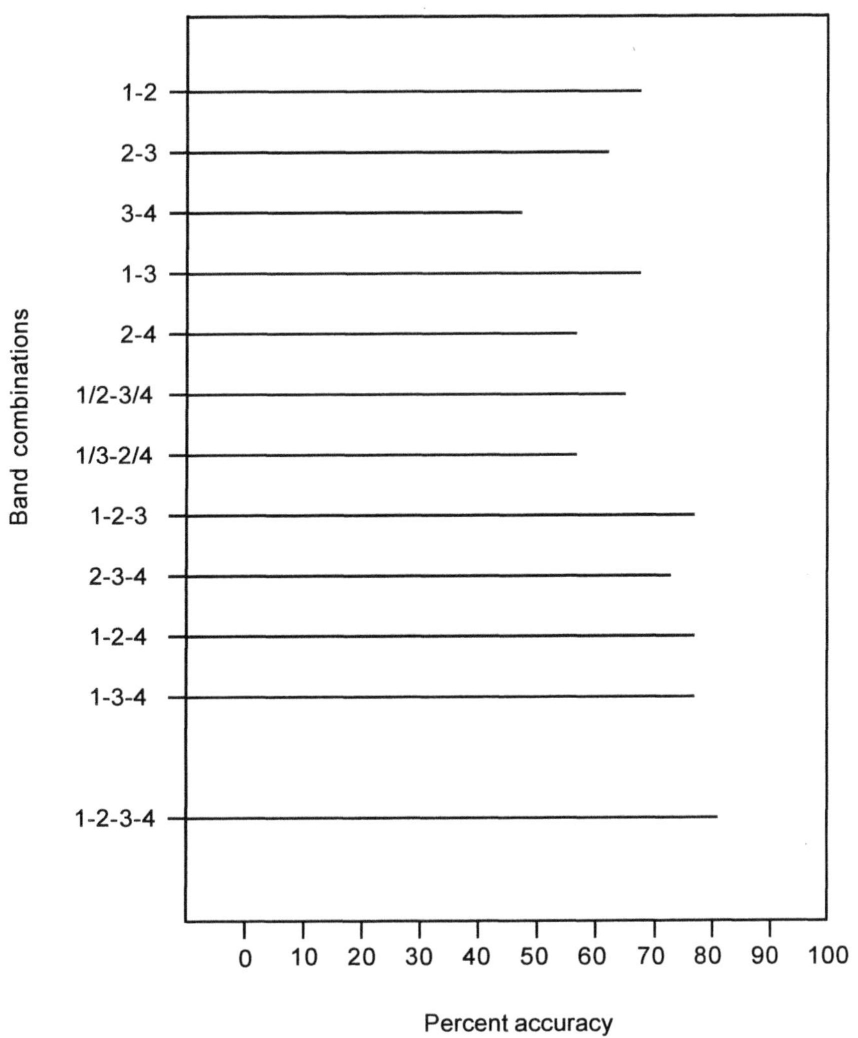

Fig. 7.26 Percentage classification accuracy of scrub.

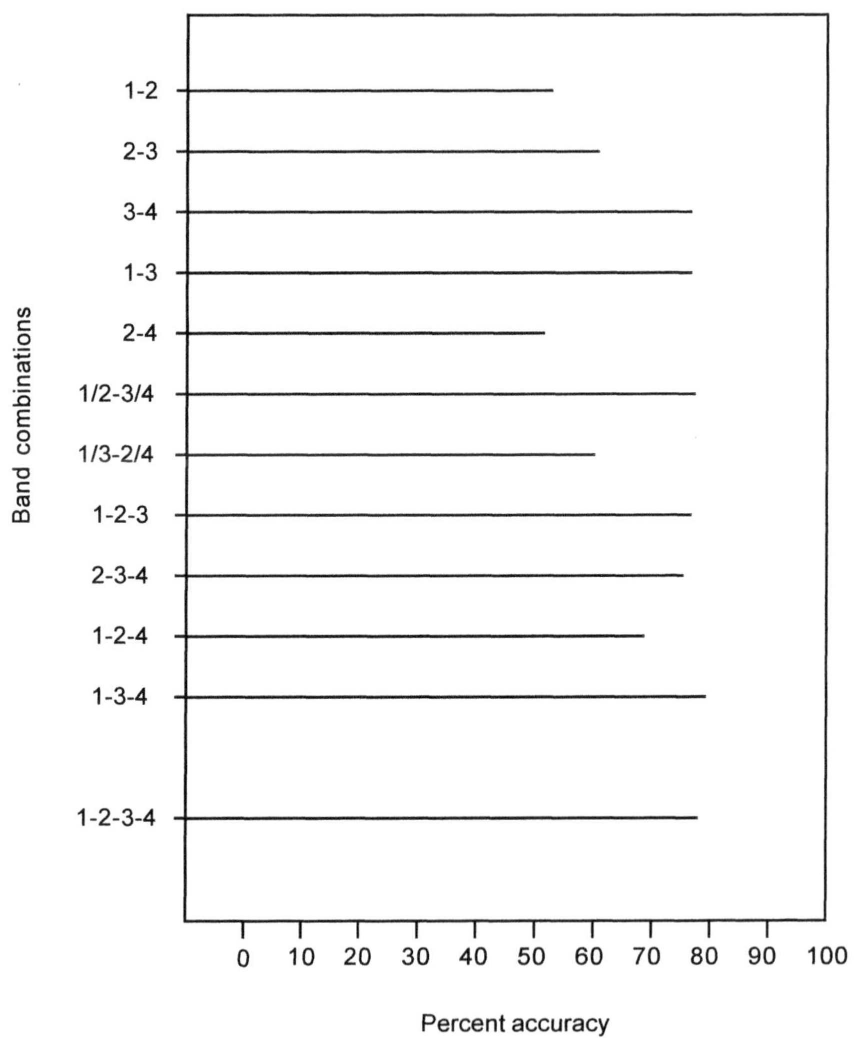

Fig. 7.27 Percentage classification accuracy of forest.

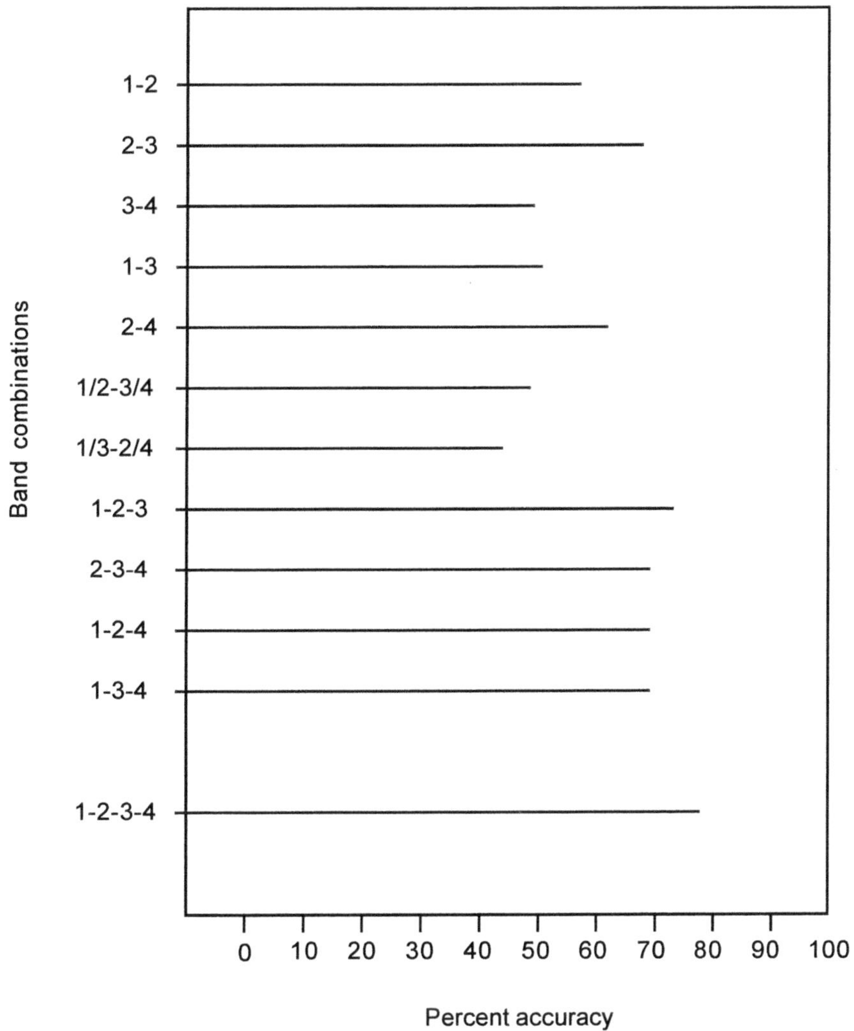

Fig. 7.28 Percentage classification accuracy of vacant land.

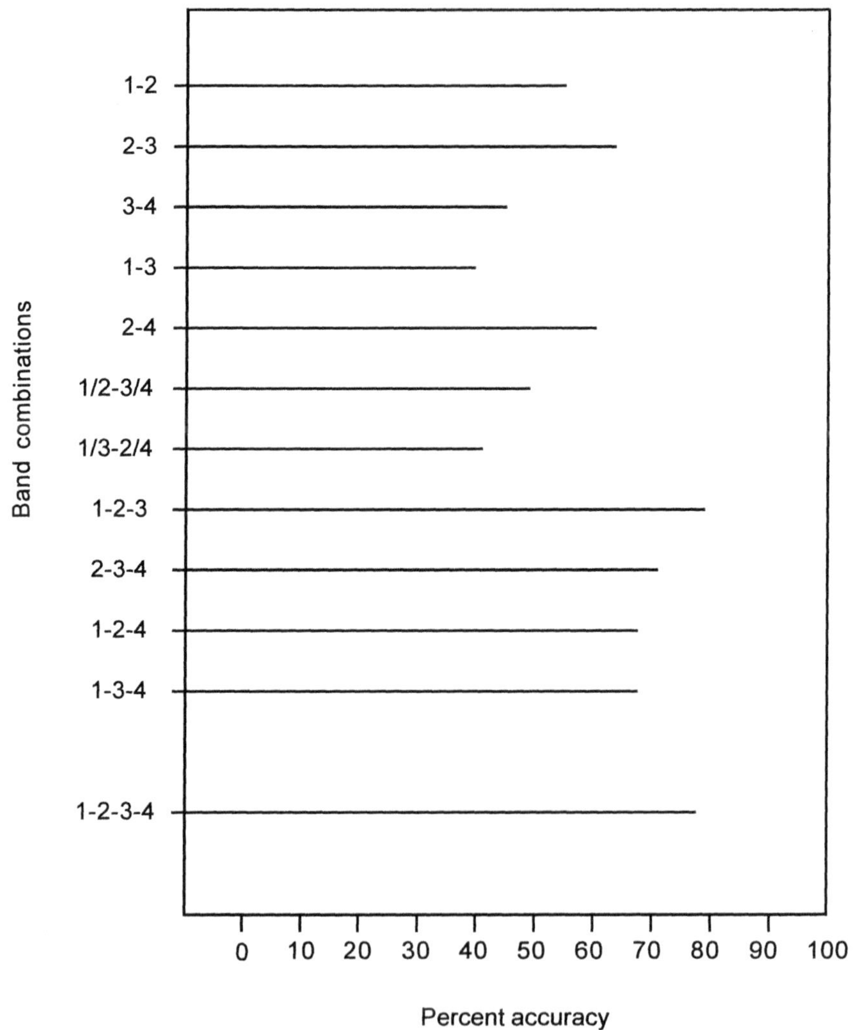

Fig. 7.29 Percentage classification accuracy of commercial land.

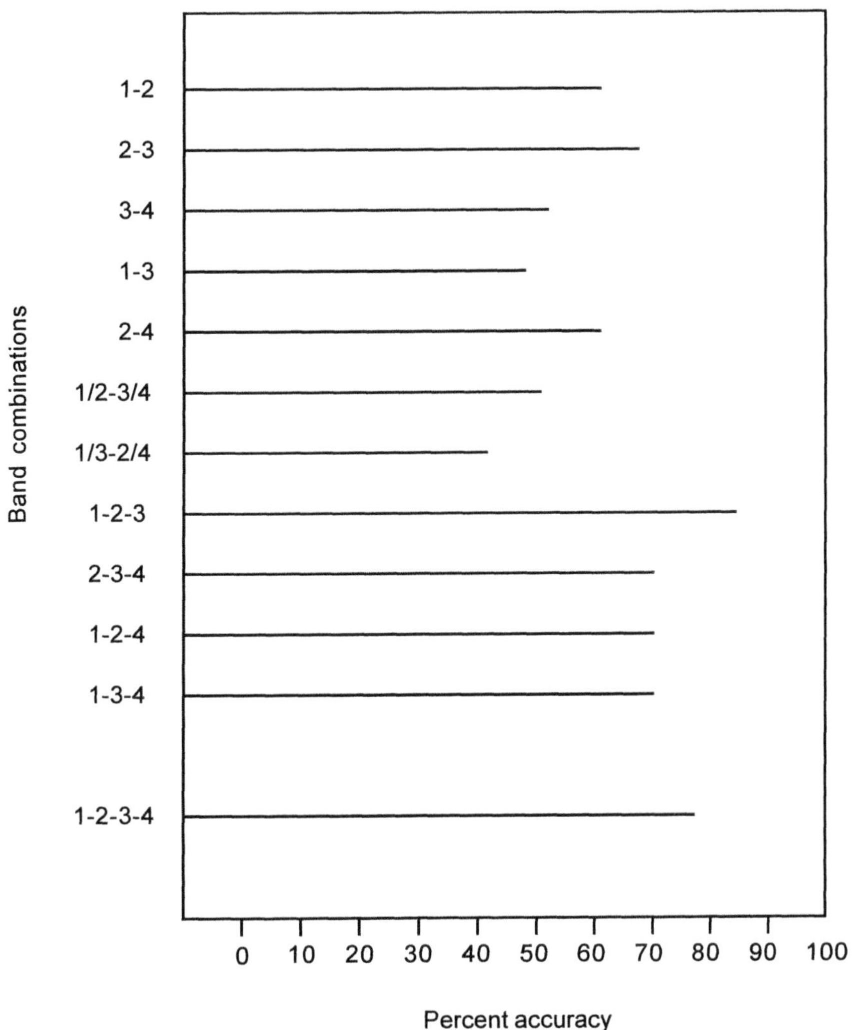

Fig. 7.30 Percentage classification accuracy of residential land.

Table 7.5 Confusion or error matrix for six classes (bands 1 and 2)

			Number of pixels classified into category					
Category type	**Number of pixels**	**% correct**	**1**	**2**	**3**	**4**	**5**	**6**
Water	1228	66.5	817	208	126	—	32	45
Scrub	964	74.06	143	714	66	—	30	11
Forest	2920	53.01	420	686	1548	—	124	142
Vacant	1916	62.01	—	57	160	1188	295	216
Industrial	1136	62.5	96	26	20	126	710	158
Residential	2884	64.0	114	52	52	272	548	1846

Overall accuracy = 61.75%

Table 7.6 Confusion or error matrix for six classes (band 1/band 2 and band 3/band 4)

			Number of pixels classified into category					
Category type	**Number of pixels**	**% correct**	**1**	**2**	**3**	**4**	**5**	**6**
Water	1228	72.14	886	128	46	56	62	50
Scrub	964	70.53	144	680	—	—	82	58
Forest	2920	81.84	48	280	2390	102	68	32
Vacant	1916	51.56	56	150	234	988	322	156
Industrial	1136	56.53	—	44	—	156	642	294
Residential	2884	58.26	134	266	62	378	364	1680

Overall accuracy = 65.76%

Table 7.7 Confusion or error matrix for six classes (bands 1, 2 and 3)

Category type	Number of pixels	% correct	Number of pixels classified into category 1	2	3	4	5	6
Water	1228	88.02	1081	86	8	—	48	—
Scrub	964	78.12	120	754	20	36	24	12
Forest	2920	80.97	40	256	2364	146	66	48
Vacant	1916	84.75	28	62	028	1624	102	72
Industrial	1136	83.89	—	10	26	52	953	194
Residential	2884	89.25	38	92	64	46	68	2574

Overall accuracy = 84.62%

Table 7.8 Confusion or error matrix for six classes (band 1, 2, 3, and 4)

Category type	Number of pixels	% correct	Number of pixels classified into category 1	2	3	4	5	6
Water	1228	94.00	1154	46	28	—	—	—
Scrub	964	80.1	12	772	92	48	24	16
Forest	2920	81.35	48	236	2376	160	44	56
Vacant	1916	80.21	12	68	48	1536	164	88
Industrial	1136	79.57	—	—	26	64	904	142
Residential	2884	82.38	16	78	60	144	208	2376

Overall accuracy = 91.58%

Table 7.9 Selection of variables and accuracy

Category	Variables	Percentage classification accuracy	95% Confidence interval
Water	Band 2 and band4	84.34	82.65
	Ratios band 1/band 2 and band 3/band 4	72.14	70.04
	Bands 2, 3 and 4	93.48	92.32
Scrub	Band 1 and band 3	79.25	77.94
	Ratios band 1/band 2 and band 3/band 4	70.53	68.12
	Bands 1, 2 and 3	78.12	75.41
Forest	Band 1 and band 3	82.08	81.34
	Ratios band 1/ band 2 and band 3/band 4	81.54	80.67
	Bands 1, 2 and 3	80.97	79.76
Vacant	Band 2 and band 3	73.38	71.72
	Ratios band 1/band 2 and band 3/band 4	51.56	49.68
	Bands 1, 2 and 3	84.75	83.92
Industrial	Band 2 and band 3	71.30	69.09
	Ratios band 1/band 2 and 3/band 4	56.53	54.09
	Bands 1 2 and 3	83.89	82.8
Residential	Band 2 and band 3	69.00	67.58
	Ratios 2 and band 3	58.26	56.74
	Bands 1, 2 and 3	89.25	88.67

Table 7.10 Overall accuracy of variables

Category	Variables	Overall accuracy (%)
02	1-2	6.75
	2-3	67.80
	3-4	64.34
	1-3	64.12
	2-4	63.05
R	1/2-3/4	65.76
	1/3-2/4	56.77
03	1-2-3	84.62
	2-3-4	80.91
	1-2-4	80.4
	1-3-4	77.44
04	1-2-3-4	91.58

8

Global Positioning System (GPS)

8.1 Introduction

The NAVSTAR Global Positioning System (GPS) is a satellite-based radio-positioning and time-transfer system designed, financed, deployed, and operated by the U.S. Department of Defence. GPS has also demonstrated a significant benefit to the civilian community who are applying GPS to a rapidly expanding number of applications. The major attractions of GPS are :

- The relatively high positioning accuracies, from tens of metres down to the millimetre level.
- The capability of determining velocity and time, to an accuracy commensurate with position.
- The signals are available to users anywhere on the globe: in the air, on the ground, or at sea.
- It is a positioning system with no user charges, that simply requires the use of relatively low cost hardware.
- It is an all-weather system, available 24 hours a day.
- The position information is in three dimensions, that is, vertical as well as horizontal information is provided.

The number of civilian users is already significantly greater than that of the military users. However, for the time being the U.S. military still operates several "levers" with which they control the performance of GPS (Bomford, 1980).Nevertheless, despite the handicap of GPS being a military system there continues to be tremendous product innovation within the civilian sector, and it is ironic that this innovative drive is partly directed to developing technology and procedures to overcome some of the constraints to GPS performance which have been applied by the system's military operators.

8.2 System Design Considerations

Development work on GPS commenced within the U.S. Department of Defence in 1973, the motivation being to develop an all-weather, 24-hour, global positioning system to support the positioning requirements for the armed forces of the U.S. and its allies. (Parkinson, B.W, 1994). The system was therefore designed to replace the large variety of navigational systems already in use, and great emphasis was placed on the system's reliability and survivability. In short, a number of stringent conditions had to be met :

- suitable for all classes of platform: aircraft (jet to helicopter), ship, land (vehicle-mounted to handheld) and space (missiles and satellites),
- able to handle a wide variety of dynamics,
- real-time positioning, velocity and time determination capability to an appropriate accuracy,
- the positioning results were to be available on a single global geodetic datum,
- highest accuracy to be restricted to a certain class of user,
- resistant to jamming (intentional and unintentional),
- redundancy provisions to ensure the survivability of the system,
- passive positioning system that does not require the transmission of signals from the user to the satellite(s),
- able to provide the service to an unlimited number of users,
- low cost, low power, therefore as much complexity as possible should be built into the satellite segment, and
- total replacement of the Transit satellite and other terrestrial navaid systems.

This led to a design based on the following essential concepts:

- A one-way ranging system, in which the satellites transmit signals, but are unaware of who is using the signal (no receiving function). As a result the user (or listener) cannot easily be: (a) detected by the enemy (military context), or (b) charged for using the system (civilian context).
- Use of the latest atomic clock and microwave transmission technology, including spread-spectrum techniques.

- A system that makes range-like measurements with the aid of pseudo-random binary codes modulated on carrier signals.
- Satellite signals that are unaffected by cloud and rain.
- A multiple satellite system which ensures there is always a sufficient number of satellites visible simultaneously anywhere on the globe, and at any time.
- Positioning accuracy degradation that is graceful.

What was perhaps unforeseen by the system designers was the power of product innovation, which has added significantly to the versatility of the GPS as a system for precise positioning and navigation. For example, GPS is able to support a number of positioning and measurement modes in order to satisfy simultaneously a variety of users, from those satisfied with general navigational accuracies (tens of metres) to those demanding very high (sub-centimetre) relative positioning accuracies. It has now so penetrated certain applications areas that it is difficult for us to imagine life without GPS!

8.2.1 Technological Advances

Rarely have so many seemingly unrelated technological advances been required to make a complex system such as GPS work. Briefly they are :

Space System Reliability : The U.S. space program had by 1973 demonstrated the reliability of space hardware. In particular, the Transit system had offered important lessons. The Transit satellites were originally designed to last 2-3 years in orbit, yet some of the satellites have operated well beyond their design life. In fact Transit continued to perform reliably for over 25 years.

Atomic Clock Technology : With the development of atomic clocks a new era of precise timekeeping had commenced. However, before the GPS program was launched these precise clocks had never been tested in space. The development of reliable, stable, compact, space-qualified atomic frequency oscillators (rubidium, and then cesium) was therefore a significant technological breakthrough. The advanced clocks now being used on the GPS satellites routinely achieve longterm frequency stability in the range of a few parts in 10^{14} per day (about 1 sec in 3,000,000 years!). This long-term stability is one of the keys to GPS, as it allows for the autonomous, synchronised generation and transmission of accurate timing signals by each of the GPS satellites without continuous monitoring from the ground.

Quartz Crystal Oscillator Technology : In order to keep the cost of user equipment down, quartz crystal oscillators were proposed (similar to those used in modern digital watches), rather than using atomic clocks as in the GPS satellites. Besides their low cost, quartz oscillators have excellent short-term stability. However, their long-term drift must be accounted for as part of the user position determination process.

Precise Satellite Tracking and Orbit Determination : Successful operation of GPS, as well as the Transit system, depends on the precise knowledge and prediction of a satellite's position with respect to an earth-fixed reference system. Tracking data collected by ground monitor stations is analysed to determine the satellite orbit over the period of tracking (typically one week). This reference ephemeris is extrapolated into the future and the data is then up-loaded to the satellites. Prediction accuracies of the satellite coordinates, for one day, at the few metre level have been demonstrated.

Spread-Spectrum Technology : The ability to track and obtain any selected GPS satellite signal (a receiver will be required to track a number of satellites at the same time), in the presence of considerable ambient noise is a critical technology. This is now possible using spread-spectrum and pseudo-random-noise coding techniques.

Large-Scale Integrated Circuit Technology : To realise the desired low cost, low power and small size necessary for much of the user equipment, the GPS program relies heavily on the successful application of VLSI circuits, and powerful computing capabilities built onto them.

8.3 GPS System Eelements

The GPS system consists of three segments (Fig. 8.1).

- The Space Segment: comprising the satellites and the transmitted signals.
- The Control Segment: the ground facilities carrying out the task of satellite tracking, orbit computations, telemetry and supervision necessary for the daily control of the space segment.
- The User Segment: the entire spectrum of applications equipment and computational techniques that are available to the users.

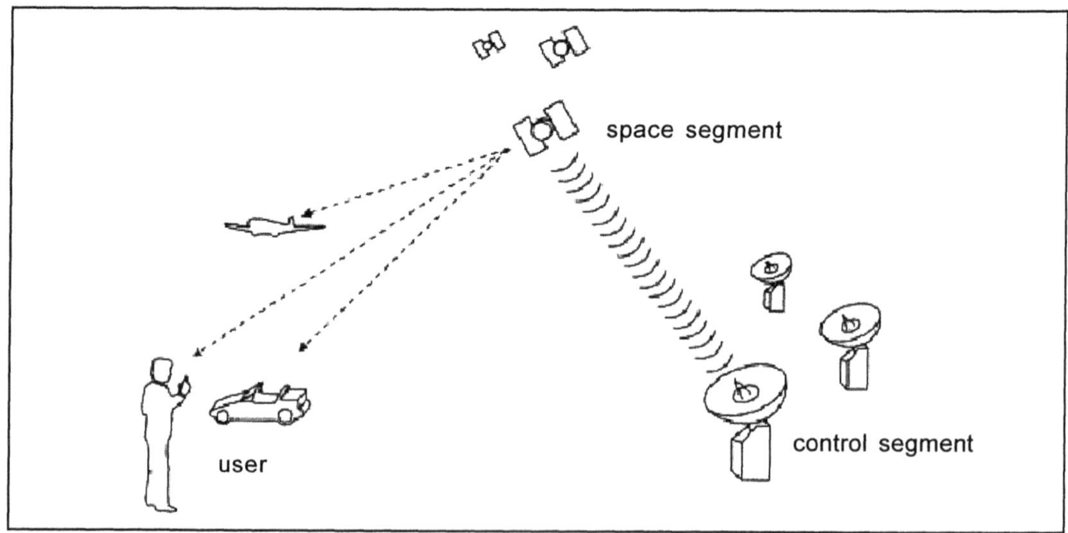

Fig. 8.1 GPS System Elements.

8.3.1 The Space Segment

The Space Segment consists of the constellation of spacecraft and the signals broadcast by them which allow users to determine position, velocity and time. The basic functions of the satellites are to :

- Receive and store data transmitted by the Control Segment stations.
- Maintain accurate time by means of several onboard atomic clocks.
- Transmit information and signals to users on two L-band frequencies.
- Provide a stable platform and orbit for the L-band transmitters.

Several constellations of GPS satellites have been deployed, and several more are planned. The experimental satellites, the so-called "Block I" satellites, were built by the Rockwell Corporation. The first was launched in February 1978, and the last of the eleven satellite series (one exploded on the launchpad) was launched in 1985. The operational series of GPS satellites, the "Block II" and "Block IIA" satellites, were also built by the Rockwell Corporation. The 20 replacement "Block IIR" series of satellites, first launched in 1997, are built by the General Electric Corporation (now the Lockheed Martin Corporation). The "Block IIF" series are still in the design phase and may, for example, incorporate an additional civilian transmission frequency. They are planned for launch from 2005 onwards. The operational satellite I.D.s are separated into three space vehicle numbering series: SVN 13 through 21 for the Block II, SVN 22 through 40 for Block IIA, and SVN 41 and above for the Block IIR satellites.

The current status of the GPS constellation and such details as the launch and official commissioning date, the orbital plane and position within the plane, the satellite I.D. number(s), etc., can be obtained from several electronic GPS information sources on the Internet. The following sections describes the general satellite orbit characteristics.

Each GPS satellite transmits a unique navigational signal centred on two L-band frequencies of the electromagnetic spectrum, permitting the ionospheric propagation effect on the signals to be eliminated. At these frequencies the signals are highly directional and so are easily reflected or blocked by solid objects. Clouds are easily penetrated, but the signals may be blocked by foliage (the extent of blockage is dependent on the type and density of the leaves and branches, and whether they are wet or dry). The signal is transmitted with enough power to ensure a minimum signal power level of -160dBw at the earth's surface (the maximum it is likely to reach is about -153dBw, (Spilker, 1995). The satellite signal consists of the following components (Fig. 8.2) :

- The two L-band carrier waves.
- The ranging codes modulated on the carrier waves.
- The so-called "navigation message".

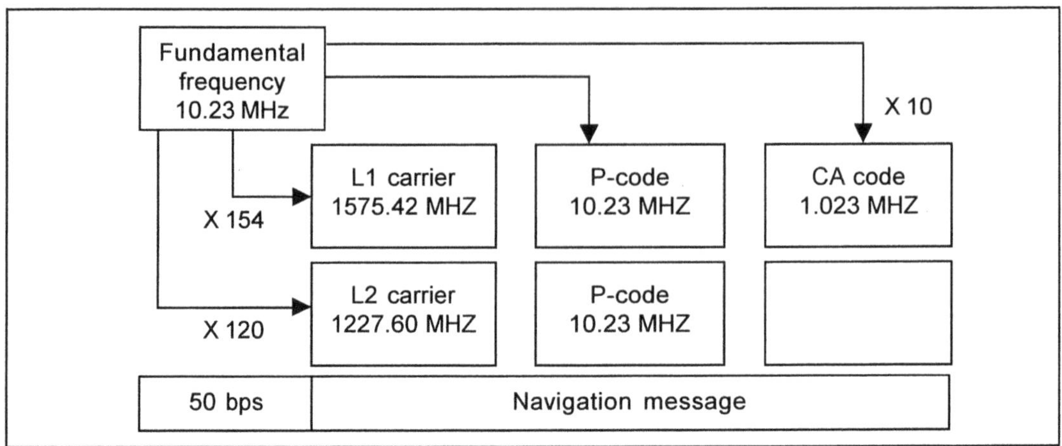

Fig. 8.2 GPS Satellite Signal Components.

Modulated onto the carrier waves are the PRN ranging codes and navigation message for the user. The primary function of the ranging codes is to permit the signal transit time (from satellite to receiver) to be determined. The transit time when multiplied by the velocity of light then gives a measure of the receiver satellite "range" (in reality the measurement process is considerably more complex). The navigation message contains the satellite orbit information, satellite clock parameters, and pertinent general system information necessary for real-time navigation to be performed. All signal components are derived from the output of a highly stable atomic clock. Each GPS satellite is equipped with several cesium and rubidium atomic clocks.

8.3.2 The Control Segment

The Control Segment consists of facilities necessary for satellite health monitoring, telemetry, tracking, command and control, satellite orbit and clock data computations, and data uplinking. There are five ground facility stations: Hawaii, Colorado Springs, Ascension Island, Diego Garcia and Kwajalein. All are owned and operated by the U.S. Department of Defense and perform the following functions:

- All five stations are Monitor Stations, equipped with GPS receivers to track the satellites. The resultant tracking data is sent to the Master Control Station.
- Colorado Springs is the Master Control Station (MCS), where the tracking data are processed in order to compute the satellite ephemerides and satellite clock corrections. It is also the station that initiates all operations of the space segment, such as spacecraft manoeuvring, signal encryption, satellite clock-keeping, etc.
- Three of the stations (Ascension Is., Diego Garcia, and Kwajalein) are Upload Stations allowing for the uplink of data to the satellites. The data includes the orbit and clock correction information transmitted within the navigation message, as well as command telemetry from the MCS.

Overall operation of the Control and Space Segments is the responsibility of the U.S. Air Force Space Command, Second Space Wing, Satellite Control Squadron at the Falcon Air Force Base, Colorado.

Each of the upload stations can view all the satellites once a day. All satellites are therefore in contact with an upload station three times a day, and new navigation messages as well as command telemetry can be transmitted to the GPS satellites every eight hours if necessary. The computation of : (a) the satellite orbits or "ephemerides", and (b) the determination of the satellite clock errors, are the most important function of the Control Segment. The first is necessary because the GPS satellites function as "orbiting control stations" and their coordinates must be known to a relatively high accuracy, while the latter permit a significant measurement bias to be reduced.

The GPS satellites travel at high velocity (of the order of 4 km/sec), but within a more or less regular orbit pattern. After a satellite has separated from its launch rocket and it begins orbiting the earth, it's orbit is defined by its initial position and velocity, and the various force fields acting on the satellite. In the case of the gravitational field for a spherically symmetric body (a reasonable approximation of the earth at the level of about 1 part in 103) this produces an elliptical orbit which is fixed in space - the Keplerian ellipse. Due to the effects of the other, non-spherical gravitational components of the earth's gravity field, and non-gravitational forces, which perturb the orbit, the actual trajectory of the satellite departs from the ideal Keplerian ellipse (Fig. 8.3).

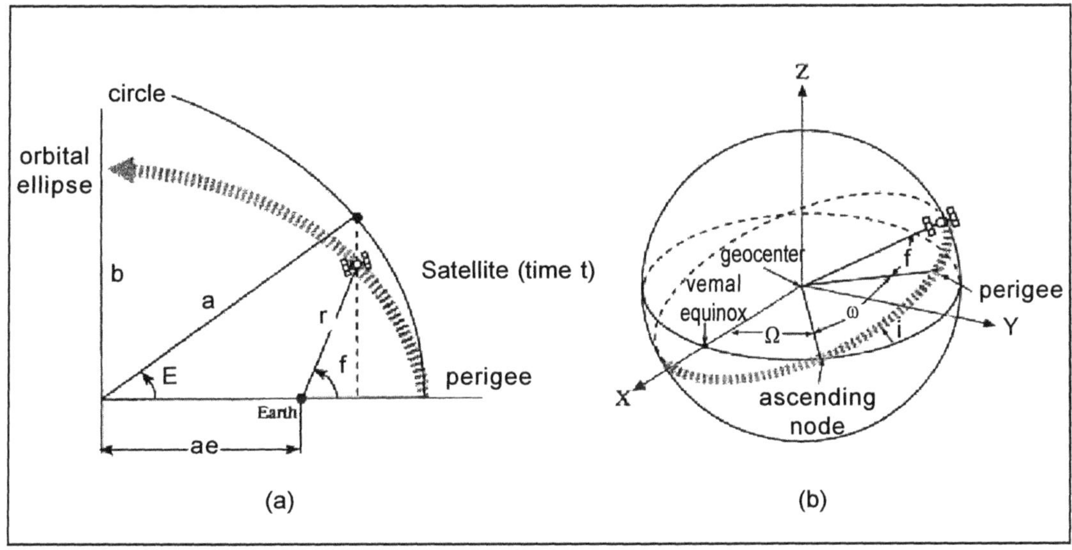

Fig. 8.3 (a) The Keplerian Ellipse, and (b) Keplerian Orbital Elements.

The most significant forces that influence satellite motion are:

- the spherical and non-spherical gravitational attraction of the earth,
- the gravitational attractions of the sun, moon, and planets (the "third body" effects),
- atmospheric drag effects,
- solar radiation pressure (both direct and albedo effects), and
- the variable part of the earth's gravitational field arising from the solid earth and ocean tides.

To determine the motion of a satellite to a high precision these perturbing forces need to be modelled accurately. If these forces were known perfectly and the initial position and velocity of the satellite were given, then the integration of the Equations of Motion would give the satellite's position and velocity at any time in the future.

However, the perturbing forces are not known to sufficient precision. An "orbit computation" process is therefore performed, in which satellite observations obtained at tracking sites of known position (in the case of GPS, the monitor stations of the Control Segment) are analysed in order to produce an orbit that is a "best fit" to the available observations. This involves the adjustment of the appropriate parameters of the orbit, possibly together with several additional force model parameters (Seebera, 1993). Determining the orbit is a complex procedure, and in the case of GPS satellites this process occurs on a continuous, automatic basis.

The product of the orbit computation process at the MCS is the satellite ephemeris (or trajectory). A satellite ephemeris may be expressed in a number of forms :

- a list of 3-D coordinates, and velocities, at regular intervals of time,
- the Keplerian elements at some reference epoch, plus their rate-of-change with time,
- a polynomial representation of the trajectory in a suitable reference system, such as along-track, cross-track, or radial components, or
- satellite position and velocity at some reference time epoch, and requiring these values to be derived for subsequent times by integrating the Equations of Motion.

The GPS broadcast ephemeris, as represented in the navigation message, is actually a combination of all of the above orbit representations (Langley, R. B, 1991b.). The orbital ephemerides are expressed in the reference system most appropriate for positioning, which is an earth-fixed reference system such as WGS84. Hence the Control Segment has the function of propagating the satellite datum, which users connect to via the transmitted satellite ephemerides.

The behaviour of each GPS satellite clock is monitored against GPS Time, as maintained by an ensemble of atomic clocks at the GPS Master Control Station. The satellite clock bias, drift and drift-rate relative to GPS Time are explicitly determined in the same procedure as the estimation of the satellite ephemeris. The clock behaviour so determined is made available to all GPS users via clock error coefficients (defining the mis-synchronization with GPS Time) in a polynomial form broadcast in the navigation message. However, what is available to users is really a prediction of the clock behaviour for some future time interval. Due to random deviations — even cesium and rubidium oscillators are not entirely predictable — the deterministic models of satellite clock error are only accurate to about 20 nanoseconds. This is not precise enough for accurate range measurement.

As the GPS system matures we can expect that the satellites will operate with greater independence from the ground-based Control Segment, without significant degradation in performance. For example the "Block IIR" and "Block IIF" satellites will have a crosslink capability enabling between-satellite communication and ranging. The satellites will talk to each other! This data will be processed to produce the ephemeris information within the space segment, with relatively little operator control having to be exercised.

8.3.3 The User Segment (The Applications)

This is the part of the GPS system with which we are most concerned — the space and control segments being largely transparent to the operations of the navigation function. Of interest is the range of GPS :

- Applications,
- Equipment, and
- Positioning strategies.

The "engine" of commercial GPS product development is, without doubt, the user applications. Each day new applications are being identified, each with its unique requirements with regards to: accuracy of the results, reliability, operational constraints, user hardware, data processing algorithms, latency of the GPS results, etc. To make sense of the bewildering range of GPS applications it may be useful to classify them according to the following :

1. Land, Sea and Air Navigation and Tracking, including enroute as well as precise navigation, collision avoidance, cargo monitoring, vehicle tracking, search and rescue operations, etc. While the accuracy requirement may be modest and the user hardware is generally comparatively low cost, the reliability, integrity and speed with which the results are needed is generally high.
2. Surveying and Mapping, on land, at sea and from the air. Includes geophysical and resource surveys, GIS data capture surveys, etc. The applications are of relatively high accuracy, for positioning in both the static and moving receiver mode, and generally require specialised hardware and data processing software.
3. Military Applications: Although these are largely mirrored by civilian applications, the military GPS systems are generally developed to "military specifications" and a greater emphasis is placed on system reliability.

4. Recreational Uses, on land, at sea and in the air. The primary requirement is for low cost instruments which are very easy to use.
5. Other specialised uses, such as time transfer, attitude determination, spacecraft operations, atmospheric studies, etc. Obviously such applications require specially developed, high cost systems, often with additional demanding requirements such as real-time operation, etc.

GPS user equipment has undergone an extensive program of development, both in the military and civilian area. In this context, GPS "equipment" refers to the combination of : hardware, software, and operational procedures or requirements.

While the military R&D programs have concentrated on achieving a high degree of miniaturisation, modularisation and reliability, the civilian user equipment manufacturers have, in addition, sought to bring down costs and to develop features that enhance the capabilities of the positioning system. The following general remarks can be made. Civilian users have, from the earliest days of GPS availability, demanded ever increasing levels of performance, in particular higher accuracy and improved reliability. This is particularly true of the survey user seeking levels of accuracy several orders of magnitude higher than that of the navigation user. In some respects the GPS user technology is being driven by the precise positioning market-in much the same way that automotive technology often benefits from car racing. Yet another major influence on the development of GPS equipment has been the increasing variety of civilian applications. For although there may exist a similar positioning accuracy requirement across many user applications, to address a particular application in the most satisfactory manner, a specific combination of hardware and software features is often required.

It is expected that the worldwide market for GPS receiver equipment will grow from about US$1 billion at the present time, to over US$8 billion by the year 2010. Market surveys suggest that the greatest growth is expected to be in the commercial and consumer markets such as ITS applications, integration of GPS and cellular phones, and portable GPS for outdoor recreation and similar activities. These could account for more than 60% of the GPS market by the turn of the century. There are at present over 100 manufacturers of GPS instruments of varying kinds.

8.3.4 The User Segment (Positioning Principles)

The basic concept of GPS positioning is that of positioning-by-ranges. The geometrical principles of positioning can be demonstrated in terms of the intersection of locii. In the two-dimensional case, a measured range to a known point constrains the position to lie on circle with the measured range as radius. In three dimensions a measured range to a known point constrains the position in 3-D space to lie on the surface of a sphere centred at the known point, with radius being the measured distance (Fig. 8.4). In the case of GPS, the distance measurement is made to a satellite with known position (coordinates are obtained from the satellite ephemeris data transmitted within the navigation message), however the principle applies to any range measuring positioning system, terrestrial or satellite-based.

In two dimensions, position can be defined as the intersection of two circles, involving distances d1 and d2 to two known points. Note that there are two possible solutions, only one of which is correct. In general one solution can be discarded rather easily through apriori knowledge of approximate position and velocity. Another possibility is to measure another range to a third point and if all ranges are measured without error the intersection of three LOPs is a single uniquely defined point.

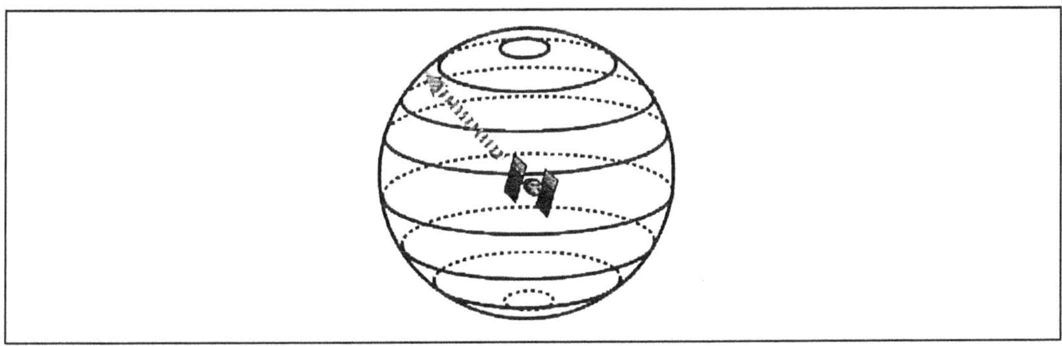

Fig. 8.4 Surfaces of Position for Range Measurements.

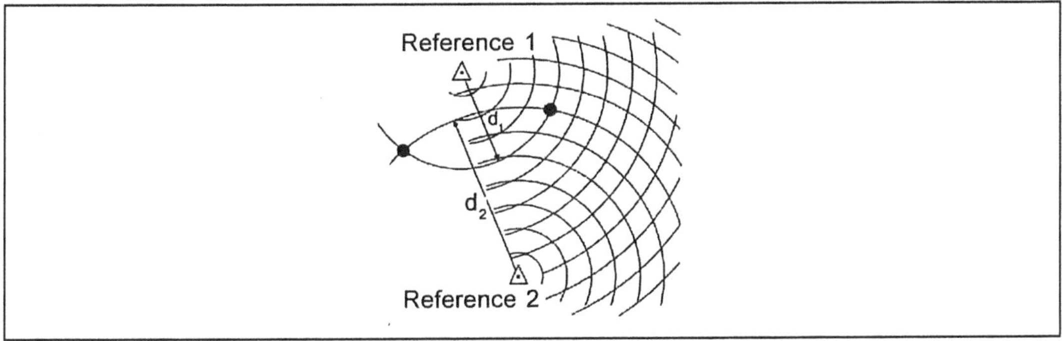

Fig. 8.5 The Intersection of Circular Lines of Position for 2-D Positioning.

In the three-dimensional case, the intersection of three spheres describes two points in space, only one of which is correct (Fig. 8.6). Hence, a minimum of three ranges are required, to three separated known points, in order to solve the 3-D position problem. The quality of the positioning solution is dependent, amongst other things, on the accuracy with which the ranges can be measured and the geometry of the intersection.

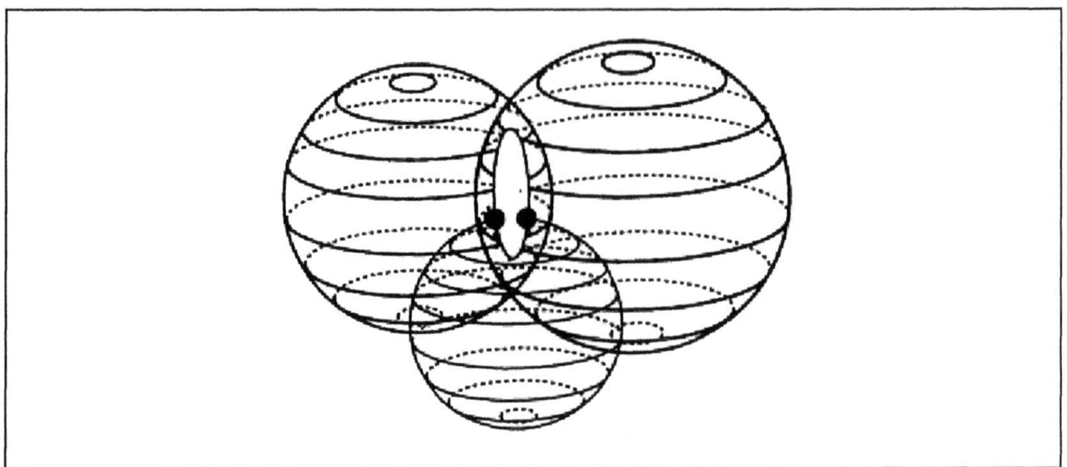

Fig. 8.6 Intersection of Surfaces of Position Based on Range Measurements.

If the point being positioned is stationary, the 2 (or 3) ranges do not need to be measured simultaneously. If the point is moving however, all ranges must be measured simultaneously (or over an interval of time during which the point has not moved by an amount greater than the uncertainty of the "fix"). Because the GPS constellation was designed to ensure at least four satellites are always visible anywhere on the earth, satellite positioning using simultaneously measured ranges is the basic positioning strategy for most navigation applications.

However, there is still the issue of how to account for measurement biases, as the technology used for making GPS range measurements does not give calibrated distance from the receiver to the satellite. Disturbing influences and errors in fact contaminate the range measurements to an unacceptable degree, and hence the basic positioning principle is modified in several ways to satisfy the varying levels of accuracies required by different applications.

8.4 GPS Satellite Constellation and Signals

8.4.1 GPS Constellation Design

The operational Block II/IIA satellite constellation was to be fully deployed by the late 1980's. However, a number of factors, the main one being the Space Shuttle Challenger disaster (28 January 1986), has meant that the GPS system only became operational in the 1990's as far as most users were concerned. Full Operational Capability was declared on 17 July 1995 — 24 Block II/IIA satellites operating satisfactorily. At an altitude of approximately 20,200km, a constellation of 24 functioning GPS satellites is sufficient to ensure that there will always be at least four satellites visible, at all unobstructed sites on the globe. Typically there are 6 to 10 satellites visible most of the day. The U.S. Department of Defense has undertaken to guarantee 24 satellite coverage 70% of the time, and 21 satellite coverage 98% of the time.

As the GPS satellites are in nearly circular orbits, at an altitude of approximately 20,200km above the earth (Fig. 8.7), this has a number of consequences :

- Their orbital period is approximately 11hrs 58mins, so that each satellite makes two revolutions in one sidereal day (the period taken for the earth to complete one rotation about its axis with respect to the stars).
- At the end of a sidereal day (23hrs 56mins in length) the satellites are again over the same position on earth.
- Reckoned in terms of a solar day (24hrs in length), the satellites are in the same position in the sky about four minutes earlier each day.
- The orbit groundtrack approximately repeats each day, except that there is a very small drift of the orbital plane to the west which is arrested by periodic manoeuvres.

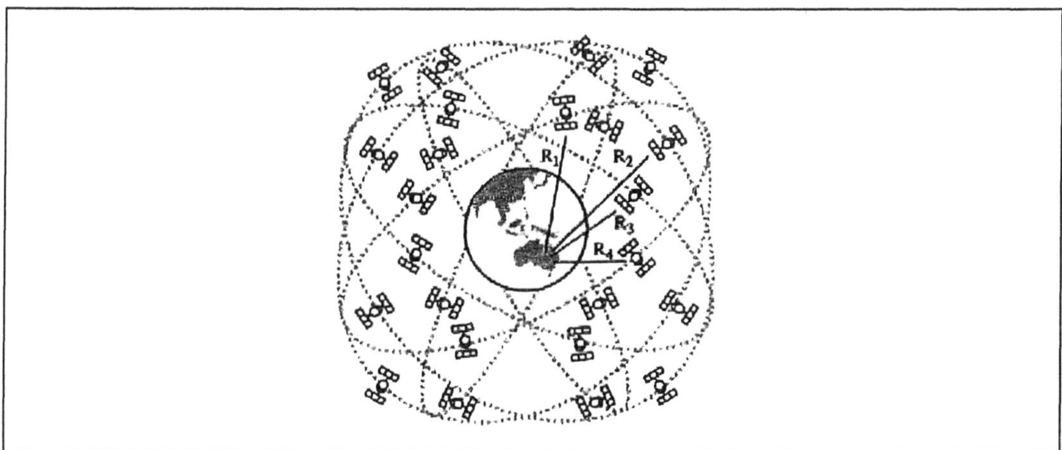

Fig. 8.7 The GPS Constellation "Birdcage".

The following general remarks can be made with regard to satellite constellation design for navigation purposes:

- The higher a satellite, the longer it is visible above the horizon (the extreme case is the geostationary satellites).
- The higher a satellite, the better the coverage due to longer fly-over passes and extended visibility of the satellite across large areas of the earth.
- The higher a satellite, the less the rate-of-change of distance, and the lower the Doppler frequency of a transmitted signal.
- The greater the angle of inclination, the more northerly the track of the sub-satellite point across the surface of the earth.
- No satellite can be seen simultaneously from all locations on the earth.

- Depending on the positioning principles being employed, there may be a requirement for observations to be made to more than one satellite simultaneously from more than one ground station.

The Block II satellites are deployed in six orbital planes at 60o intervals about the equator, with each containing four satellites. The satellites can be moved round their orbits if it becomes necessary to "cover" for a failed satellite. The orbital planes are inclined at an angle of 55^0 to the equatorial plane.

As the satellites are at an altitude of more than three times the earth's radius, a particular satellite may be above an observer's horizon for many hours, perhaps 6-7 hours or more in the one pass. At various times of the day, and at various locations on the surface of the earth, the number of satellites and the length of time they are above an observer's horizon will vary. Although at certain times of the day there may be up to 12 satellites visible simultaneously, there are nevertheless occasional periods of degraded satellite coverage (though naturally their frequency and duration will increase if satellites fail). "Degraded satellite coverage" is generally defined in terms of the magnitude of the Dilution of Precision (DOP) factor, a measure of the quality of satellite geometry. The highest the DOP value, the poorer the satellite geometry. For example, if all the visible satellites are located in the same part of the sky, the intersections of the SOPs will be very obtuse.

8.4.2 GPS Signal Components

The basis of the GPS signal are the two L-band carrier signals. These are generated by multiplying the fundamental frequency f_o (10.23MHz) by 154 and 120, yielding the two microwave L-band carrier waves L1 and L2 respectively. The frequency of the two waves are: $f_{L1} = f_o \times 154 = 1575.42$ MHz, and $f_{L2} = f_o \times 120 = 1227.6$ MHz. These are radio frequency waves capable of transmission through the atmosphere over great distances, but which cannot penetrate solid objects. Note that all GPS satellites transmit carrier waves at the same two L-band frequencies (unlike the GLONASS system, where a different frequency is assigned to each satellite. However, the L-band carrier waves themselves carry no information, and must be modified (or modulated) in some way. In the Global Positioning System the L-band carrier waves are modulated by two ranging codes, and the navigation message. The two distinct GPS ranging codes are :

- The C/A code (sometimes referred to as the "clear/access" or "coarse/acquisition" code), sometimes also referred to as the "S code".
- The P code (the "private" or "precise" code) was designed for use only by the military, and other authorised users.

The L1 carrier was designed to be modulated with both the P and C/A codes, whereas the L2 carrier would be modulated only with the P code. Under the policy of Anti-Spoofing the P code is encrypted through modulation by a further secret code (the "W code") to produce a new "Y code". Both carrier signals contain the navigation message.

The C/A and P (or Y) codes provide the means by which a GPS receiver can measure one-way ranges to the satellites. These codes have the characteristics of random noise, but are in fact binary codes generated by mathematical algorithms and are therefore referred to as "pseudo-random-noise" or PRN codes. Both the C/A and P code generating algorithms are known, and are based on a simple Tapped Feedback Shift Register scheme. One C/A code is assigned to each GPS satellite (the PRN code number is often used as the satellite I.D.). Each C/A code is a 1023 "chip" long binary sequence, generated at a rate of 1.023 million chips per second (that is, 1.023 MHz), thus the entire C/A code sequence repeats every millisecond. The P code is a far more complex binary sequence of 0's and 1's, being some 267 days long with a chipping rate at the fundamental frequency f_0 (10.23 MHz). The resolution of this code (length of the P code chip) is ten times the resolution of the C/A code. Instead of assigning each satellite a unique code, as is the case with the C/A code, the P code is allocated such that each satellite transmits a one week portion of the 267 day long PRN sequence, restarting the code sequence at the end of each week.

To measure one-way range (from satellite to receiver), a knowledge of the ranging code(s) is required by the user's receiver. Knowing which PRN code is being transmitted by a satellite means that a receiver can generate a local replica of the same code sequence. These PRN codes possess a very important attribute : a given C/A (or P or Y) code will fully correlate with an exact replica of itself only when the two codes are aligned, and has a low degree of correlation with other alignments.

As the same P (or Y) code is synchronously modulated on both carrier waves, any difference in signal transit time of the same PRN sequence is due to the retardation of the two L-band signals by a different amount as they travel through the ionosphere. The effect of the ionosphere on signal propagation is essentially a function of signal frequency, hence measurement on both frequencies is a very effective way of overcoming the ionospheric signal delay.

8.4.3 The Civilian - Military Relationship

Although GPS is a military navigation system the civilian sector represents an important (and rapidly growing) user group that has increasingly lobbied the U.S. Government in order to influence: (a) the direction of GPS system development; (b) official GPS policy concerning enhancement and control; and (c) the design of follow-on systems to GPS for the 21st century. Several policy decisions have already been made which impact on GPS performance. Some of these actions were agreed to during the early system design phase, while others were made only after much of the present system had been deployed :

- Two PRN ranging codes are implemented. The C/A code is intended for general, civilian use, while the P code was reserved for military and other authorised users. Because of the P code's higher measurement resolution it was expected that the accuracy of positioning using the P code would be much better than that possible using the C/A code. However, it was found that the performance of C/A code positioning was often no worse than that of P code positioning by a factor of two. (The latest C/A code tracking technology permits ranging quality almost as good as P code ranging and hence, all other things being equal, positioning accuracy close to that expected from the processing of P code ranges should be possible.)

- These two levels of positioning performance were designed into the GPS system from the very beginning. The positioning service based on using C/A code ranging data is referred to as the "Standard Positioning Service" (SPS), while the service based on P code ranging data is known as the "Precise Positioning Service" (PPS).
- As a result of the demonstration of a surprisingly good level of SPS accuracy, the policy of "Selective Availability" (SA) was endorsed in order to artificially widen the gap between the two levels of positioning. SA is an intentional degradation of the accuracy of GPS horizontal positioning to 100m and vertical positioning at the 160m (at the 95% confidence level) for SPS users. SA is implemented through an encryption of the navigation message whereby a part of the transmitted ephemeris and satellite clock data is falsified (the so-called "epsilon" effect) and the satellite clock is "dithered" (the so-called "delta" effect). SA therefore affects the precision of all measurements, code and carrier phase. SA does not affect PPS users who have user equipment able to decipher the correct ephemeris and clock error data.
- Any GPS hardware manufacturer is able to construct a P code ranging receiver (the P code PRN generation algorithm is published). The non-military market for P code receivers was always assumed to be very small, and one that could be controlled by the U.S. Government through the issuance of "export licences", etc. However, a significant demand by the land surveying market for dual-frequency phase tracking GPS receivers (code-correlating phase tracking receivers need a knowledge of the P code PRN in order to make a carrier phase measurement) led to an expansion in the production of P code capable positioning equipment.
- Under another policy known as "Anti-Spoofing" (AS), access is denied to the P code modulated on both L-band frequencies. AS was implemented on 31 January 1994, through the encryption of a secret "W code" onto the P code. The rationale behind this decision was that by keeping the military PRN code secret, an enemy of the U.S. could not jam the signal using a ground-based transmitter, nor "spoof" military GPS receivers by transmitting a false P code signal from a satellite. However, several GPS receiver manufacturers have developed proprietary techniques for making dualfrequency measurements even in the presence of AS.

- Dual-frequency observations will lead to more accurate positioning results than single frequency observations, because the ionospheric bias can be eliminated from the code range measurements. The fact that the C/A code is only modulated on the L1 carrier is therefore an intentional design decision to ensure that the SPS service cannot deliver the accuracy that the PPS service can, even with SA off.

It must be emphasised that the situation as far as the policies on SA and AS is under almost continuous review, particularly since the Presidential Decision Directive on the "U.S. National GPS Policy" was released in March 1996.

8.4.4 Why is the GPS Signal so Complicated?

GPS was designed as a complex military navigation system, which would also be used by civilians, hence we can summarise the reasons why the signal is so complicated as:

(a) The requirement for GPS to be multi-user system is most easily satisfied if it designed as a listenonly, or self-positioning system. An unlimited number of users could be supported because no return signal was necessary, however, the measurements and positioning procedures must then be based on one-way (satellite-to-receiver) signals.

(b) The very important requirement for real-time positioning influenced a number of design criteria related to the satellite signals:

- Simultaneous measurements from many satellites, but all signals are at the same frequency (though Doppler-shifted) — need to identify signals by use of different codes
- Unambiguous range measurements were required — need to determine signal delay
- Satellite positions are needed — broadcast ephemerides

Hence the PRN codes were used to distinguish the different satellite signals. The measurement of a "range" to a particular satellite required that a comparison be made of the PRN code generated within the receiver (corresponding to the satellite in question) to the PRN code sequence contained within the incoming satellite signal.

(c) The requirement for GPS to support high accuracy positioning also had implications for satellite signal design:

- High frequency modulation — P code at 10 MHz
- Dual-frequency signal — permits ionospheric delay estimation
- Microwave carrier frequency — 1.2 to 1.6 GHz

(d) The anti-jamming requirement could be partly met through the use of spread-spectrum codes. This also fulfilled the requirement for a low power signal that could still be detected even though it was below the ambient signal noise level. By ensuring that some of the PRN codes are secret (known only to military and a few authorised users), jamming or "spoofing" of signals was made much more difficult. (However, it is possible to jam a GPS receiver if the jamming source is within several metres of the receiver.)

(e) As the system was intended as a "dual-use" technology, satisfying both military and civilian users, this had several implications the most important of which was that the civilians were not to have the same level of accuracy as the military users :

- The use of two PRN codes — P and C/A code
- Restriction on the dual-frequency signal for civilians — C/A code only on the L1 frequency
- Microwave carrier frequency — 1.2 to 1.6 GHz

8.4.5 GPS Satellite Ephemerides

How are the GPS satellite ephemerides computed? As the forces of gravitational and non-gravitational origin perturb the motion of the GPS satellites, the coordinates of the satellites in relation to the WGS84 reference system must be continually determined through the analysis of tracking data. In the case of the GPS broadcast ephemeris, this procedure is a three-step process (Langley, R. B, 1991a)

- An off-line orbit determination is performed through the analysis of tracking to generate a "reference orbit". This is an initial estimate of the satellite trajectories computed from about one week's of tracking from the five Control Segment monitor stations.
- An on-line daily updating of the reference orbit using a Kalman filter as new data are added. This provides the current estimates of the satellite orbit which is used to predict the future orbit.
- The ephemeris is estimated for 1 to 14 days into the future. To obtain the necessary broadcast information, curve fits are made to 4 to 6 hour portions of the extrapolated ephemeris, and hourly orbit parameters determined.

Note that these parameters are not true Keplerian elements as they only describe the ephemeris over the interval of applicability and not for the whole orbit. (Although only intended for use during the transmission period, they do, however, adequately describe the orbit over intervals of 1.5 to 5 or more hours, with a graceful degradation in accuracy.). The user can derive the earth-centred, earth-fixed WGS84 Cartesian coordinates of the GPS satellite from the broadcast orbital parameters, using an algorithm described in (Vandierendonck, A. J., et.al, 1980) and implemented in every GPS receiver.

8.5 GPS Measurements

8.5.1 The Transmitted Signal

The signal that actually leaves a GPS satellite antenna is a combination of the three components: carrier wave, ranging codes and navigation message. The generation of the signal to be transmitted is carried out in a number of steps, and relies on the fact that all the components are derived by multiplying or dividing the fundamental frequency (Fig. 8.2). There are two distinct procedures for the combination of signal components: (a) BPSK modulation of binary sequences onto the carrier waves, and (b) modulo-2 addition of binary sequences.

Biphase Shift Key Modulation (BPSK) is the technique used to added a binary signal to a sine wave carrier. This amounts to causing a 180° phase shift in the carrier each time the binary sequence undergoes a transition from "0" to "1", or "1" to "0". The P code plus navigation message is modulated on both the L1 and L2 carriers, while the C/A code plus navigation message is only modulated on the L1 carrier.

An example of binary-to-binary modification of codes is the modulo-2 addition of the navigation message data to the C/A code. Because of the difference in frequency between the C/A code and the message stream this has the effect of inverting 20 C/A code binary "states" whenever the data bit of the navigation message is equal to "1". Conversely, when the data bit is "0" the C/A code sequence remains unaffected. The same satellite message is also modulated onto the P code sequence using this modulo-2 addition procedure. There are two main types of measurements that can be made on the GPS signals (Langley, R. B, 1993) :

- range observations based on the PRN codes, sometimes referred to as "code range" or "code phase", and
- carrier phase observations, which are more precise range-type measurements, but which have a much higher degree of "ambiguity" than the code ranges.

8.5.2 The GPS Range Measurements

The PRN codes are accurate time marks that permit the receiver's navigation computer to determine the time of transmission for any portion of the satellite signal. Before considering how ranging is carried out we need to describe, in general terms, how the incoming satellite signal is processed within a GPS receiver. (Parkinson, 1994 and Kaplen 1996)

Neglecting any discussion on extraneous noise (in relation to which PRN codes have special properties), and assuming that the only satellite signals received at the antenna originate from one GPS satellite, the following simplistic procedure is carried out within a receiver "channel". The L1 carrier modulated by the C/A code is converted to a signal of lower frequency and then mixed with a locally on a different time scale (due to nonsynchronization of receiver clock to GPS Time and the travel time of the signal from the satellite to antenna) to that of the incoming C/A code. As soon as the incoming signal and the receiver C/A code sequences are aligned within the receiver the ones and zeros of the two codes cancel, leaving the received carrier modulated only by the binary navigation message.

The extraction of the code range, or more precisely the determination of the amount by which the receiver generated PRN code must be shifted to align it with the incoming signal, is carried out with the aid of a PRN code correlator in some form "delay-lock loop" scheme. How accurately is this carried out? The C/A code has a sequence rate of 1.023 Mbps, corresponding to a resolution of about 300m (speed of light divided by the frequency). The P (or Y) code, on the other hand, has a socalled "chip" rate of 10.23 Mbps, and hence an effective resolution of about 30m. As a rule-of-thumb, alignment of the incoming and receiver generated codes is generally possible to within about 1-2% of the chipping rate, hence the measurement precision of C/A code ranging is of the order of 3-5m, and for P code ranging it is of the order of 0.3-0.5m. However, under the policy of "Anti-Spoofing", the Y code is encrypted and is therefore not available to civilian applications.

It should be noted, however, that modern "narrow correlator" C/A code measurement technology as implemented in several "top-end" GPS receivers has demonstrated ten times better correlation performance for the C/A code correlation than that quoted above.

8.5.3 The GPS Carrier Phase Measurements

The wavelengths of the carrier waves are very short (approximately 19cm for L1 and 24cm for L2) compared to the C/A and P code wavelengths. Assuming a measurement resolution of 1-2% of the wavelength, this means that carrier phase can be measured to millimetre precision compared with a few metres for C/A code measurements. Unfortunately a phase measurement is ambiguous as it cannot discriminate one (L1 or L2) wavelength from another. In other words, time of transmission information for the L-band signal cannot be imprinted onto the carrier wave as is done using PRN codes (this would be possible only if the PRN code frequency was the same as the carrier wave, rather than 154 or 120 times lower in the case of the P code, and 1540 or 1200 times lower for the C/A code). The basic phase measurement is therefore an angle in the range 0o to 360o. It is nevertheless the basis for high precision GPS positioning.

8.5.4 Ranging Using PRN Codes

Consider for a moment a perfect system, where all satellite clocks are synchronized to the same time system: the GPS Time (GPST). Furthermore, the ground receiver's clock also maintains the same synchronization, and none of the clocks drift from this GPST scale. Now suppose the satellite starts transmitting its L1 carrier (modulated with the combined C/A PRN code and navigation data). At the same instant, the receiver begins generating the C/A PRN code corresponding to that particular satellite. Under these circumstances the satellite and receiver generated C/A codes would be output in unison. When the satellite signal is received, however, it will be lagging the receiver generated code due to the signal transit time. Multiplying the time offset required to align the two codes (in effect determining the signal transit time) by the speed of light yields the satellite to receiver distance.

Measuring ranges simultaneously in this fashion to three satellites would fix one's position at the intersection of three spheres of known radii (the satellite ranges), centred at each satellite whose coordinates can be calculated from the navigation message. In reality the situation is more complex :

- GPS receivers are equipped with crystal clocks that do not keep the same time as the more stable satellite clocks (the satellite clocks can be nearly synchronised to GPST using the clock correction model transmitted in the navigation message). Consequently each range is contaminated by the receiver clock error. This range quantity is therefore referred to as pseudo-range, and in order for the user to derive position from pseudo-range data, the receiver equipment is required to track (a minimum of) four satellites, and solve for four unknown quantities: the three-dimensional position components and the receiver-clock offset (from GPST). This is the basis of GPS real-time navigation, and why GPS could be considered an example of a time-differenceof- arrival system.
- There is in fact a 300 km "ambiguity" in the C/A code pseudo-range measurements (300 km is the approximate length of the C/A code sequence). That is, all measured "distances" appear to have a range of 0 to 300 km. This ambiguity is resolved in a number of ways, but the easiest to assume that if the approximate receiver position is known to within say 100 km, the "missing" component of the distance can be determined, and hence the raw pseudo-range measurement can be corrected for this ambiguity to obtain the full satellite-receiver distance.
- Ranging (and hence receiver position determination) can be carried out using the C/A code or the P code. P code ranging can be done on the combination of the two frequencies, hence eliminating the bias due to ionospheric refraction. Furthermore, the C/A code is "coarser", and hence the C/A derived ranges are subject to greater measurement "noise". The absence of a C/A code on L2 is intentional, as one of the accuracy limitations of the GPS system for the general class of civilian users.
- As previously mentioned, this differentiation between ranging codes, and the formulation of policies for their use (in peacetime and in times of global emergencies), is responsible for the provision of two GPS services: The Precise Positioning Service based on P or Y code (dual-frequency) ranging, and the Standard Positioning Service based on single frequency C/A code ranging.

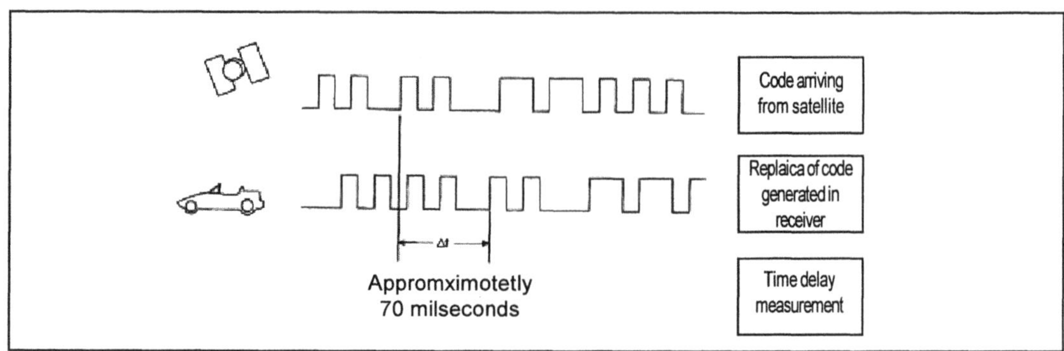

Fig. 8.8 One-Way Ranging Using PRN Codes.

8.5.5 An Observation Model of the Pseudo-Range

The observation equation for a receiver-clock-biased range (Langley, R. B. 1991d)is

$$p = \rho + \varepsilon_{rc}(t_r) \cdot c \qquad \text{..........(8.1)}$$

where c is the velocity of electromagnetic radiation in a vacuum (or simply the "velocity of light"), erc is the receiver clock error caused by the receiver oscillator (assume satellite clock time is "true" time) at time of reception t_r, p is the measured range and ρ is the true "geometric" range. Each observation made by the receiver can be parameterised as follows :

$$(x^s - x)^2 + (y^s - y)^2 + (z^s - z)^2 = (\rho - \varepsilon_{rc} \cdot c)^2 \qquad \text{..........(8.2)}$$

where x^s, y^s, z^s is the coordinate of the satellite and x, y, z is the coordinate of the receiver. Note that the time argument has been discarded (satellite coordinates are time-dependent, and so are the receiver coordinates if the receiver is moving).

As the 3-D coordinate of the satellite is known, then each measurement p contains four parameters which may be considered unknown: the 3-D coordinate of the receiver (x, y, z) and the receiver clock error (ε_{rc}). By making four measurements, to four different satellites, the following system of equations is obtained:

$$(x^{s1} - x)^2 + (y^{s1} - y)^2 + (z^{s1} - z)^2 = (p^1 - \varepsilon_{rc} \cdot c)^2$$
$$(x^{s2} - x)^2 + (y^{s2} - y)^2 + (z^{s2} - z)^2 = (p^2 - \varepsilon_{rc} \cdot c)^2$$
$$(x^{s3} - x)^2 + (y^{s3} - y)^2 + (z^{s3} - z)^2 = (p^3 - \varepsilon_{rc} \cdot c)^2$$
$$(x^{s4} - x)^2 + (y^{s4} - y)^2 + (z^{s4} - z)^2 = (p^4 - \varepsilon_{rc} \cdot c)^2 \qquad \text{..........(8.3)}$$

which has a unique solution **(see, for example [15]).** If more than four measurements are made the method of Least Squares, or other estimation techniques, can be used to determine the optimum solution. Does the receiver clock error have to be estimated at each epoch? That depends upon such factors as :

- How well the clock error is estimated.
- How often the position solution is carried out.
- The stability of the clock.

Under the assumptions that (a) there is a range measurement uncertainty of about 1 metre attributable to receiver "noise", and (b) the receiver is equipped with a quartz crystal clock (stability of 0.1 nanoseconds/second), then the uncertainty of the clock time after 30 seconds is as great as the range measurement error. Hence, once the clock error were determined, it would have to be independently reestimated at least every 30 seconds, otherwise it would dominate the range error budget. This is one of the reasons why this procedure could not be used if pseudo-range measurements to different satellites were not made "simultaneously" (here, within 30 seconds of each other), so that the clock error could be assumed to be a constant.

This is the conventional strategy used for pseudo-range-based GPS positioning: the receiver clock error is simply treated as an additional unknown and the navigation problem can be considered as 4-D estimation. However, the GPS satellite clock error is assumed to be a known quantity, and parameters defining this clock error are transmitted in the navigation message. The ionospheric delay is also modelled using transmitted parameters. An estimate of the tropospheric delay may be obtained from a simple tropospheric model such as Hopfield. All other biases are assumed to be insignificant compared to the measurement noise.

8.5.6 Coping with the Satellite Clock Bias

As the satellite clock error is the largest source of GPS measurement bias it deserves closer study. Under the assumption that the satellite clock error is an unknown quantity, the observation equation for such a satellite-biased range :

$$p = \rho + \varepsilon^{sc}(T^s) \, . \, c \qquad \text{..........(8.4)}$$

ε^{sc} is the satellite clock error caused by the satellite oscillator not being synchronized to "true" time (GPST). p is the measured range, ρ is the true range and T^s is the time of transmission. Each observation made by the receiver can be parameterised as in equation (1.1), except for the replacement of the term ε_{sc} for ε_{rc} :

$$(x^s - x)^2 + (y^s - y)^2 + (z^s - z)^2 = (p - \varepsilon_{sc} \, . \, c)^2 \qquad \text{..........(8.5)}$$

The 3-D coordinate of the satellite signal transmitter (x^s, y^s, z^s) is known, hence in the case of three range observations there are six unknowns in the system: the 3-D coordinate of the receiver (x_{r1}, y_{r1}, z_{r1}) and the three satellite clock error terms (ε^{sc1}, ε^{sc2}, ε^{sc3}). Hence, at first glance, six satellite-biased range observations are required to solve this positioning problem. It is not, however, possible simply to make observations to more satellites as each new observation introduces a new satellite clock parameter. There are two options for overcoming this dilemma.

It is possible to take advantage of the fact that all observations made to a particular satellite are biased by the same amount (if made at the same time, or close enough together so that the satellite clock error can be assumed to have not changed by a significant amount). If three range observations are made from another station, whose coordinate is known (x_{r2}, y_{r2}, z_{r2}), then it is possible to obtain a system of six equations in six unknowns:

$$(x^{s1} - x_{r1})^2 + (y^{s1} - y_{r1})^2 + (z^{s1} - z_{r1})^2 = (p_{r1}{}^{s1} - \varepsilon^{sc1} \,.\, c)^2$$
$$(x^{s2} - x_{r1})^2 + (y^{s2} - y_{r1})^2 + (z^{s2} - z_{r1})^2 = (p_{r1}{}^{s2} - \varepsilon^{sc2} \,.\, c)^2$$
$$(x^{s3} - x_{r1})^2 + (y^{s3} - y_{r1})^2 + (z^{s3} - z^{r1})^2 = (p_{r1}{}^{s3} - \varepsilon^{sc3} \,.\, c)^2 \quad \text{..........(8.6)}$$
$$(x^{s1} - x_{r2})^2 + (y^{s1} - y_{r2})^2 + (z^{s1} - z^{r2})^2 = (p_{r2}{}^{s1} - \varepsilon^{sc1} \,.\, c)^2$$
$$(x^{s2} - x_{r2})^2 + (y^{s2} - y_{r2})^2 + (z^{s2} - z^{r2})^2 = (p_{r2}{}^{s2} - \varepsilon^{sc2} \,.\, c)^2$$
$$(x^{s3} - x_{r2})^2 + (y^{s3} - y_{r2})^2 + (z^{s3} - z^{r2})^2 = (p_{r2}{}^{s3} - \varepsilon^{sc3} \,.\, c)^2$$

for which a unique solution can be obtained. In a conceptual sense this is the basis of differential GPS positioning, although in reality there are several possible implementations.

The other strategy for accounting for satellite clock error is for the GPS operators to periodically determine the clock error. As the satellite clocks have significantly better long-term drift characteristics than the receiver clocks, a suitable clock error model could be a time polynomial :

$$\varepsilon^{sc} = a_0 + a_1 (t - t_{oc}) + a_2 (t - t_{oc})^2 \quad \text{.........(8.7)}$$

where : a_0 is the clock bias term

a_1 is the clock drift term

a_2 is the clock drift-rate

t is satellite clock time

t_{oc} is some reference epoch for the definition of the coefficients

What is actually available to users via the navigation message is a prediction of the satellite clock error behaviour for some time into the future (24 hours or more). Such deterministic models of satellite clock error are accurate to about 20 nanoseconds, or about six metres in equivalent range, depending upon the time since last navigation message update. Hence the measured pseudo-ranges may be corrected for satellite clock error, and then the observation model in equation (8.1) can be used.

Selective Availability complicates matters because it is a further artificial "dithering" of the satellite clock causing an additional several tens of metres error in the pseudo-range measurement.

8.6 GPS Instrumentation

The following components of a generic GPS receiver can be identified (Fig. 8.9):

- ***Antenna and Preamplifier :*** Antennas used for GPS receivers have broadbeam characteristics, thus they do not have to be pointed to the signal source like satellite TV receiving dishes. The antennas are compact and a variety of designs are possible. There is a trend to integrating the antenna assembly with the receiver electronics.

- ***Radio Frequency Section and Computer Processor :*** The RF section contains the signal processing electronics. Different receiver types use somewhat different techniques to process the signal. There is a powerful processor onboard not only to carry out computations such as extracting the ephemerides and determining the elevation/azimuth of the satellites, etc., but also to control the tracking and measurement function within modern digital circuits, and in some cases to carry out digital signal processing.
- ***Control Unit Interface :*** The control unit enables the operator to interact with the microprocessor. Its size and type varies greatly for different receivers, ranging from a handheld unit to soft keys surrounding an LCD screen fixed to the receiver "box".
- ***Recording Device :*** in the case of GPS receivers intended for specialised uses such as the surveying the measured data must be stored in some way for later data processing. In the case of ITS applications such as the logging of vehicle movement, only the GPS-derived coordinates and velocity may be recorded. A variety of storage devices were utilised in the past, including cassette and tape recorders, floppy disks and computer tapes, etc., but these days almost all receivers utilise solid state (RAM) memory or removable memory "cards".
- ***Power Supply :*** Transportable GPS receivers these days need low voltage DC power. The trend towards more energy efficient instrumentation is a strong one and most GPS receivers operate from a number of power sources, including internal NiCad or Lithium batteries, external batteries such as wet cell car batteries, or from mains power.

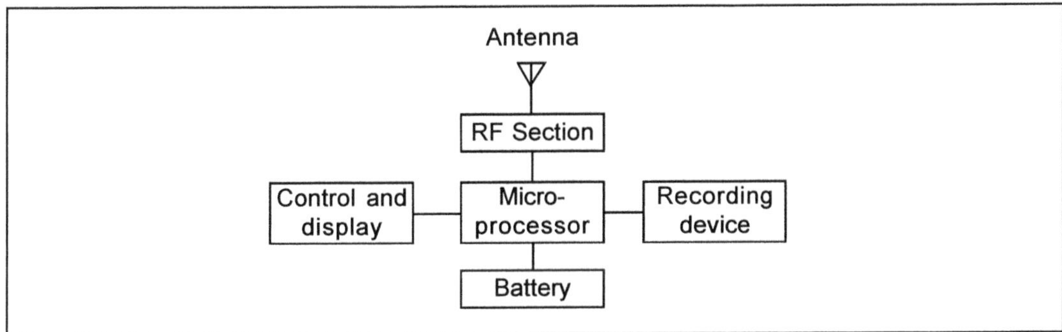

Fig. 8.9 The Generic GPS Receiver.

The antenna and RF technology components are briefly discussed below. For further details the reader is referred to (Wells, D. E, et.al. 1987).

8.6.1 Antennas

The task of the antenna is to convert the energy of the arriving electromagnetic waves into an electric current that can be processed by the receiver electronics. There are a number of special considerations as far as antenna design is concerned:

- the antenna must be able to pick and discriminate very weak signals,
- the antenna may need to operate at just the L1 frequency, or at both the L1 and L2 frequencies,
- as the signals are right-hand circularly polarised, the GPS antenna must also be right-hand circularly polarised,
- antenna gain pattern that enhances the ability of the RF section to discriminate against multipath signals (such as left-hand circularly polarised signals),
- a stable electrical centre,
- low cost, and
- reliable.

There have been several types of GPS antennas used:

- monopole or dipole configurations,
- quadrifilar helices,
- spiral helices,
- microstrip,
- choke ring, and other multipath resistant designs

In short, the antennas are required to be rugged, simple in construction, have stable phase centres, be resistant to multipath, have good gain and pattern coverage characteristics. The microstrip antenna is almost universally used for navigation applications.

8.6.2 Signal Processing Principles for Code-Correlating Receivers

The RF section of a GPS receiver converts the incoming (preamplified) signal to a signal of a more manageable frequency. This intermediate frequency is obtained by mixing the incoming signal with a pure sinusoidal signal generated by the local oscillator (the quartz "clock"). The frequency of this beat frequency is the difference between the original (Doppler-shifted) received carrier frequency and the local oscillator. The intermediate or beat frequency is then processed by the signal tracking circuitry.

There are several classes of signal processing techniques that can be employed to make observations, as well as several proprietary implementations of tracking technologies. This is particularly the case for instruments which track carrier phase. It is beyond the scope of this chapter to discuss in detail the electronic circuitry and the reader is referred to Kaplen 1996, and Spilkar 1995. As our interest is the pseudo-range measuring instruments, we need only consider the issues of the "code-correlating" signal processing technique, and some brief details of the processing architecture within the receiver, and in particular the form of the tracking "channel(s)". Code-correlating receivers employ tracking loops to extract the useful measurements from the beat signal. A typical GPS receiver contains two types of tracking loops :

- the delay-lock, or code-tracking, loop, and
- the phase-lock, or carrier tracking, loop.

The delay-lock loop is used to make the alignment of the PRN code sequence (C/A or P code) that is contained in the signal from a satellite with an identical PRN code generated within the receiver. A correlator in the delay-lock loop continuously cross-correlates the two code streams, time shifting the receiver generated stream until alignment is obtained. The time shift is then converted to a pseudo-range measurement. Once the code-tracking loop is aligned, the PRN code can be stripped from the satellite signal. The resulting signal then passes to the phase-lock loop where the satellite message is extracted. Once the local oscillator is locked onto the satellite signal it will continue to follow the variations in the phase of the carrier as the satellite-receiver distance changes.

A GPS receiver may be described as continuous or switching depending on the type of channel(s) it has. A continuous-tracking receiver has dedicated hardware channels, and each channel tracks a single satellite, maintaining continuous code and/or phase lock on the signal. Each channel is controlled and sampled by the receiver's microprocessor with input/output operations being performed fast enough so that tracking is not disturbed. Continuous-tracking receivers may enjoy a signal-to-noise advantage over switching receivers in that the satellite signal is continuously available and may be more frequently sampled. A further advantage is a potential redundancy capability. Should one of the hardware channels fail, it may still be possible to obtain sufficient data to determine a position. One disadvantage of a multi-channel receiver is that the differences in signal path delay in the channels, the so-called inter-channel biases, must be well calibrated. This is the most common architecture used in GPS receivers today.

A switching receiver has hardware channels which sequentially sample the incoming signal from more than one satellite. There may be a single channel for all satellites, or each channel may track, say, two satellites. Code and/or carrier phase tracking for the individual signals is controlled by software (or, more usually, the "firmware") within the microprocessor. As a result, greater demands are placed on the microprocessor in a switching receiver and its programming is necessarily more complex — in effect hardware complexity is exchanged for software complexity. If the cost of hardware components is a significant factor in determining the selling price of a receiver then in principle, the cost of a switching receiver should be cheaper than a continuous receiver (this is certainly the case for the low cost GPS navigation units).

There are three basic kinds of switching receivers, distinguished by the time required to sequence through the signals tracked by a particular channel. A multiplexing channel is one for which the sequencing time to sample all satellites assigned to the channel is equal to 20 milliseconds, the period of one bit in the satellite navigation message. The sampling can be arranged so that no message bit boundary is spanned by any tracking interval. In this way, the messages from all satellites phase tracked by the channel can be read simultaneously. A multiplexing channel can be used to obtain both L1 and L2 data by alternating between the frequencies every 20 milliseconds. If a channel switches between signals at a rate which is asynchronous with the message bit rate, the channel is referred to as a sequencing channel. A fast sequencing channel is one which takes the order of a second or so to sequence through the signals. A slow sequencing channel may take several seconds or even minutes. A single sequencing channel would lose bits in a particular satellite message during those intervals spent sampling the signals from other satellites. Consequently, sequencing receivers may have an extra hardware channel just for message decoding. Alternatively, the navigation message must be decoded before the receiver starts the tracking cycle for real-time positioning (note that the message only changes once an hour).

The configuration of channels is selected so that, for example in the case of a navigation receiver, a minimum of four satellites can be tracked at the same time. This may call for a single switching channel, or two switching (between two satellite signals) channels, or even five channels (one used for calibrating the other four channels, and decoding the navigation message).

8.6.3 Trends In GPS Instrumentation

It is impossible to precisely predict trends in GPS instrumentation. The task is no easier if we focus our attention only to a specific market segment, such as positioning hardware for ITS applications. Nevertheless, speculation based on R&D activities being undertaken at present gives us some clues :

- The third generation of GPS receivers presently available already exhibit significant gains in miniaturisation, reduction in power consumption, and portability, over the earlier models. We can expect this trend to continue.
- The choice confronting GPS manufacturers is whether to maintain a broad-based development program, and market a product that is versatile enough to satisfy many applications (including the provision of interchangeable components such as antennas, RF units, memory, etc., to accomplish this), or to focus on specialist applications (military, navigation, differential navigation, surveying, kinematic, etc.). To date, most manufacturers are split between those with products for the high precision surveying market, and those focussing on the low cost, high volume navigation market. Only a few address all markets.

- There have been many predictions of low cost GPS receivers. However there is an enormous range of prices from the "top-of-the-line" surveying receivers to the "bare-bones" GPS "engines" implemented on one or more chips. GPS boardsets with basic navigation functionality are available at less than one hundred U.S. dollars each (when purchased in high volumes).
- There is a trend towards product differentiation, with many different configurations of tracking channels, data recording options, and, in particular, software options. Although some of these trends may be due to manufacturers wanting to give their product "an edge" in the marketplace, it is equally valid to suggest that this is in response to the different demands (some very specialist) of the market. However, even in the case of complex ITS systems, it is possible to identify the "basic GPS component", and distinguish it from the "add-ons" .
- An exciting marriage is possible between satellite navigation and satellite communication, combining real-time positioning with instantaneous transmission of position. GPS receivers may be fitted to many different platforms (some unmanned, for example rail rolling stock and ship cargo containers), and their locations remotely monitored at a central site via a satcom link. In addition, for land navigation these could include electronic map displays to aid the driver. What is clear is that for many applications the navigation data provided by a GPS receiver will merely be the first link in a complex information system.
- New tracking electronics, such as "narrow-correlator" technology, would improve pseudo-range measurement precisions, as would a combination of phase and pseudo-range measurement. Both are, however, still relatively expensive technologies and are not found on "bare-bones" GPS boardsets. Further electronic tracking refinements will lead to more multipath-resistant receivers.

9

Fundamentals of GIS

9.1 Introduction

To answer apparently simple geographical questions like : What is the population of a particular city? What are the characteristics of the soils in a particular land parcel? Are there any trends in the patterns of earthquakes in India which could help predict future quakes? How has the distribution of urban and rural population changed between the past two censuses? All these questions are pertinent and require the proper and accurate data derived from several sources, and be integrated into a consistent form. The art, science, engineering and technology required to answer these geographical questions constitute called Geographical Information System (GIS). GIS is a generic term denoting the use of computers to create and depict digital representations of the Earth's surface.

Roger Tomlin the father of Canada GIS is credited with visualising the need for computers to perform certain simple but labour-intensive tasks associated with Canada Land Inventory.

Tomlin (1990) saw that if a map could be represented in digital form, then it would be easy to make measurements of its basic elements, specifically the areas assigned and the tedious hand-measurement of area by conducting dots on transparent overlays of known dot density. Tomlin's cost-benefit analysis showed that computerisation would be cost effective, despite the enormous costs involved and the primitive nature of the computers of the time.

It is, however, important to note that the earlier investigators also realised the importance of such a computerised analysis of the spatial data. David Bickmore (1934), the primary GIS innovator, has discovered that, GIS stems from the benefits of automating the map production process. Once information of any kind is in digital form, it is much easier to manipulate, copy, edit, and transmit. Ray Boyle invented the "free pencil " digitiser, and by 1964 Bickmore and Boyle set up the Oxford system for high quality digital cartography (Rhind 1988). At that time, major mapping agencies, including the US and other military bodies began the lengthy and often rocky process of automation. Widespread achievement of the benefits of automated cartography had to await the development of suitable mechanisms for input, display, and output of map data, but the necessary devices map digitiser, interactive graphics display device and plotter, respectively became available at reasonable cost by the early to mid 1970's and from then onwards an increasing number of organisations set out to convert all their maps into a computerised form.

The world as composed of a set of largely independent layers and spatial elements, each representing some component of the environment, and some set of environmental concerns as well. These layers might include groundwater, natural vegetation, soil and other resources. McHarg (1996) was the foremost proponent of this view, and his group at the University of Pennsylvania applied it in a long series of exemplary studies. Although the initial idea was strictly manual, the computerisation of these ideas in a layer-based raster GIS was a simple step, and many systems owe their origins to McHarg's simple model (Tomlin 1990).

9.2 Roots of GIS

GIS has its roots in the stimulus provided by the development of remote sensing, in the late 1960s and early 1970s, as a potentially cheap and effective source of earth observations. While many of the techniques for processing remote sensing data are highly specialised, more general GIS techniques become important in order to combine information derived from remote sensing with other collateral information (Star et. al 1997). Today many GIS include extensive functionality for image processing, and all types of remote sensing are increasingly the primary data source, particularly for detection of landscape change, natural resource management and environmental monitoring and management. Similarly, GIS has many roots of evolution like map production process; one root lies in landscape architecture and environmentally sensitive planning; another root in urban and demographic data analysis; the roots of remote sensing merging with vector GIS; the roots of representational issues of space and time in GIS; the root of large scale data integration around a common data model; possibility of storing large number of layers of information on temporal changes to handle an image within a relational database environment etc, and so on.

If GIS has so many apparently independent roots, what brought them together, and why has the umbrella term 'GIS' become so widely accepted? First, there are obvious commonalities. For example, the representation of topology invented for the Dual Independent Map Encoding (DIME) system at the US Bureau of the Census, is almost identical to that incorporated in CGIS and in Australian work. The methods of raster (grid cell data) processing and storage used in remote sensing systems are almost conceptually identical to those used by the systems that have implemented McHarg's multi-layer view of the World. Second, it was easy from the viewpoint of the software engineering paradigms of the 1970s and 1980s to integrate functions around common representations. Once a raster or vector data model had been established, functions that process the data model in different ways were easy to add, so as to build large-scale integrations of image processing functions around a common raster representation.

Geographical Information is information about geography, that is, information tied to some specific set of locations on the Earth's surface including the zones of atmosphere. 'Spatial' is often used synonymously with or even in preference to, 'geographical' in this context, although in principle it might be taken to include information that is tied to frames other than the earth's surface, such as, human body (as in medical imaging) or a building. Because of this difficulty, the term 'geospatial' has become popular recently, notably in the context of the US National Spatial Data Infrastructure, the Canadian National Geospatial Infrastructure, and the UK National Geospatial Data Framework.

Today, the term GIS tends to be applied whenever geographical information in digital form is manipulated, whatever be the purpose of that manipulation. Thus using a computer to make a map is referred to as 'GIS'. This entails using the same computer to analyse geographical information and to make future forecasts using complex models of geographical processes. The earth images collected by remote sensing satellites are geographical data, but the systems that process the images are not to be called GIS as long as they remain confined to this particular form of data in such cases. 'GIS' tends to be reserved for systems that integrate remotely sensed data with other types, or process data that have already been cleaned and transformed. Similarly, an atmospheric scientist or oceanographer will tend to associate 'GIS' with the system used more for multidisciplinary work and policy studies, and other software environments for modelling and analysis within the confines of one's own discipline.

9.3 Overview of Information System

GIS might provide the medium for studying one or more of the fundamental issues that arise in using digital information technology to examine the surface of the earth or any related systems. The Resource Information System (RIS) for agricultural management, for instance, has to be considered multidimensional with attribute dimension, spatial dimension and temporal dimension. Geographic Information System (GIS) offers capabilities of integrating multisector, multilevel and multiperiod database. GIS is a computerized database system for capture, storage, retrieval, analysis, and display of spatial data. It is a general-purpose technology for handling geographic data in digital form, and satisfying the following specific needs, among others :

(i) the ability to preprocess data from large stores into a form suitable for analysis, including operations such as reformatting, change of projection, resampling, and generalisation.

(ii) direct support for analysis and modeling, so that form of analysis, calibrations of models, forecasting, and prediction are all handled through instructions to the GIS.

(iii) post processing of results including such operations as reformatting, tabulation, report generation, and mapping.

In all these operations, the typical GIS user now expects to be able to define requirements and interact with the system through a "user-friendly" intuitive interface icons and desktop metaphors (Mark 1986). The function of an information system is to improve one's ability to make decisions. An information system is the chain of operations that takes us from planning the observation and collection of data, to storage and analysis of the data, and to the use of the derived information in some decision-making process. This brings us to an important concept that a map is kind of information system. A map is a collection of stored, analysed data, and the information derived from this collection is used in making decisions. Fig. 9.1 shows the simplified information system overview.

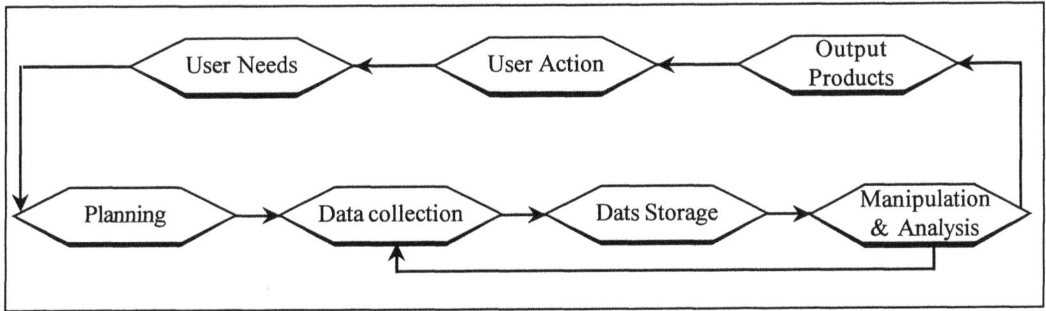

Fig. 9.1 Simplified information system overview.

9.4 The Four Ms

There are mainly four key activities that any urban planners or scientists or resource managers and others use geographic information for. They observe and measure environmental parameters and develop maps which portray characteristics of the earth. They monitor changes in our surroundings in space and time. In addition, they model alternatives of actions and process operation in the environment. These, four activities are Measurement, Mapping, Monitoring and Modelling termed as key activities which can be enhanced by the using information systems technologies through GIS. Fig 9.2 explains these four Ms.

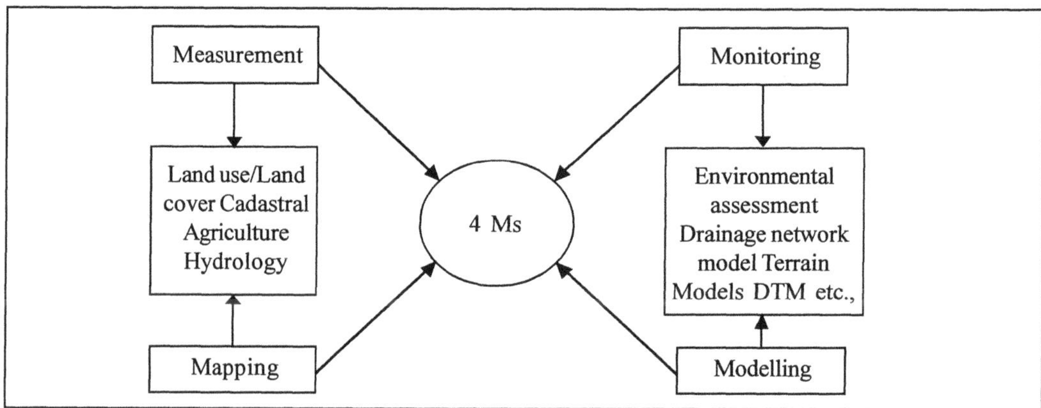

Fig. 9.2 Schernatic representation of Four Ms: Measurment, Mapping, Monitoring and Modelling.

GIS technology is more different from traditional mapping and map analysis. GIS is based on a mathematical framework of primitive map analysis operations analogous to those of traditional statistics and algebra. From this perspective, GIS forms a toolbox for processing maps and fundamental concepts for spatial measurement. It provides a foundation for advanced analytic operations involving spatial analysis and measurement. Most of GISs contain analytic capabilities for reclassifying and overlaying maps. Any GIS system for the measurement of areas, distances, angles and and so on requires two components, namely, a standard measurement unit and a measurement procedure.

Another major function of GIS capability is the study of environmental surroundings and the monitoring of environmental parameters (Burrough et. al 1988). Although analytical models have been linked to GIS for spatial measurement and resource assessment, the cross fertilisation between the modules of modelling, measurement and automated mapping allows the GIS user to monitor the environment and the earth systems.

In principle, it is possible to make a clear distinction between GIS and digital cartography. Mapping technology or digital cartography deals with map features and with associated attributes of colour, symbology, name of annotation, legends, neatlines and north arrows. GIS includes the capabilities for storing, editing, and handling the relationships of attributes with their spatial entities along with the capabilities of digital cartography. A map, an ultimate product of digital cartography or GIS, is a very persuasive form of data display and a computer drawn map carries the authority of a powerful technology.

GIS applications now span a wide range, from sophisticated analysis and modelling of spatial data to simple inventory and management. They also dictate the development directions of much of the industry. However, several vendors have chosen to concentrate on the niche market for environmental applications and to emphasise support for environmental modelling. GRASS is a significant public domain GIS software developed by USA with substantial capabilities for modelling.

9.5 Contribution Disciplines

GIS is convergence of technological fields and traditional disciplines. GIS has been called an " enabling technology " because of the potential it offers for the wide variety of disciplines dealing with spatial data. Many related fields of study provide techniques which make up GIS. These related fields emphasise data collections while GIS brings them together by emphasising integration, modelling and analysis. Thus GIS often claims to be the science of spatial information. Fig. 9.3 shows the technical and conceptual development of GIS. The contributing disciplines for the evolution of a GIS (Burrough, 1998) are geography, cartography, remote sensing, surveying and photogrammetry, computer science technology, mathematics, and statistics.

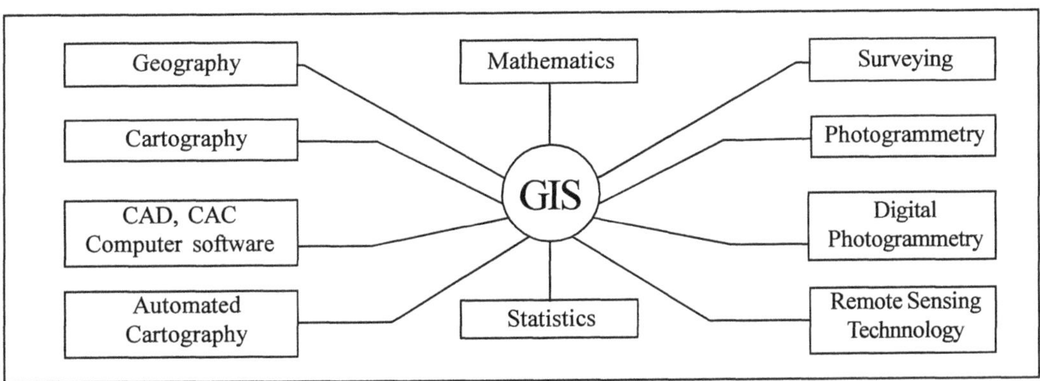

Fig. 9.3 GIS : the result of linking parallel developments in many separate spatial data processing disciplines.

Geography is broadly concerned with understanding the world and man's place in it. Geography has a long tradition in spatial analysis, and provides techniques for conducting spatial analysis and a spatial perspective on research. *Cartography* is concerned with display of spatial information. It is now the main source of input data for GIS (maps) and has a long tradition in the design of maps which is an important form of output from GIS.

Remote Sensing is becoming an important source of geographical data by providing digital images derived from space and the air. Remote sensing provides techniques for data acquisition and processing anywhere on the globe at a low cost, and consistent update potential. While integrated with GIS, remotely sensed imagery can be merged with other data in a GIS providing real-time spatial information. The first part of this book enlightens the concepts and the potential utility of remote sensing.

Surveying and Photogrammetry provide high quality data on positions of cadastral objects like land parcel and building, and topography. Aerial photogrammetry deals with the photographs taken by an aerial camera on board aircraft at different altitudes. Aerial photogrammetry is one of the most powerful data-capturing techniques for the creation of GIS spatial database. The relevant data can be extracted from the aerial photographs of various scales (Fig. 9.4), and may be used as input for GIS. Digital orthophotos provide the source of digital data. These products are scanned airphotos that have been rectified to eliminate displacement caused by variable elevation of the ground surface and the tilt of the camera. Properly registered with other digital data sets, these images can be used directly as backdrops for vector data or to provide a basemap for onscreen digitising. The user may abstract information on land use, vegetation type and other aspects of the landscape from the photograph.

Curran (1989) identifies six characteristics of aerial photographs of immense value as a data source for GIS. They are (i) wide availability, (ii) low cost, (iii) wide area views, (iv) time - freezing ability, (v) high spectral and spatial resolution, and (vi) three-dimensional perspective.

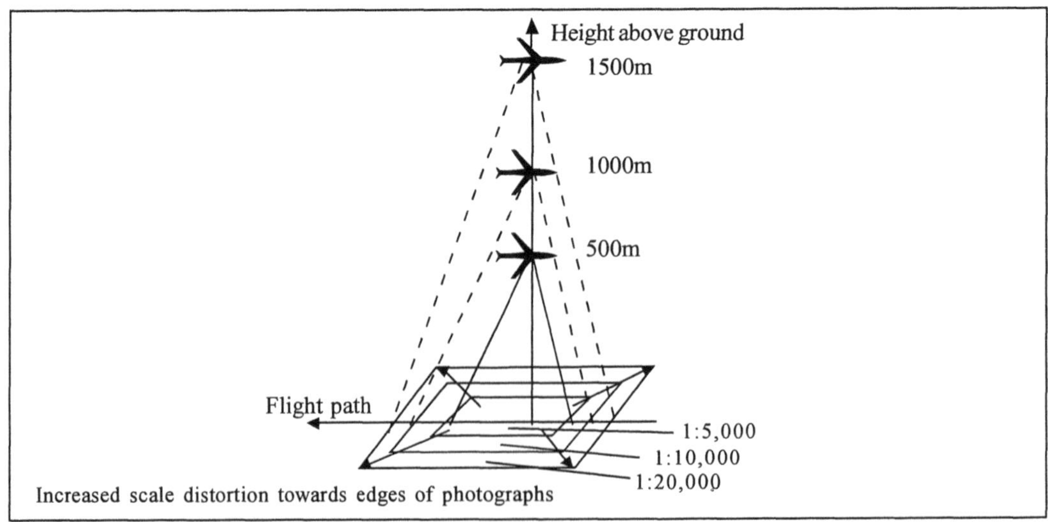

Fig. 9.4 Varying scale on Aerial photographs.

Computer Assisted Design (CAD) provides software, techniques for data input, display and visualisation, and representation, particularly in 3-dimensions. Advances in computer graphics provide hardware and software for handling and displaying graphic objects. Data Base Management System (DBMS) contributes methods for representing data in digital form and procedures for system design and update. Artificial intelligence (AI) uses the computer to make choices based on available data in a way that is seen to enhance the human intelligence and decision-making. Using AI, computer can act as an "expert" in such functions as designing maps, generalising map features and classification.

Several branches of mathematics, especially geometry and graph theory, are used in GIS system design and analysis of spatial data. Statistics is used to build models and perform spatial data analysis in GIS. Statistics is also important in understanding issues of error, quality and uncertainty in GIS data.

Availability of large quantities of spatial data in the form of digital aerial photograph, digital remote sensing imagery, advancement of computer hardware, software and software development, increasing demand of spatial information for management, and infrastructure development parameters, lead to have a system to handle all these requirements. In order to handle such data to meet these demands, to store, retrieve, handle, analyse, manipulate, and display the results, it requires computer based system. Such a system is Geographical Information System (GIS).

9.6 GIS Definitions and Terminology

GIS are decision support computer based systems for collecting, storing, presenting and analysing geographical spatial information. These systems are spatially referenced databases giving users the potentiality to control queries over space, and usually through time. GIS is much more advanced than Computer Aided Design (CAD) or any other spatial data system. The basic output of GIS or spatial data analysis system is a map. The need to analyse maps to compare and contrast patterns of earth related phenomena, is confirmed by the long standing tradition of doing so with traditional maps.

Many geographical phenomena are best described scientifically as fields. Good examples are topographic elevations, air temperatures, and soil moisture content. A 2-dimensional field may be defined as any single valued function of location in a 2-dimensional space and discrete fields, with nominal dependent variables. It appears that any geographical phenomenon can be represented either as a field or as a collection of digital objects. For example, a set of states or revnue or administrative units like mandals within a country would commonly be represented in a GIS as a set of areal objects, or as a set of linear objects that form their boundaries. Fields can be digitally represented by vector approaches, but are often represented by raster data structures.

9.6.1 Geographical Entities

Spatial analysis is a technology which typically requires two types of information about spatial objects : attribute, and geographical or locational information. Given the diversity of analytical perspectives within GIS it is difficult to define spatial data analysis. The results of such analysis depend upon the spatial arrangement of events (Goodchild et. al 1992). Events may be represented as geographical entities associated with the attributes. The following sections highlight these two types of information and their relationships. To the query, where the particular object is with respect to any coordinate system, the answer is the spatial or geographical or locational information. To the query, what that particular object is, the answer is the attribute data. Alternatively, it can be noted that the GIS consists of two types of data : spatial and attribute. The structure of data and various models of representation and the management of spatial data are given in chapter 10, the attribute data and management are provided in the chapter 11.

'Entities' are things in the real world. 'Objects' are things in the digital world. Digital objects and associated attributes and values represent geographical entities. The distinction between entity and objects makes explicit the difference between things and their representations in a formal system. The entity or the field model is more appropriate and are particularly interesting for topographic elevations. Topographic data normally are represented in GIS as fields, either through grid based digital elevation models (DEMs) or as triangular tessellations. Robinson (1958) identified four kinds of geographical quantities. They are point, line, area, and volume. There are three kinds of cartographic symbols : point, line, and area. Robinson discusses 2-dimensional data on 'Mapping quantitative point, line, and area data', and separates volume data under the title 'Mapping 3-dimensional data'.

The frequency of geographical entities with indistinct boundaries has been known for some time; yet vector GIS is tuned to represent entities with crisp boundaries, whereas raster GIS does not represent entity boundaries at all. Thus, formal methods for the representation of geographical entities with uncertain or graded boundaries is an important new area of study in GIS (Burrough and Frank 1996). Fuzzy set theory represents a possible approach to modelling entities with graded boundaries, but it has problems. Geographical entities and classification of geographical entities into categories is a well-known process both in everyday thinking and in scientific work. Various subfields of geography have developed elaborate classifications for landforms, vegetation assemblages, and settlements. In brief, we can note that locational information about the spatial objects of concern are generally described by means of their position on a map or geographical coordinate systems.

Map Features are holding the spatial information of the geographic feature entities, such as, the spatial location like latitude, longitude, x, y, z, shape of points like churches and tram stops, lines like roads and creeks, and polygons like blocks of land and parks.

9.6.2 Attributes

Attributes are the characteristics of the map features, and holding of the descriptive information about the geographic features. Attributes are the non-spatial data associated with time and area entities. They are considered characteristics of entity (Lawrini and Thompson, 1992). The GIS attributes are represented using colours, textures, and linear or graphic symbols like the gardens. The parks are shaded green, the church locations are designated using the special symbol, the bus routes are drawn with a specific line width and style, as broken lines of 12 points width, Contour lines are brown in colour, and so on. The actual value of the attribute that has been measured (sampled) and stored in the database is called attribute value. A classical example of attribute data associated with spatial entities of the environs of Hussain Sagar in Hyderabad may tell us that, a point represents a hotel, a line represents the road and area represents the boundaries of the lake. Each spatial entity may have more than one attribute associated with it , that is, a point representing the hotel may have a number of rooms, standard of accommodation and other related information.

Broadly speaking two types of attributes may be distinguished : primary attributes and secondary attributes. Socioeconomic characteristics, and physical properties of objects are some of the examples of primary attributes. Flows of information levels, districts, capitals, and mandal names are considered secondary attributes.

9.6.3 Topology

In GIS, topology is the term used to describe the geometric characteristic of objects which do not change under transformations and are independent of any co-ordinate system (Berrhardsen, 1992). The topological characteristics of an object are also independent of scale of measurement (Chrisman, 1997). Topology, as it relates to spatial data, consists of three elements, namely, adjacency, containment and connectivity (Burrough, 1986). Topology may be defined as constituting those properties of geometrical figures that are invariant under continuous deformation (Mc Donnell and Kemp, 1995). Broadly, topology can be explained any two ways. Firstly, topologically spatial relationships with the entity which are learned by human beings at a very early age. Secondly, topology consists of metric aspects of spatial relations, such as, size, shape, distance and direction. Many spatial relations between objects are topological in nature, including adjacency, containment and overlap.

Adjacency and containment describe the geometric relationships which exist between area features. Areas can be described as being adjacent when they share a common boundary. For example, boundary of the area of municipal corporation of Hyderabad and Secunderabad is common, or may be adjacent. Containment is an extension of the adjacency that describes area features which may be wholly contained within another area feature, such as, an island within a lake. Connectivity is a geometric property used to describe the linkages between line features, like road network. The geometric relationship between spatial entities and corresponding attributes are very crucial for spatial analysis and integration in GIS.

In other cases, metric properties, such as, distance or direction, expressed either quantitatively or qualitatively, may determine the meanings of various terms, for example, both 'north of' any 'near' normally refine the 'disjoint' topological relation, and are ill-defined for non-disjoint entities. Spatial relations between disjoint entities, which neither touch nor overlap, are characterised by a system of distinctions that is essentially independent of the system used to describe and classify spatial relations for non-disjoint entities. Some of the spatial relations between disjoint objects are distance, direction, and reference frames.

Distance may be pure Euclidean distance. In natural language 'hedge' words such as 'about' are often associated with approximate numerical distance. Distance may be given in qualitative rather than metric terms, dividing distance into just three categories : 'at', 'near', and 'far'. Direction may be either qualitative or quantitative. Direction is an orientation specified relative to some reference frame. Directional relations are thought of as being between points. Directions are not so straightforward between spatially-extended entities, since a large range of directions may exist, between any point in one entity and any point in the other. Reference frames are used in discourse and spatial reasoning. Geographically, in many cultures, a reference frame based on cardinal directions seems, dominant for geographical spaces, whereas viewer-centered or object-centered reference frames often dominate over bodily or tabletop ('manipulatable') spaces and entities.

9.6.4 Congnitive Models

Both entities and fields exist in cognitive models. Entities are typically conceptualised as being organised by dimensionality in points, lines, areas, volumes. Entities often have indistinct boundaries, a fact which is at odds with typical GIS representation schemes. Entities are also categorised, and since many aspects of nature from a continuum, categories may be relatively arbitrary and thus subject to disparate cultural differences. Spatial relations, on the other hand, seem to be very similar in disparate cultures and languages. Congnitive spatial relations are predominantly topological but metric factors such as distance and direction often refine the relations and characterise prototypical relations.

There is a very real sense in which all representations are cognitive. Mathematics is, after all a formalisation of how at least some people think. The cognitive view of spatial relations, however, emphasises the importance of human subjects being testing, preferably under laboratory-controlled conditions, in defining the nature of the spatial representations that are needed for geographical information systems and spatial analysis. Many geographical distributions, such as, those of soil variables are inherently complex, revealing more information at higher spatial resolution apparently without limit (Mandelbrot, 1982). Geographical data modelling is the process of discrimination that converts complex geographical reality into a finite number of database records or objects. Objects have geographical expression as points, lines, and areas, and also possess descriptive attributes, for example, collection of water samples from different wells, in which the location of the well creates point object (with the address of the location) and associated water quality parameter is its attribute value. A major difficulty arises in the case of the object models. Many geographical objects have inherently fuzzy spatial extents. One common solution to this problem is to allow objects to have multiple representations which depend on the scale. For example, a river might be a single line at scales smaller than 1:50,000, but a double line at larger scales. Six field models are in common use in GIS: (i) irregular point sampling, (ii) regular point sampling, (iii) contours, (iv) polygons, (v) cell grid modelling, (vi) triangular network models. The object models are commonly used to represent man-made facilities. For example, an underground pipe more naturally represents as a linear object than as a value in a layer. Pipes can cross each other in object model, but cause problems in a field model.

9.7 GIS Queries

As a decision support system, a GIS must provide potential for aspatial (nonspatial) and spatial queries, as well. The answer to aspatial kind of queries does not require the spatial locations of the geographical features involved. See the question : "calculate the number of churches in Hyderabad," or "Compute the percentage of grass in Hyderabad." To answer questions like this, a GIS as well as a number of statistical and spreadsheet packages, don't require the stored value of latitude and longitude, or x, y, z coordinates.

Spatial queries carried out spatial operations and links data sets using location as the common key. Location condition trends pattern modelling. See the question: "Calculate the number of Greek Orthodox Churches in the area surrounded by the roads, SN Colony Road, Rajbhavan Road, and other roads." To answer a question like this, the GIS must know the spatial location of particular map features, in this case, the roads as line map features and churches as point map features.

9.8 GIS Architecture

According to the definition proposed by Marble and Peuquet (1983), GIS deals with space-time data, and often but not necessarily, employs computer hardware and software. GIS can be understood as the subsystem nature within the framework of a main system. According to these investigators, GIS has the following generic subsystems:

(i) A data input subsystem which is also called data capture subsystem
(ii) A data storage and retrieval subsystem
(iii) A data manipulation and analysis subsystem
(iv) A reporting subsystem.

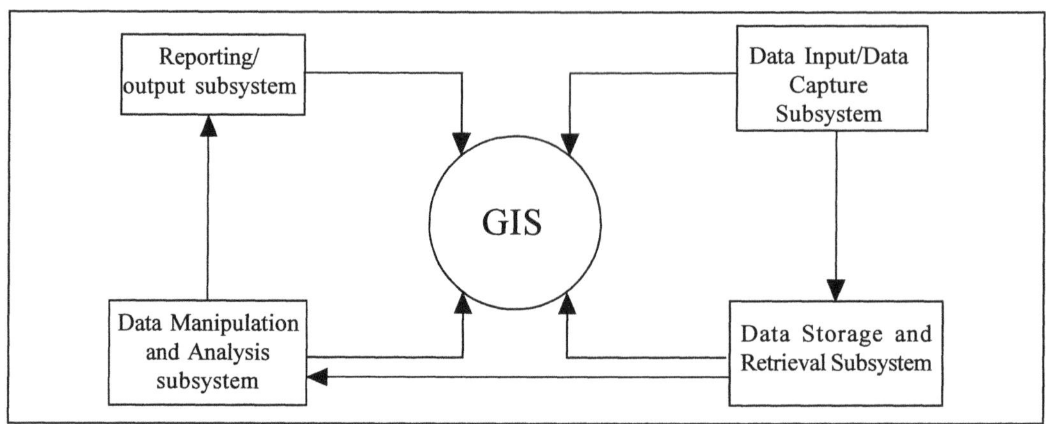

Fig. 9.5 Subsystem nature of GIS (structural prespective).

Each of the subsystems has been described in terms of functions that the respective subsystem performs. The data input/capture subsystem provides operational functions for acquiring data. The data management or data storage and retrieval subsystem stores and retrieves the data elements. The manipulation and analysis subsystem handles the transformation of data from one form to another and derivation of information from the data. The fourth subsystem output/reporting subsystem provides a way for the user to see the data in the form of diagrams, maps, and/or tables. Fig. 9.5 shows the architechture of all subsystems of GIS from a structrual perspective.

9.8.1 Components of a GIS

Geographical Information Systems have three important components, namely, computer hardware, sets of application software modules, and a proper organisational setup. These three components need to be in balance if the system is to function satisfactorily. GIS run on the whole spectrum of computer systems ranges from portable personal computers to multi-user supercomputers, and are programned in a wide variety of software packages. Systems are available that use dedicated and expensive work stations, with monitors and digitising tables built in. In all cases, there are a number of elements that are essential for effective GIS operations. These include (Burrough, 1986):

(i) the presence of a processor with sufficient power to run the software
(ii) sufficient memory for the storage of large volumes of data
(iii) a good quality, high resolution color graphics screen and
(iv) data input and output devices, like digitisers, scanners, keyboards, printers and plotters.

The general hardware components of a GIS include control processing unit which is linked to mass storage units, such as, hard disk drives and tape drives, peripherals such as digitiser or scanner, printer or plotter and Visual Display Unit (VDU). Fig. 9.6 shows the major hardware components of a GIS.

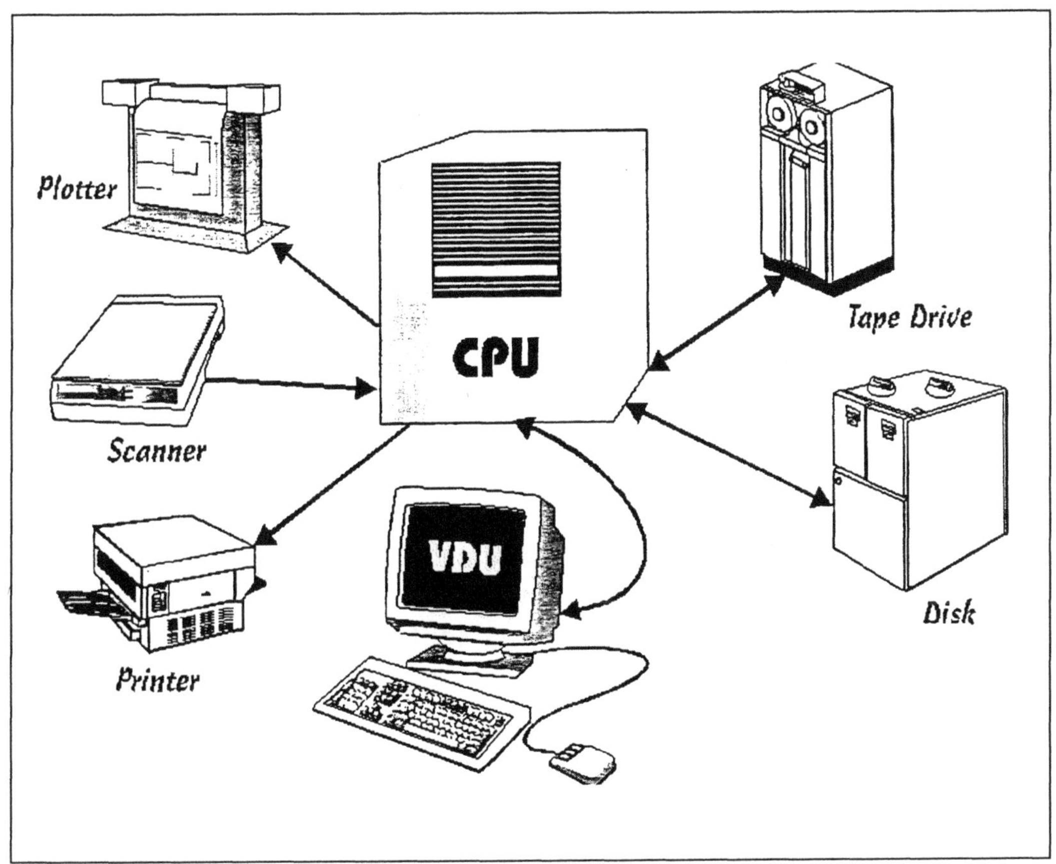

Fig. 9.6. Hardware components of GIS.

There are a number of essential software elements that must allow the user to input, store, manage, transform, analyse and output data. Therefore, the software package for a GIS consists of four basic technical modules. These basic modules are : (i) data input and verification, (ii) data storage and database Management (iii) data transformation and manipulation, and (iv) data output and presentation. The GIS software package should have the capabilities performing all the four GIS subsystems. The GIS hardware and software govern the way in which geographical information can be processed but they donot themselves guarantee that any particular GIS will be used effectively.

9.8.2 GIS Work Flow

There are five essential elements that a GIS must contain. They are data acquisition, preprocessing, data management, manipulation and analysis, and product generation. For any application of GIS, it is important to view these elements as a continuing process. Fig. 9.7 shows the work flow process of GIS in procedural perspective. Data acquisition is the process of identifying and gathering the data required for any given application. This typically involves a number of procedures. One procedure might be to gather new data by preparing large-scale maps of natural vegetation from field observations. Other procedures for data acquisition may include locating and acquiring existing data, such as, maps, aerial and ground photography, and data acquired by satellite sensing systems.

Collection, input, and correction operations concerned with receiving data into the system include manual digitising, scanning, keyboard entry of attribute information, and online retrieval from other database systems. It is at this stage that a digital map is first constructed. The digital representation can never be of a higher accuracy than the input data, although the mechanisms for its handling will frequently be capable of greater precision than that achieved during data collection. The essential preprocessing procedures include: (a) format conversion, (b) data reduction and generalisation, (c) error detection and editing, (d) Merging of points into lines, and lines into polygons, (e) Edge matching and tiling, (f) Rectification/registration, (g) Interpolation, and (h) Interpretation.

The functions of database management govern the creation of an access to the database itself. These functions provide consistent methods for data entry, update, deletion, and retrieval. Modern database management systems isolate the users from the details of data storage, such as, the particular data organisation on a mass storage medium. A modern Database Management System (DBMS) is used to create GIS database, that is, attribute database.

Storage and retrieval mechanisms include the control of physical storage of the data in memory, disk or tape, and mechanisms for its retrieval to serve the needs of the other three components. In a disaggregate GIS this data storage may be physically more from the rest of the system, and may meet the database requirements. This module includes the software structures used to organise spatial data into models of geographic reality.

The development of new derived data layers, which may form the input to further analysis, is an important function of any GIS. The list of data manipulation and analysis operations are, (i) reclassification and aggregation, (ii) Geometric Operations: as

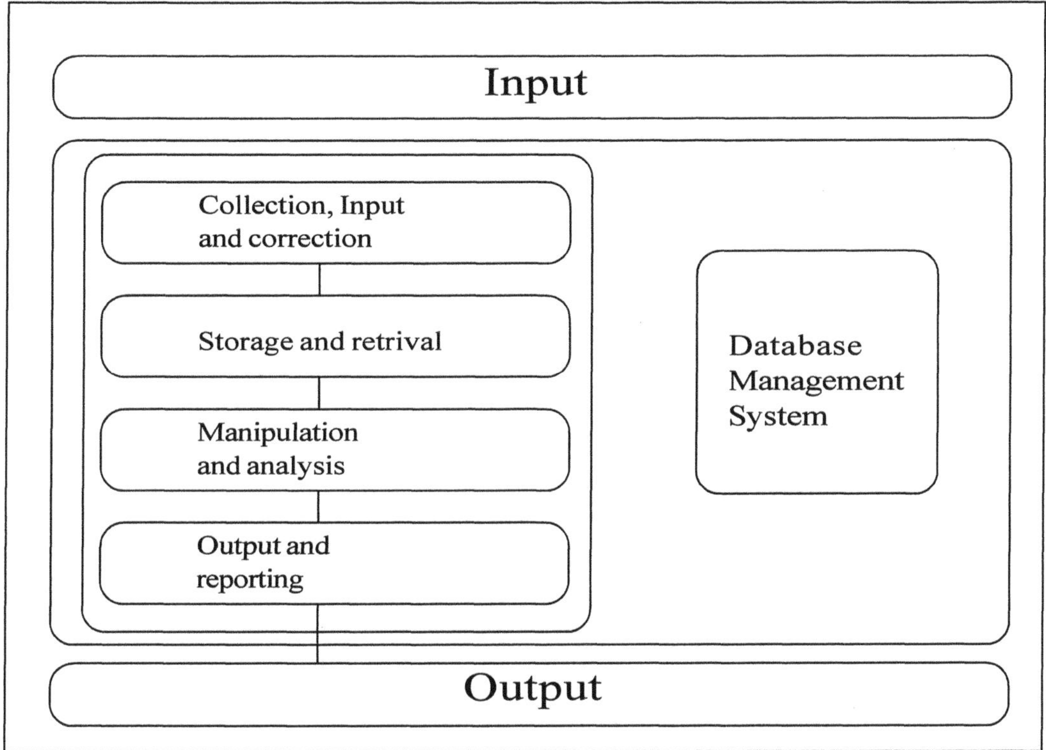

Fig. 9.7. Workflow process of GIS (Procedural perspective)

rotation, translation and scaling, rectification, and registration, (iii) Controlled determination, (iv) Data structure conversion, (v) Spatial operations of connectivity and neighborhood operations, (vi) Measurement of Distance and Direction, (vii) Statistical analysis as descriptive statistics regression, correlation, and cross-tabulation, and (viii) Modelling. This operation or subsystem represents the whole spectrum of techniques available for the transformation of the digital model by mathematical means. A library of data-processing algorithms is available for the transformation of spatial data, and incorporated in new visual maps. Using these techniques it is possible to deliberately change the characteristics of the data representation in order to meet theoretical requirements. It is equally possible to mishandle or unintentionally distort the digital map at this state.

Product generation is the phase where final outputs from the GIS are created. These output products might include statistical reports, maps, and graphics of various kinds. Some of these products are softcopy images and hardcopy.

9.9 Theoretical Models of GIS

Reference should now be made to the existing models of GIS operation, which are broadly similar in nature. These may be considered in two main groups, namely, (i) the functional elements of GIS, and (ii) the fundamental operations of GIS. It will be seen that these approaches generally make little or no reference to the very important process as which have been outlined above. It is due to their failure to address these issues that they are unable to offer much help to those who would seek to apply GIS in new situations. In particular, certain socioeconomic phenomena need special treatment. Conceptual models tend only to address aspects of the operation or composition of GIS systems. They make no statement about the nature of the data representation. The components in these models are basically analogous to the main software components in any general-purpose systems. Bracken and Webster (1989) suggest an alternative classification which recognies three major components in its characterization of GIS. They are the problem-processor model, database model and interface model. However, this is still an explicitly software-oriented approach to understanding GIS.

9.9.1 Functional Elements of GIS

Bracken and Webster (1987) outlined four functional elements to address the GIS technology. They are database approach, the process-oriented approach, an application oriented approach, and toolbox approach.

Database approach stresses the ability of the underlying data structures to contain complex geographical data. The process-oriented approach focuses on the sequence of system elements used by an analyst running an application. An application oriented approach defines GIS based on the kinds of information manipulated by the system and the utility of the derived information produced by the system while the toolbox approach emphasis as the software components and algorithms that should be contained in a GIS.

9.9.2 Fundamental Operations of GIS

This approach considers the functions which GIS is able to perform. The operations discussed in this section fall entirely within the manipulation-and-analysis subsystem referred to above, and are thus wholly internal to the GIS. The fundamental classes of operations performed by a GIS have been characterised as 'map algebra' (Tomlin and Berry, 1979; Berry, 1982, 1987; Tomlin, 1991) in which context primitive operations of map analysis can be seen as analogous to traditional mathematical operations. The 'classes of analytical operation' are divided into reclassification, overlay, distance/connectivity measurement and neighbourhood characterisation of the data. These operations can be identified as follows :

(i) Reclassification operations transform the attribute information associated with a single map coverage.

(ii) Overlay operations involve the combination of two or more maps according to boolean conditions and may result in the delineation of new boundaries.

(iii) Distance and connectivity measurement include both simple measure of inter-point distance and more complex operations such as the construction of zones of increasing transport cost away from specified locations, and

(iv) Neighbourhood characterisation involves the values to a location both summary and mean measures of a variable, and include smoothing and enhancement filters. Sequences or such manipulation operations have become known as 'cartographic modelling'.

9.10 Theoretical Framework for GIS

This discussion is based on an analysis of the way in which data are transformed and held as a digital model of the external world. The geographic data-processing system outlined is not intended to be a description of any specific software system, but is a model of the processes which may operate with digital geographic data. The idea of data representation used here should not be confused with work on specific data structures, either spatial, such as, vector, raster, triangulated irregular network, and quadtree, or attribute such as, hierarchical and relational which are mainly technical issues. Actual software systems may be identified which perform all or some of the principal transformations to a greater or lesser degree, but only those containing some capacity for input, manipulated and output digital spatial data in some form are considered to be 'GIS' in the present context.

The ground work for this approach has been laid by cartographers seeking to understand the relationships between the world and the map as a model. We have seen how the digital map is related to the analog map, and we must now modify and extend the existing theoretical structure to replace the paper map with a digital map sitting within a GIS. The process of analog map production may be modelled as a series of transformations between the real world, raw data, the map, and the map image (Fig. 9.8).

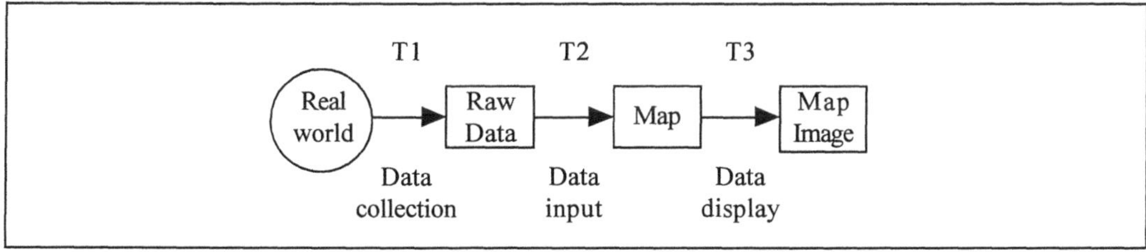

Fig. 9.8 Transformation stages in the tradional cartographic process.

The significance of these transformations is that they control the amount of information transmitted from one stage to the next. The cartographer's task is to devise the very best approximation to an 'ideal' transformation involving a minimum of information loss. A detailed explanation of the 'transformational' view of digital cartography is given in Clarke (1990).

In the context of GIS, we may add an additional transformation stage which sits entirely within the GIS (Fig. 9.9.). The sequence of transformation stages illustrated in the figure forms the basis for the discussion of GIS techniques later in this book. In the first transformation (T_1), data are selected from the real word, as for example, surveying measurements or census data. These are then input for the GIS in some form (T_2) to provide the basis for its digital map representation of the real world. Within the system, a vast range of manipulation operations are available to further transform the data and store the results (T_3), and these may be communicated as tabulated or graphic images by means of a hardcopy or screen (T_4). It is worth noting that each of these transformation stages may actually involve several physical operations on the data. For example, T_1 may involve both collection and aggregation, and T_3 will almost always consist of a whole series of data-processing operations. If the 'thematic and spatial' charateristics noted by Berry (1987) are to be understood, it is necessary for us to carefully consider the way in which these four transformations may effect the digital representation of the real-world objects.

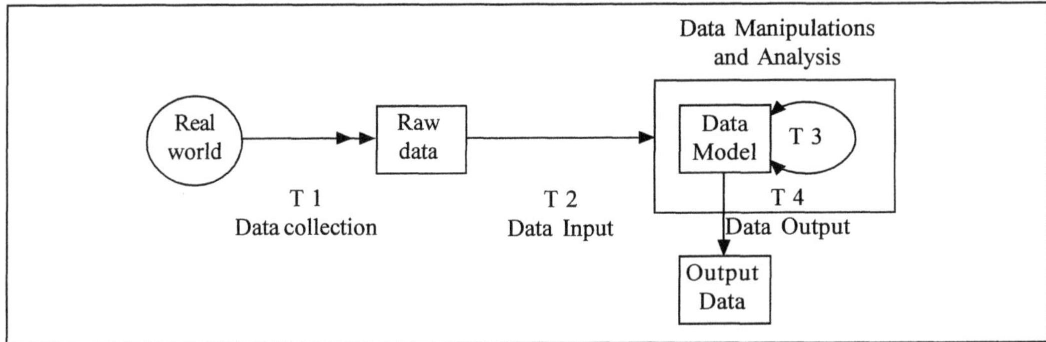

Fig. 9.9 A transformation based view of GIS operation.

9.11 GIS Categories

For more than two decades, the GIS applications were dominated by the so-called "map geometry", that is GIS data with only x, y, coordinates. If an entity or a point/line is represented by two values, for example, x-coordinate y-coordinate, thus it is said to be 2-D GIS. Efficient GIS products are offered by vendors which, besides the two-dimensional x, y-geometry, intergrate the "third-dimension" using the "normal case" Digital Terrain Model (DTM) equation $z = f(x, y)$ for the height, or any other predefined simple equation for an attribute like soil type. This is called 2.5 D GIS.

In the 2.5-D case, for the same pair of x, y co-ordinates only one spatial location (point's height) or attributes or attribute value (soil type) exist, and this can be calculated using a predefined formula.

Users' perception is three-dimensional. Therefore, there is an existing need for 3-D data, but in this case the data handling is more difficult, and requires more complex algorithms. In the 3-D case, for the same pair of x, y co-ordinates a number of spatial locations (points) with a different z-co-ordinate could exist because the z co-ordinate is calculated independently from the x, y pair. Hence, these systems are suitable for any kind of terrain modelling. Also, in cases where the third dimension is referred to an attribute, the actual attribute value is indepedent from the x, y spatial location and it is calculated using complex mathematics. 4 -D GIS or Temporal GIS is the fourth category of GIS. In this case, the fourth dimension namely, time, is introduced. For this reason, the kind of GIS applications depending on time are regraded as temporal GIS systems.

9.12 Levels/Scales of Measurement

So far, we have examined the fundamental considerations and, various components of GIS. The entities associated with them, a set of coordinates that allow to locate, the positions and the corresponding descriptions of each location, can be called attribute value.

All these spatial features or entities, contain information not only about how they occupy space but also about what they are and how important they are. The additional nonspatial information that helps us to describe the objects we observe in space comprises the feature's attributes. The character of attribute data themselves can influence the utility of data sets in GIS analysis. One characteristic which is of considerable importance is the scale of measurement used to record and report the data. There is already a well-established measurement frame work for nearly all forms of data including geographic data. These are called levels of geographic data measurement (Fig. 9.10). They illustrate the levels of measurements in terms of commonly used geographic features.

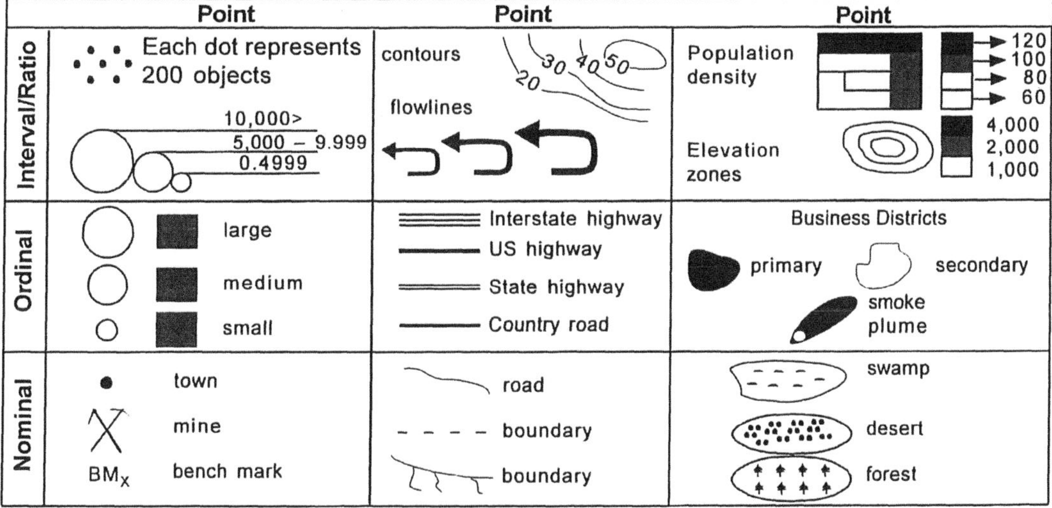

Fig. 9.10 Levels of geographic data measurments.

The four commonly refered levels of measurement are, (i) nominal scale, (ii) ordinal scale, (iii) interval scale, and (iv) ratio scale. Nominal scales are those variables which are described by name, with no specific order, for example, Landuse, parks, residential areas and so on. These are 'named' data. The system allow us to make statements about what to call the object, but it does not allow direct comparisons between one named object and another.

Table 9.1 Scale (level) of measurement

Data	Unit of measurement	Scale/level
Hotel name	Text	nominal
Status of Hotel	Three Star	ordinal
Average Tariff	in Rupees	Interval
Size of the Hotel	m^2	Ratio

Ordinal variables are those variables which are lists in discrete classes but with an inherent order, for example, classes of streams, like I order and II order. In the interval scale of measurement, the numbers are assigned to the items measured. Data measured can be compared as in the case of ordinal scale of measurement and example have a natural sequence of temperature measured in degrees. The ratio variables have the same characteristic as interval variables, but they have natural zero or a starting point. Rainfall per month is an example of ratio variable. Table 9.1 shows an example of level (scale) of measurement.

10
Spatial Data Modelling

10.1 Introduction

Burrough (1986) observed that the human eye is highly efficient at recognising shapes and forms but the computer needs to be instructed exactly how spatial patterns should be handled and displayed. Computers require precise and clear instructions on how to turn data about spatial entities into graphical representations. The process is the second stage in designing and implementing a data model. At present there are two main approaches in which computers can handle and display spatial entities. They are the raster and vector approaches. The data structures that have little to do with the graphic representation of cartographic objects are simple lists, ordered sequential files and indexed file systems. These three systems are discussed in the next chapter under attribute database management.

The human mind is capable of producing a graphic abstraction of space and objects. This representation is actually quite sophisticated if we use computers to handle graphic devices. A map appears as a graphic device which contains an implied set of relationships about the spatial elements, such as, monuments, roads/rivers, and parks. Lines are connected to other lines and together are linked to create areas or polygons. The lines are related to one another in space through angles and distances. Some are connected, but others are not. Some polygons have neighbours, but others are isolated. The list of possible relationships that can be contained on a graphic diagram is virtually endless. From this endless relationships among objects, there should be a way to find and represent each object and relationships by means of a set of rules. These rules then assist the computer to recognise all the points, associated lines, and areas to represent something on the earth. The representation may be with respect to explicit locations related to other objects within space, absolute and/or relative location, proximity of each object and many other relationships. In order to extract all such information, we need to create a language, known as language of spatial relationships through spatial modelling. Spatial modelling is very much useful in understanding the geographical problems. In general, spatial modelling in GIS can be split into two parts: a model of spatial form and a model of spatial processes. The model of spatial form represents the structure and distribution of features in geographical space, while the interaction between these features are considered in spatial data processing models.

10.2 Stages of GIS Data Modelling

The construction of models of spatial form can be taken as a series of stages of data abstraction. By applying this abstraction process the GIS designer moves from the position of observing the geographical complexities of the real world to one of simulating them in the computer. This process involves,

(i) Identifying the spatial features from the real world that are of interest in the context of an application.

(ii) Representing the conceptual model by an appropriate spatial data model. This involves choosing between one of the two approaches: raster or vector.

(iii) Selecting an appropriate spatial data structure to store the model within the computer. The spatial data structure is the physical way in which entities are coded for the purpose of storage and manipulation.

Fig. 10.1 provides an overview of the stages involved in creating a GIS data model. At each stage in the model-building process, we move further away from the physical representation of a feature in reality and closer to its abstract representation in the computer. In this chapter, the definition of entities and graphical representation of the surface features in the computers are considered along with the different spatial data models and structures available. The modelling of more complex features and the difficulties of including the third and fourth dimensions in a GIS model are also presented.

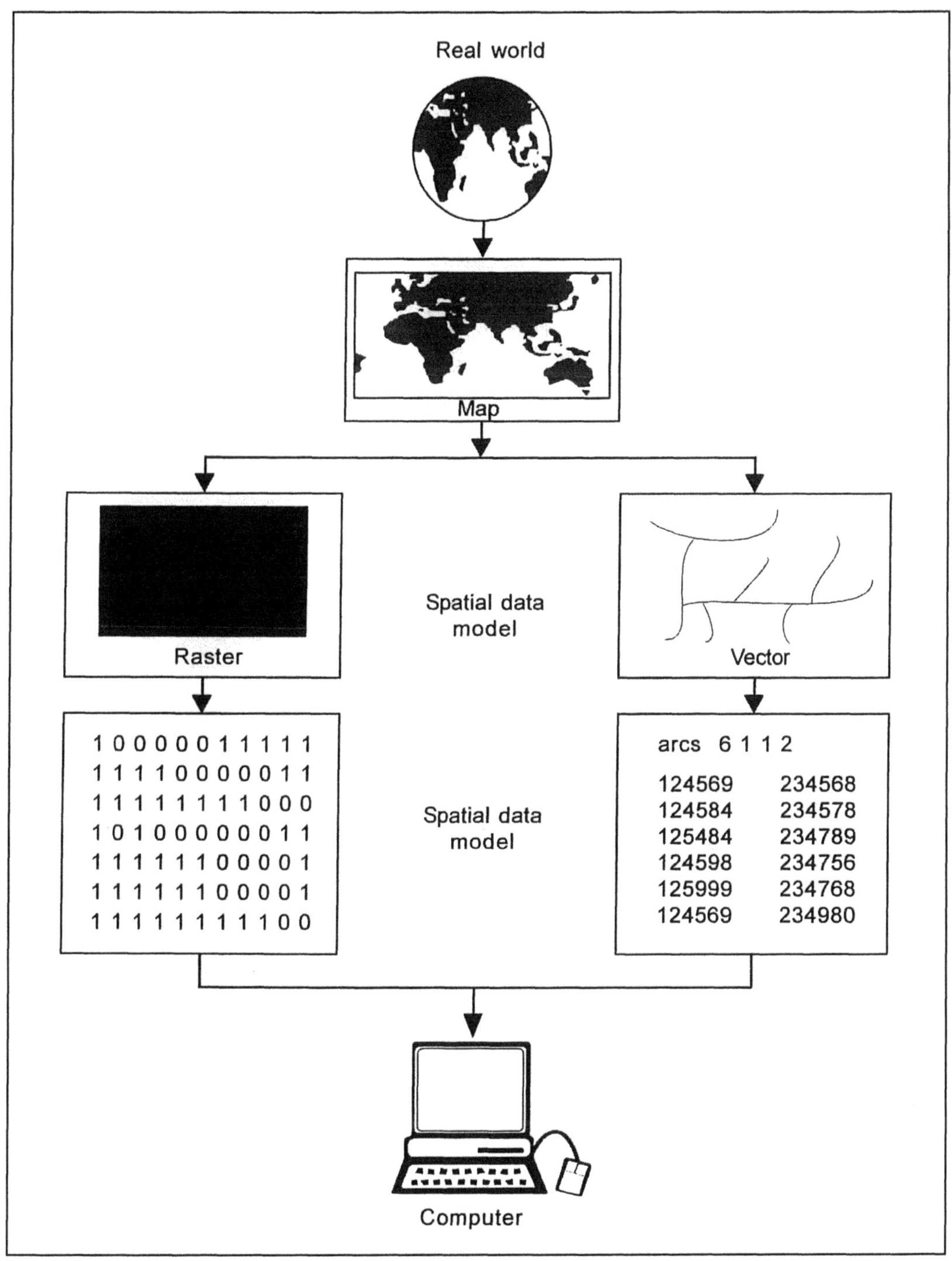

Fig. 10.1 Stages in creating a GIS model.

10.3 Graphic Representation of Spatial Data

An entity is the element in reality. It is a phenomenon of interest in reality that is not further subdivided into phenomena of the same kind. For example, a city can be considered an entity. A similar phenomena stored in a database are identified as entity types. All geographical phenomena can be represented in two dimensions by three main entity types : points, lines, and areas. Fig. 10.2 shows how a spatial data model could be constructed using points, lines, and areas. Fig. 10.2 also introduces two additional spatial entities : networks and surfaces. These are an extension of the area and line concepts.

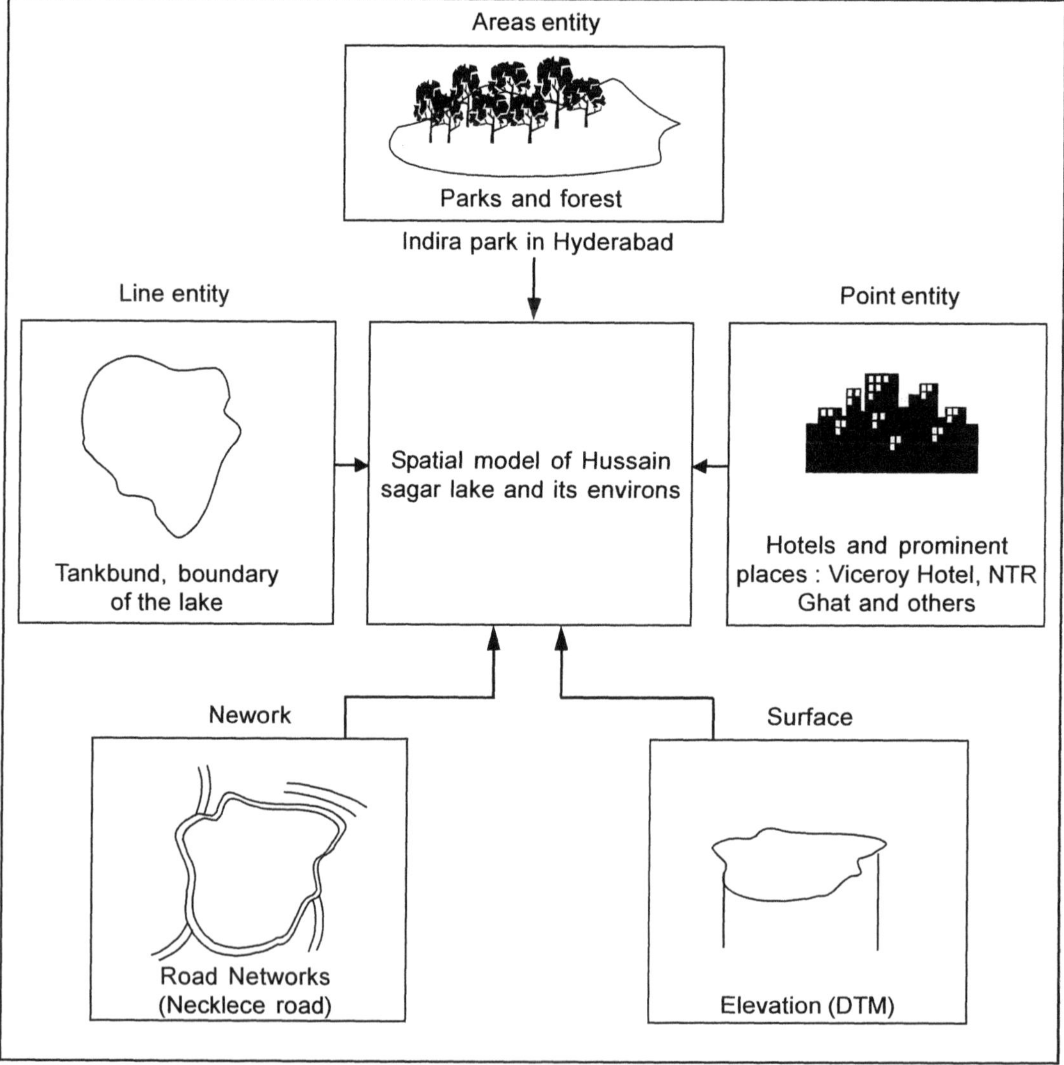

Fig. 10.2 Spatial entity Data model.

A surface entity is used to represent continuous features or phenomena. For these features there is a measurement or value at every location, as in the case of elevation, temperature and population density. This makes representation by a surface entity appropriately. The continuous nature of surface entities distinguishes them from other entity types (points, lines, areas, and networks) which are discrete, that is, either present or absent at a particular location.

A network is a series of interconnecting lines along which there is a flow of data, objects or materials, for example, the road network, along which there is a flow of traffic to and from the areas. Another example is that of a river, along which there is a flow of water. Others not visible on the land surfaces, include the sewerage and telephone systems considered network type of entities.

The dynamic nature of the world poses two problems for the entity-definition phase of a GIS project. The first is how to select the entity type that provides the most appropriate representation for the features being modelled. Is it best to represent a forest as a collection of points (representing the location of individual trees), or as an area (the boundary of which defines the territory covered by the forest)? The second problem is how to represent changes over time. A forest, originally represented as an area, may decline until it is only a dispersed group of trees that are better represented by using points.

The definition of entity types for real-world features is also hampered by the fact that many real-world features simply do not fit into the categories of entities available. An area of natural woodland does not have a clear boundary as there is normally a transition where trees are interspersed with vegetation from a neighbouring habitat type. In this case, if we wish to represent the woodland by an area entity, where do we place the boundary? The question is avoided if the data are captured from a paper map where a boundary is clearly marked, as if someone has already made a decision about the location of the woodland boundary. But is this the true boundary? Vegetation to an ecologist may be a continuous feature (which could be represented by a surface), whereas vegetation to a forest is better represented as series of discrete area entities.

Features with 'fuzzy' boundaries, such as the woodland, can create problems for the GIS designer and the definition of entities, and may have an impact on later analysis. Deciding which entity type should be used to model a real-world feature is not always straightforward. The way in which individuals represent a spatial feature in two dimensions will have a lot to do with how they conceptualise the feature. In turn this will be related to their own experience and how they wish to use the entity they produce. An appreciation of this issue is central to the design and development of all GIS applications.

There are two fundamental methods of representing geographical entities. They are (i) Raster method, and (ii) Vector method.

10.3.1 Raster Data Representation

In raster representation, the terrain is divided into a number of parcels or quantised the space into units. A parcel or a unit is called a grid cell. Although a wide variety of raster shapes like triangles or hexagons are possible, it is generally simpler to use a series of rectangles, or more often squares, called grid cells. Grid cells or other raster forms generally are uniform in size, but this is not absolutely necessary. For the sake of simplicity, we will assume that all grid cells are of the same size and that, therefore, each occupies the same amount of geographic space as any other.

Raster data structures do not provide precise locational information because geographic space is now divided into discrete grids, as much as we divide a checkerboard into uniform squares. Instead of representing points with their absolute locations, they are represented as a single grid cell (Fig. 10.3). This stepped appearance is also obvious when we represent areas with grid cells. All points inside the area that is bounded by a close set of lines must occur within one of the grid cells to be represented as part of the same area. The more irregular the area, the more stepped the appearance.

In grid-based or raster GIS, there are two general ways of including attribute data for each entity. The simplest is to assign a single number representing an attribute like a class of land cover, for each grid cell location. By positioning these numbers, we, ultimately, are allowing the position of the attribute value to act as the default location for the entity. For example, if we assign a code number of 10 to represent water, then list this as the first number in the X or column direction, and the first in the Y or row direction, by default the upper left grid cell is the location of a portion of the earth representing water. The larger the grid cell, the more land area is contained within it --a concept called resolution. The coarser the resolution of the grid, the less we know about the absolute position of points, lines, and areas represented by this structure.

Raster structures, especially square grid cells, are pieced together to represent an entire area. Raster data structure may seem to be rather undesirable because of the lack of absolute locational information. Raster data structures have numerous advantages over other structures. Notably, they are relatively easy to conceptualise as a method of representing space. Remotely sensed data acquired by a sensor is one of the well known example of raster data representation. In fact, the relationship between the pixel used in remote sensing and the grid cell used in GIS allows data from satellites to be readily incorporated into raster-based GIS without any changes. A characteristic feature of grid-based systems is that many functions, especially those involving the analysis and modelling of surfaces and overlay operations, are simple to perform with this type of data structure. The major disadvantages of the raster data structure are a reduced spatial accuracy, decrease of the reliability of area and distance measures, and the need for large storage capacity associated with having to record every grid cell as a numerical value.

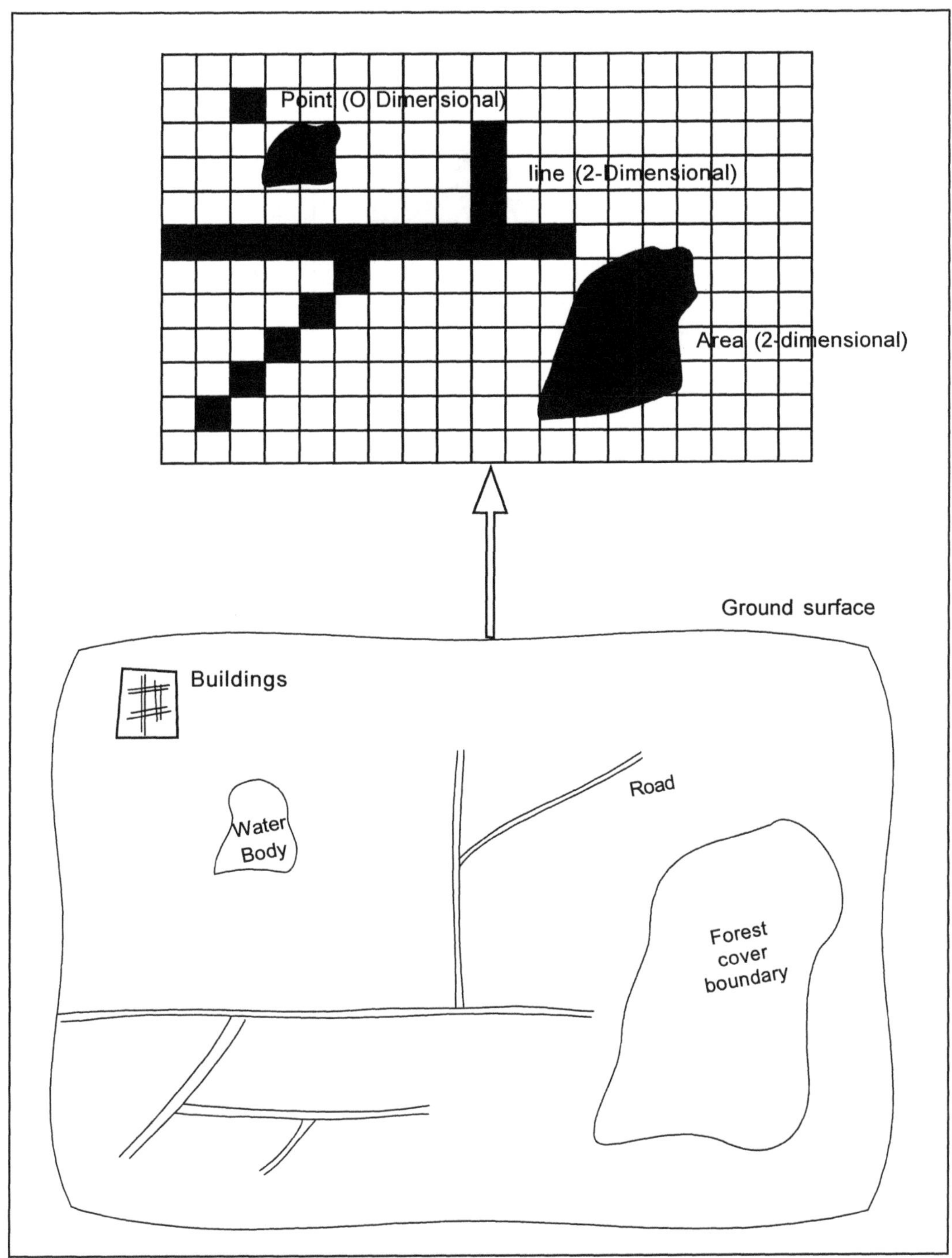

Fig. 10.3 Raster Graphic Data Representations.

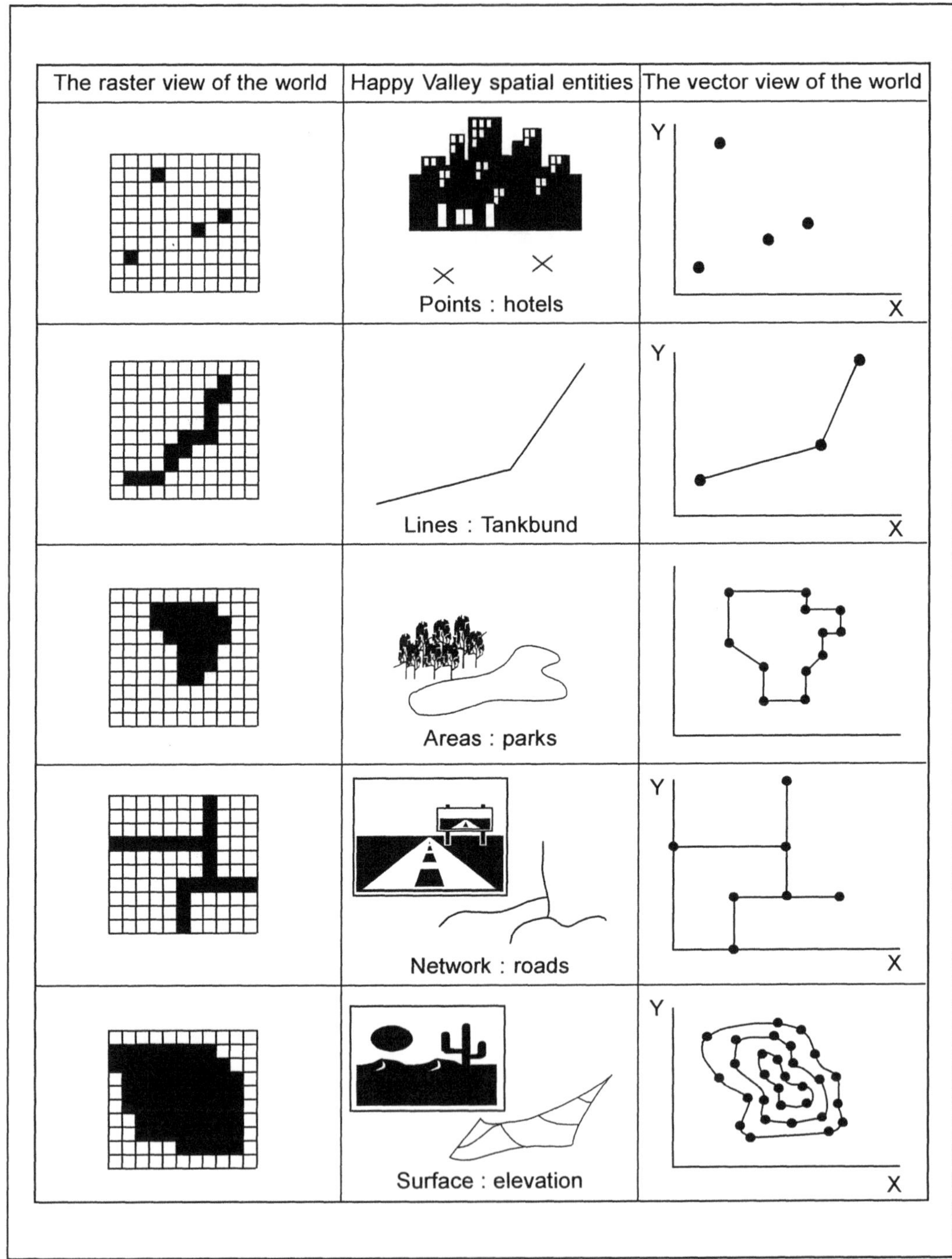

Fig. 10.4 Raster and vector spatial data models.

The raster spatial data model is one of a family of spatial data models described as tessellations (Demers, 1999). In the raster world individual cells are used as the building blocks for creating images of point, line, area, network, and surfaces. Fig 10.4 shows how a range of different features represented by the five different entity types can be modelled using the raster approach. Hotels are modelled by single and discrete cells, the tankbund is modelled by linking cells into lines, the forest by grouping cells into blocks, and the road network by linking cells into networks. The relief of the area has been modelled by giving every cell in the raster image an altitude value. In Fig. 10.4 the altitude values have been grouped and shaded to give the appearance of a contour map.

10.3.2 Vector Data Representation

The second method of representing geographic space, called vector, allows us to give specific spatial locations explicitly. In this method it is assumed that geographic space is continuous, rather than being quantised as small discrete grids. This perspective is acquired by associating points as a single set of coordinates (X and Y) in coordinate system, lines as connected sequences of coordinate pairs of points, and areas as sequences of interconnected lines whose first and last coordinate points are the same (Fig. 10.5). Anything that has a single (X, Y) coordinate pair not physically connected to any other coordinate pair is a point (zero-dimensional) entity.

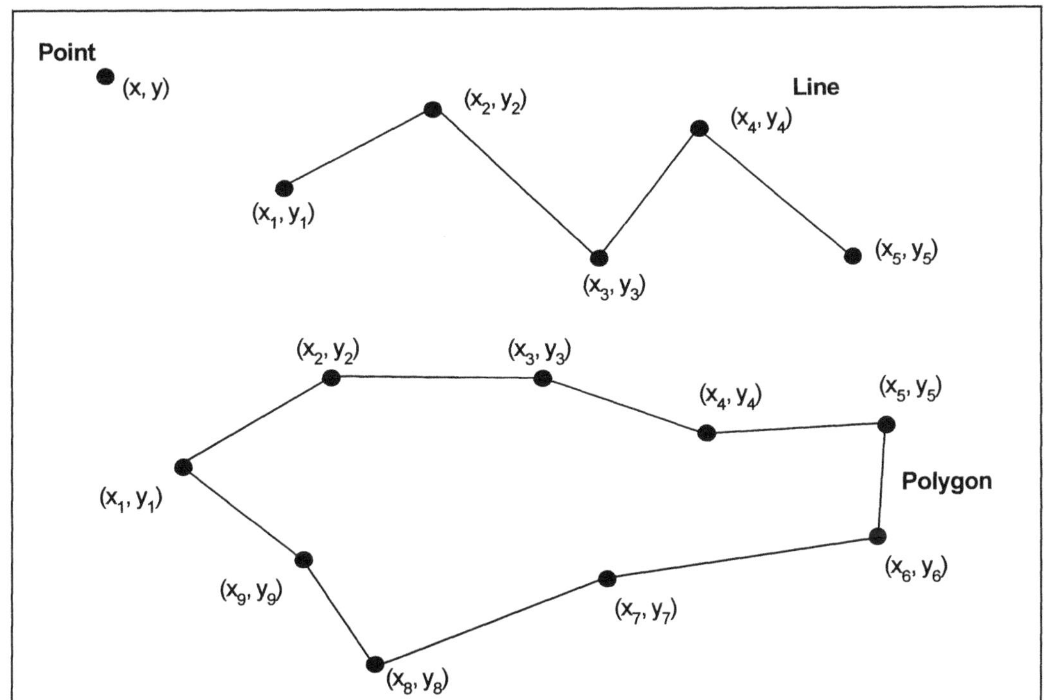

Fig. 10.5 Vector graphic data representation.

A vector spatial data model uses two-dimensional Cartesian (x, y) coordinate system to store the shape of a spatial entity. In the vector world the point is the basic building block from which all spatial entities are constructed. The simplest spatial entity, the point, is represented by a single (x, y) coordinate pair. Line and area entities are constructed by connecting a series of points into chains and polygons. Fig. 10.4 shows how the vector model has been used to represent various features. The more complex the shape of a line or area feature, the greater the number of points required to represent it. Selecting the appropriate number of points to construct an entity is one of the major problems in vector based GIS data representation. In the vector data model, the representation of networks and surfaces is very complex and closely linked to the way the data are structured for computer encoding.

The representation of the vector data is much more representative and generally, we combine the entity data with associated attribute data kept in a separate file through a database management system, and then link them together. It means that the entity data and corresponding attribute data in the form of tables can be stored and linked through a software linkage.

In vector data structures, a line consists of two or more coordinate pairs, again storing the attributes for that line in a separate file. This is explained in the next section under vector models. For straight lines, two coordinate pairs are enough to show location and orientation in space. More complex lines will require a number of line segments, each beginning and ending with a coordinate pair. For complex lines, the number of line segments must be increased to accommodate the many changes in angles. The shorter the line segments, the more exactly will they represent the complex line. Thus we see that although vector data structures are more representative of the locations of objects in space, they are not exact but are still an abstraction of geographic space.

10.3.3 Spatial Data Models

Spatial data structures provide the information that the computer requires to reconstruct the spatial data model in digital form. Although some lines act alone and contain specific attribute information that describes their character, other more complex collections of lines called networks add a dimension of attribute characters. Thus not only does a road network contain information about the type of road or similar variables, but it will also indicate, that travel is possible only in a particular direction. This information must be extended to each connecting line segment to advise the user that movement can continue along each segment until the attributes change-perhaps until a one-way street becomes a two-way street. For example, one node might indicate the existence of a stop sign, a traffic signal, or a sign prohibiting U-turns. All these attributes must be connected throughout the network so that the computer knows the inherent real-world relationships that are being modelled within the network. Such explicit information about connectivity and relative spatial relationships is called topology.

Like line entities area entities can be produced in the vector data structure. By connecting pairs of coordinates into lines and organising the lines into a looping form, where the first coordinate pair on the first line segment is the same as the last coordinate pair on the last line segment, we create an area or polygon. As with point and line entities, the polygon will also have associated with it a separate file that contains data about the attributes or characteristics of the polygon. Again, this convention improves the simple graphic illustration of area entities, making it possible for them to represent in a better way the abstraction of area patterns we observe on the earth's surface.

To store such a huge quantity of GIS data in vector and/or raster, a number of models are developed. Raster models are based on grid cells and vector models are based on the vectors in the form of coordinate pairs of points, lines and areas. Each one of these models has its advantages and disadvantages and hence the selection of model depends upon number of parameters. These models are discussed in the following sections.

10.4 Raster GIS Models

The simplest approach of structuring spatial data is to use grid cells to represent quantised portions of the earth which is called GRID based GIS or raster GIS. In the raster GIS, a range of different methods are used to encode a spatial entity for storage and representation in the computer. Fig. 10.6 shows the most straight forward

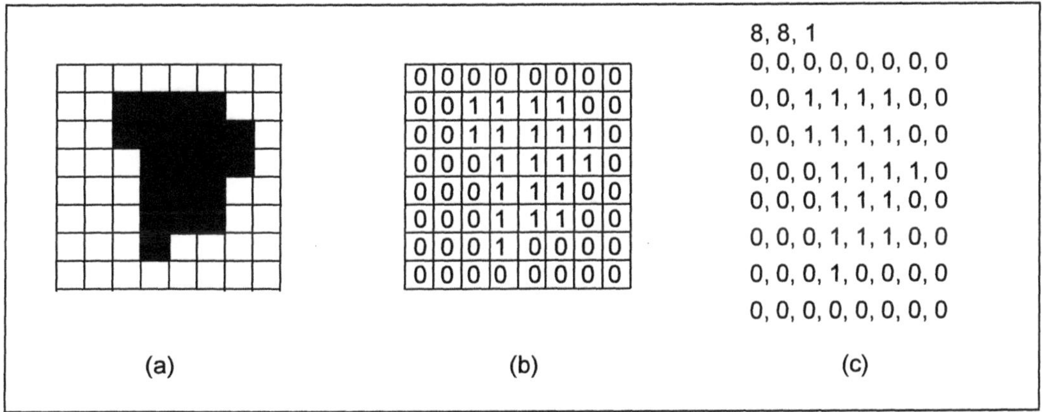

Fig. 10.6 A simple Raster Data structure (a) entity model; (b) cell values and (c) file Structure

method of coding raster data. The cells in each line of the image are mirrored by an equivalent row of numbers in the structure. The line of the file structure indicates the number of rows, the number of columns and the maximum cell value in the image. In the example shown in the Fig. 10.6, it can be seen that there are 8 rows, 8 columns and the maximum cell value is 1. The remaining cells are filled with 0. It indicates that the entity is not present. If the cell fills with '1' it indicate that the entity is present. Fig. 10.4 shows five different raster entities, so five separate data files would be required, each representing a different layer of spatial data. However, if the entities do not occupy the same geographic location (or cells in the raster model), then it is possible to store them all in a single layer, with an entity code given to each cell. This code informs the user which entity is present in which cell. Fig. 10.7 shows how different land uses can be coded in a single raster layer. The values 1, 2 and 3 have been used to classify the raster cells according to the land use present at a given location. The value 1 represents residential area; 2, forest; and 3, farm land.

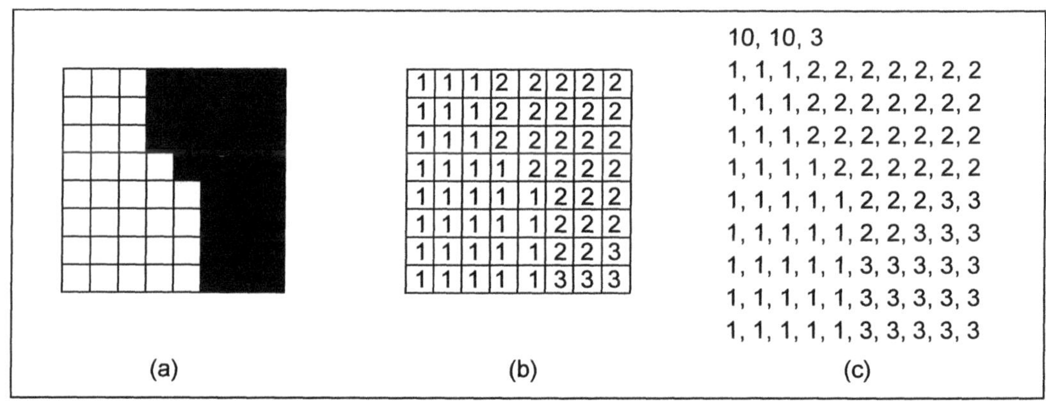

Fig. 10.7Coding of features in Raster GIS.

One of the major problems with raster data sets is their size, in that a value must be recorded and stored for each cell in an image. Thus, a complex image made up of a mosaic of different feature such as a soil map with 20 distinct classes requires the same amount of storage space as a similar raster map showing the location of a single forest. To address this problem a range of data compaction methods have been developed. These include run length encoding, block coding, chain coding and quadtree data structures. One of the simplest data structures is a raster or cellular organisation of spatial data. In a raster structure, a value for the parameter of interest, elevation in meters above datum, and land use class form a specified list. Plant biomass in grams per square meter, is developed for every cell in any array over space. A set of cells are located by coordinates, and each cell is independently addressed with the value of an attribute. The simplest raster data structure consists of an array of grid cells. Each grid cell is referenced by a row and column number and it contains a number representing the type of value of the attribute being mapped. Raster representation assumes that the geographical space can be treated as though it were a flat Cartesian surface. Each pixel or grid cell is then associated with a square parcel land. The resolution or scale of raster data is the relation between the cell size in the database and the size of the cell on the ground. The use of this type of model mainly related to the volume of data size of memory required. Data storage requirements can be considerably reduced by chain codes, run-length codes, quadtrees, and block codes.

10.4.1 Simple Raster Arrays

The horizontal dimension of the simplest raster, along the rows of the array, is often oriented parallel to the east-west direction for convenience. Following the conventional practice in image processing, raster elements in this direction along the rows of the array are sometimes called samples, and numbered from the left (or west) margin. Positions in the vertical direction, aligning with the columns of the array, are often numbered starting from the top (or northern) boundary. This numbering scheme comes from the computer graphics field, in which displays are often painted on the computer screen or printer from the top down. Thus, the origin of the raster is frequently the upper left corner. This location is considered position (1, 1) in some systems of notation, and position (0, 0) in others.

Note that this referencing system for cells in a raster is different from more traditional georeferencing systems, such as, latitude-longitude in which one specific point on the Earth's surface (such as, the point where the prime meridian crosses the equator) is the origin. It is also different form the Universal Transverse Mercator system, where (in the northern hemisphere) the origin of the coordinate system is in the lower left corner, which is similar to a conventional cartesian system. Often, the distances between cells in the raster are constant in both the row and column directions; In other words, the cells in the raster are square. In this case, it is natural to store the data on a computer in a two-dimensional array.

10.4.2 Hierarchical Raster Structures

Consider a set of digital elevation data values, where the fundamental data are stored on a 50 meter square grid (that is, each cell representing a square that is 50 meters on a side). Rather than storing this information as a single layer in GIS, we shall store it in several interrelated layers. One layer corresponds to the original 50 meter interval raster data. A second layer consists of data resampled to a 100 meter interval. Each cell in the 100 meter layer is the algebraic average of four cells in the 50 meter layer. A third spatial averaging process, decreasing the spatial resolution at each "higher" layer, until at the highest layer we might have a single pixel, whose elevation value is the numerical average of all the data in the original 50 meter layer. This is called a pyramidal data structure (Fig. 10.8) since we can imagine each of the derived layers stacked on top of previous layers in the space of a pyramid (Star and Estes, 1990). If each higher level layer has pixel that are exactly twice as wide as the earlier one (thus four times the area), as in our example, this is called a quadtree data structure.

The size of the raster cell in a dataset is sometimes confused with the minimum mapping unit, that is, the smallest element we can uniquely represent in our data. However, raster cell size and minimum mapping unit are not quite the same. Choosing an appropriate minimum mapping unit for a study is a very important decision in the design phase of a project.

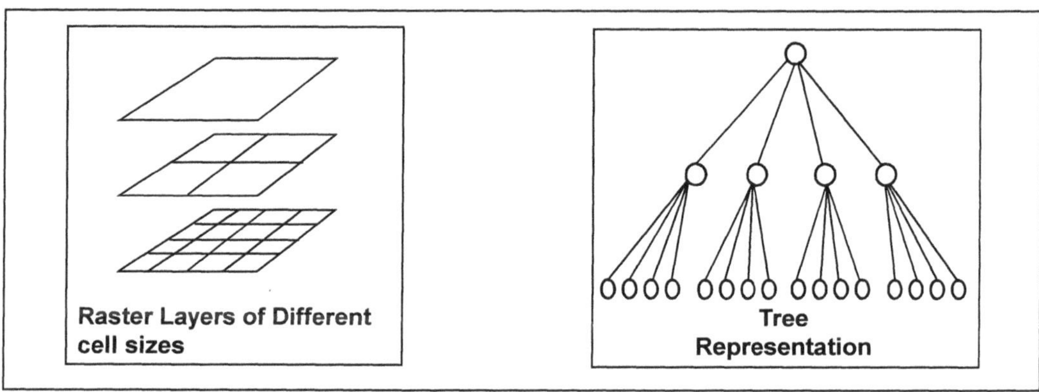

Fig. 10.8 Hierarchical Data Structures.

10.4.3 Types of Raster GIS Models

The grid based GIS spatial data can be stored, manipulated, analysed, and referenced basically in any one of the three methods/models. These three models (Burrough, 1983) are: GRID/LUNAR/MAGI model, IMGRID model and MAP model. All of these models use the grid cell values, their attributes, coverages and corresponding legends. These models are developed depending upon the requirements from time to time. Based on the applications of interest, availability of softwares and other related information, any one of the above models can be selected for the execution of a particular GIS project. There are a number of ways of forcing a computer to store and reference the individual grid cell values, their attributes, coverage names and legends. Fig. 10.9 shows these three raster models for managing multiple sets of grid coverages. These models can be understood by considering a checkerboard. Red indicates water and black, land. Perhaps you could think of a checkerboard, with its red and black squares. If each of these squares is taken to represent a simple map of land cover we have produced a simple coverage. But the problem is, how are the attributes of our landcover physically connected to these grid squares. We can pick up the entire checkerboard because it is a physically connected structure. Likewise, when we pick up a thematic map, it also represents all the different changes in the theme as a single, connected object. The similarity between the checkerboard as a single unit of play for a game and the map as a single unit of storage for spatial information is natural. All these problems can be resolved if we use any one of the above mentioned models as discussed in the following paragraphs.

GRID Model

The first and foremost model for the representation of raster data is the GIRD model. The method of storing, manipulating, and analysing the grid based data was first conceptualised by an attempt to develop GRID model. Burrough (1983) used this approach, because each of those early GIS systems used this model. Fig. 10.9 (a) illustrates the GRID model. In this method, each grid cell is referenced and addressed individually and is associated with identically positioned grid cells in all other coverages, rather like a vertical column of grid cells, each dealing with a separate theme. Comparisons between coverages are therefore performed on a single column at a time. For example, to compare soil attributes in one coverage with vegetation attributes in a second coverage, land use/land cover attributes in a third coverage, each X and Y location must be examined individually. So a soil grid cell at location must be examined individually. So a soil grid cell at location X10-Y10 will be compared to its vegetation counterpart and third layer land use/land cover at location X10-Y10. You might be able to envision this by imagining a geological core in which each rock type is lying directly on top of the next, and to get a picture of the entire study area, it will be necessary to put a large number of cores together.

The advantage of this model is that computational comparison of multiple themes or coverages for each grid cell location is relatively easy. This is a reasonable approach and has proven successful. The main disadvantage is that it limits the efficient examination of relationships of themes to one-to-one relationships within the spatial framework. In other words, it is more inconvenient to compare groups in one coverage to groups in another coverage because each grid cell location must be addressed individually. Second disadvantage is more storage space for the cell data and the representation is vertical rather than horizontal, which would more closely resemble our notion of maps.

IMGRID Model

With a slight modification of the checkerboard analog, the second basic raster data model, that is the IMGRID data model, can be illustrated (Fig. 10.9 (b)). This model is also used in the early GIS system (Burrough, 1983). Let us assume that the red squares on checkerboard map serve to contain a single attribute, rather than just a theme. Instead, we can use the number 1 (red squares) to represent water and 0 (black squares) to indicate the absence of water. How can we represent a thematic map of land use that contains, say four categories, namely, recreation, agriculture, industry, and residences? Each of these four attributes would have to be separated

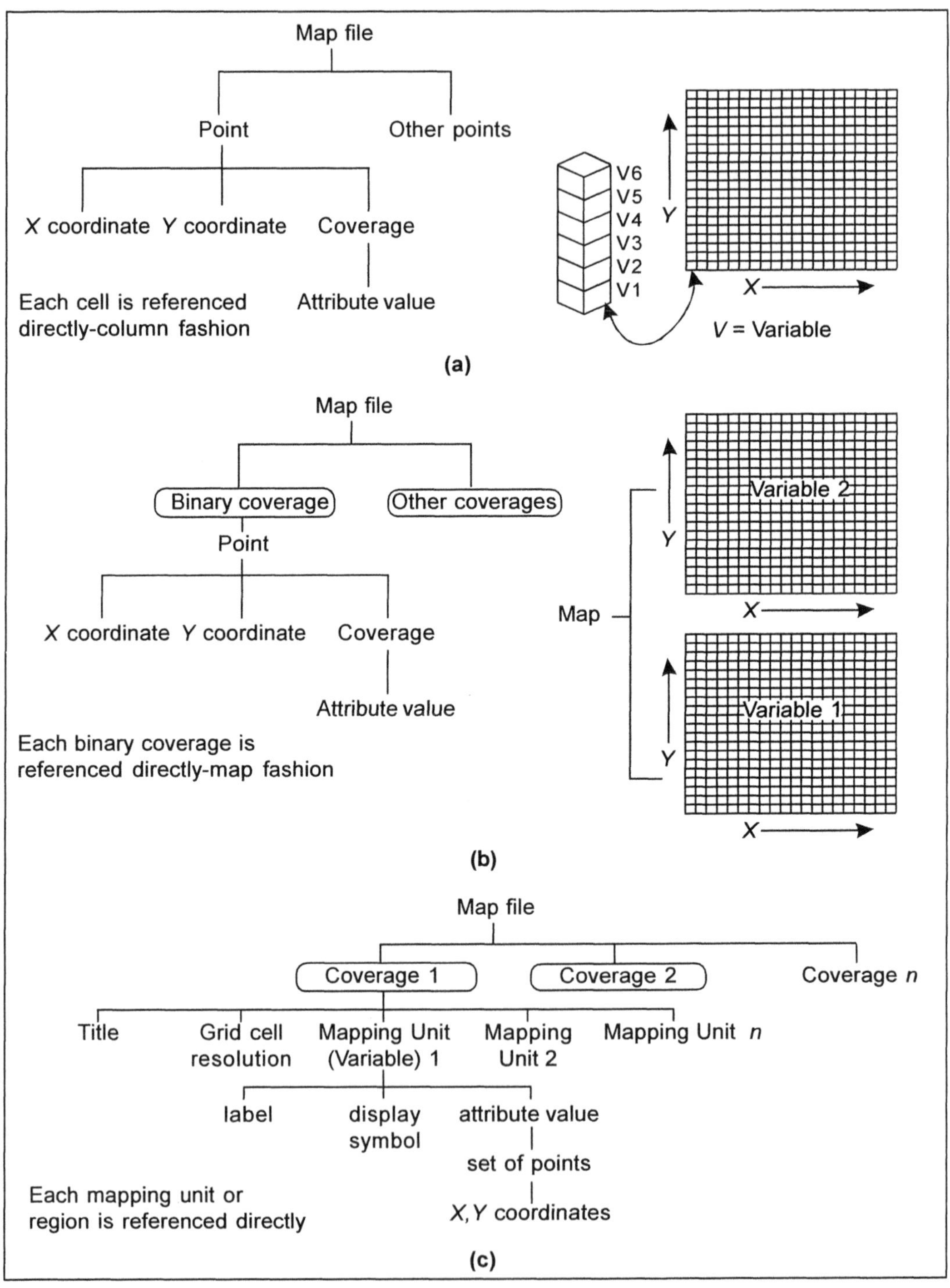

Fig. 10.9 Three types of Raster GIS models.

out as an individual layer. One layer would stand for agriculture only, with 1's and 0's representing the presence or absence of this activity for each grid cell. Recreation, industry, and residences would be represented in the same way, with each variable referenced directly, rather than referencing the grid cell as we did in the GRID/LUNAR/MAGI data model. Finally, the coverages would be combined vertically, or in column fashion, to produce a single theme or coverage, much as red, yellow, green, and blue printing plates are combined to create a single color image.

IMGRID system has two major advantages. First, we have a contiguous object that more closely resembles how we think about a map. That is, our primary storage object is a two-dimensional array of numbers, rather than a column of numbers for different themes. Second, we reduce the numbers that must be contained in each coverage to 0's and 1's. This will certainly simplify our computations and will eliminate the need for map legends. Since each variable is uniquely identified, assigning a single attribute value to a single grid cell is possible, and this is a third advantage. Let us assume that a given grid cell partly occupies agriculture and partly recreation and each of these attributes of land use theme is separated out. In such a case, we may encounter difficulties when creating our final thematic coverage if multiple values occur in individual cells. To avoid such problems, we must be able to ensure that each grid cell has only a single value for each variable.

The IMGRID model is seems to be more intuitive from a map abstraction viewpoint, and requires us to be very specific about the attributes to be contained in each coverage. But it offers the advantage of using the coverage as the direct object of reference for the computer. Its limitations stem primarily from the problem of data explosion. Imagine for a moment that you have a database composed of 50 themes. Each theme must be separated out into binary (0's and 1's) coverages on the basis of individual attributes within each theme. Suppose that there is an average of 10 categories for each theme. To represent this rather modest database, you will need a total of 10 × 50 or 500 coverages. Although available storage devices can certainly manage such volumes, you need to manage and keep track of examining this approach further. Imagine how many values must be modified and recoded to create a new theme. For example, to combine 10 binary coverages to create a new thematic coverage with 10 categories, you would have to separate the thematic coverage into 20 new binary coverages each. Thus, for a simple operation you had to combine 10 grid cell values, and to create additional thematic coverage it is necessary to produce 10 new values of 0 and 1 for each variable. This is a rather tedious approach.

MAP Model

The third raster GIS model Map Analysis Package (MAP) model developed by C. Dana Tomlin (Burrough, 1983) formally integrates the advantages of the above two raster data structure methods. In this data model (Fig. 10.9 (c)) each thematic coverage is recorded and accessed separately by map name or title. This is accomplished by recording each variable, or mapping unit, of the coverage's theme as a separate number code or label, which can be accessed individually when the coverage is retrieved. The label corresponds to a portion of the legend and has its own symbol assigned to it. In this way, it is easy to perform operation on individual grid cells and groups of similar grid cells, and the resolution changes in value require rewriting only a single number per mapping unit, thus simplifying the computations. The overall major improvement is that the MAP method allows ready manipulation of the data in a many-to-one relationship of the attribute values and the sets of grid cells.

The MAP data model is compatible to almost all computer systems from its original mainframe version to Macintosh and PC versions and modern UNIX-based workstation versions. It can be used as a teaching version of GIS as it is very flexible and also becomes a major module in commercial GIS packages like ARC/INFO.

Although raster GIS systems have traditionally been developed to allow single attributes to be stored individually for each grid cell, some have evolved to include direct links to existing database management systems. This approach extends the utility of the raster GIS by minimising the number of coverages and substituting multiple variables for each grid cell in each coverage. Such extensions to the raster data model have also allowed direct linkage to existing GIS systems that use a vector back and forth from raster to vector. The user can operate with all the advantages of both the data structures. The conversion process is often quite transparent, allowing the user to perform the analyses needed without concern for the original data structure. This feature is particularly important because it is strengthening the relationship between traditional digital image processing software used to manipulate grid cell-based, remotely sensed data and GIS software. Many software systems already have both sets of capabilities, and still more are likely in the future. Together with the linkage with existing statistical packages, we are rapidly approaching the systems that operate with a superset of spatial analytical techniques, resulting in a maturing of automated geography.

10.4.4 Compact Raster Data Models

In execution of any GIS related projects, huge quantity of raster data has to be stored, retrieved, manipulated, and analysed. This involves a number of thematic coverages to be stored in the disk of the computer system. The common methods of storing raster data with substantial savings in disk space require a data model similar to the MAP data model. The compact methods of storing raster data allow groups of grid cells to be accessed directly, because the compactions generally operate by reducing the information content of these groups of cells to the absolute minimum needed to represent them as a unit. Compact methods for storing raster data certainly operate under the storage and editing subsystem of a GIS, but they can also be applied directly during the input phase of the GIS operation (chapter 12). Based on the nature of the GIS data and existence of available facilities , all the compact methods are grouped as (a) run-length codes, (b) raster chain codes, (c) block codes,

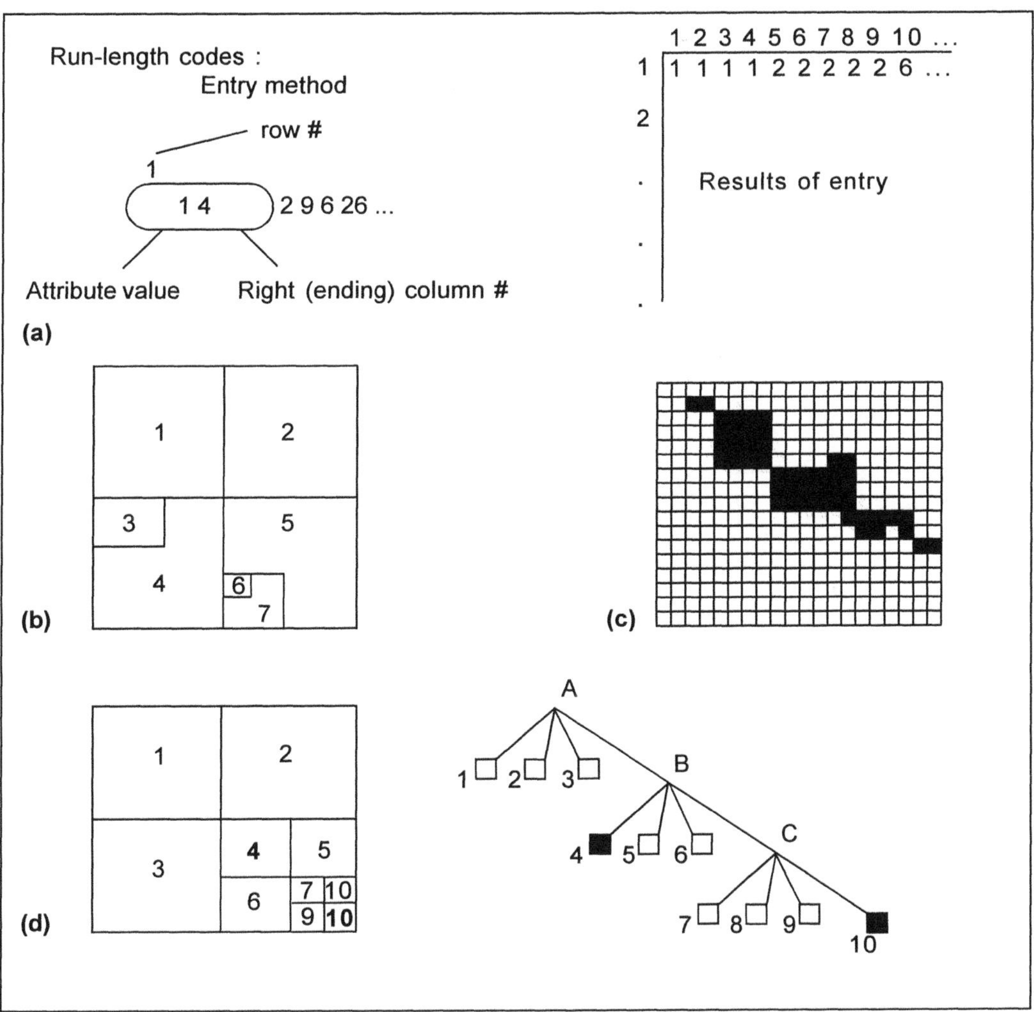

Fig. 10.10 Methods of compacting raster data to preserve storage.

and (d) the unique structure called quadtrees. Fig. 10.10 illustrates how these four methods are used to store in the raster data in order to save the disk space.

Run-Length Coding

The first method of compacting raster data is a process called run-length codes. In the raster data, each grid cell has a numerical value corresponding to a category of data on the map that must be put (generally typed) into the computer. For example, for a map of 500 × 500 grid cells, 2,50,000 numbers have to the typed into the computer. As you begin typing, you will quickly see patterns emerging from the data that present oppurtunities for reducing your workload. Specifically, there are long strings of the same number in each row. Think how much time you could save if for a given row, you could just tell the computer that starting at column 8 all the numbers are 1s, representing some map variable, until you get to column 56, then at column 57 the numbers are 2s until the end of the row. Indeed, you could also save a great deal of space by simply giving starting and ending points for each string and the value that should be stored for that string. This method of storing the data is called run-length coding.

This technique reduces data volume on a row by row basis. It stores a single value where there are a number of cells of a given type in a group, rather than storing a value of each individual cell. Fig. 10.10 (a) shows a run-length encoded version of the forest cover. The first line in the file represents the dimensions of the matrix (10 × 10) and the number of entities present. In the second and subsequent lines of the file, the first number in the pair (either 1 or 0 in this example) indicates the presence or absence of the forest. The second number indicates the number of cells occupied by the forest. Therefore the first pair of numbers at the start of the second line tell us that no entity is present in the first 10 cells of the first row of the image. The main disadvantage of this method of storing data is that the operation is on a row-by-row basis.

Raster Chain Codes

The chain coding method of data reduction works by defining the boundary of the entity. The boundary is defined as a sequence of unit cells starting from a cell and returning to a given origin. The direction of travel around the boundary is usually given using a numbering system (for example, 0 = North, 1 = East, 2 = South and 3 = West). Fig. 10.10 (b) shows how the boundary cells for the forest would be coded using this method. Here, the directions are given in letters (N, S, E, and W) to avoid any possible mistake. The first line in the file structure tells us that the chain coding started at cell 4, 3 and there is only one chain. On the second line the first letter in each sequence represents the direction and the number of cells lying in this direction. The raster chain method of storing data is based on X and Y position, a grid cell value for the entire area, and the directional vectors. Usually the vectors include nothing more than the number of grid cells and the vector direction based on a simple coding scheme, 0, 1, 2, and 3 could indicate north, south, east, and west respectively.

Block Codes

The third method of storing the grid-based data for reducing the storage is block codes. The block codes method is a modification of run-length codes. Instead of giving starting and ending points, plus a grid cell code, select a square group of cells and assign a starting point, the centre or a corner, pick a grid cell value, and tell the computer how wide the square of grid cells is, based on the number of cells. Block coding is also called a two-dimensional run-length code. Each square, group of grid cells, including individual grid cells, can be stored in this way with a minimum group of numbers. Block coding methods are a very effective method of reducing the storage space for most thematically layered digital data in a GIS.

Fig. 10.10 (c) shows how the simple raster map of the forest cover has been subdivided into a series of hierarchial square blocks. Ten data blocks are required to store data about the forest image. These are seven unit cells, two four-cell squares and one nine-cell square. Coordinates are required to locate the blocks in the raster matrix. In the example, the top left-hand cell in a block is used as the locational reference for the block.

Quadtrees

The final method of compact storage is a rather difficult approach. Still at least one commercial system called Spatial Analysis System (SPANS), from Tydac, and one experimental system called Quilt are based on this scheme. Like block codes, quadtrees operate on square groups of cells. In this the entire map is successively divided into uniform square groups of grid cells with the same attribute value. Starting with the entire map as entry points the map is then divided into four quadrants (NW, NE, SW, and SE). If any of these quadrants is homogeneous containing grid cells with the same value, that quadrant is stored and no further subdivision is necessary. Each remaining quadrant is further divided into four quadrants, again NW, NE, SW, and SE. Each quadrant is examined for homogeneity. All homogeneous quadrants are again stored, and each of the remaining quadrants is further divided and tested in the same way until the entire map is stored, as square groups of cells, each with the same attribute value. In the quadtree structure, the smallest unit of representation is a single grid cell.

One of the advantages of this raster model is that each cell can be subdivided into smaller cells of the same shape and orientation. This unique feature of the raster data model has produced a range of innovative data storage and data reduction methods that are based on quadtree works on the principle of recursively subdividing space. The most popular of these is the area or region quadtree. The area quadtree works on the principles of recursively subdividing the cells in a raster image into quads (or quarters). The subdivision process continues until each cell in the image can be classed as having the spatial entity either present or absent within the bounds of its geographical domain. The number of subdivisions required to represent an entity will be a trade-off between the complexity of the feature and the dimensions of the smallest grid cell. The quadtrees principle is illustrated in Fig. 10.10 (d) where the division of the region of the image is mainly based on the resolution of the system as minimum mapable unit. Therefore the systems based on quadtrees are called variable resolution systems because they can operate at any level of quadtree subdivision. Thus users can decide how fine the resolution needs to be for various manipulations and applications. In addition, because of the compactness of storage from this method, a very large database, perhaps of a continental or even global scale, can be stored in a single system.

The major difficulty with the quadtree structure is in the method by which it separates the grid cells into regions. In block codes, the decision was based entirely on the existence of homogeneous grid cells, regardless of where they were located on the map. With quadtrees, the subdivision is preset to the four quadrants (NW, NE SW, SE), resulting in some otherwise homogeneous regions lying in two or more different quadrants. This results in computational difficulties for analysis of shape and pattern that must be overcome through rather complex computational methods. GIS software using the quadtree data model operates under workstation and PC platforms and use multiple operating systems. Such programs are in use worldwide and offer some interesting opportunities, especially to those who need very large databases.

10.5 Vector GIS Models

Vector data structures allow the representation of geographic space in an intuitive way reminiscent of the familiar analog map. The geographic space can be represented by the spatial location of items or attributes which are stored in another file for later access. Fig. 10.5 shows how the different entity, namely, points, lines, and areas can be defined by coordinate geometry. Like the raster spatial data model, there are many potential vector data models that can be used to store the geometric representation of entities in the computer.

A point is the simplest spatial entity that can be represented in the vector world with topology. A point requires to be topologically correct with respect to a geographical reference system which locates it with respect to other spatial entities. To have topology a line entity must consist of an ordered set of points a locus of number points, (known as an arc, segment, or chain) with a defined start and end points (nodes). Knowledge of the start and end points gives a line direction. For the creation of topologically correct area entities, the data about the points and lines used in its construction, and a knowledge of how these are connected to define the boundary, are required. The combination of points gives the line entity and the combination of points and line segments forms an area entity.

The simplest vector data structure that can be used to reproduce a geographical image in the computer is a file containing (x, y) coordinate pairs that represent the location of individual point features. Fig. 10.11 shows such a vector data structure for a car park near Hussain Sagar lake in Hyderabad. Now, how a closed ring of coordinate pairs defines the boundary of the polygon, is clear. The limitations of simple vector data structures start emerging when more complex spatial entities are considered. There are several ways in which vector data structures can be put together into a vector data model by which the relationships between variables in a single coverage or among variables in different coverages can be defined. The two basic types of vector data models are (i) spaghetti model, and (ii) topological model.

10.5.1 Spaghetti Model

The simplest vector data structure that can be used to reproduce a geographical image in the computer is a file containing (x, y) coordinate pairs that represent the location of individual point features. Fig. 10.12 is essentially a one-for-one translation of the graphical image or a map which is also termed as the conceptual model. Let us consider a conceptual model in which an analog map covering each graphic object is shown in Fig. 10.12. Each graphic object can be represented with a piece of spaghetti. Each piece of spaghetti acts as a single entity. The shortest spaghetti can be represented as a point, collection of a number of point spaghettis for a line entity and collections of line segments that come together at the beginning and ending of surrounding areas form an area entity. Each entity is a single, logical record in the computer, coded as variable length strings of (x, y) coordinate pairs. Let us assume that two polygons lie adjacent to each other in a thematic coverage. These two adjacent polygons must have separate pieces of spaghetti for adjacent sides. That is, no two adjacent polygons share the same string of spaghetti. Each side of polygon is uniquely defined by its own set of lines and coordinate pairs. In this model of representing vector data, all the spaghetties are recorded separately for polygons. But in the computer they should have the same coordinates.

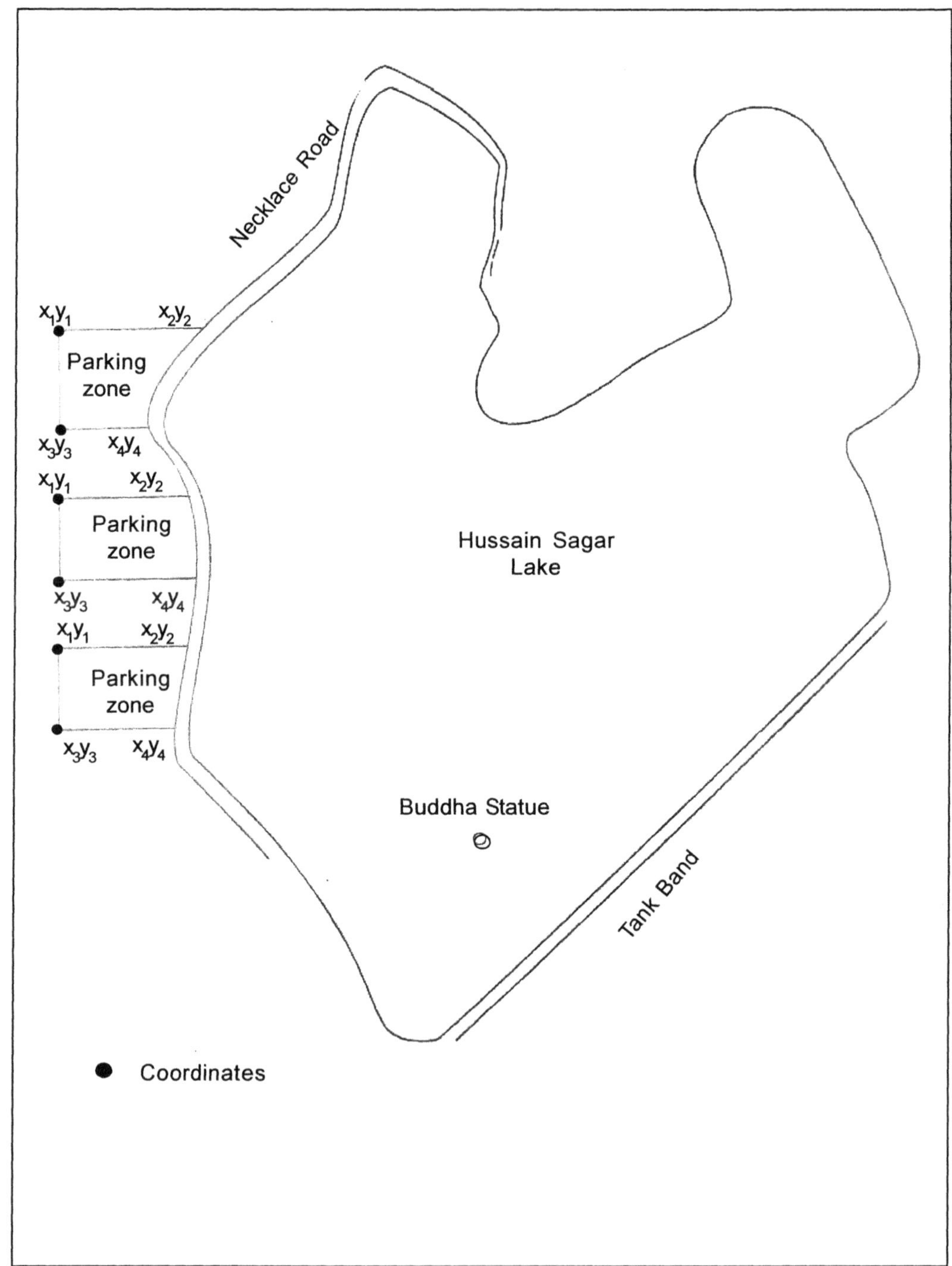

Fig. 10.11 Vector Data Structure for a car park near Hussain Sagar Lake.

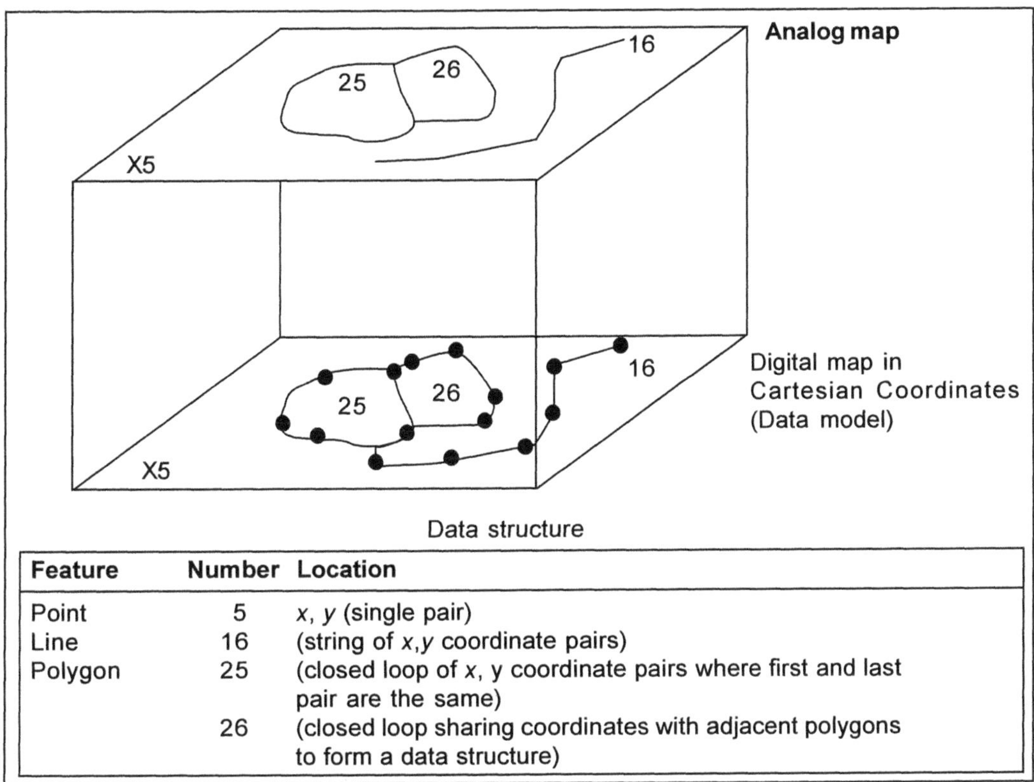

Feature	Number	Location
Point	5	*x*, *y* (single pair)
Line	16	(string of *x*,*y* coordinate pairs)
Polygon	25	(closed loop of *x*, y coordinate pairs where first and last pair are the same)
	26	(closed loop sharing coordinates with adjacent polygons to form a data structure)

Fig. 10.12 Spaghetti Vector Data Model.

A result of this lack of explicit topology with enormous computational overheads, makes measurements and analysis difficult. Because it so closely resembles the analog map, the spaghetti model is relatively efficient as a method of cartographic display and is still used quite often in computer aided cartography when analysis is not the primary objective. The representation is quite similar to that found in many plotting devices, making the translation of the spaghetti model to the plotter language easy and efficient. Plotting of spaghetti data models is usually quite fast compared with some others. The characteristic feature of spaghetti model is recording of the coordinates of all the points associated with all the polygons. This record of explicit reference information is known as a point dictionary (Burrough, 1986). The data structure in Fig. 10.12 shows how such an approach has been used to store data for the different zones of the project area.

However, this model would not address any information about the linkage between lines. Linkages would be implied only when the lines are displayed on the computer screen. In the same way, a series of polygons created using either the simple data structure or a point by point approach may appear connected on the screen when in fact the computer sees them as discrete entities unaware of the presence of neighbouring polygons. Therefore the spaghetti model does not consider the topology. But any vector model would be called a perfect vector GIS model, if and only if it has the functional capabilities of spatial database management, attribute database management, and linkage mechanism between these two databases to exhibit the topological relationships. For the representation of line networks and adjacent island polygons, a set of instructions is required to inform the computer where one polygon or line is with respect to its neighbours. Topological data structures and linkage mechanism contain this information. There are numerous ways of providing topological structures in a form that the computer can understand. The topological data structures and the management of huge quantities of topological information are explained under topological models.

10.5.2 Topological Models

In order to use the data manipulation and analysis subsystem more efficiently and obtain the desired results, to allow advanced analytical techniques on GIS data and its systematic study in any project area, much explicit spatial information is to be created. The topological data model incorporates solutions to some of the frequently used operations in advanced GIS analytical techniques. This is done by explicitly recording adjacency information into the basic logical entity in topological data structures, beginning and ending when it contacts or intersects another line, or when there is a change in the direction of the line. Each line then has two sets of numbers: a pair of coordinates and an associated node number. The node is the intersection of two or more lines, and its number is used to refer to any line to which it is connected. In addition, each line segment, called a link, has its own identification number that is used as a pointer to indicate the set of nodes that represent its beginning and ending polygon. These links also have identification codes that relate polygon numbers to see which two polygons are adjacent to each other along its length. In fact, the left and right polygon are also stored explicitly, so that even this tedious step is eliminated. This design feature allows the computer to know the actual relationships among all its graphical parts to identify the spatial relationships contained in an analog map document. Fundamentally, the topological models available in GIS ensure (a) that no node or line segment is duplicated, (b) that line segments and nodes can be referenced to more than one polygon, and (c) that all polygons can be adequately represented. Fig. 10.13 shows one possible topological data structure for the vector representation. To understand the topological vector data structure, let us consider a network with 8 nodes encoded as n1 to n8. The links

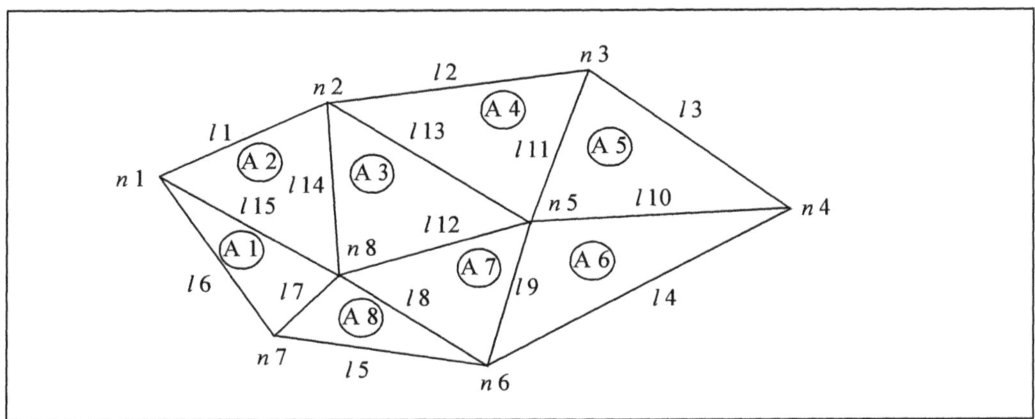

Fig. 10.13 Topological vector Data Model.

joining all these nodes are encoded as *l* 1 to *l* 14 and the polygons created by all these line segments/links are coded as A1 to A8. The creation of this structure for complex area features is carried out in a series of stages. Burrough (1986) identifies these stages as identifying a boundary network of arcs (the envelope polygon), checking polygons for closure, and linking arcs into polygons. The area of polygons can then be calculated and unique identification numbers attached. This identifier would allow nonspatial information to be linked to a specific polygon. Table 10.1(a) provides the spatial data base along with the coordinate file (Table 10.1(b)) of all the nodes, and the corresponding attribute information can be given to each point, line and polygon by keeping the identification numbers.

There are a number of topological vector data models. Out of the available models, three models are very common in use. These three models are : (a) GBF/ DIME model created by US Department of Commerce, Bureau of the Census, 1969 (b) TIGER model (Marx, 1986) and (c) POLYVERT (Peuquet, 1984).

GBF/DIME Topological Vector Model

The best-known topological data model is the GBF/DIME (Geographical Base File/Dual Independent Map Encoding) model created by the US Bureau of the Census to automate the storage of street map data for the decennial census (US Department of Commerce, Bureau of the Census, 1969). GBF/DIME models were designed to incorporate topological information about urban areas for use in demographic analyses (Cooke, 1987) and were created by graph theory. In this case the straight-line segment ends when it either changes direction or intersects another line, and the nodes are identified with codes. In addition to the basic topological model, the GBF/DIME model assigns a directional code in the form of a 'From node and a To node,' that is, a low-value node to a high-value node in the sequence. The useful feature of this type

Table 10.1(a) Spatial Database of Topological Data (Topological file)

Link No.	Left node	Right node	Left poloygon	Right polygon
*l*1	n1	n2	0	A2
*l*2	n2	n3	0	A4
*l*3	n3	n4	0	A5
*l*4	n6	n4	A6	0
*l*5	n7	n6	A8	0
*l*6	n1	n7	A1	0
*l*7	n7	n8	A1	A8
*l*8	n8	n6	A7	A8
*l*9	n5	n6	A7	A6
*l*10	n5	n4	A5	A6
*l*11	n3	n5	A4	A5
*l*12	n8	n5	A3	A7
*l*13	n2	n5	A3	A4
*l*14	n2	n8	A2	A3

Table 10.1(b) Database related to coordinates of nodes (coordinate file)

node No.	x - coordinate	y - coordinate
n1	x1	y1
n2	x2	y2
n3	x3	y3
n4	x4	y4
n5	x5	y5
n6	x6	y6
n7	x7	y7

of data structure is that the associated attribute data can be accessed by means of geographic coordinates. This approach makes it easy to check for missing nodes during the editing process. If, for instance, you want to see whether a polygon is missing any links, simply match the 'to node' of one line to the 'from node' of the preceding link. If the nodes do not completely surround an area, it means a node is missing.

An additional useful feature of the GBF/DMIE system is the creation of the files for both the street address and coordinates for each node and links. The disadvantage of such a model is the slowest possible way to search for records in a computer. The model would also back the geographical specificity of the entities. Since there is no particular order in which the line segments occur in the system, to search for a particular line segment, the program must perform a tedious sequential search of the entire database. The GBF/DIME system, is based on the concept of graph theory. It does not matter whether the line connecting any two points is curved or straight. Thus, a side of a polygon serving to indicate a curved lake boundary would be stored not as a curved line but rather as a straight line between two points, with the resulting model lacking in geographic specificity.

TIGER Topological Vector Model

TIGER stands for Topologically Integrated Geographic Encoding and Referencing system. This model does not depend upon the graph theory designed for use in the 1990 US census. In this system, points, lines, and areas can be explicitly addressed, and therefore census blocks can be retrieved directly by block number rather than by relying on the adjacency information contained in the links. Real-world features such as meandering streams and irregular coastlines are given a graphic portrayal more representative of their true geographic shape. Thus TIGER files are more generally used in research which is not related to census.

POLYVRT Topological Vector Model

POLYVRT developed by Peucker and Chrisman (1975) and later implemented at the Harvard Laboratory for Computer Graphics was called the POLYVRT (POLYgon con VERT) model. In this method of representing vector data, each type of geographic entity is stored separately. These separate objects are then linked in a hierarchical data structure with points relating to lines, which in turn are related to polygons through the use of pointers. Each collection of line segments, is collectively called chains in this explicit directional information in the form of To-From nodes as well as left-right polygons (Fig. 8.13).

This structure is also called arc-node structure and objects in the database are structured hierarchically. In this system, points are the elemental basic components. Arcs are the individual line segments that are defined by a series of x-y coordinate pairs. Nodes are at the ends of arcs and form the points of intersection between arcs. There may be a distinction made between nodes at the ends of lines, and points that are not associated with lines. Polygons are areas that are completely bounded by a set of arcs. Thus, nodes are shared by both arcs and contiguous polygon. Several commercial geographic information systems use forms of this arc-node data structure. POLYVRT model has the advantage of retrieving selective and specific entity types like points, lines, or polygons based on their codes. One more advantage of POLYVRT model is that the chains which are a combination of a number of individual lines forming a polygon, can be accessed directly saving time for searches.

POLYVRT has the following advantages. It allows to store and retrieve specific entity types and identify them based on their codes. The corresponding attributes data can also be retrieved based on these codes. Since a polygon can be stored with indirect line segments, individual line segments are straight as nodes and with these nodes as coordinate pairs, each entity can be accessed, retrieved, stored, manipulated, and analysed selectively.

POLYVERT chain lists bounding polygons that are explicitly stored and linked through pointers to each polygon code. The size of the database is largely controlled by the number of polygons, rather than by the complexity of the polygon shapes. This makes storage and retrieval operations more efficient, especially when highly complex polygonal shapes found in many natural features are encountered. The major drawback of POLYVRT is that it is difficult to detect an incorrect pointer for a given polygon until the polygon has actually been retrieved, and even then you must know what the polygon is meant to represent.

10.5.3 Shape File

Advancement of computer technology in terms of database management techniques, speed of the processor, and massive storage capacity of the devices, leads to the development of a newer and nontopological structure called the shape file (Peuquet, 1984). The shape file structure is the file structure that stores the geometry, topographic information, and attribute information of the geographical features in a dataset file. The geometry and shape of the feature comprise a set of vector coordinates and topology corresponding to their attributes.

The shape file generally has lower processing overhead than its topological counterpart. This results in faster drawing speed and editing, allows for the handling of overlap and noncontiguous features, reduces disk space requirements and makes the files themselves easier to read and write. Shape files are usually three separate and distinct types of files: main files, index files, and database tables. The main file is a direct access, variable record length file that contains the shape as a list of vertices. The index file contains character length and offset information for locating the values, and a database table which contains the attributes that describe the shapes.

10.5.4 Compact Vector Data Models

The data of raster models, discussed in the previous section, can be compacted to reduce storage space in a number of ways. Similarly, compacting vector data models are developed to reduce the storage space. Although vector data models are generally more efficient at storing large amounts of geographic space, it is still necessary to consider reductions. In fact, a simple codification process developed more than a century ago by Sir Francis Galton (1884) is relatively similar to the compaction technique in vector data storage. There are two schemes of compacting vector data models : Galton's schemes, and Freeman-Haffman chain codes (Fig. 10.14).

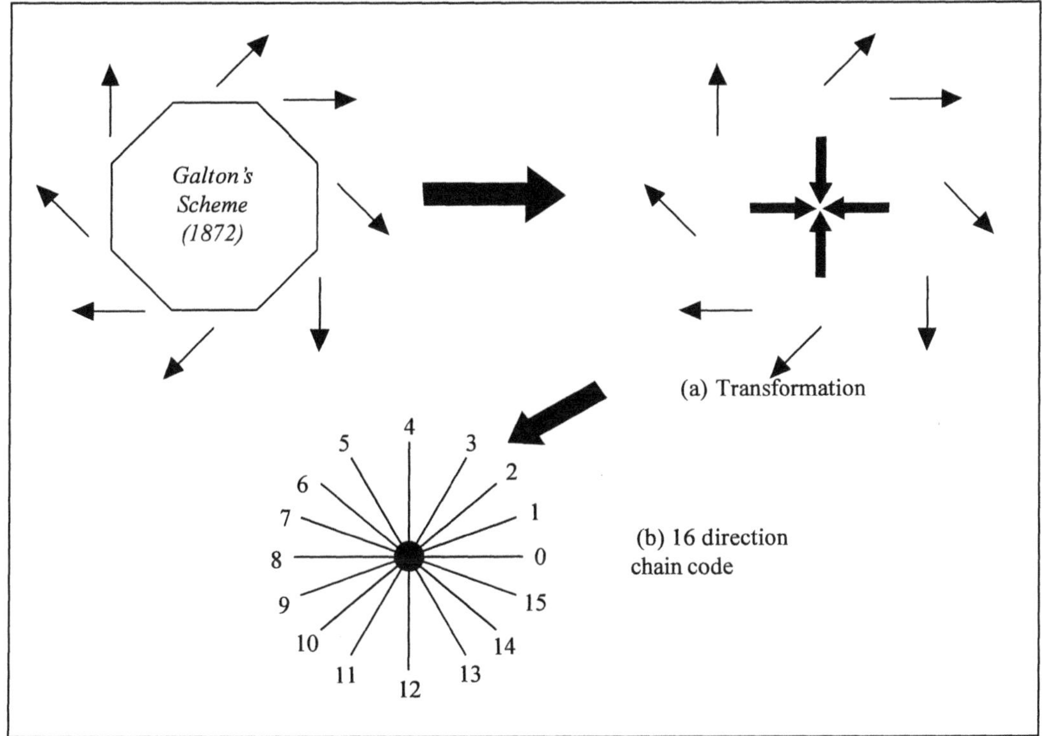

Fig. 10.14 Chain codes for compacting vector data.

In the Galton's Scheme, we have to apply eight numerical values, one for each of the cardinal compass direction and one for each of the intermediate, northeast, south east, southwest, and northwest (Fig. 10.14(a)). The second coding scheme is known as Freeman-Hoffman chain codes. Eight unique directional vectors are assigned the numbers 0 through 7. As Galton had done for ground navigation on his journeys, the Freeman-Hoffman method assigned these vectors in the same four cardinal directions and their diagonals. By assigning a length value to each vector, individual line entities can be given a shorthand to show where they begin, how long they are, in which direction they are drawn, and where the vector changes direction. There are many variations of this scheme including increasing the codes to 16 (Fig. 10.14(b)) or even 32 values rather than 8, to enhance accuracy. But the result is the same reduced storage for vector database.

Although the chain code models produce significant improvements in storage, they essentially compact spaghetti models and contain no explicit topological information. This limits their usefulness to storage, retrieval, and functions, because of the analytical limitations of nontopological data structures. In addition, the way the lines and polygons are encoded as vectors, performing coordinated transformations, especially rotation, leads to heavy cost computing overhead. Chain code models are good for distance and shape calculation, because much of this information is already part of the directional vectors themselves. In addition, because the approach is very similar to the way vector plotters operate, the models are efficient for producing rapid plotter output.

10.6 Comparison of Raster and Vector Models

The traditional advantages and disadvantages of raster versus vector spatial data structures have been documented by Kenndey and Meyers (1997). The basic issues include data volume, retrieval efficiency, data accuracy, data display, correctness to perturbation, and data manipulation, efficiency, and processing capabilities. Comparisons of data volume between raster and vector systems are entirely dependent upon the database elements, as well as considerations of accuracy and precision. A detailed comparisons between raster model and vector model are discussed in the Table 10.2.

Table 10.2 Comparison of raster and vector Data models

Raster model	Vector model
Advantages	**Advantages**
1. It is a simple data structure. 2. Overlay operations are easily and efficiently implemented. 3. High spatial variability is efficiently represented in a raster format. 4. The raster format is more or less required for efficient manipulation and enhancement of digital images.	1. It provides a more compact data structure than the raster model. 2. It provides efficient encoding of topology, and, as a result, more efficient implementation of operations that require topological information, such as, network analysis. 3. The vector model is better suited to supporting graphics that closely approximate hand-drawn maps.
Disadvantages	**Disadvantages**
1. The raster data structure is less compact. 2. Topological relationships are more difficult to represent. 3. The output of graphics is less aesthetically pleasing because boundaries tend to have a blocky appearance rather than the smooth lines of hand-drawn maps. This can be overcome by using a very large number of cells, but it may result in unacceptably large files.	1. It is a more complex data structure than a simple raster. 2. Overlay operations are more difficult to implement. 3. The representation of high spatial variability is inefficient. 4. Manipulation and enhancement of digital images cannot be effectively done in the vector domain.

11

GIS Data Management

11.1 Introduction

Management of GIS data consists of storing a variety of data categorised under two types, entity (spatial data) and attribute (aspatial) data in a way that permits us to retrieve or display any combinations of these data after analysis and manipulation. In order to perform these operations, the computer is able to store, locate, retrieve, analyse and manipulate the raw data derived from a number of sources by using representational file structures. In other words, each graphical identity must be stored explicitly, along with its attributes, so that, we can retrieve and select the correct combinations of entities and attributes in a reasonable time. GIS database comprises spatial or entity or graphical database, nonspatial or attribute database, and a linkage mechanism for their topology, to show the relationship between the spatial data and attribute data for further analysis.

An entity (either a point, or a line, or an area) has both spatial and attribute data to describe it. Spatial data can be known as "where things are" data and attribute data the 'what things are' (Ian Heywood et. al 1998). For example, a point entity, the Charminar, a monument in Hyderabad, has the reference in terms of a latitude and longitude, and to accompany this there would be an attribute data about the nature of the real world feature that the point represents. More clearly, in this example,

Entity type is the point
Spatial data are longitude and latitude
Attribute data is the monument, Charminar.

Nonspatial (attribute) data can be stored in any conventional databases, whereas spatial data, which is the dominant data in GIS, should have the database which is capable of handling spatial data.

A spatial database describes a collection of entities, some of which have a permanent location on some global and dimensional space. Normally, there is a mixture of spatial and aspatial entity types. Spatial entity types have the basic topographical properties of location, dimension, and shape, while aspatial entity types do not have location. Thus before the analysis can be done, the 'additional' data need to be specified and incorporated in the geographical database. To manage the GIS data it is useful to examine the concepts of database management systems.

In this chapter, apart from the fundamental concepts and components of DBMS, some basic file structures and database structures that enable large amounts of data to be organised, stored, searched, and analysed, basic concepts and models involved in the representation of space and its objects by graphic data structures, are discussed. These fundamental considerations allow to develop more comprehensive GIS data models to link a set of cartographic data with their attributes.

11.2 Data Base Management Systems

There are many definitions of a DBMS. Dale and McLaughlin (1988) define a DBMS as a computer program to control the storage, retrieval and modification of data (in a database). Stern and Stern (1993) consider that a DBMS will allow users to join, manipulate or otherwise access the data in any number of database files. A DBMS must allow the definition of data and their attributes and relationships, as well as providing security, and an interface between the end users and their applications and the data themselves. The functions of a DBMS can be (i) File handling and file management (for creating, modifying, or deleting the database structure), (ii) adding, updating, and deleting records, (iii) the extraction of information from data, (iv) maintenance of data security and integrity, and (v) application building.

The overall goal of management of GIS database is to provide users with access without having to learn the details of the database itself. In effect, the database management system hides many of the details, and thus provides a higher-level set of tools for users. In the GIS data management field, two types of distinct data are important, one is logical data and the other is physical data. The way in which data appear to a user is called a logical view of the data, and the physical data includes the details of data organisation as it actually appears in memory or on a storage medium.

The functions that a GIS should be able to perform include data input, data storage, management, transformation analysis and output. The data management functions necessary in any GIS facilitate the storage, organisation, and retrieval of data using a database management systems. A DBMS is a set of computer programs for organising information at the core of which will be a database. DBMS is helpful in a number of ways like payrolls, bibliographies, travel agency booking systems, and students enrolment. DBMS can also be used in handling both the graphical and non-graphical elements of GIS data. An ideal GIS DBMS should provide support for multiple users and multiple databases for efficient utility of GIS for allied applications.

In general, there are two approaches to use DBMS in GIS. The first approach is the total solution in which all spatial and aspatial data are accessed through the DBMS to check whether they fit the assumptions imposed by DBMS designer. The second approach is mixed solution in which some data are accessed through the DBMS because they fit the model well. These systems usually adopt a dual database system, one for spatial data managed by database systems specially designed for spatial data, and the other for aspatial data managed by a DBMS.

11.2.1 Functions of DBMS

A database management system is the software that permits the users to work efficiently with the data. The essential functions of the system must provide the means to define the contents of a database, insert new data, delete old data, ask about the database contents, and modify the contents of the database, updating of data, minimisation of redundancy, and physical data independence, security and integrity. Some of these functions are described in the following paras.

Security : Not all users should have all modes of access to a database. Those without proper knowledge or proper authority, should not have the liberty to modify the contents of the database. Database management software allows a user to access data efficiently without being concerned with its actual physical storage implementation, and allows degrees of protection in terms of what a user may see, and what a user is permitted to do. Security refers to the protection of the data against accidental or intentional disclosure to unauthorised persons and protection against unauthorised access, modification, or destruction of the database.

Integrity : Integrity in DBMS is the ability to protect data from systems problems through a variety of assurance measures like range checking, backup, and recovery. A DBMS checks elements as they are entered to enforce the necessary structural constraints of the internal data. Users are forced to enter only those data fields that are required.

Synchronisation : This refers to forms of protection against inconsistencies that can result from multiple simultaneous users. A mechanism is required, so that when one user is about to remove something from the collection, the other user is either warned or prevented from accessing the information until the first has committed to the transaction. This means we must be able to support simultaneous access to the database by multiple users when required, as well as logically view the database as arbitrary subsets of the entire physical database.

Physical data independence : The underlying data storage and manipulation hardware should not matter to the user. The hardware could be changed without users having any awareness of the change. This independence permits us to change hardware as needs and technology change, without rewriting the associated data manipulation software. Data independence implies that data and the application programs that operate on them are independent, so that either may be changed without affecting the other.

Minimisation of redundancy : In a database, storing values that are dependent on other stored values without explicitly keeping track of the dependencies can lead to disruption of the database. At the same time, storing and manipulating the dependencies, in addition to the data itself, increases the difficulties of working with the data. Redundancy in a database is generally not desirable.

Efficiency : Efficient data storage, retrieval, deletion, and updating are dependent upon many parameters. In the creation of a spatial database, it is necessary to provide modes of access for retrieval of both spatial and nonspatial information. Efficient data-retrieval operations are largely dependent upon the volume of the data stored, the method of the data encoding, the design of database structure and complexity of the querry. These operations affect the necessary calculations as well as the types and amounts of requests to be made of the database management systems.

The functions of data management permit the efficient use of a database and the entry points to hardware and software facilities. A modern GIS possesses a number of qualities that are common to all database management systems. The storage, retrieval, deletion and updating of large data sets are an expensive process. These are the essential management functions for any database and must be carried out efficiently regardless of the physical storage device or database location.

11.2.2 Components of DBMS

The interaction with database systems is to perform the following broad types of tasks :

(a) Data definition
(b) Storage definition
(c) Database administration, and
(d) Data manipulation

The first three tasks are most likely to be performed by the database professional, while the fourth will be required by a variety of user types possessing a range of skills and experience as well as variable needs or requirements in terms of frequency and flexibility of access. To meet these tasks, a specialised database system is built with the components described in Fig. 11.1. It shows schematically various DBMS components. To retrieve the required data from the database, mapping must

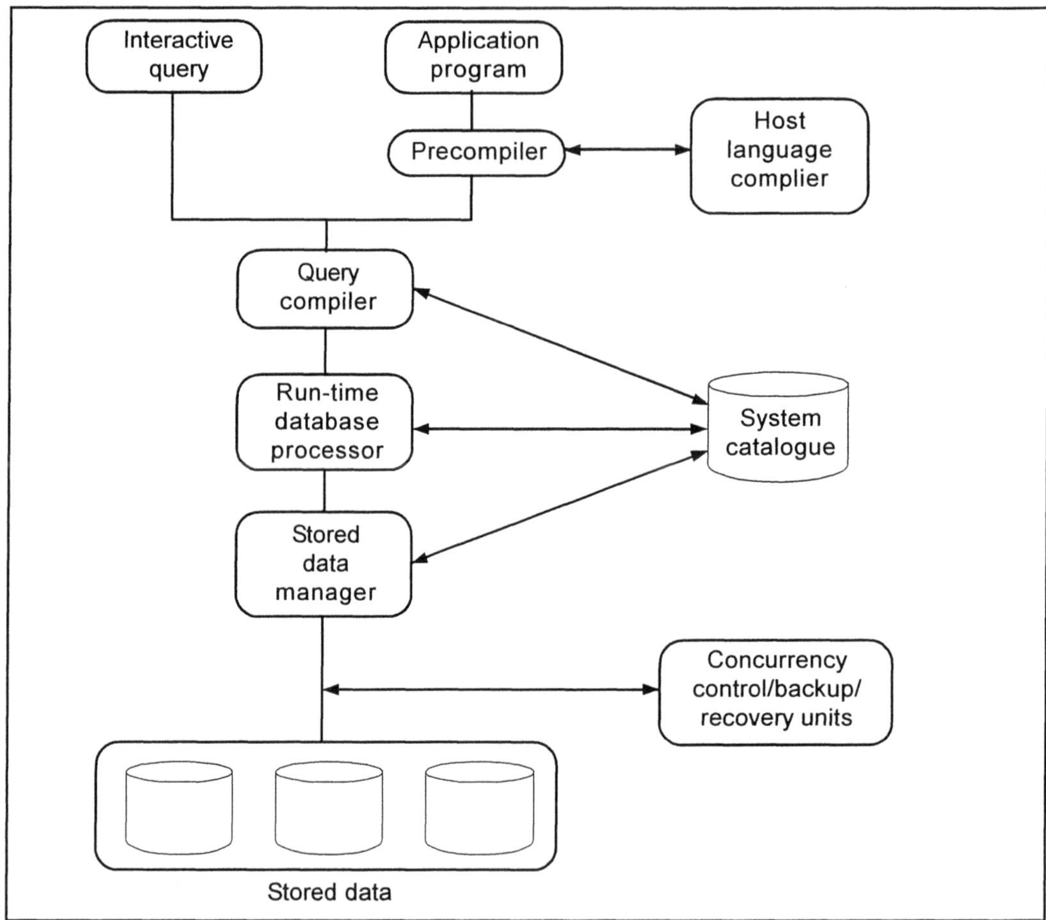

Fig. 11.1 Schematic diagram of DBMS components used to queries.

be made between the high-level objects in the querry language statement and the physical location of the data on the storage device. These mappings are made using the system- catalogue. Access to DBMS data is handled by the stored data manager, which is called the operating system for control of physical access to storage devices. The DBMS has a querry complier which may call the querry optimiser to optimise the code, so that the performance on the retrieval is improved. The logical item of interaction with a database is the transaction, which broadly means to create, modify and delete.

11.3 GIS Data File Management

Traditional computer file structures allow storing, ordering, and searching of pieces of data from a DBMS. Database structures, composed of combination of various file structures and other graphic data structures, allow complex methods of managing data and analysing multiple thematic layers to be used for a particular GIS.

The storage and management of non- spatial/attribute data is a well established technology and is analogous to filing a system. Files are nothing more than a simple accounting system that allows the machine to keep track of the records of data you give it and retrieve these records in any order you wish. Much of what we do in GIS consists of storing entity and attribute data in a way that permits us to retrieve any combination of these objects. This requires the computer, using a representational file structure, to be able to store, locate, retrieve, and cross-reference records. In other words, each graphical entity must be stored explicitly, along with its attributes, so that we can select the correct combinations of entities and attributes in a reasonable amount of time. There are three basic computer file structures: simple lists, ordered sequential files, and indexed files.

11.3.1 Simple List

The simplest file structure is called a simple list consisting of data like names and addresses in a separate index card for each name in a file. Rather than organising the names in any formal order, however, the cards are placed in the order in which they are entered. The only advantage with such a file structure is that to add a new record, you simply place it behind all the rest. Clearly, all the cards are there, and an individual name can be located by examining the cards, but the lack of structure makes searching very inefficient. Suppose your database contains 200,000 records. If your basic file structure is a simple, unstructured, and unordered structure, you may have to search 200,000 cards to find what you are looking for. If it takes, for example, 1 second to perform each search, it will require you to perform as many as (n + 1)/2 seconds, or nearly 28 hours of searching for one point. In contrast, a computer based database management system (DBMS) allows us to extract information, not only by name, but also according to a selection of the other pieces of information in each record given as addresses we could make a search to find out who lived where or to identify all individuals with a given age. These operations would be intolerably tedious using a filing cabinet, which is only indexed for a single field of data like the name of the owner. The computer information systems are based around a digital model, which may be manipulated rapidly to perform the task required.

11.3.2 Ordered Sequential Files

Ordered sequential files are based on the use of alphabetic characters. The data can be arranged in recognisable sequences against which individuals can be compared. The normal search strategy is a sort of divide-and-conquer approach. A search is begun by dividing the file in half and looking first at the item in the middle. If it exactly matches the target combination of numbers or letters, the search is done; if not, the item of interest is compared to each of its neighbours to determine whether the alphanumeric combination is lower or higher. If it is lower, the half containing higher numbers or letters is searched in the same way. If it is higher, the half containing lower numbers or letters is searched by the divide-and-conquer method. This method of arranging data avoids usage of much time for searching the desired data record. The search strategy is based on the key attributes themselves. In GIS, as in many other situations, the items you want to search are points, lines, and areas, primarily based on their coded numbers. Each point, line, and area entity will have often been assigned to it a number of descriptive attributes. Typically, a search will consist of finding the entities that match a selected set of attribute criteria. Thus you might ask the GIS to find all study plots in excellent condition for subsequent display or analysis. Because of the possibly large numbers of attributes linked to each entity, a more efficient method of search will be necessary if we are to find specific entities with associated, cross referenced attributes. Our search method otherwise will rapidly deteriorate into an exhaustive search of all attributes associated with all entities -- the same tedious process employed with the simple list file structure (Burrough, 1983). In short, we need an index to our directory much like the Yellow Pages you would use to find a particular type of store.

11.3.3 Indexed Files

Indexed files are far superior to the above two methods of storing data as these files are created based on the index or code. Indexed files can be created as direct files and/or inverted files. Files, in direct indexed files record themselves are used to provide access to other pertinent information. Let us explain creation and development of indexed files by considering hydrogeomorphological mapping using GIS technique. If you want search ground water potential zones of a particular terrain element from the database created for hydrogeomorphology of the terrain, then the computer will invoke explicit file information, perhaps a code, that tells the exact location of entities bearing the code for ground water potential zones. The program search can now be directed to those specific locations or record numbers by creating an index that directly relates the codes for these zones to their locations in the file, and zones that do not meet this rule will be ignored.

Further improvements in search speed can be obtained if a formal index is created for a selected attribute to be searched. We could, for example, create an index of type of zone attributes (excellent, very good, poor, very poor) and have this index associated with specific entities and their computer locations, analogous to the setup. Because the index is based on possible search criteria, rather than the entities themselves, the information is literally inverted, in that the attributes are the primary search criteria and the entities rely on them for selection. For this reason we call this an indexed inverted file structure.

To create an inverted file structure requires an initial sequential search of the data to order the entities on the basis of the attributes of interest. This search, has three requirements. First, it requires search criteria before hand. Second, it requires recalculation of the index from the original data. Third, use of sequential search methods are needed to obtain the information, if you forget the search criteria. This third condition of database design often causes problems in GIS work because sequential searches are both difficult and time-consuming.

11.3.4 Building GIS Worlds

The previous chapter has explored the way in which different spatial data models (raster and vector) and their respective data structures are used to store entities. In this section, let us consider some of the fundamental methods used to construct computer based GIS worlds by grouping these entities together. There are, in general, four options to build or construct GIS real world model, namely, LCGU based GIS, layer based GIS, feature based GIS and object oriented GIS. The most common method of structuring the geography of the real world in the computer is to use a layered approach.

LCGU based GIS : The least common geographical units (LCGU) also known as Integrated Terrain Units (ITU) integrate all pertinent spatial data records into a single set of all classes.

Layer based GIS : This approach is the most widely used. The layering approach allows users to visualise a map database as a set of registered map separations. This layered approach is illustrated in Fig. 11.2. Each layer is thematic and reflects either a particular use or a characteristic of the land use. For example, layer one describes land use characteristic and layer two, soils. The layered approach to structuring spatial information characterised by the use of thematic maps may be derived from the analysis of remote sensing data to show different features of the area. The concept of breaking the real world into a series of thematic layers is the process that is used to develop the first map overlay analysis (McHarg, 1969). The logical extension of the layer concept is the use of tiles. This approach breaks down

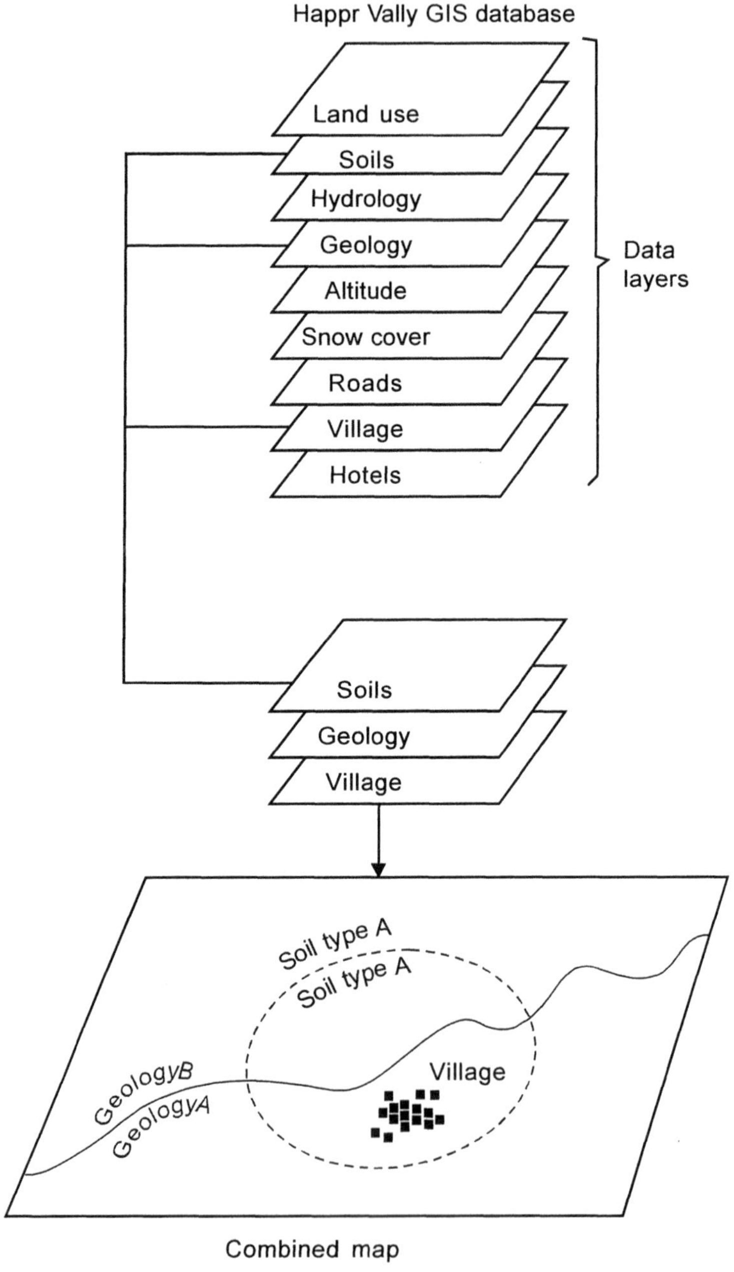

Fig. 11.2 The layer-based approach.

geographical space into a series of regular or irregular units that can be reassembled through the use of a coordinate system. In this model, GIS requires that data are broken down into a set of logical terrain units to assist with physical display of data.

Feature based GIS : This is a new approach having the trappings of object oriented GIS, spatial analysis functionality from raster models and power from vector models.

Object oriented GIS : This is a relatively new approach compared to structuring geographical space views of the real world as a set of individual objects and groups of objects. This object-oriented approach integrates individual geographical entities and attributes into semantic objects with inheritable properties. This approach is based on object-oriented programming. Fig. 11.3 shows how such an approach might be used to structure some of the information for Hussain Sagar lake environs. Notice that features are not divided into separate layers but grouped into classes and hierarchies of objects. The object oriented GIS approach, has the characteristic feature of collecting, managing, and maintaining geographical (size, shape and location), topological (slope and aspect), and behaviour of objects as related to one another.

The objec orient GIS has particular advantages for spatial modelling over the layered GIS. However, it does cause some problems associated with implementation into a workable GIS.

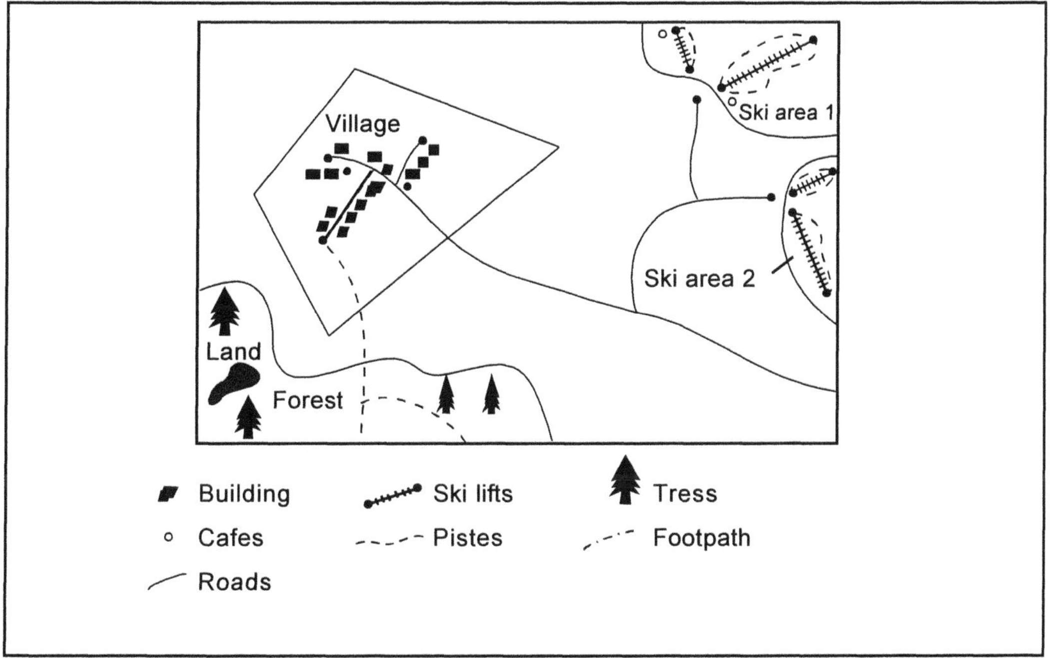

Fig. 11.3 Object-oriented GIS model.

11.4 Database Models

Organisation of data related to utilities, natural resources, automated mapping, facilities management, environmental management, natural resources management and decision support applications have adopted GIS. An important issue for successful implementation of many of these new systems is their ability to integrate GIS with existing internal databases. Many of these databases are relational, as the relational database approach for data management is being implemented in most of the GISs. There are a number of database models by which GIS real world model is built. Amongst those that have been used for the management of attribute data are hierarchical, networks and relational database management systems.

Martin (1996) has named the hierarchical and network systems together as navigational database management systems. The navigational database models are conventional DBMS technologies, and they may be thought of as direct graphs and flow charts. That means, the relationships between different tuples are often displayed as links in a diagram. These links are called pointers and the relationships are linked through a series of pointers. Since almost all existing and most widely used GIS softwares like ARC/INFO are considered RDBMS technologies, an emphasis is given to RDBMS in this book. The relational database model has become more popular than the navigational database models in recent years.

While the navigational model may provide a faster response time for predefined queries, the relational model is extremely easy to explain to a user, and is well suited to adhoc queries. Further, the relational query languages may often be easier to learn than those for navigational database systems.

The purpose of a data model is to provide a common computationally meaningful medium for use by system developers and users. For developers, the data model provides a means to represent the application domain. For the users, it provides a description of the structure of the system, independent of specific items of data or details of the particular implementation. A clear distinction should be made between data models upon which database systems like a relational model and a semantic data model, are built. This can be explained by means of entity-attribute relationships. The three currently most important data modelling approaches are record-based, object-based and object-relational based on the entity-attribute relationship.

11.4.1 Hierarchical Database Models

When the data have a parent/child or one-to-many relation it is called a hierarchical model. Hierarchical systems of data organisation are well known to environmental science, being the methods used for plant and animal taxonomies, soil classification, and so on. Hierarchical systems have the advantage that they are easy to understand and they are easy to update and expand They provide quick and convenient means of data access, and they are good for data retrieval, if the structure of all possible queries can be known beforehand. Disadvantages of hierarchical database structures are that large index files have to be maintained, and certain attribute values may have to be repeated many times, leading to data redundancy, which increases storage and access costs. Fig. 11.4 shows how the database for a university administration can be created using hierarchical DBMS.

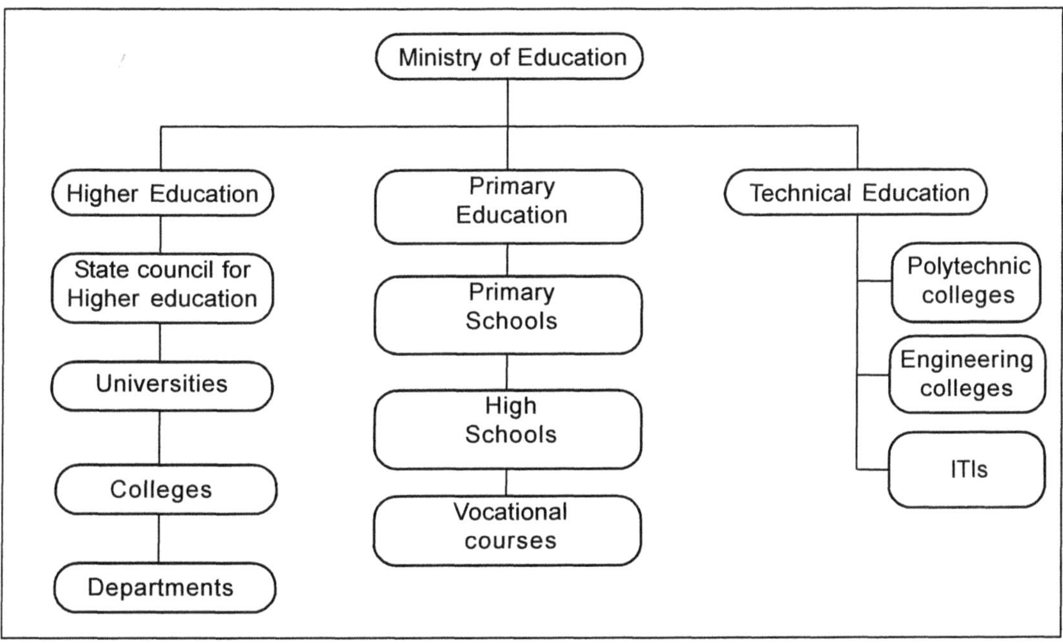

Fig. 11.4 Hierachical Database Model for Education.

Burrough (1983) cited a classic example of how a system based on a hierarchical structure fails to comply with a user's request because the information needed is not even included in the system. Clearly the rigid structure of the hierarchical system also would make it difficult to restructure to allow further searches of this nature. Beyond this severe limitation, the hierarchical structure creates large, often cumbersome index files, frequently requiring repeated entries of important keys to perform its searches. This adds substantially to the amount of memory needed to store the data and sometimes contributes to slow access times as well.

Searches performed in hierarchical data structures are restricted to the branching network itself. In graphic databases, users frequently need to jump to different portions of the database to acquire entity information based on queries of attributes. Since attribute and entity data may very well be located in different locations, the creation of a hierarchical structure requires that direct links be made between search criteria and the graphic devices used to illustrate the locations in space.

Many GIS databases have, in addition to one-to-one and many-to-one relationships, many-to-many relationships, in which a single entity may have many attributes, and each attribute is linked explicitly to many entities. This structure allows users to move from data to data item through a series of pointers. The pointers indicate the relationships among data items.

11.4.2 Network Systems

In data structures much more rapid linkage is required for graphic features where adjacent items in a map or figure need to be linked together even though the actual data about their coordinates may be written in very different parts of the database. Networks systems fulfil this need. Ring pointer structures are very useful ways of navigating around complex topological structures. Network systems are very useful when the relations or linkages can be specified beforehand. They avoid data redundancy and make good use of available data. The disadvantages are that the database is enlarged by the overhead of the pointers, which in complex systems can become quite a substantial part of the database.

Network systems are generally considered to be an improvement over hierarchical structures for GIS work because they are less rigid and can handle many-to-many relationships. As such, they allow much greater flexibility of search than do the hierarchical structures. Also, unlike the hierarchical structure, they reduce redundancy of data like coordinate pairs. Their major drawback is that in very complex GIS databases, the number of pointers can get quite large, coming to comprise a substantial portion of storage space. In addition, although linkages between data elements are more flexible, they must still be explicitly defined with the use of pointers. The disadvantages of large numbers of pointers can be avoided by using another database structure, such as, relational database structures.

11.4.3 Relational Database Models

An understanding of the database models for the management and handling of attribute data in GIS is essential for effective use of the systems available. It is important to understand how both spatial data and attribute data are structured within GIS to enable you to specify appropriate and achievable questions for your GIS to answer, and to implement them effectively (Abel, 1989). The way in which the spatial and attribute data are linked in an individual system is also an important area where considerable differences are seen between systems.

Database systems provide the engines for GIS. In the database approach, the computer acts as a facilitator of data storage and sharing. It also allows the data to be modified and analysed while in the store. For a computer system to be an effective data store, it must have the confidence of its users. Database management systems have grown out of file management systems that perform basic file handling operations, such as, sorting, merging, and report generation.

The records in a file are all of the same record type, containing a fixed set of fields (attributes). A relational database is a collection of tabular relations, each having a set of attributes (Cood, 1970, 1982). The data in a relation are structured as a set of rows and is called tuples consisting of a list of values, one for each attribute. An attribute has associated with it a domain from which its values are drawn. Most current systems require that values are atomic in that they cannot be decomposed as lists of further values, so a single cell in a relation cannot contain a set, list or array of values. This limits the possibilities of the pure relational model for GIS. There are two schema upon which the entire database depends and they are relation schema and database schema.

The relation schema is usually declared when the database is set up and then remains relatively unaltered during the life-span of the system. A database schema is a set of relation schema and a relational database is a set of relations possible with some constraints. Fig. 11.5 shows the example of a database schema and each row in this database is called a tuple.

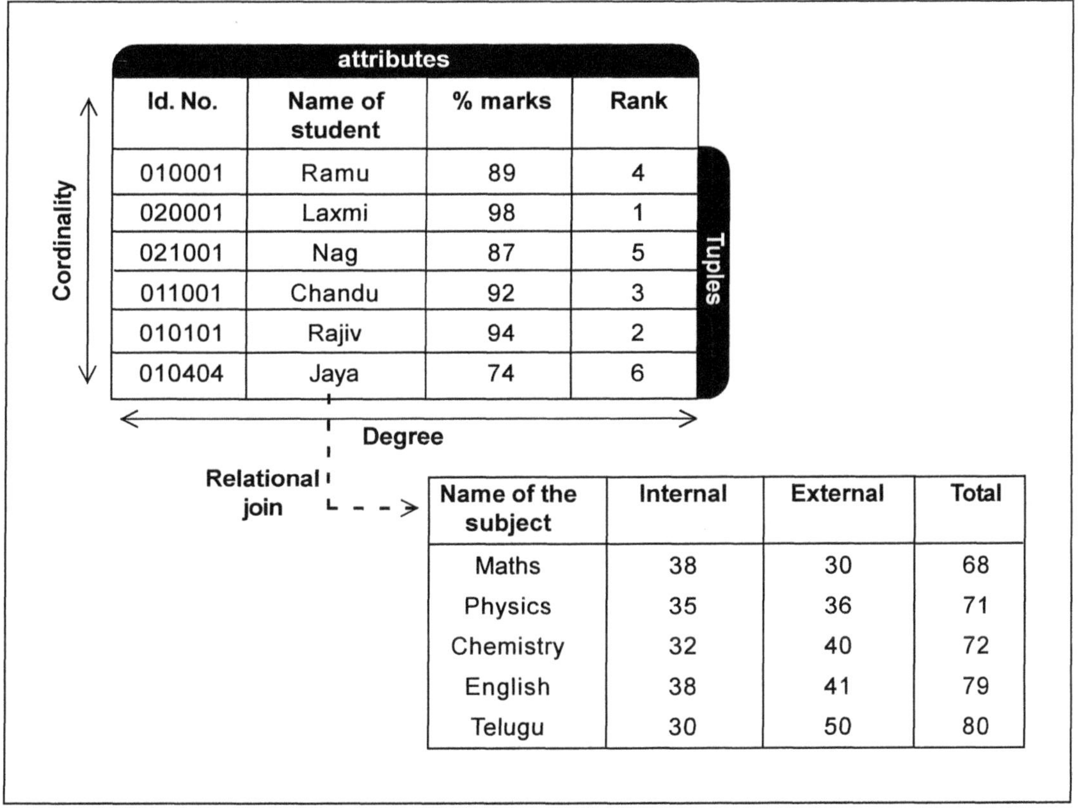

Id. No.	Name of student	% marks	Rank
010001	Ramu	89	4
020001	Laxmi	98	1
021001	Nag	87	5
011001	Chandu	92	3
010101	Rajiv	94	2
010404	Jaya	74	6

Name of the subject	Internal	External	Total
Maths	38	30	68
Physics	35	36	71
Chemistry	32	40	72
English	38	41	79
Telugu	30	50	80

Fig. 11.5 Relational database structure and tables.

The primitive operations that can be supported by a relational database are the traditional set of operations of union, intersection and difference, along with the characteristically relational operations of projecting, restrict, and joining and dividing. The structure of these operations and the way that they can be combined are provided by relational algebra (Cooke and Maxfield, 1967).

Relational systems are based on a set of mathematical principles called relational algebra that provides a specific set of rules for the design and function of these systems. Because relational algebra relies on a set theory, each table of relations operates as a set, and the first rule is that a table can't have any row (tuple) that completely duplicates any other row of data. A single column or even multiple columns can be used to define the search strategy and this search criterion is called a primary key for searching the other columns in the database (Date, 1985). No primary key row can have missing values because missing row values could result in permitting duplicate rows to be stored, thus violating our first rule.

Keeping the organisational tasks and user's needs, we need to match data from one table to corresponding (same row) data in another table by the use of a linking mechanism called a relational join. Because of the predominance of the use of relational systems in GIS, and the large databases produced for GIS, this process is a common one and you must pay close attention to it. Any number of tables can be "related". The process is one of matching the primary key (column) in one table to another column in a second table. The column in the second table row to which the primary key is linked is called a foreign key. Again, the related row to values are assumed to be in the same positions, to ensure that they correspond. This link means that all the columns in the second table are now related to the columns in the first table. In this way, each table can be kept simple, making accounting easier. You can link a third table by finding a key column in the second table that acts as the primary key to a corresponding key column (now called the foreign key) in the third table. The process can continue, connecting each of the simple tables to allow for quite complex searches while maintaining a very simple, well-defined and easily developed set of tables. This approach eliminates the confusion found in database development using network systems. In the process of joining one table to the other table of the database, each table must have at least one column in common. This sometimes creates the redundancy in relations joins. A set of rules called normal forms has been established to reduce the redundancy. There are three basic normal forms.

The first normal form states that the table must contain columns and rows, and because the columns are going to be used as search keys, there should be only a single value in each row location. The second normal form requires that every column that is not a primary key be totally dependent upon the primary key. This simplifies the tables and reduces redundancy. The third normal form, which is related to the second normal form, states that the columns are not the primary keys must "depend" upon the primary key, whereas the primary key does not depend upon any nonprimary key. In other words, you must use the primary key to find the other columns, but you don't need the other columns to search for values in the primary key column. Again, the idea is to reduce redundancy to ensure that the smallest number of columns are produced.

The rules of the normal forms were summed up by Kent (1983) when he stated that each value of a table represents something important "about the primary key, the whole primary key, and nothing but the primary key". For the most part, these rules are highly useful and should be rigorously enforced. The terminology described in the above paras is summerised and presented in Table 11.1.

Table 11.1 Relational database terminology

Paper version	File version	RDBMS
Table	File	Relation
Row	Record/case	Tuple
Column	Field	Attribute
Number of columns	Number of fields	Degree
Number of rows	Number of cases	Cardinality
Unique ID	Primary key	Index
	Possible values	Domain

The terminology of relational databases can be confusing, since different software vendors have adopted different terms for the same thing. Table 11.1 illustrates the relationship between relational database terminology and the traditional table or simple computer file. Plate 1 applies this terminology to the table suggested for the Shivannagudem watershed and shows a relational database structure development for the data for creation of hydrogeomorphological map and groundwater potential zones.

11.4.4 Standard Query Language (SQL)

The data in a relational database are stored as a set of base tables with the characteristics described above. Other tables are created as the database is queried and these represent virtual views. The table structure is extremely flexible and allows a wide variety of queries on the data. Queries are possible on one table at a time (for example, you might ask 'which hotels have more than 14 beds?' or 'which hotels are of a luxury standard?'), or on more than one table by linking through key fields (for instance, 'which passengers originating from India are staying in luxury hotels? Queries generate further tables, but these new tables are not usually stored. There are few restrictions on the type of queries possible.

With many relational databases querying is facilitated by menu systems and icons, or 'query by example' systems. Frequently, queries are built up of expressions based on relational algebra, using commands such as SELECT (to select a subset of rows), PROJECT (to select a subset of columns), or JOIN (to join tables based on key fields). SQL has been developed to facilitate the querying of relational databases. The advantages of SQL for database users are its completeness, simplicity, pseudo English language style and wide application. However, SQL has not really developed to handle geographical concepts such as 'near to', 'far from', or 'connected to'.

The availability of SQL is one of the advantages of the relational database model. Additionally, the model has a sound theoretical base in mathematics and a simple logical data model that is easy to understand. The relational database model is also easy to understand. The relational database model is more flexible than either of the previously used hierarchical or network models. However, the model will always produce some data redundancy and can be slow and difficult to implement. There are also problems with the handling of complex objects such as those found in GIS (and CAD and knowledge-based applications) as there is a limited range of data types, and difficulties with the handling of time. Seaborn (1995) considers that many of the limitations of relational databases in GIS stem from the fact that they were developed to handle simple business data, not complex multidimensional spatial data. However, they have been widely adopted and successfully used. Relational databases are predominantly used for the handling of attribute data in GIS. For example, ARC/INFO maintains an attribute table in relational database software, using a unique ID to link this to spatial data. Some commonly used relational DBMS are INFO used in ARC/INFO ; ORACLE used in ARC/INFO, Geovision ; D Base used in PC ARC/INFO and other PC based GIS.

11.5 Storage of GIS Data

As discussed a number of times in the earlier chapters, the GIS data consists of two different types of data, namely, spatial and attribute. The spatial data may be raster or vector depending upon the source and nature of the data used in a particular project under study. The attribute data, which is in the form of tables, text and descriptive associated with the spatial data conceptually, spatial data is location based whereas the attribute data is the description of that location. Therefore, the GIS should provide a storage space for both of these data products either individually or separately and the design of this kind of system is such that these data products are retrieved as and when the analyst needs further analysis. In addition, the GIS should have a capability of linking spatial as well as attribute through a software in order to establish the topology. In principle, there are three different files to be created for any GIS project, a topological data file, coordinate file, and file with attribute data.

The differing emphasis placed by GIS system designers on the database approach for storage of digital map coordinates, and corresponding attribute data has led to the development of two different approaches to implementation of relational database in vector GIS. These two approaches are hybrid data model and integrated data model. The benefits of using an integrated architecture are considerable, allowing a uniform treatment of all data by the DBMS and thus not consiging the spatial data to a less sheltered existence outside the database, where integrity, concurrency and security may not be so rigorously enforced. It seems only logical that two technologies, it linked through software, would provide the best of both the modes.

11.5.1 The Hybrid Data Model

The starting point for this approach is the data storage mechanisms which are optimal for locational information and are not optimal for attribute/thematic information. The GIS software manages linkages between the cartographic files (operating system files), and the DBMS (attribute data) during different map processing operations, such as, overlay. From an initial situation where each GIS supported only one DBMS, a clear trend is now being established to provide access to multiple DBMS systems, with vendors of GIS software and others developing generic relational DBMS interface. The hybrid GIS model is shown in Fig. 11.6. In this model, the topological data and coordinate data are stored in one file system and the corresponding attribute table data are in another file system. For example, the ARC in ARC/INFO GIS software contains the topological and coordinate data files and INFO contains attribute tables. These two data files can be linked or hybridised by means of a software linkage mechanism.

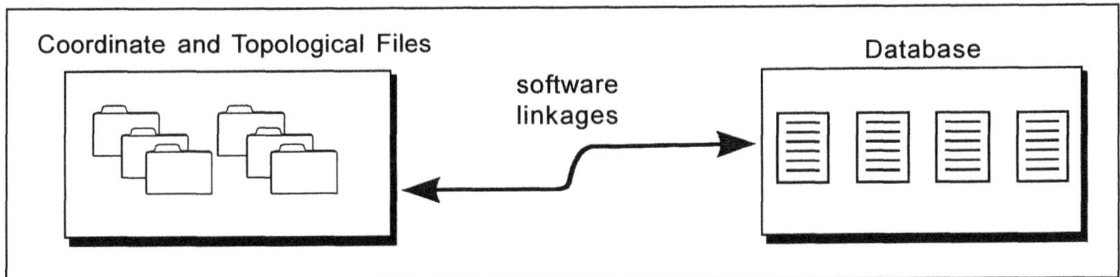

Fig. 11.6 Hybrid vector geographic information system (GIS).

To implement the hybrid vector GIS model, the coordinate and topological data required for graphics are stored as a separate set of files (Fig. 11.6). The attribute tables carrying all the necessary attribute data for each graphic entity are also stored separately within existing DBMS software. Linkage is performed by storing identification codes as a column data (primary key) in the attribute data base.

11.5.2 The Integrated Data Model

The second major type of GIS model is the integrated data model which is more closely integrated with the database management system than in the hybrid system (Guptill, 1987). The integrated data model approach is also described as the spatial data base management system approach, with the GIS serving as the query processor sitting on top of the database itself. The integrated data model has a number of implications in terms of the special characteristics of spatial data. From the data base viewpoint, it is possible to store both the coordinates and the topological information required to characterise digital cartographic elements using a design based on Codd's Normal Forms. (x, y) coordinate pairs for individual vertices along line segments are stored as different rows in a data base table. To achieve satisfactory retrieval performance it has been found necessary to store coordinate strings in long or 'bulk data' columns in tables. Handling of large spatial databases is the need to convert 2-D coordinate information into 1-D spatial keys that can be stored as data base table columns. These can then be indexed in the normal way and used for fast retrieval of map elements contained within or overlapping a specified geographical search area.

There are two ways of storing coordinate information as relational tables. The first records individual (x, y) coordinate pairs and polygon terminator and vertices as individual atomic elements or a row in a database. It is very difficult to search any element because each element must be recomposed from its atomic format to grate whole polygons or groups of polygons. Fig. 11.7 shows the integrated GIS with which it can be understood how a single database can be configured to certain separate files for entities and attributes.

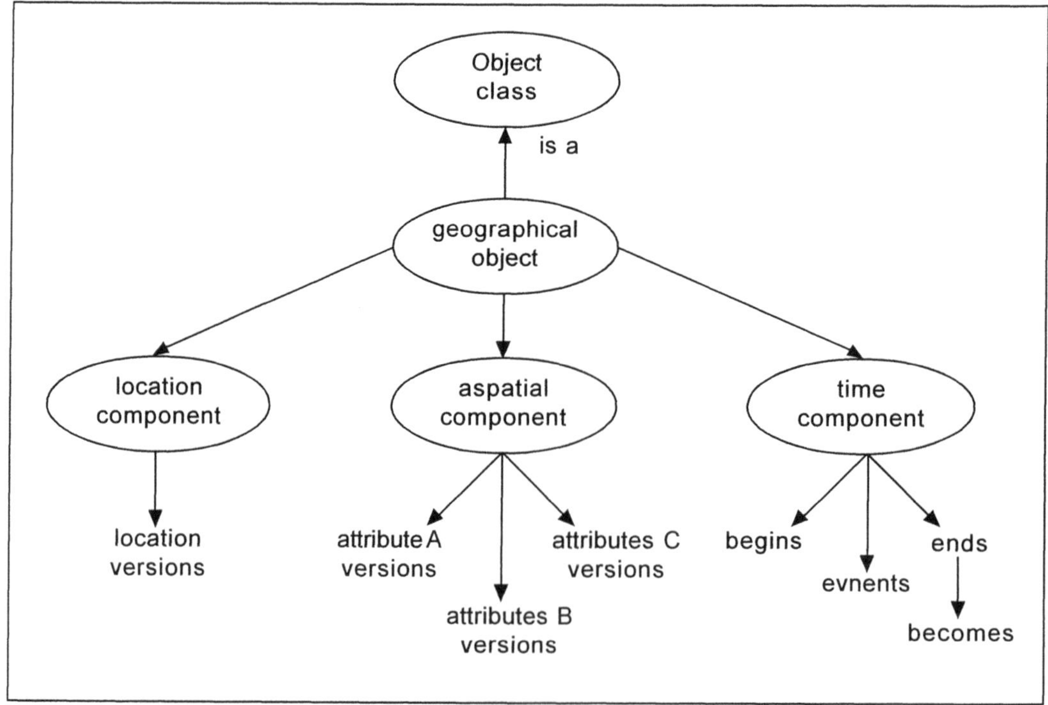

Fig. 11.7 Integrated GIS model for Data Storage.

11.6 Object Based Data Models

Extensions of the integrated model to incorporate spatial query language functions have recognised that, in the final analysis, it is not sufficient merely to hold data on map elements in the database. For GIS purpose it must also be possible to access the operations to be performed on these elements. This led to the concept of the object-oriented approach in spatial data handling. In a general definition, an object is an entity that has a state represented by the values of local variables and a set of operations or methods that operate on the object. The major example of the object-oriented approach in the GIS field is the INTERGRAPH TIGRIS system, which utilises object-oriented programming, rather than an object-oriented interface. Further work in this area also includes user interface aspects, data modelling, and query modelling. It is clear, however, that an exploration of the potential of object-oriented DBMS system, as the organising framework for spatial data bases, may have to wait until they move further from the 'proof-of-concept' stage towards widespread availability. The hierarchy of the object classes is shown in Fig. 11.8. Object based data models

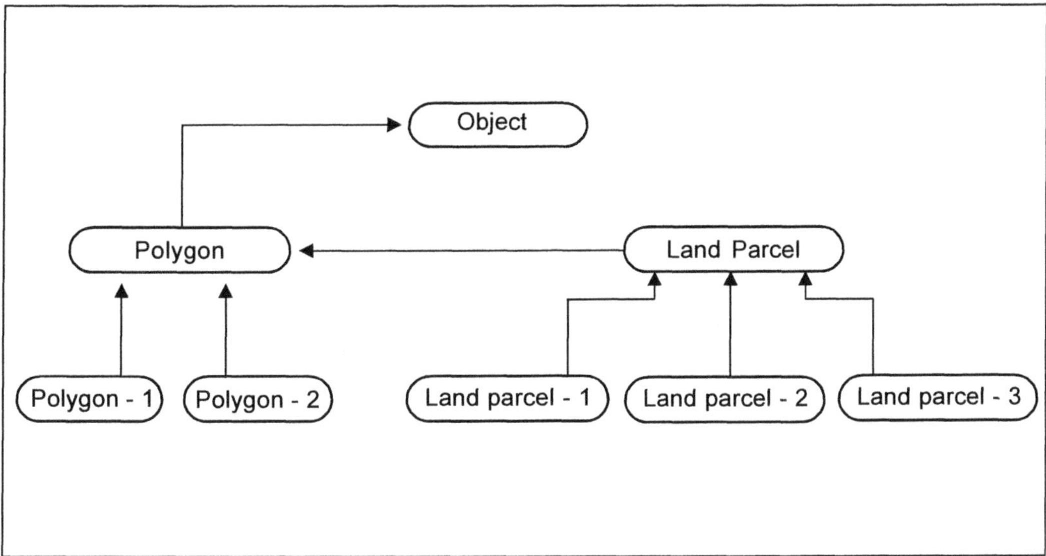

Fig. 11.8 Object-oriented geographic informaiton sytem (GIS) Model.

for GIS purpose are based on entities, attributes, and the relationship between entities and attributes with respect to the locations, temporal changes and both location and time. In the following sections, an attempt is made to provide a priliminary study for creating object based data models on the basis of entity-attribute relationships, location based representations of spatio-temporal data, and time based representations of spatio-temporal data and a combined approach of representing spatio-temporal data.

11.6.1 Entity-Relationship-Attribute Model

The primary components of an object-based model are its objects or entities. The entity-relationship attribute (ERA) model and the object-oriented (OO) models are the two main object-based modelling approaches. The ERA approach is attributed to Chen (1976) and has been a major modelling tool for relational database systems for about 20 years. In the ERA approach, an entity is a semantic data modelling construct and is something that has an independent and uniquely identifiable existence in the application domain. Entities are describable by means of their attributes (for example, the name, boundary, and population of a district). Entities have explicit relationships with other entities. Entities are grouped into entity types, where entities of the same type have the same attribute and relationship structure. Fig. 11.9 shows an ERA diagram representing the structure of these data in the example database. Entity types are represented by rectangles with offshoot attributes and connecting edges showing relationships.

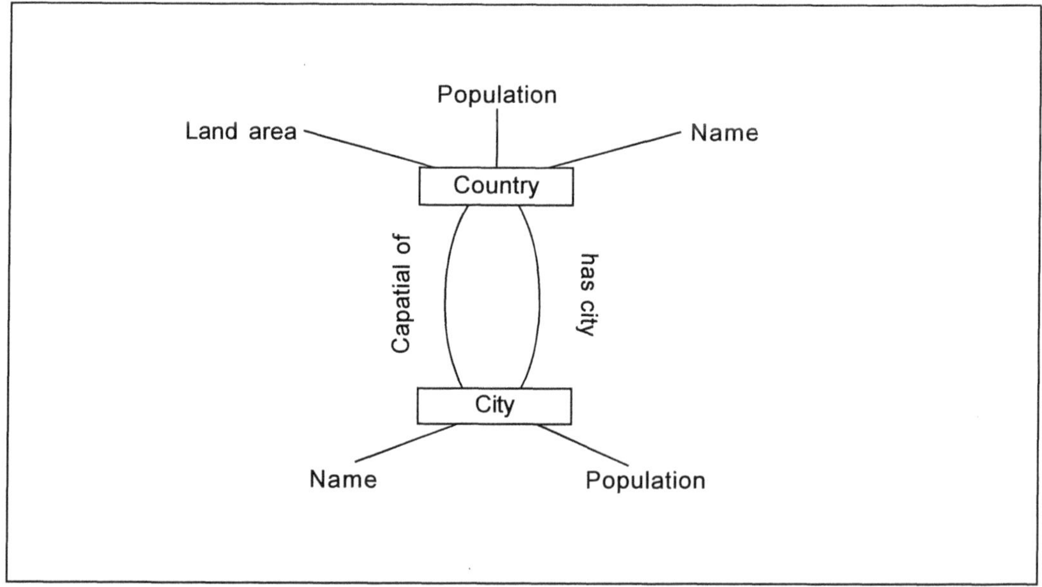

Fig. 11.9 Example of an ERA diagram.

The purpose of this type of ERA model is to serve a user community as a data store for a particular range of applications. Regarding geospatial applications, relational databases have fallen short of effectively achieving that purpose for two main reasons:

(i) the relational model has not provided a sufficiently rich set of semantic constructs to allow users to model naturally geospatial application domains,

(ii) relational technology has not delivered the necessary performance levels for geospatial data management.

11.6.2 Location-Based Representations for Spatio-Temporal Data

The only data model available within existing GIS which can be viewed as a spatiotemporal representation is a temporal series of spatially-registered 'snapshots', as shown graphically in Fig. 11.10. This is not an intended representation but is rather a convenient redefinition within a standard GIS database organisation of what is stored

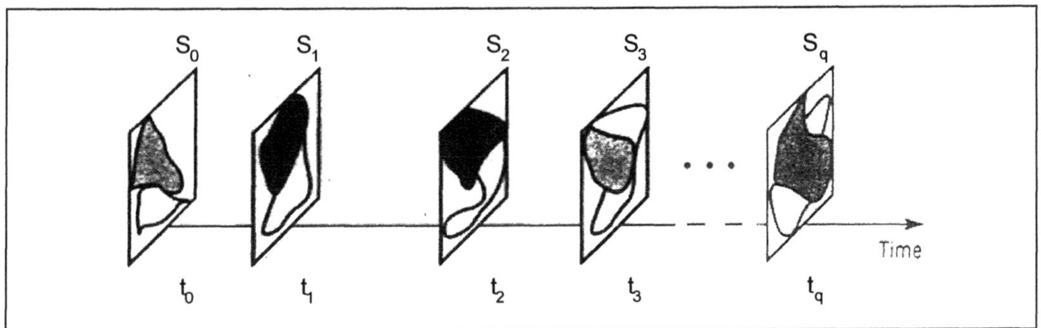

Fig. 11.10 The 'snapshot' approach for representing spatio-temporal data ; each 'snapshot' S_i, represents the state for a given point in time, t_i.

in the individual thematic layers. The 'snapshot' approach for space-time representation usually employs a grid data model, although a vector model can also be used. Instead of storing all information relating to a given thematic domain like elevation or landuse within a single layer, a layer holds information relating to a single thematic domain at a single known time. Data are thus recorded over a series of discrete temporal intervals. The distinguishing feature of the snapshot representation is that a 'world state map' S_i at each given point in time t_i is stored as a complete image or snapshot. Everything is included regardless of what has or has not changed since the previous snapshot, and the temporal distance between snapshots is not necessarily uniform. There are, however three drawbacks inherent in this approach (Peuquet and Duan, 1995) :

(i) The data volume increases enormously when the number of snapshots increases since each snapshot is a complete map of the entire region. This necessitates storage of a significant amount of redundant data, since in most cases, the spatial changes in two consecutive snapshots are only a small portion of the total data volume.

(ii) The changes of spatial entities that accumulate between two points in time are stored implicitly in the snapshots and can only be retrieved via a cell-by-cell (or vector-by-vector) comparison of adjacent snapshots. This process can be very time consuming.

(iii) Any individual change that occurs cannot be exaclty determined. Chrisman warned against the use of snapshots on the basis of the last two characteristics alone since, volume problems can be overcome with greater hardware storage capacity (Chrisman 1994).

A modification of the grid model that allows the time and place of individual changes like events to be recorded was proposed within a GIS context by Langran (1990) and was implemented on a prototype basis (Peuquet and Qian 1996). This model is also used in electronic circuitry design analysis. Instead of recording only a single value for a single pixel a variable-length list is associated with each pixel. Each entry in the list records a change at that specific location denoted by the new value and the time at which the change occurred. This is shown in Fig. 11.11 in which each new change for a given location is added to the beginning of the list for that location. The result is a set of variable-length lists referenced to grid cells. Each list represents the event history for that cell location sorted in temporal order. The world state for the entire area is easily retrieved as the first value stored in all the locationally-referenced lists. In contrast to the snapshot representation, this representation stores only the change related to specific locations and avoids storing redundant information of values for locations which remain unchanged.

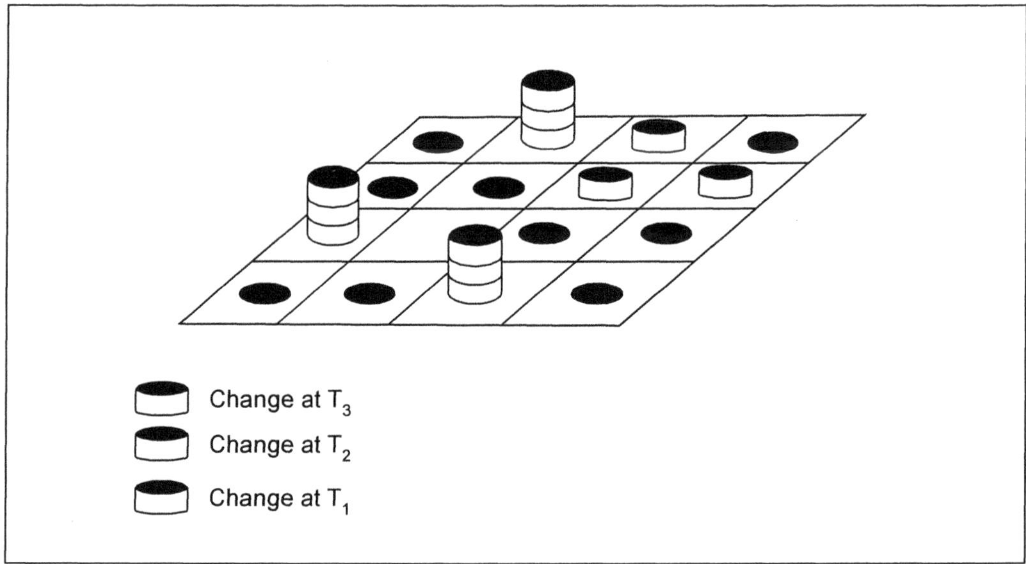

Fig. 11.11 The temporal grid approach for represnting spatio-temporal data.

11.6.3 Entity-Based Representations for Spatio-Temporal Data

Several spatio-temporal models have also been proposed which explicitly record spatial changes through time as they relate to specific geographical entities instead of locations (Hazelton 1991, Kelmelis 1991; Langran 1992). This is an extension of the topological vector approach. As such, they track changes in the geometry of entities through time. These spatio-temporal models rely on the concept of amendments, where any changes subsequent to some initial point in time in the configuration of polygonal or linear entities are incrementally recorded. The first of these models was proposed by Langran (1989b). It relies on what she describes as 'amendment vectors'. As a simple graphic example Fig. 11.12 shows the historical sequence for a small portion of the roadways in a growing urbanised area.

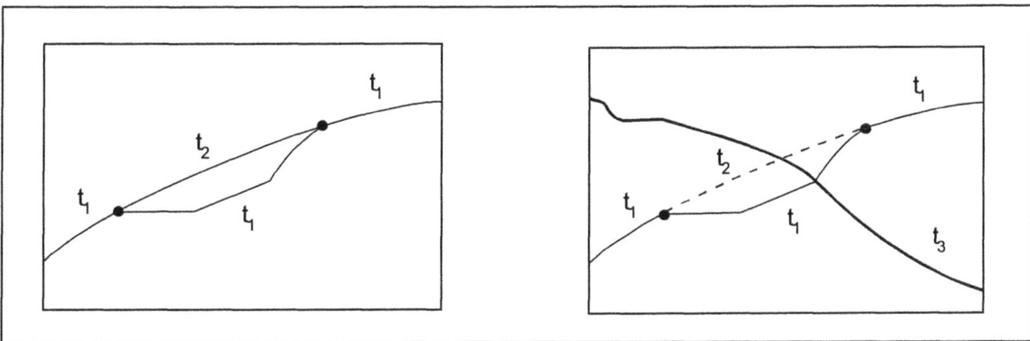

Fig. 11.12 The 'amendment vector' approach.

The thin black line shows the original configuration of a road at time t_l. At some later time t_2, the route of the original road was straightened. Note that this modification required cutting the original line at two points, designating the piece of the original route between those two points as obsolete, and inserting a new line segment between the same two points to represent the new portion of the road. This results in four line segments where there was only one before the update. At a still later time, t_3, a new road is built and entered into the database which has an intersection point along the realigned segment of the first road. The time, t_n, when the change occurred is recorded as an attribute of each vector. This organisation allows the integrity of individual entities like lakes and roads. Components of those entities like boundary lines and the vector topology are to be explicitly maintained over time. The amendment vector approach also has the advantage of being able to represent asynchronous changes to entity geometries. This capability, however, comes at a significant cost. As time progresses and the number of amendment vectors accumulate, the space-time topology of these vectors become increasingly complex.

The object-oriented approach to representation has been proposed by a number of researchers as a means of handling this problem. The object-oriented approach was originally developed as a generally applicable data representation method for software design and implementation. The key to the object-oriented approach is the idea of storing, as an integral unit, all components that define a particular thing, as a *concept* like a single bank transaction and geographical entity. This is known as *encapsulation*. The object-oriented approach thus provides a cohesive representation that allows the identity of objects as well as complex interrelationships, to be maintained through time. Rules for determining how to split or merge entities can be stored as part of the entity and entity class definitions.

11.6.4 Time-Based Representations for Spatio-Temporal Data

Spatio-temporal representations that use time as the organisational basis have also been proposed recently. These also suggest maintaining the explicit storage of temporal topology as an adjunct to location and entity-based representations in a temporal GIS. In the time-based representation proposed by Peuquet and Duan, shown diagrammatically in (Fig. 11.13 (a,b)), all changes are stored as a sequence of events through time. The time associated with each change is stored in increasing temporal order from an initial, stored 'world state'. Differences between stored times denote the temporal intervals between successive events. Changes stored within this timeline or 'temporal vector' can relate to locations, entities, or to both. Such a timeline, then, represents an ordered progression through the time of known changes from some known starting 3-5 or moment (t_0) to some other known, later point in time (t_n). Each location in time along the timeline (with its temporal location noted as t_0, t_1, ..., t_n) can have associated with it a particular set of locations and entities in spacetime that changed at that particular time and a notation of the specific changes.

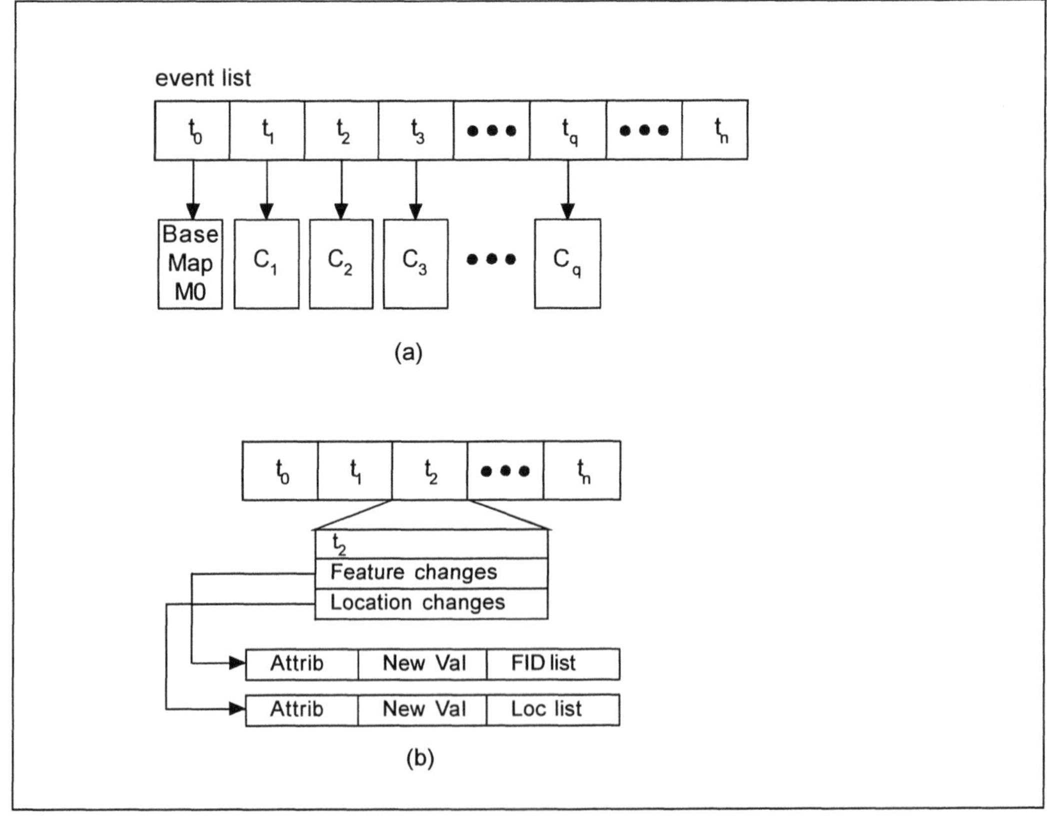

11.13 A time-line, or 'temporal vector'.

With this type of time-based representation, the changes relating to times are explicitly stored. This type of representation has the unique advantage of facilitating time-based queries. Adding new events as time progresses is also straightforward, and are simply added to the end of the timeline.

11.6.5 A Combined Approach for Spatio-Temporal Representation

Associating additional temporal information with individual entities provides a means of recording entity histories, and thereby allows histories of entities and types of entities to be easily traced and compared. Similarly, associating temporal information with locations allows the history of individual locations and sets of locations to be traced and compared. Locational overlay operations can also be used in a location-based spatiotemporal representation to characterise locations on the basis of collocation of multiple changes or types of change. Since these different types of representation provide differing perspectives of the data, each facilitates a different class of queries.

11.7 Temporal Topology

Peuquet (1994) has proposed three basic types of temporal relationships. They are:

(i) Association between elements within a given temporal distribution at a given temporal scale;

(ii) Combination of elements from different temporal distributions, and

(iii) Transformations between temporal scales.

The first type of basic temporal relationship includes metric and topological relations. Given that time is 1-dimensional, there is only a single temporal metric, namely, temporal distance. Temporal distance is the quantitative measurement of the interval between any two given points in time. Temporal distance can be used to denote the length of an event, that is, its duration, the duration between two events, or the duration of some continuous state.

Temporal topology is the expression of relationships between events strictly in an ordering sense. In this type of relation the timeline itself can be viewed as elastic with events recorded upon it as knots or singularities. In terms of analysis this also allows for examination of different sequences of events that may occur at varying time scales.

The second basic type temporal relationship includes those that combine different types of temporal distributions. These are the temporal overlay relationships that act as Boolean set operators (intersection, union, negation). This type of relationship allows temporal co-occurrence (or non co-occurrence) of different states or events over specific temporal intervals to be examined.

The third basic type of temporal relationship, temporal scale change, includes generalisation and extrapolation over a specific temporal distribution. This type of relationship involves combining events, episodes, and states in the case of generalisation. In the case of extrapolation events, episodes, and states may be inferred in order to fill in an expanded temporal sequence. The method of temporal generalisation (or extrapolation) used is dependent upon the given temporal measurement used in 4 seasons, 12 months, or 365 days each equal to one year. The method may also vary from scale to scale.

11.8 Organisational Strategy of DBMS in GIS

The storage of nonspatial attribute data is a well established technology, quite apart from GIS. In its simplest form, it is analogous to a filing system, allowing information to be extracted from the database via some organizational structure. A traditional filing cabinet imposes an alphabetical structure on the data, allowing us to retrieve many pieces of related information from a record card indexed by an alphabetic name. We know where in the system to find objects referenced by a particular name, thus speeding up the search for information, in an address record system, for instance.

In contrast, a computer-based database management system (DBMS) allows us to extract information, not only by name, but also according to a selection of the other pieces of information in each record. Given an address, we could search to find out who lived there, or to identify all individuals with a given age. These operations require a tedious using of filing cabinet, which is only indexed for a single field of data like the name of the owner. As we have seen already, information systems are based around a digital model, which may be manipulated rapidly to perform the task required.

There are a range of organisation strategies for DBMS packages, each of which imposes different models on the data, affecting the ways in which it can be extracted. Date (1986) cites network, hierarch, inverted list, and relational structures as the four major categories. Software tools for the manipulation of data in these structures include methods for traversing the structure itself trying to find all property owners with the same postal sector in their address, or finding all doctors serving patients on a particular register. The most popular structure for recent GIS software design has been the relational model (hence the term relational DBMS, or 'RDBMS'). A RDBMS consists of a series of tables, linked by key fields, which allows the establishment of relations between the entries in different tables and impose no navigational constraints on the user. The divisions between these tables are 'invisible' to the user, making this structure ideally suited to applications where the nature of queries is not known in advance. Whereas hierarchic structures, for example, embody rigid relationships between objects in the database, relational systems allow the establishment of temporary links between tables sharing common fields (Aronoff, 1985). This serves to reduce redundancy and makes the database more flexible. The key fields in the geographic RDBMS are typically the unique identifiers of the point, line, and area objects to which the attribute data in the tables is relate. Healey (1991) notes the recent dominance of relational approaches, but stresses the advantages of hybrid approaches and object-orientation.

Date (1986) summarizes the problems with the traditional approach to data management as:

- Redundancy (the unnecessary repetition or duplication of data);
- High maintenance costs
- Long learning times and difficulties in moving from one system to another
- The possibility that enhancements and improvements to individual files of data will be made in an *ad hoc* manner
- Data-sharing difficulties
- A lack of security and standards, and
- The lack of coherent corporate views of data management.

In addition, the data storage mechanisms may be inflexible, creating difficulties in dealing with *ad hoc* 'one-off' queries. Oxborrow (1989) also identifies the problem of modelling the real world with traditional data management methods—many are simply not suitable. The data should be structured in such a way that they represent features from the real world and the relationships between them. Traditional methods of storing data cannot represent these relationships, so a database approach is necessary.

12

Data Input and Editing

12.1 Introduction

So far this book has presented the characteristics of spatial and attribute data and considered how to represent and structure them in the computer. This chapter introduces the method for getting data into the computer by a process known as data encoding. The first step in developing the database for a geographical information system is to acquire the data and to place them into the system. GIS must be able to accept a wide range of different kinds and formats of data for input. There may be times when a given user generates his own datasets. This, however, is, rare. Even so, getting data is one of the greatest problems relating to operation and costs. Kennedy and Guinn (1975) described the importance of the data in automated spatial information systems as follows. While models which use the data are important to support the decision-making activities of those who use the system, a large portion of the investment will be in obtaining, converting, and storing new data. Spatial data, on the other hand, can be obtained from many different sources and in a number of formats. These data can be input for GIS using a number of different methods. For instance, maps may be input by digitising, scanning or direct file transfer. Aerial photographs may be scanned into GIS and satellite images are downloaded from digital media. This chapter looks in detail at the range of methods available to get data into a GIS, the methods of storing the data, detection of errors in the digital data base and editing them before using the data for GIS data processing for any specific application like urban and regional planning, watershed management, natural resources management and a wide range of environmental management and assessment studies.

12.2 The Data Stream

The data to be input for GIS are typically acquired in a diverse variety of forms. Some data come in graphic and tabular forms. These would include maps and photographs, records from site visits by specialists, related to non-spatial information from both printed and digital files (including descriptive information about the spatial data, such as date of compilation, and observational criteria). Other data come in digital form. These would include digital spatial data such as computer records of demographic or land ownership data, magnetic tapes containing information about topography and remotely sensed imagery. The data to be input for GIS are of different forms. These include key board entry or key coding, digitising, scanning and digital data. The process of data encoding and editing is often called as data stream. Fig. 12.1 shows the stages in the data stream, which is the basic frame work of this chapter.

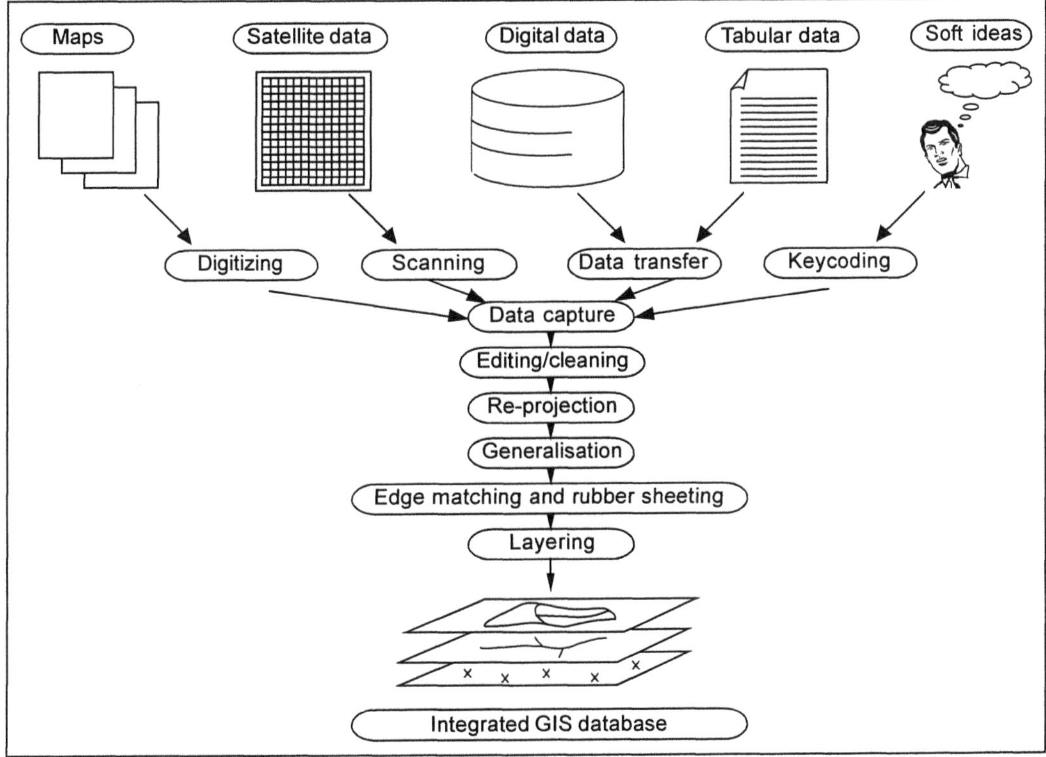

Fig. 12.1 The conceptual view of data stream in GIS.

Often these data will require manual or automated preprocessing prior to data encoding. For example, tabular records may need to be entered into the computer system. Aerial photographs might require photo interpretation to extract the important spatial objects and their relative location, a digitising process to convert the data to digital form, and numerical rectification algorithms to convert the locations of significant features to a standard georeferencing system. Computer programmers may need to help move digital datasets from one computer to another. Airborne scanner data might require thematic classification and rectification before the data are suitable for entry into GIS.

Other kinds of spatial data might be described as continuous fields, where we can theoretically calculate or measure a value at any location. Examples of this class of data are descriptions of elevation, plant biomass, or population density and many kinds of demographic variables. Finally, an important class of spatial data involves dividing a portion of the Earth's surface into relatively homogeneous discrete regions. A political map is perhaps the most common example of this type, since it subdivides a portion of the Earth into countries, states, and so forth. A similar way to subdivide the Earth would be to develop classes of land cover and land use, indicating the boundaries of the different classes and the characteristics within the boundaries.

There are a number of other important attributes of a dataset, including information about accuracy and precision as well as the density of observations used to develop the entire dataset. Regarding the latter, we may develop a large dataset based on a small number of measurements and then use numerical models to infer values at many other locations. This is commonly the case for elevation datasets as well as detailed demographic information.

From a very simple point of view, we can distinguish two different families of datasets, and thus have an idea of the kinds of effort each requires before it is ready for use in a geographic information system. Existing datasets are those that are already compiled and available in some form. We, of course, do not minimize the fact that they may require a great deal of effort to make them appropriate for a particular use. In contrast, there are many circumstances where we must develop or generate the dataset ourselves. In the second case while we may have complete control over the data gathering process, we generally have much more work to do.

12.2.1 Existing Datasets

The data in analogue or digital form need to be encoded and is compatible with the GIS being used. This would be a relatively straight forward exercise if all GIS packages used the spatial and attribute data models. However, there are different approaches to the handling of spatial and attribute data. There is a great deal of spatial information available in the public domain for some parts of the world, if you know where to begin to look. In India, much spatial data collected by government agencies, including maps, photographs, and many kinds of digital data, are considered in the public domain. Thus there are effectively no restrictions on access to this data except the maps of restricted zones. In contrast, in many other parts of the world, spatial data are considered the proprietary resource of the agency that collected the data, or may be controlled for economic or security reasons.

The most common form of spatial data is a map. Maps of various kinds are in common use for many kinds of spatial analyses. National agencies in many developed countries have systematic collections of map products at various scales, and programs for distributing and maintaining these resources. When appropriate maps are available, a digitising process permits us to extract the information from the flat map and place it into a digital computer. For example Table 12.1 shows the various data types and corresponding sources of availability of these data products in India.

Table 12.1 Data types and their source in India

Data Type	Data Source
Topography	GSI, NATMO, SOI
Digital Elevation Model	
Digital Terrain Data	
Land Use and Land Cover	
Ownership and Political Boundaries	SOI, Land Records
Transportation	R & D, A.P.S.R.T.C.
Hydrography	Ministry of Water Resources State & Central Ground Water Board
Socioeconomic and Demographic data	
Census Tract Boundaries	Department of Census
Demographic Data	Bureau of Economics & Statistics
Soils	NATMO
Wetlands	NATMO
Remotely Sensed Data	National Remote Sensing Agency

GSI - Geological Survey of India

NATMO - National Atlas Thematic Mapping Organisation

SOI - Survey of India

APSRTC -Andhra Pradesh State Road Transportation Corporation

12.2.2 Creation of Data

There will be times and circumstances in which it is necessary to develop your own datasets. Existing information resources may not be relevant to the problem, or perhaps they are not sufficiently current. There may also be datasets whose validity is unknown, thus forcing us to collect our own data either to test the existing datasets or to replace them. These occasions have a particular advantage in that they give us the opportunity to design the data acquisition program to meet our needs. The usual disadvantage of such a program is, unfortunately, the expense of designing, implementing, and managing the data-gathering task.

Constructing new datasets involves fieldwork of many kinds. Maps of terrain or the location of certain cultural features may need to be created. Details of plant and animal populations, such as, noting the occurrences of different species or determining the age of distributions within a population may be required for an environmental report. Knowledge of the general trends in groundwater, elevation in an area, as well as changes through the annual cycle of discharge and recharge, may be needed to site a water well or a waste disposal facility. Sampling design is one of the important elements of any data gathering plan where decisions are made about how to gather the data of interest.

12.3 Data Input Methods

Before explaining the input methods, it is necessary to make a distinction between analogue (non-digital) and digital sources of spatial data. Analogue data are normally in paper form and include paper maps, tables of statistics and hardcopy aerial photographs. All these forms of data need to be converted to digital form before use in a GIS. Digital data like remote sensing data are already in compute-readable formats and are supplied on diskette, magnetic tape or CD-ROM or across a computer network. All data in analogue form need to be converted to digital form before they can be input into GIS. There are four methods of data input which are widely used: keyboard entry, manual digitising, automatic digitisation, and scanning. Digital data must be downloaded from their source media and may require reformatting to convert them to an appropriate format for the GIS being used. Reformatting or conversion may also be required after analogue data have been converted to digital form. For example, after scanning a paper map, the file produced by the scanning equipment may not be compatible with the GIS, so it needs reformatting. For both the analogue and digital data, keyboard entry method, manual digitising and automatic digitising and scanning methods are very important as detailed below.

12.3.1 Keyboard Entry

Keyboard entry, often referred to as keycoding, is the entry data into a file at a computer terminal. This technique is used for attribute data that are available only on paper. This technique can be mixed with digitising process for the creation of GIS database as discussed in chapter 15 for the land use/land cover database for Hyderabad city. The attribute data, once in digital format, are linked to the relevant map features in the spatial database using identification codes. There are unique codes that are allocated to each point, line and area feature in the dataset.

The coordinates of spatial entities like point, line and area features can be encoded by keyboard entry. This method is used when the coordinates of these spatial entities are known and there are not too many of them. If the coordinates are more in number, this data can be encoded using digitising. The procedure of keyboard entry can be used to enter land record information. This method leads to obtain very high level of precision data by entering the actual surveying measurements. This method is used for the development of cadastral information system.

12.3.2 Manual Digitising

Manual digitising is the most common method of encoding spatial features from paper maps. It is a process of converting the spatial features on a map into a digital format. Point, line, and area features that form a map, are converted into (x, y) coordinates. A point is represented by a single coordinate, a line by a string of coordinates, and, when one or more lines are combined with a label point inside an outline, then an area (polygon) is identified. Thus digitising is the process of capturing a series of points and lines. Points are used for two different purposes: to represent point features or to identify the presence of a polygon. Manual digitising requires a table digitiser that is linked to a computer work station (Fig 12.2). To achieve good results, the following steps are necessary. Before, discussing these steps, the description of digitisers is provided for the beginners in this field of technology. Digitisers are the most common device for extracting spatial information from maps and photographs. The position of an indicator as it is moved over the surface of the digitising tablet is detected by the computer and interpreted as pairs of x, y coordinates. The indicator may be a pen-like stylus or a cursor. Frequently, there are control buttons on the cursor which permit control of the system without having to turn attention from the digitising tablet to a computer terminal. The current most popular digitiser is contemporary tablets using a gird of wires embedded in the tablet to a generate magnetic field which is detected by the cursor. The accuracy of such tables are typically better than 0.1mm which is better than the accuracy with which the average operator can position the cursor. Sometimes the functions for transforming coordinates are built into the tablet and used to process data before it is sent to the host (Fig 12.2).

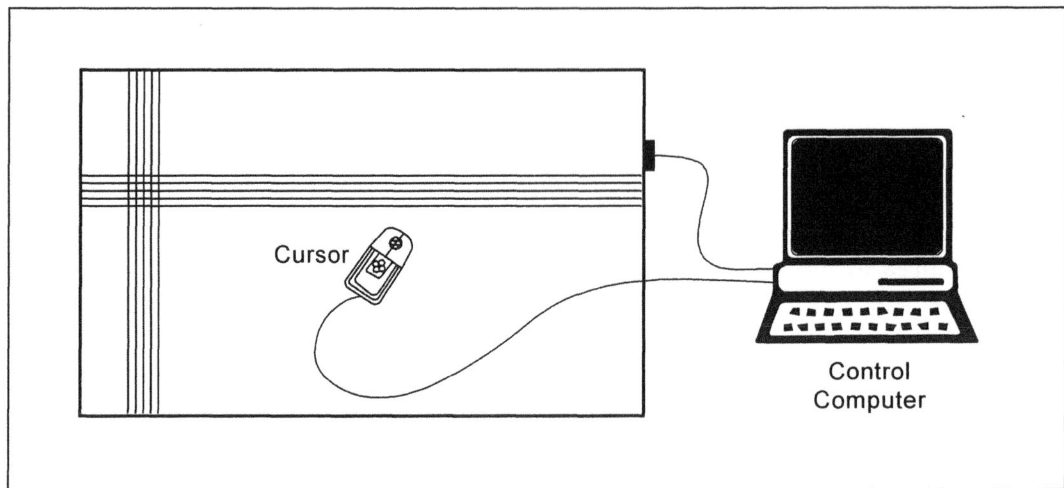

Fig. 12.2 Contemporary tablets using a grid of wires embedded in the tablet to generate a magnetic field which is detected by the cursor.

The Digitising Operation

The map is affixed to a digitising table. Three or more control points are to be identified and digitised for each map sheet. These points should be those that can be easily identified like intersections of major streets and prominent landmarks. The points are called reference points or tics or control points.

The coordinates of these control points will be known in the coordinate system to be used in the final data, such as, latitude and longitude. The control points are used by the system to calculate the necessary mathematical transformations to convert all coordinates to the final system. The more the control points, the better the accuracy of digitisation. Digitising the map contents can be done in two different modes: point mode and stream mode.

Point mode is the mode in which the operator identifies the points to be captured explicitly by pressing a button, and **stream mode** is the mode in which points are captured at set time intervals, typically 10 per second, or on movement of the cursor by filed distance. Most digitising is currently done in point mode.

Problems with Digitising Maps

The problems that come during the process of converting the maps into digital mode through the process of digitisation vary from one CAD operator to another. It depends upon the experience and skill of the operator and density of points, lines and polygons of the map. The accuracy of the output of the digitisation also depends upon the selection and distribution of the control points. Some of the commonly occurred problems during the digitisation of any paper map are as follows:

(i) Paper maps are unstable; each time the map is removed from the digitising table, the reference points must be re-entered when the map is affixed to the table again.

(ii) If the map has stretched or shrunk in the interim, the newly digitised points will be slightly off in their location when compared to previously digitised points.

(iii) Errors occur on these maps, and these errors are entered into the GIS data base as well.

(iv) The level of error in the GIS database is directly related to the error level of the source maps.

(v) Maps are meant to display information, and do not always accurately record vocational information.

For example, when a train, a stream and a road all go through a narrow mountain pass, the pass may actually be depicted wider than its actual size to allow for the three symbols to be drafted in the pass (Fig. 12.3).

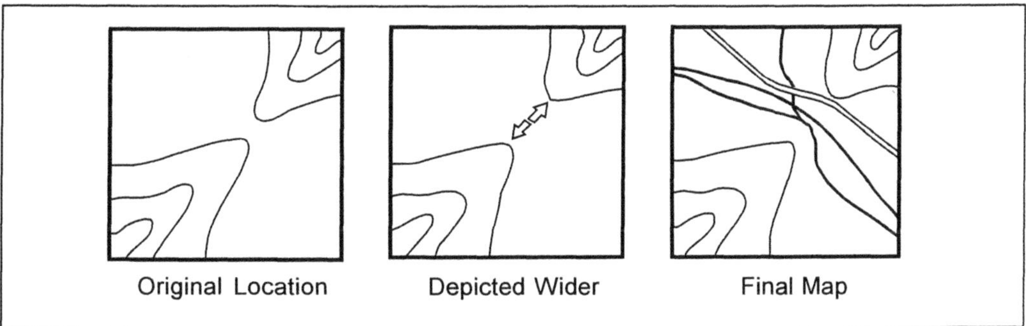

Fig. 12.3 Maps do not always accurately record locational information.

Discrepancies across map sheet boundaries can cause discrepancies in the total GIS database. For example, roads or streams do not meet exactly when two map sheets are placed next to each other (Fig.12.4). User error causes overshoots, undershoots (gaps) and spikes at intersection of lines (Fig. 12.5).

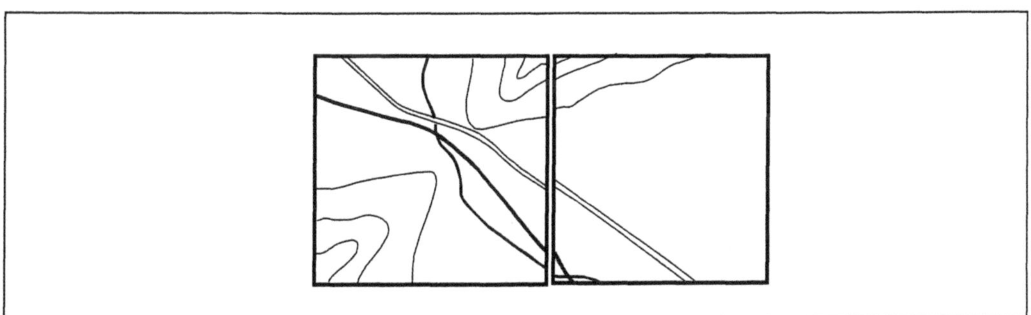

Fig. 12.4 Discrepancies across map sheet boundaries.

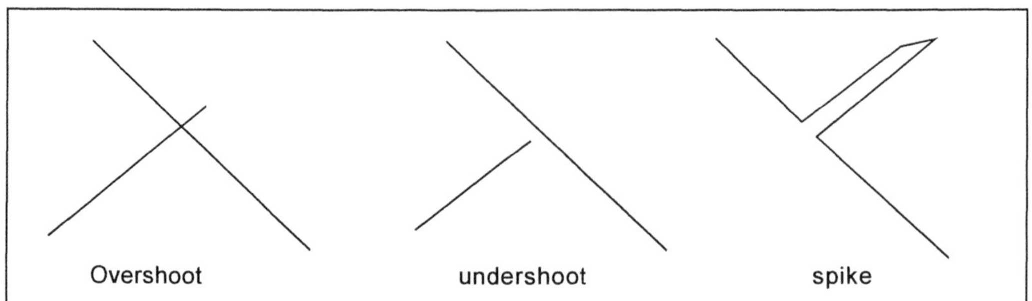

Fig. 12.5 Overshoots, undershoots and spikes.

The accuracy of the digital data is directly affected by the quality of the paper map. The maps should be in good physical condition and the features should be clearly visible on the maps. Determine how the maps are to be digitised and establish a standard sequence of procedure. For example, all arcs digitised before points. Establish a sequence for digitising features and map sheets, so you can easily track which portions of the database have already been digitised. Also establish standard naming conventions for naming different features and types of features. Map preparation will help minimise problems at the digitising stage as well as later, during the editing phase. Overall, the goal is to minimize the number of questions the person digitising will have to stop and resolve.

Guidelines for Digitisation

After digitising or importing data from some other source, make sure that the data in that class is free of spatial errors, that is,

(i) All features that were to be digitised, were really digitised (no missing data).
(ii) All features that are there, should be there (no extra data). The features are in the right place and the arcs have the correct shape.
(iii) All polygons have one, and only one, label point
(iv) All features are within the outer boundary.

This means that all the arcs outlining regions, are in the right place, have the right shape and connect to the outer boundary. Also, each polygon should have a label point with a unique identifier so that the code identifying its use, such as, agriculture, urban, and forest, can be associated with it. These relationships are known as topological relationships. They make those relationships digitally explicit, which we can see when looking at a map. Creating topology for a class also creates the data-base table containing the descriptive data for the class. Care should be taken to minimise errors by following the guidelines listed below :

(i) Locate TIC registration points and assign them a unique number. These must be known points for which you can get real-world coordinates. Once established, the same TIC points will be recorded and used for each separate manuscript.

(ii) Make a new boundary, slightly larger than the real one and then extend the internal lines just beyond it. This ensures that data you digitise will completely fill your study area. Later, the real boundary of the study area will be used to clip all the data extending beyond the edges.

(iii) When lines don't clearly intersect, indicate a definite intersection by specifically marking node points on the manuscript. This helps establish consistent intersections.

(iv) Make sure all polygons close and have a single label point with a unique identifier or identification code.

12.3.3 Scanning and Automatic Digitising

Scanning is the most commonly used method of automatic digitising. Scanning is an appropriate method of data encoding when raster data are required, since this is the automatic output format from most scanning software. Thus scanning may be used as a background raster dataset for the over-plotting of vector infrastructure data, such as, pipelines and cables.

A scanner is a piece of hardware (Fig. 12.6) for converting an analogue source document to a digital raster format (Jackson Woodsford, 1997). There are two types of scanners, (i) Flatbed scanner and (ii) rotating drum scanners. The cheapest scanners are small flatbed scanners, and high quality and large format scanners are rotating drum scanners in which the sensor moves along the axis of rotation.

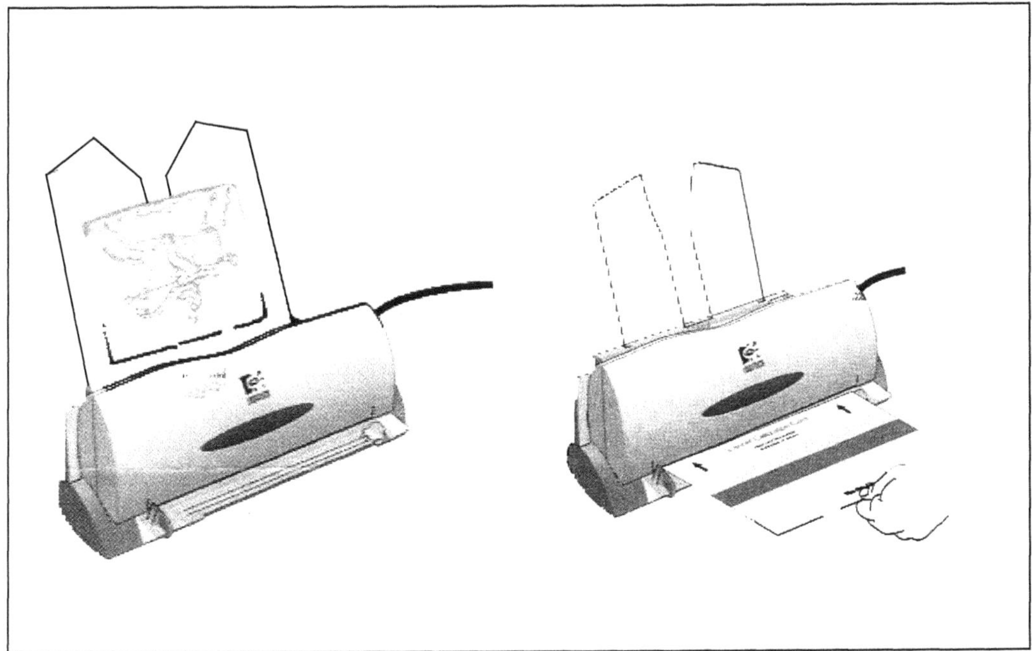

Fig. 12.6 Abokos colour scanner.

A digital image of the map is produced by moving an electronic detector across the map surface. The size of the map area viewed by the detector and scanning should be processed or edited to improve the quality and convert the raster to vector after online digitisation. The accuracy of the scanned output data depends on the quality of the scanner, the quality of the software used to process the scanned data, and the quality of the source document. A very important feature that a GIS user should observe after scanning the paper map is the occurrence of splines, which is black appearance on the scanned output. This can be removed by using a process called thinning.

The resolution of the scanner used affects the quality and quantity of output data. The cheaper flat-bed scanners have resolutions of 200-500 mm whereas the more expensive drum scanners use resolutions of 10-50 mm. The higher the resolution, the larger the volume of the data produced. Fig. 12.7 shows scanned output of paper map and the output of automatic digitisation.

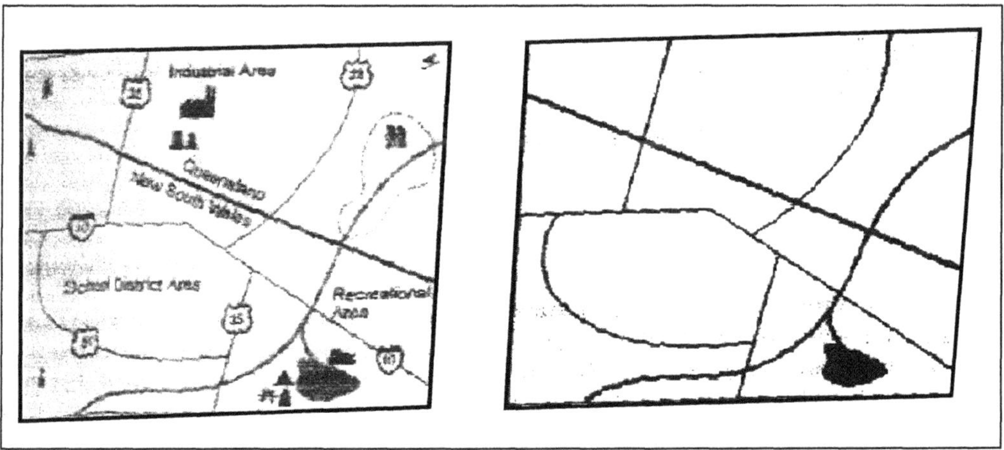

Fig. 12.7 Scanned output of paper map and the output of automatic digitisation.

12.4 GPS For GIS Data Capture

The explosion in interest in GIS as the management's tool has been accompanied by the development of a number of enabling technologies, one of the most important of which is the GPS (Global positioning system) (Lange, A.F and Gilbert, C, 1999). While GIS technology offers tremendous capabilities for more informed management decision making, sending competent decisions depends upon having reliable data. This section describes how GPS works and how to obtain reliable data using it. The term 'GPS' is used interchangeably here which satellite-based navigation systems in general, and the ground-based geographical data collection instruments, in particular. The fundamental components of GPS are in terms of basic segments: space, control and user. The space segment consists of the orbiting satellites making up the constellation composed of 24 satellites, and the design of the orbits, and the spacing of the satellites in the orbital planes. Most users will have six or more satellites available at all times. The control segment oversees the building, launching, orbital positioning, and monitoring of the system. The user segment comprise the users making observations with GPS receivers. The civilian GPS user community has increased dramatically in recent years because of the availability of low-cost portable GPS receivers.

12.4.1 Capturing Coordinate Data

Several key issues need to be explored when considering whether GPS is an appropriate tool for capturing co-ordinate data for a GIS database. First and foremost is the need to determine the position accuracy requirements. If the data were to be used for site-specific analysis that requires position accuracy to be within a meter, high-quality code-based differential GPS receivers would be necessary. If still higher accuracy were required, to the order of 10 centimeters or better as for property boundary mapping, then phase differential GPS techniques would be required (Hurn, 1989).

Every GIS database must be referenced to a base map or base data layer, and the reference datum of the various data layers must be uniform. Ideally, the database should be referenced to a very accurate base map. If however, the base map is 1:125000 or smaller, there could be problems when attempting to view the true spatial relationships between features digitised from such a map and features whose coordinates were captured with GPS. This can be a real problem if the GIS analyst decides to use a particular GIS data layer that was originally generated using small-scale base maps as a base to which all newly generated data are referenced (Weibel and Heller, 1991). The best way to avoid such an incompatibility is to develop an accurate base data layer, based on geodetic control and photogrammetric mapping.

12.4.2 Advantages of GPS

Broadly, GPS has three advantages.

(i) GPS may be used to identity or define the geographical co-ordinates associated with satellite imagery. GPS is used to reduce distortions and to improve the positional accuracy of these images. When three or more distinctive points (the more the better) can be located both on a satellite image and on the ground, GPS receivers can be used to collect accurate geographical co-ordinates at these locations. The rest of the images can then be located both on a satellite image and on the ground. GPS receivers can be used to collect accurate geographical coordinates at these locations. The rest of the image can then be adjusted so that it provides a better match to the real-world coordinates.

(ii) GPS can be used in the ground truthing of satellite images. When a particular satellite image has a region of unusual or unrecognised reflectivity or back scatter, the co-ordinates of that region can be loaded into a GPS receiver.

(iii) GPS has developed into a cost effective tool for updating GIS or computer-aided design (CAD) systems. Using GPS to collect data is analogous to digitising a map by moving a mouse or digitising pack over a map. The users of GPS equipment simply move along the surface of the earth and the geographical co-ordinates.

However, GPS is an excellent tool for data collection in many environments where the user can generally see the sky and is able to get close to the objects to be mapped.

12.4.3 GPS Data Creation

The final stage of the process of GPS -based data collection for GIS is to transfer the data from the field device to the target database. In the interests of preserving valuable memory, most GPSs store data internally in their own proprietary formats. This data transfer is most often accomplished by running a translation program that will convert the data from the compact internal storage format to the database interchange format of the user's choice.

12.5 Data Editing

The storage and editing subsystem of GIS provides a variety of tools for storing and maintaining the digital representation of study area. It also provides tools for examining each coverage for mistakes that may have crept into our preparations. The input data that is encoding may consist of a number of errors derived from the original data source as well as errors that have been introduced during the encoding process. There may be errors in coordinate data as well as in accuracies and uncertainty in attribute data. Before successfully using the methods of data analysis for any specific application, it is better to intercept errors before they contaminate the GIS database. The process of detecting and removing errors through editing is called ***cleaning***. The author's experience on the execution of a number of operational projects made the author put all the errors into 3 groups : (i) entity errors, (ii) attribute errors, and (iii) entity-attribute errors. They occur during the execution of the project using vector based GIS and attribute and entity attribute agrement errors occur in the use of both raster and vector based GIS. Data editing and cleaning of GIS database are covered under three subheads : (a) Detecting and correcting errors (b) Data reduction and generalisation and (c) edge matching and rubber sheeting.

12.5.1 Detecting and Correcting Errors

In any information system, facilities must be provided to detect and correct errors in the database. Different kinds of errors are common in different data sources. To illustrate some of the common varieties, we will discuss some of the errors that must be detected when generating a vector dataset, whether developed from a manual digitisation or an automated digitisation.

Polygonal areas, by definition, have closed boundaries. If a graphic object has been encoded as a polygon (rather than a vector or point), the boundary must be continuous. Software should be able to detect that polygon which is not closed. The causes for this kind of error include encoding error (the object is a vector, rather than a polygon) and digitising error (either points along the boundary of the polygon, or connections between the points, are missing).

With some systems, the boundaries between adjacent polygons may have to be digitised twice – once for each polygon and either side of the shared vector. This also happens on more advanced systems when the document cannot be digitised in a single session. Such systems create slivers and gaps. Slivers occur when the adjacent polygons apparently overlap, and gaps occur when there is an apparent space between adjacent polygons. Similar problems occur when several maps are required to cover the area of interest. In this case spatial objects may be recorded twice because of overlapping regions of coverage. An arc-node database structure avoids this problem, since points and lines may be recorded and then referenced by many different spatial objects. Errors of several kinds can appear in the polygon attributes. An attribute may be missing or may be inappropriate in a specified context. The latter kind of error is often related to problems in coding and transcription, such as, recording an elevation value into a land use data layer. Software to detect such errors is vital.

Care may be taken while digitising, otherwise it is not possible to make the lines connect perfectly. So making this spatial data usable really means making coordinate data error free and topologically correct. Errors in input data may be derived from three main sources: errors in the source data, errors introduced during encoding, and errors propagated during data transfer, construction, and conversion. For example, there may be subtle errors in a paper map source used for digitising. There may be printing errors in paper records as a source data. Topology makes explicit the relationships between geographic features within any given class. This process helps to identify errors that may exist in data. Some of the common errors that topology construction can identify are : arcs that do not connect to other arcs, polygons that are not closed, polygons that have no label point or too many label points, and/or user-Ids that are not unique.

Constructing topology helps identify these errors because when constructing topology arc intersections are created. The arcs that make up each polygon is identified and a label point is associated with each polygon. Until topology was constructed, no polygon existed and arcs that crossed each other were not connected at a node since there was no intersection. It is essential to build topology in order to have spatial relationships between features. Each feature, namely, point, segment, and region, is assigned a unique feature, Id. There will be corresponding record for each feature in the attribute file. CLEAN deals with only segments and regions, whereas BUILD performs operations on point, segment, and region. Moreover before performing the BUILD, the data should be CLEANed.

Most of the digitising errors can be identified by comparing the plot and the source. Examples of map errors that may arise during the construction of topology, digitisation, entering the data by means of scanning and keyboard entry, and encoding, are presented in Table 12.2

Table 12.2 Common errors in GIS database

Error	Description
Missing entities	missing points, lines or boundary segments
Duplicate entities	points, lines or boundary segments that have been digitised twice
Mislocated entities	points, lines or boundary segments digitised in wrong place.
Mislocated labels	unidentified polygons
Duplicate labels	two or more identification labels for the same polygon.
Artifacts of digitising	undershoots, overshoots, wrongly placed nodes, loops and spikes
Noise	Irrelevant data entered during digitising, scanning or data transfer.

Pseudo nodes occur where a single line connects itself (an island) or where only two arcs interest. Pseudo nodes do not indicate an error. There can be pseudo nodes representing an island or an intermediate point having an attribute data attached to it. A dangling node refers to the unconnected node of a dangling segment. Every segment begins and ends at a node point. So if a segment does not close properly (undershoot), or was digitised past an intersection (overshoot), it will register a dangling node. There can be instances where dangling nodes are representing some real-world feature. It is always better to have overshoots. It is much easier than editing an undershoot. Fig 12.8 shows some of the examples of spatial error in vector data.

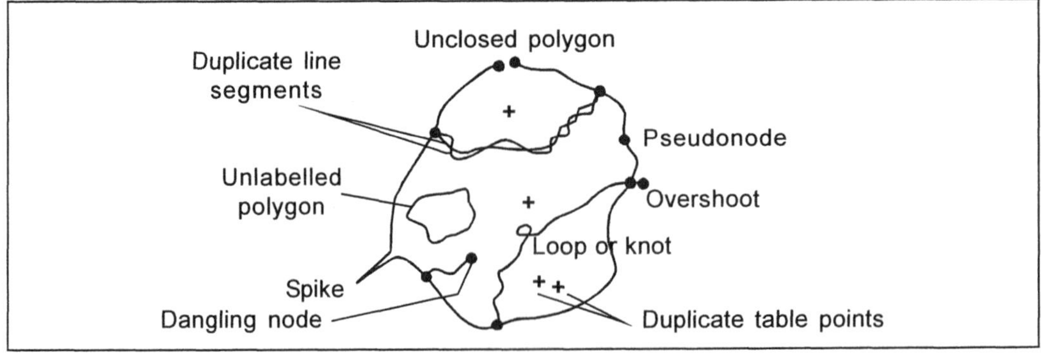

Fig. 12.8 Examples of commonly committed errors in vector data.

Fixing errors is one of the most important steps in constructing GIS data base. Unless errors are fixed correctly, your area calculations, any analysis and subsequent maps will not be valid. For example, polygons that do not have a label point cannot have descriptive attributes attached to them, and if a polygon is not closed, it will 'leak' into surrounding polygons when trying to shade in it. Fixing errors simply means that missing data is added, and bad data is removed and replaced by the correct data. For example, Table 12.3 explains the appropriate action for the errors during the construction of a GIS database.

Table 12.3 Error and Corresponding action to remove that error

Error	Action
Missing Segment	Draw it
A gap between two Segments	Indicate which arc to extend or which node to move
An overshoot	Delete node if needed
An Undershoot	Merge nodes or extend segment

Editing the spatial characteristics of a class alters the topology of the class. In such cases, always reconstruct topology to reestablish the spatial relationships. After reconstructing the topology, repeat the correction methods until the database is free from errors.

12.5.2 Data Reduction and Generalisation

An input record may require data reduction of various kinds. For example, field crew records of tree-crown diameter may contain several measurements at different points around the circumference of the tree. In a given application, we may need to average the measurements of a single tree, and enter the average into our database. For another example, existing datasets may record vegetation by species, and our requirements may be satisfied by simply recording by species (deciduous versus coniferous trees).

A more complex form of data reduction involves changes of scale in spatial data. We may need to assemble property ownership records in an area, and the original surveyor's records may be more detailed than we require. There are two obvious options: either accept the level of detail (and thus, incorporate a greater volume of data than necessary, with the attendant increased processing and storage costs), or develop a less precise representation from the original source data. The latter is called generalisation. For vector data, consider a locus of points, connected by straight line segments, which describe a coastline. In this way, reduce the total number of points by a factor of two, and in the process, reduce the level of details in the dataset.

A more sophisticated approach would involve modeling the behavior of a modest number of successive points by a numerical algorithm. For example, a polynomial curve of specified order can be least-squares fitted to a sequence of data points, and then a smaller number of points along the path of the fitted polynomial recorded in their place. Another alternative is to eliminate points from the original dataset that are too close together for a specified purpose, or fall along a straight line (within a specified tolerance). The former procedure is a common operation in many computer graphics systems, when points in the original dataset are too close to resolve on the desired graphics device. In a plotter, for example, the limiting resolution could be a simple function of the width of a pen and the repeatability of the plotter's positioning machinery. By repeating these processes, depending on the dataset and the desired resolution, we can derive a lower-resolution representation of the original vector dataset (Fig. 12.9). This approach is often termed as thinning.

For continuous raster data (such as millimeters of rainfall per year or elevation), we can similarly create a more generalized dataset in two steps. First, compute the average value of the attribute in a two-by-two neighbourhood. Second, record this average value in a new raster cell at the geographic location of the point shared by the four original raster cells. This kind of procedure, called resampling, can also be used when the required new raster cells are not of a length that is an integer multiple of the initial cell length. For nominal and ordinal raster data, rules for aggregation must be developed. Common aggregation rules include determining the majority or plurality class in the averaging neighbourhood.

12.5.3 Edge Matching and Rubber Sheeting

Sometimes, any given project area under study, extends across two or more map sheets like waste land mapping of Andhra Pradesh, small differences or mismatches between adjacent map sheets may need to be resolved. Normally, each map sheet would be digitised separately and then adjacent sheets joined after editing, projection, transformation, generalisation, and scaling. The process of joining is known as edge matching. Many vector GIS systems also allow you to separately stores potions of database as large, predefined subsections for archival purposes. This process is called tiling. This tiling process is commonly used to reduce the volume of the data needed for the analysis of extremely large database. Edge matching is a process of operating on more than one tile at a time to ensure that there is a correct match between the two tiles for entities for entries that extend across the tile boundaries.

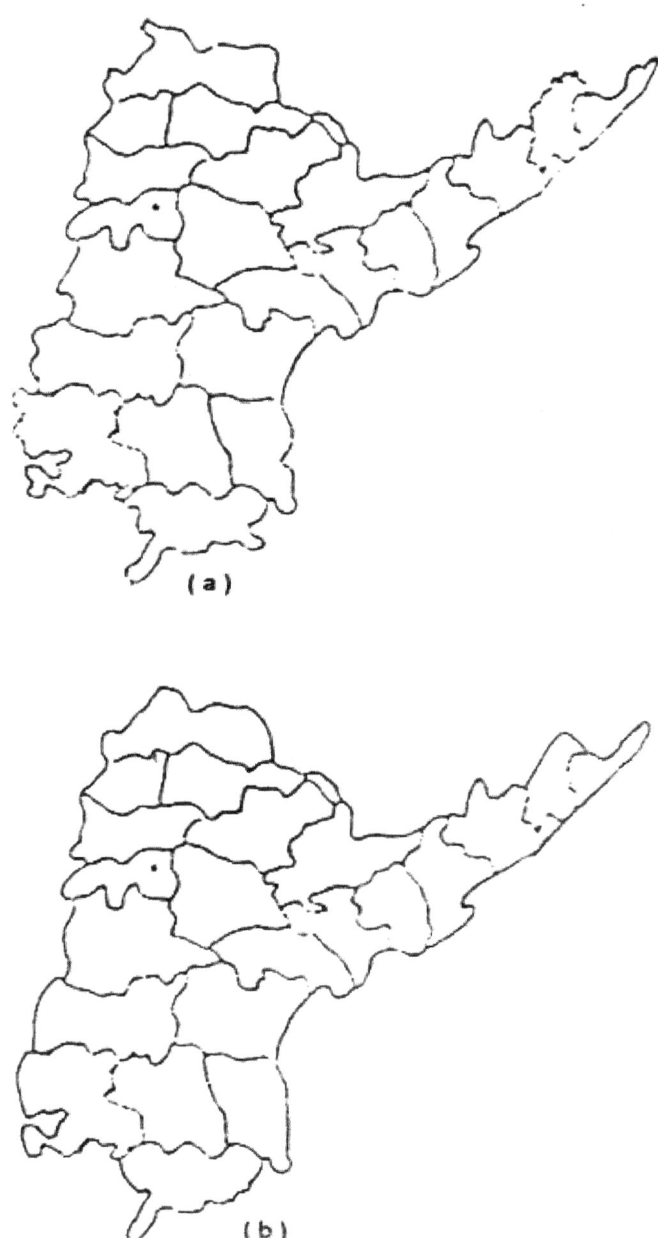

Fig. 12.9 Generalisation (a) original vector data
(b) generalised vector data.

Consider the example sketched in Fig. 12.10 where a region we wish to digitise into a GIS dataset lies across the boundaries of two map sheets. In this example, we see a water body and roads, both of which lie across the artificial boundary between the two map sheets. One option is to place both map sheets on the digitising tablet or scanning system at the same time, after manually creating a mosaic of the two. This is often impractical, both because the resulting composite sheet is too large for the available digitising equipment and storage facilities, and because this can destroy the maps for other uses. The more common procedure is to digitise or scan each sheet separately.

Frequently, when the two map sheets are digitised or scanned separately, features that cross the boundary do not align properly. This distortion can come from several causes. Even when maps are printed with no discernible error, the physical size of the map can change with temperature and humidity, and this can be a significant problem with maps printed on paper. Errors at the margins can also be caused by georeferencing errors during the digitising process, extrapolations and numerical round-off errors in the georeferencing algorithms, accuracy errors in the digitising tablet itself, and slivers and gaps caused by overlapping map coverage.

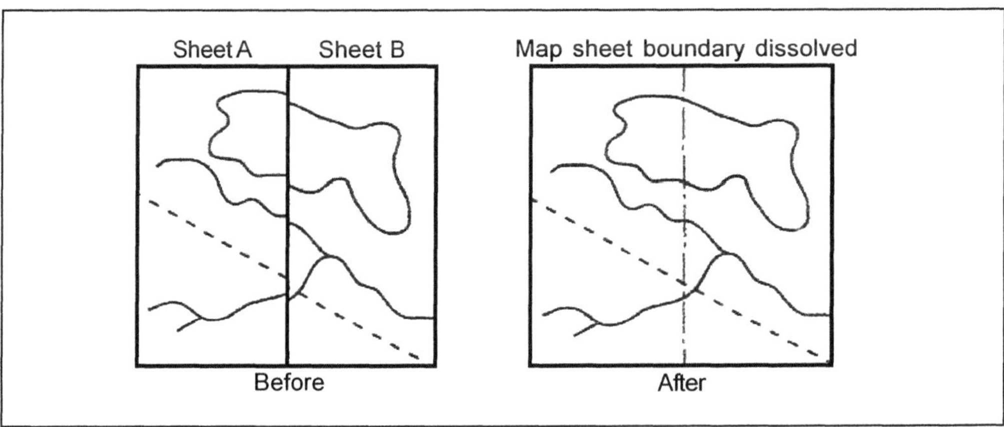

Fig. 12.10 Edge matching means.

Under perfect circumstances, problems relating to map shrinkage would be solved before entering the data into the GIS database. However, this is not always possible. There are two families of adjustments to correct these errors at the edges between map sheets or between different digital data files. In the first, the analyst manually adjusts the locations of points and vectors to maintain the continuity of the dataset. A graphics device is used to display the general area of the boundary between data sets, while the analyst uses "cartographic license" to manually adjust vectors that cross the boundary (Fig.12.10). In the second family of adjustments, automated are derived to reduce the edge effects. Line attributes and the spatial distribution of the lines on either side of the boundary are first matched. Then, appropriate locations from the line on each side are modified slightly to match exactly at the boundary. The latter is much like the adjustment a surveyor applies to balance traverse data (Anderson and Mikhail, 1985).

Certain data sources may give rise to internal distortion within individual map sheets. This is especially true of data derived from aerial photographs as the movement of the aircraft and the distortion caused by the camera lens can cause internal inaccuracies in the location of features within the image. These inaccuracies remain even after transformation and reprojection. These problems can be rectified through a process known as rubber sheeting or conflation. Rubber sheeting involves stretching the map in various directions as if it were drawn on a rubber sheet. Objects on the map that are accurately placed are tacked down and kept still while others that are in the wrong location or have wrong shape are stretched to fit with the control points, which are known points identified on the ground and on the image. The coordinates of these control points may be determined from field observation using GPS. Fig. 12.11 illustrates the process of rubber sheeting.

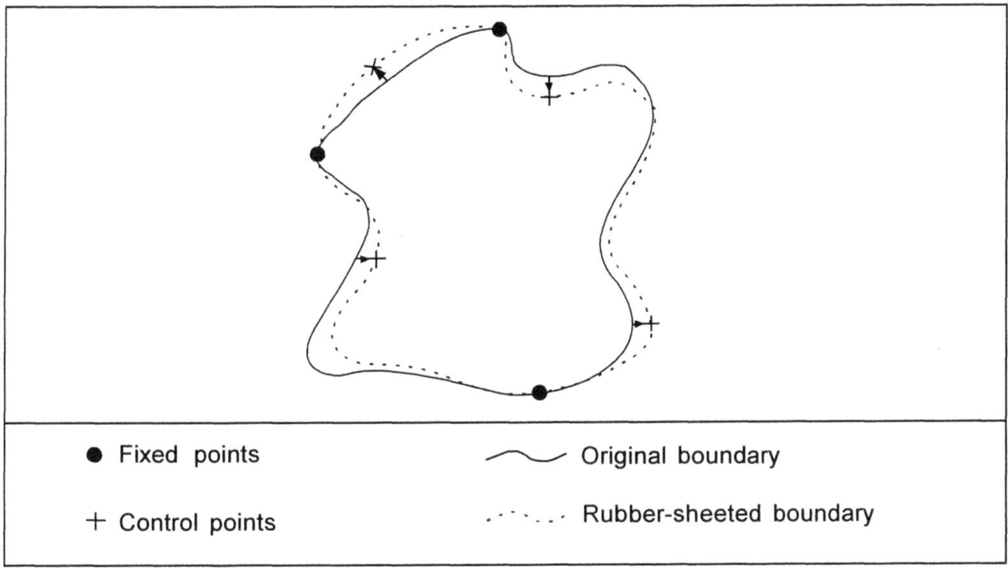

Fig. 12.11. Rubber sheeting.

13
Data Quality Issues

13.1 Introduction

GIS and remote sensing can bring enormous benefits by its ability to handle spatial data in ways that are precise, rapid, and sophisticated. But remote sensing and GIS user community paid little attention to the problems caused by error, inaccuracy and imprecision in spatial and nonspatial data sets. Certainly, there is an awareness that all data suffers from inaccuracy and imprecision, but their effects on GIS problems and solutions are not considered in great detail. This situation has changed substantially in recent years as the quality of output is an important component for any mapping activity. It is now recognized that error, inaccuracy and imprecision can 'make or break' many types of GIS projects and any GIS output without checking of quality is almost worthless. The word 'error' is used here in its widest sense to include the statistical concept of error, meaning variation. Therefore, the data quality check has become a very important component of any GIS project.

The meaning of 'quality' depends upon the context in which it is applied. The term is commonly used to indicate the superiority of a manufactured item or to attest to a high degree of craftsmanship or artistry. Quality is rather difficult to define for data. Unlike manufactured products, data do not have physical characteristics that allow quality to be easily assessed. Quality is a function of intangible properties such as 'completeness' and 'consistency'. More specifically, data quality refers to the relative accuracy and precision of a particular database used for any given application. Thus quality of data has become a concern among the user community for the following reasons: (a) increased data production by private sector, (b) increased use of GIS as a decision-support tool, and (c) increased reliance on secondary data sources.

These trends have contributed to reappraisal of the responsibilities of data producers and consumers for data quality. Responsibility for assessing whether a database meets the needs of a particular application has, therefore, shifted to the consumer who is in a position to make such an assessment. This is referred to as determining 'fitness-for-use'. From the perspective of the data producer, quality refers to the difference between the actual characteristics of the product and the relevant specifications that define it. Information on quality is immensely useful in managing the production process, particularly if the results of quality analysis point back to suspect sources of data.

Peter Fisher (1999) discusses alternative models of uncertainty. The traditional scientific concept of measurement error, which accounts for differences between observers or measuring instruments, turns out to be far too simple as a framework for understanding quality in geographical data. If agreement can be reached on how to measure and express data quality, then this information should be made available to users, preferably by storing it as a part of, or in conjunction with, the database. With adequate information available on data quality, it is possible to determine its effects on the results of GIS analysis, and to reflect on the uncertainties present in the base data relating to the decisions made with GIS.

In this chapter we first review the dimensions of the accuracy problem, and discuss the premises and assumptions on which the remainder of the chapter is based. The second section reviews the models that are available for analysing and understanding errors in spatial data. The chapter concludes with a summary of the current state of the art and an agenda for future research and development.

13.2 Components of Data Quality

Geographical observations describe phenomena with spatial, temporal, and thematic components. Space, which defines geographical location, is the dominant member of this troika. Although poorly accommodated in conventional geospatial data models, time is critical to an understanding of geographical phenomena, not as entities that exit at some location, but as events that appear and disappear in space and time. A second problem is that geographical phenomena are not really about space, but about themselves.

Data quality components are the key elements to explain the accuracy and precision of data and the results of the data analysis. Data quality components can be differentiated in space, time, and theme. There are several components of quality, namely accuracy, precision, resolution, consistency, and completeness. These terms are used almost interchangeably with reference to spatial data but we will need to establish more exact definitions.

Accuracy is the degree to which information on a map or in a digital database matches true or accepted values. Accuracy is an issue pertaining to the quality of data and the number of errors present in a dataset or map. In discussing a GIS database, it is possible to consider horizontal and vertical accuracy with respect to geographic position, as well as attribute, conceptual, and logical accuracy.

Precision refers to the level of measurement and exactness of description in a GIS database. Precise locational data may measure position to a fraction of a unit. Precise attribute information may specify the characteristics of features in great detail. Surveyors may make mistakes or data may be entered into the database incorrectly. The level of precision required for particular applications varies greatly. Engineering projects, such as, road and utility construction, require very precise information measured to the millimeter or tenth of an inch. High precision does not indicate high accuracy nor does high accuracy imply high precision. But high accuracy and high precision are both expensive. GIS practitioners are not always consistent in their use of these terms. Sometimes the terms are used almost interchangeably and this should be guarded against.

The resolution of a data set defines the smallest object or feature that is included or discrenable in the data. Scale and resolution are intimately related because there is a lower limit to the size of an object that can be usefully shown on a paper map. This limit is often assumed to be 0.45 mm as a rule of thumb, so the effective resolution of a 1:1000 map is about 1000 × 0.5 m, or 50 cm. The standards of most mapping agencies are common map scales assuming 0.5 mm resolution as shown in Table 13.1. Consistency refers to the absence of apparent contradictions in a database. Completeness refers to the relationship between the objects in the database and abstract universe of all such objects.

Table 13.1 Scale and Resolution for some common map scales

Scale	Effective resolution	Minimum resolvable area
1 : 1250	62.5 cm	
1 : 10000	5 m	
1 : 24000	12 m	
1 : 50000	25 m	0.0625 ha
1 : 100000	50 m	0.25 ha
1 : 250000	125 m	1.56 ha
1 : 500000	250 m	6.25 ha
1 : 1000000	500 m	25 ha
1 : 10000000	5 km	2500 ha

13.3 Accuracy

A useful starting point for discussing accuracy is the entity-attribute value model, which serves as the conceptual basis for most database implementations of real-world phenomena. According to this model, 'entities' represent real-world phenomena, such as, streets, districts, and hazardous waste sites, and 'attributes' specify the relevant properties of these objects, such as, width and number of lanes, while 'values' give the specific qualitative and quantitative measurements pertaining to a particular attribute. In this model error is defined as the discrepancy between the encoded and actual value of a particular attribute for a given entity. Accuracy is the inverse of error. This model can be used to define spatial, temporal, and thematic error for a particular entity as, the discrepancies in the encoded spatial, temporal, and thematic attribute values.

This definition is useful because geospatial data are always acquired with the aid of a model that specifies, implicitly or explicitly, the required level of abstraction and generalisation relative to real-world phenomena (Fig. 13.1). Accuracy is a relative measure rather than an absolute one, since it depends upon the intended form and content of the database. Different specifications can exist for the same general types of geospatial data. To judge the fitness-for-use of the data for some applications, one must not only judge the data relative to the specification, but also consider the limitations of the specification itself. Accuracy of the GIS data can be discussed in terms of spatial accuracy, temporal accuracy, thematic attribute data accuracy, and conceptual accuarcy.

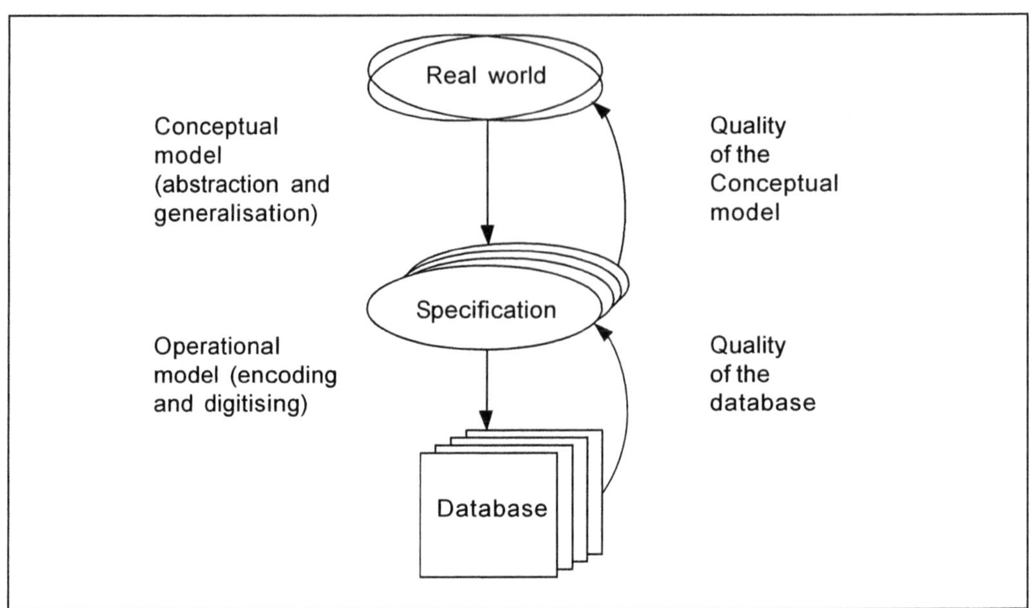

Fig. 13.1 The mediating role of the database specification in assessing data quality.

13.3.1 Spatial Accuracy

Spatial accuracy also called 'positional accuracy' refers to the accuracy of the spatial component of a database. Measurements of spatial accuracy depends upon the dimensionality. This applies to both horizontal and vertical positions. Accuracy and precision are the functions of the scale. This means that when we see a point on a map we have its "probable" location within a certain area. This applies to lines also to make us beware of the dangers of false accuracy and false precision, that is, reading locational information from map to levels of accuracy and precision beyond which they are created. This is a very great danger in computer systems that allow users to pan and zoom to an infinite number of scales. Accuracy and precision are tied to the original map scale and do not change even if the user zooms in and out.

For points, error is usually defined as the discrepancy (normally Euclidean distance) between the encoded location and the location as defined in specification. Error can be measured in any one of, or in combinations of, the three dimensions of space. The most common measures are horizontal error (distance measured in x and y simultaneously) and vertical error (distance measured in z) (Fig 13.2). Various metrics have been developed to summarise spatial error for sets of points. One such metric is mean error, which tends to zero when 'bias' is absent. Bias refers to a systematic pattern of error like the error arising from map misregistration. When bias is absent, error is said to be random. Another common metric is root mean squared error (RMSE), which is computed as the square root of the mean squared errors. RMSE is commonly used to document vertical accuracy for digital elevation models (DEMs). RMSE is a measure of the magnitude of error but it does not incorporate bias since the squaring eliminates the direction of the error. There is a close analogy between classical approaches to error in the location of a point. Horizontal error is a 2-dimensional extension of the classical error model in which error in position is defined in terms of a bell-shaped probability surface (Goodchild 1991a). Errors in lines arise from the errors in the points that define those lines. However, as these points are not randomly selected the errors present at points cannot be regarded as somehow typical of errors present in the line.

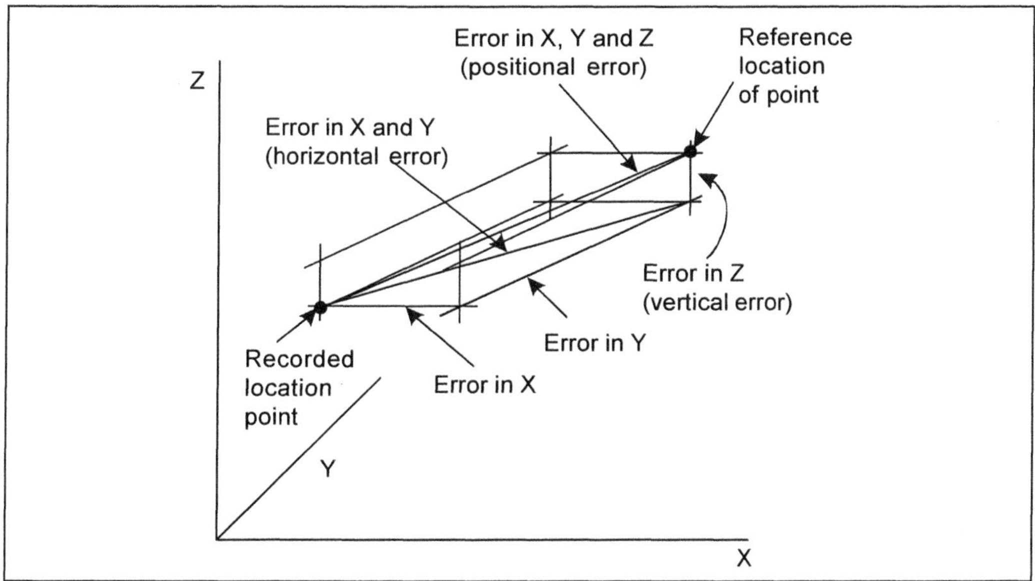

Fig. 13.2 Measuring components of spatial error.

Error is usually defined for lines using some variant of the epsilon band. The epsilon band is defined as a zone of uncertainty around and encoded line within which there is a certain probability of observing the actual line. The epsilon may be interpreted as if zone as a uniform sausage idicates the distribution of error is uniform (Chrisman, 1982) as shown in Fig. 13.3.

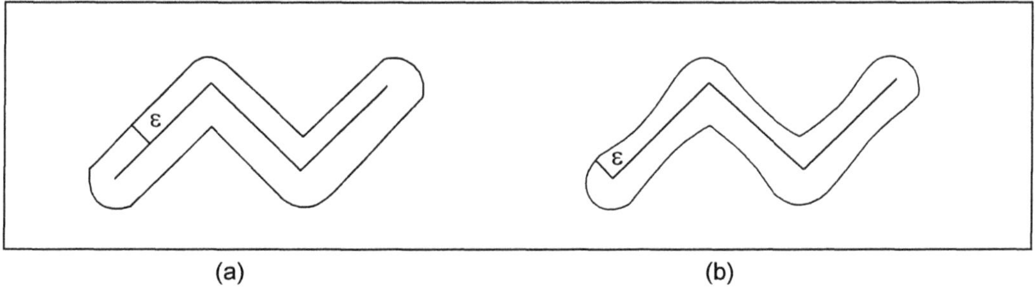

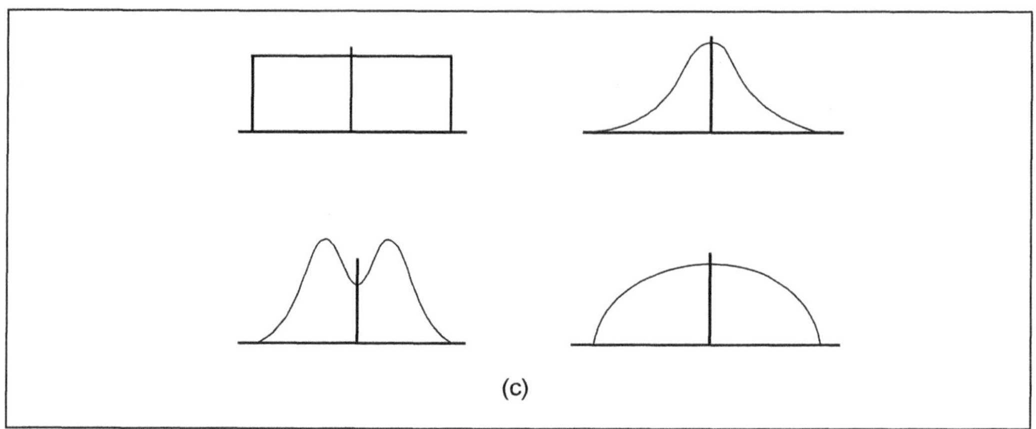

Fig. 13.3 Early models of the epsilon bond.

13.3.2 Temporal Accuracy

Temporal accuracy and currentness are two distinct concepts (Thapa and Bossler 1992). Temporal accuracy refers to the agreement between 'encoded' and 'actual' temporal coordinates. Currentness is an application-specific measure of temporal accuracy. A value is current if it is correct in spite of any possible time-related changes in value. Thus currentness refers to the degree to which a database is uptodate (Redman 1992). To equate temporal accuracy with currentness is to state, in effect, that to be temporally accurate a database must be uptodate. Clearly this is not the case since a database can achieve a high level of temporal accuracy without being current and is dependant upon the availability of historical data.

Assessment of temporal accuracy depends upon the ability to measure time objectively using a standard temporal coordinate system. Another impediment to the measurement of temporal accuracy is that time is often not dealt with explicitly in geospatial databases. Temporal information is often omitted, except in databases designed for explicitly historical purpose.

13.3.3 Attribute Accuracy

Metrics of thematic accuracy or attribute accuracy vary with measurement scale and is a function of scale. For quantitative attributes, metrics are similar to those used to measure spatial accuracy for point features like RMSE. For categorical data, most of the research into quality has come from the field of classification accuracy assessment in remote sensing. Accuracy assessment is based on the selection of a sample of point locations and a comparison of the land cover classes assigned to these locations by the classification procedure with the classes observed at these locations on a reference source usually called 'ground truth' (Aronoff 1985).

The non-spatial data linked to location may also be inaccurate or imprecise. Inaccuracies may result from mistakes of many sorts. Non-spatial data can also vary greatly in precision. Precise attribute information describes phenomena in great detail. For example, a precise description of a person living at a particular address might include gender, age, income, occupation, level of education, and many other characteristics. An imprecise description might include just income, or just gender.

13.3.4 Conceptual Accuracy

GIS depends upon the abstraction and classification of real-world phenomena. The users determine what amount of information is used and how it is classified into appropriate categories. Sometimes users may use inappropriate categories or misclassify information. For example, classifying cities by population size would probably be an ineffective way to study fertility patterns. Failing to classify power lines by voltage would limit the effectiveness of a GIS designed to manage an electric utilities infrastructure. Even if the correct categories are employed, data may be misclassified. A study of drainage systems may involve classifying streams and rivers by "order," that is where a particular drainage channel fits within the overall tributary network. Individual channels may be misclassified if tributaries are miscounted. Yet some studies might not require such a precise categorization of stream order at all. All they may need is the location and names of all streams and rivers, regardless of order.

13.4 Precision and Resolution

Precision refers to the amount of detail that can be discerned. It is also known as granularity or resolution. The later term is commonly used in GIS and related fields, and adopted here to avoid confusion with the statistical concept of precision as observational variance. Resolution is also limited because geospatial databases are intentionally generalised. Generalisation includes elimination and merging of entities, reduction in detail, smoothing, thinning, and aggregation of classes. Generalisation is inevitable because, geospatial databases can encompass only a fraction of the attributes and their relationships that exist in the real world.

Resolution affects the degree to which a database is suitable for a specific application. The resolution of the database must match the level of detail required in the application. Resolution is also important because it plays a role in interpreting accuracy. For example, two databases may have approximately equal spatial accuracy levels, but if their spatial resolutions are significantly different, then the accuracy levels do not denote the same level of quality. One would generally expect accuracy and resolution to be inversely related, such that a higher level of accuracy will be achieved when the specification is less demanding. The concept of spatial resolution, thematic resolution, and temporal resolution are discussed in the following sections.

13.4.1 Spatial Resolution

The concept of spatial resolution is well developed in the field of remote sensing (chapter 5). It is defined in terms of the ground dimensions of the picture elements, or pixels, making up a digital image. This defines the minimum size of objects on the ground that can be discerned. The concept is applicable without modification to raster databases. For vector data, the smallest feature that can be discerned is usually defined in terms of rules for minimum mapping unit size which depend upon map scale.

Spatial resolution is related to, but distinct from, the concept of the spatial sampling rate. Resolution refers to the fineness of detail that can be observed while the sampling rate defines the ability to resolve patterns over space. For remotely sensed images, resolution refers to the pixel size (ground area resolved) and sampling rate to the space between pixels. Thus in theory one could mix high spatial resolution with low sampling rate (small pixels with large gaps between them) or low spatial resolution with high sampling rate (large pixels that overlap). Normally, resolution and sampling rate are approximately equal.

13.4.2 Temporal Resolution

Temporal resolution refers to the minimum duration of an event that is discernible. It is affected by the interaction between the duration of the recording interval and the rate of change in the event. A shorter recording interval implies higher temporal resolution, just as a faster film has given us the ability to photograph quickly-moving objects. For geospatial data, the situation is more complicated because interactions between spatial and thematic resolutions must also be considered.

There is a clear distinction between resolution and sampling rate in the temporal domain. Sampling rate refers to the frequency of repeat coverage while resolution refers to the time collection interval for each measurement. Geosynchronous satellites are capable of much higher sampling rate than sun-synchronous satellites (repeat coverage several times per minute vs several times per month). Resolution, however, is a function of the time required to obtain spectral reflectance data for one pixel.

13.4.3 Thematic Resolution

In the thematic domain, the meaning of resolution depends upon measurement scale. For quantitative data, resolution is determined by the precision of the measurement device. For categorical data, resolution is defined in terms of the fineness of category definitions. Land cover classification systems define the level of detail in taxonomic definitions in terms of the spatial resolving power of the remote sensing system. This illustrates the interdependence between space and theme when extracting spatial information (land cover class boundaries) from thematic information (spectral reflectance data).

13.5 Consistency

Consistency refers to the absence of apparent contradictions in a database. For geospatial data the term is used primarily to specify conformance with certain topological rules (Kainz 1995). These rules vary with dimensionality; for example, only one point may exist at a given location, lines must intersect at nodes; and polygons are bounded by lines. Elimination of topological inconsistences is usually a prerequisite for GIS processing, such that most databases are topologically 'cleaned' before being released.

Topological consistency is one aspect of consistency in the spatial domain. Spatial inconsistencies can also be identified through redundancies in spatial attributes. For example, an entity might have the value 'Ongole' for the attribute 'state' but the value 'Ranga' for the attribute 'district'. This is inconsistent since there is no Ranga district in Ongole (Andhra Pradesh state). In this case redundancy is partial. Non-redundancy implies that there is independence between two attributes, such that meaningful consistency constraints do not exist (Redman 1992).

Little work has been done on consistency in the temporal domain, although a framework for temporal topology has been developed. For example, since at a given location only one event can occur at one time, an inconsistency exists if a different entity appears at the same location on two maps of the same date. Since events have a duration, this idea can be extended to identify events that exhibit temporal overlap. In the thematic domain, the ability to identify inconsistencies requires a level of redundancy in thematic attributes, for example, the three sociodemographic variables 'population', 'mean household size' and 'total number of households'. Of course, the identification of an inconsistency does not necessarily imply that it can be corrected or that it is possible to identify which attribute is in error. Thus consistency is appropriately viewed as a measure of internal validity.

13.6 Completeness

Completeness refers to relationship between the objects in the database and the 'abstract universe' of all such objects. Selection criteria, definitions, and other mapping rules used to create the database are important determinants of completeness. The abstract universe can be defined in terms of a desired degree of abstraction and generalisation. There are different types completeness. They are feature or entity completeness, attribute completeness, and value completeness.

Feature or entity completeness can be defined over space, time or theme like a database that depicted, the locations of historical monuments in the state of Andhra Pradesh placed on the register of Department of Tourism Development in 1997. This database would be incomplete if it includes only buildings were in Hyderabad district. This may be said that the database is incompleteness in space since Hyderabad district covers only a portion of Andhra Pradesh. If all the buildings placed in the register by 1990, it is said to be incompleteness in time since historical monuments may have been added after 1990. If the database consists only monuments of buildings, then the incompleteness in theme is due to the omission of other types of historical monuments. Feature or entity completeness may be explained by means of data completness and model completeness.

Data completeness is used to assess data quality which is application-independent. 'Model completeness' refers to the agreement between the database specification and the abstract universe that is required for a particular database application. Model completeness is application-dependent and therefore an aspect of fitness-for-use. It is also a component of 'semantic accuracy' (Salge 1995). The second type of completeness is 'attribute completeness'. This can be identified as the degree to which all relevant attributes of a feature have been encoded. A final type of completeness is 'value completeness' which refers to the degree to which values are present for all attributes (Brassl et al 1995).

13.7 Sources of Error in GIS

Spatial and attribute errors can occur at any stage in a GIS project. These errors may arise during the derivation of spatial entities. Encoding these entities in the computer system, forms the use of data in analysis. In addition, these errors may be present in source data, arising during data conversion and , arising during data manipulations and processing, and can be produced during the presentation of results.

There are many sources of error that may affect the quality of a GIS dataset. Some are quite obvious but others can be difficult to discern. Few of these will be automatically identified by the GIS itself. It is the user's responsibility to prevent them. For example, smooth changes in boundaries, contour lines, and the stepped changes of choropleth maps are "elegant misrepresentations" of reality. In fact, these features are often "vague, gradual, or fuzzy" (Burrough 1986). There is an inherent imprecision in cartography that begins with the projection process and its necessary distortion of some of the data, an imprecision that may continue throughout the GIS process. Recognition of error and importantly what level of error is tolerable and affordable must be acknowledged and accounted for by GIS users. Burrough (1986) divides sources of error into three main categories: (a) Obvious sources of error, (b) errors resulting from natural variations or from original measurements, and (c) errors arising through processing. Table 13.2 shows the various errors caused during the executions of the GIS project.

Table 13.2 : Sources of Errors

(i)	Obvious sources of error Age of data Areal coverage Map scale Density of observations Relevance Format Accessibility
(ii)	Error resulting from natural variations or from original measurements Positional accuracy Accuracy of content Qualitative and Quantitative Variation in data Natural variation Data entry/output faults
(iii)	Error arising through processing Numerical error in computer Faults due to topological analyses Misuse of Logic Problems associated with map overlay Classification and generalisation problems Interpolation

Generally errors of the first two types are easier to detect than those of the third, because errors arising through processing can be quite subtle and may be difficult to identify. Burrough further divided these main groups into several subgroups. Group (i) errors include topics that are more obvious and easy to check. Group (ii) contains subtle sources of error that can often be detected while working intimately with the data. Group (iii) is perhaps the most important because it includes the mistakes, errors and miss apprehensions that can arise as a result of carrying out certain kinds of processing. This group of errors are the most difficult to spot because they require an intimate knowledge of not only the data, but also the data structures and the algorithms used.

13.8 Modelling Errors

With the basic introduction to accuracy and error in the previous section, we can now consider the specific issue of quality in spatial data. This is not a simple and straightforward extension; as we will see, spatial data require a different and more elaborate approach.

The objective is to find an appropriate model of the errors or uncertainties that occur in spatial data. A model is taken here to mean a statistical process whose outcome emulates the pattern of errors observed in real digital spatial data. The first subsection considers models of error in the positioning of simple points; the second section looks at the ways this can be extended to more complex line or area features. For simple scalar measurements, conventional error theory is based on the Gaussian distribution. In effect, what we seek is the equivalent of the Gaussian distribution for the ensembles of measurements referred to here as spatial databases.

13.8.1 Point Data Error Models

The closest analogy between the classical theory of measurement and the problem of error in spatial data concerns the location of a single point. Suppose, we wish to determine the location accurately in coordinates of a single street intersection. It is possible to regard the coordinates as two separate problems in measurements, subject to errors from multiple sources. If the coordinates are in fact, determined by the use of a transparent roamer on a topographic sheet, then this is a fairly accurate model of reality. Our conclusion might be that both coordinates had been determined to an accuracy of 100 m, or 2 mm on a 1:50,000 sheet.

If the errors in both coordinates are represented by Gaussian or normal distributions, then we can represent accuracy in the form of a set of ellipses centered on the point. If the accuracies are the same in each coordinate and if the errors are independent, then the ellipses become circles, again centered on the point. This gives us the circular normal model of positional error, as the two-dimensional extension of the classic error model. The error in position of point can be visualized as a probability density forming a bell-shaped surface centered over the point. Using the model we can compute the probability density forming a bell-shaped surface centered over the point. Using the model we can also compute the probability that the true location lies within any given distance of the measured location, and express the average distortion in the form of a standard deviation. A commonly used summary measure is the Circular Map Accuracy Standard, defined as the distance within which the true location of the point lies 90% of the time, and equal to 2.146 times the standard deviation of the circular normal distribution. These indices are incorporated into many of the widely followed standards for map accuracy.

13.8.2 Line and Area Data Error Model

In vector database lines are represented as sequences of digitised points connected by straight segments. One solution to the problem of line error would therefore be to model each point's accuracy, and to assume that the errors in the line derived entirely from errors in the points. Unfortunately this would be inadequate for several reasons. First, in digitising a line a digitiser operator tends to choose points to be captured fairly carefully, selecting those that capture the form of the line with the greatest economy. It would therefore be incorrect to regard the points as randomly sampled from the line, or to regard the errors present in each point's location as somehow typical of the errors that exist between the true line and the digitised representation of it.

Second, the errors between the true and digitised line are not independent, but instead tend to be highly correlated. If the true line is to the east of the digitised line at some location along the line, then it is highly likely that its deviation immediately on either side of this location is also to the east by similar amounts. Much of the error in digitised lines results from misregistration, which creates a uniform shift in the location of every point on the map. The relationship between true and digitised lines cannot therefore be modelled as a series of independent errors in point positions.

One common method of dealing with this problem is the concept of an errors band, often known as the Perkal epsilon band (Perkal, 1966; Blakemore, 1984; chrisman, 1982). The model has been used in both deterministic and probablistic forms. In the deterministic form, it is proposed that the true line lies within the band with probability 1.0, and thus never deviates outside it. In the probablistic form, on the other hand, the band is compared to a standard deviation, or some average deviation from the true line. One might assume that a randomly chosen point on the observed line had a probability of 68% of lying within the band, by analogy to the percentage of the normal curve found within one standard deviation of the mean.

Some clarification is necessary in dealing with line errors. In principle, we are concerned with the differences between some observed line, represented as a sequence of points with intervening straightline segments, and a true line. The gross misfit between the two versions can be measured readily from the area contained between them, in other words, the sum of the areas of the spurious or silver polygons. To determine the mismatch for a single point is not so simple, however, since there is no obvious basis for selecting a point on the true line as representing the distorted version of some specific point on the observed line. Most researchers in this field have made a suitable but essentially arbitrary decision, for example, that the corresponding point on the true line can be found by drawing a line from the observed point that is perpendicular to the observed line (Keefer, smith, and Gregoire, 1988). Using this rule, we can measure the linear displacement error of any selected point, or compute the average displacement along the line.

13.8.3 Models for Dot and Pixel Counting

One of the traditional methods of measuring area from a map involves placing an array of grid cells or dots over the area to be measured and counting. In principle this process is similar to that of obtaining the area of a patch from an image by counting the pixels that have been classified as belonging to or forming the patch. The accuracy of the area estimate clearly depends directly upon the density of the pixel size, but so does the cost of the operation; therefore it would be useful to know the precise relationship in order to make an informed judgment about the optimum density or size.

The literature on this topic has been reviewed by Goodchild (1980). In essence two extreme cases have been analysed, although intermediates clearly exist. In the first, the area consists of a small number of bounded patches, often a single patch. In the second, it is highly fragmented of area is on the order of one or two. We refer to these as Case A and B, respectively. Case A was first analysed by Frolov and Maling (1969), and later by Goodchild (1980). The limits within which the true area is expected to lie 95% of the time expressed as a percentage of the area being estimated, are given by:

$$1.03\ (kn)^{1/2}\ (SD)^{-3/4} \qquad \text{..........(13.1)}$$

where n is the number of patches forming the area; k is a constant measuring the contortedness of the patch boundaries as the ratio of the perimeter to 3.45 times the square root of area (k ~ 1 for a circle); S is the estimated area; and D is the density of pixels per unit area.

The results for Case B follow directly from the standard deviation of the binomial distribution, since this case allows us to assume that each pixel is independently assigned. The limits within which the true area is expected to lie 95% of the time, again as a percentage of the area being estimated, are given by:

$$1.96\ (1 - S/S_0)^{1/2}\ (S/D)^{1/2} \qquad \text{..........(13.2)}$$

where S_0 is the total area containing the scattered patches. Error rises with the 3/4 power of cell size in Case A, but with the 1/2 power in Case B. Thus a halving of cell size will produce more rapid improvement in error in Case A, other things being equal, because only cells that intersect the boundary of the patch generate error in case A, whereas all cells are potential sources of error in Case B, irrespective of cell size.

13.9 Error Evaluation by Graphical Methods

The graphic display should allow a data distribution and its reliability to be displayed independently or jointly. Complex graphic design issues arise in trying to display data and reliability together, so users can observe correlations in the patterns. Mac Eahren (1994) offers three possibilities for joint display of data and reliability: (i) side-by-side images; (ii) composite images; and (iii) sequenced images. Major data quality issues that are adopted in using GIS for application development are, (i) metadata issues, (ii) graphic design issues, (iii) error analysis issues, and (iv) user satisfaction issues. Visualisation of data is a demanding display problem (Robertson 1991) which becomes even more demanding when the display must also include error and uncertainty and address the characteristics of spatial data.

13.9.1 Metadata Issues

Information on how the spatial data are collected, the sampling design, and whether any compilation or processing steps are performed on the data, is usually minimal or missing. This is a serious problem for error analysis, as, without this information, there is little basis on which to proceed. Several issues remain on how such metadata can be effectively stored with the data and maintained as the data evolve. Fortunately that standards effort mentioned having some impact and geographical datasets are appearing with more documentations.

13.9.2 Graphic Design Issues

Graphic detection and evaluation of spatial data error and uncertainty create particular problems for graphic design. These problems require a representation of space of linkage of a spatial displays to a spatial representation, so that spatial distribution of errors can be known. These errors may be regular, random and clustered in space. Graphic displays need to allow for both implicit and explicit displays of uncertainty.

13.9.3 Error Analysis Issues

As Tufte (1983) points out, graphics gather their power from content and interpretation beyond the immediate display of numbers. Thus by good graphic designing and association, effective detection and evaluation, are highly dependent upon effective error analysis. Plotting the data can work as an error detection device because we often have some expectation about the pattern we see. Deviations from this pattern suggest framework, from which departures can be determined. These may include (i) a known or postulated distribution for a set of observations; (ii) a hypothesised or assumed relationships; (iii) an expected set or range of values; and (iv) an independent (and more accurate) set of observations. These models and frameworks can range from being simple and inexpensive to being complex and expensive. In standard statistics, errors and their significance are characterised by their distance from the central values. This has some limitations as a method for detecting errors since outliers may in fact be unusual values and not errors. However, statistical methods make clear that for detection to be possible, we must first establish some expected distribution for values. In the case of spatial data we can add departures from assumed stationarity of mean or stationarity of dependence as the basis for detection of possible errors.

Other error detection methods require ground truth data or other sources of higher accuracy for their computation. Root mean square error (RMSE) measures the error between a mapped point and a measured ground position. A limitation of this measures is that the error standard deviation is spatially invariant. There is no information about variation in positional accuracy at individual points. Errors and uncertainty in spatial data are not static. New errors and uncertainty can occur as data are subjected to geographical information processing operations. Detection is thus an ongoing process which should be continually informed by metadata.

In the evaluation of errors or uncertainties we assume that having detected them, the users wish to assess their significance. We need to know the context of use, and a model and possibly a hypothesis to determine significance. Cross-validation is one common method used to assess statistical prediction. Repeating this over many deleted subsets allows an assessment of the variability of prediction error.

Fuzzy classifiers have recently become source of error descriptions and error models (Burrough). In a raster representation fuzzy classifiers provide a means of describing uncertainty by associating each pixel with a vector of class memberships. Goodchild et al (1993) describe an error model based on the vector of probabilities for a pixel's class membership. These methods can create quite large processing and/or large storage overheads.

13.9.4 User Satisfaction Issues

User satisfaction issues relate to the packaging around of the graphic error analysis tools. The interface to these tools should, as always, be intuitive and easy to use. The ideal graphic displays are those which are simple, relevant, and unambiguous. Uncertainty in the data should not be mapped to an uncertainty in the graphics in a way which requires the user to search hard or spend a long time interpreting the results. For example, will users be satisfied with depiction of the existence (location of errors or uncertainty), or will they desire more extensive information, such as, the rank, magnitude, or significance of errors? Computing the magnitude and significance of errors or uncertainty, requires more processing of the data while the display of quantitative values requires some very thoughtful consideration of the visual display.

To summarise the state of the art, we have good techniques for describing and measuring :

- The accuracy of location of a single point
- The accuracy of a single measured attribute
- The probability that a point at a randomly chosen location anywhere on a map has been classified
- The effects of digitising error in measures of length and area
- The propagation of errors in raster based area class maps through GIS operations such as overlay, and
- The uncertainty in measures of area derived from dot counting.

14 Data Analysis and Modelling

14.1 Introduction

The analysis and modelling subsystem is the very heart of the GIS. This is, however, the most abused subsystem of GIS. The abuses range from attempts to compare noncomparable nominal spatial data with highly precise ratio spatial data to statements about the causative nature of spatially corresponding phenomena made without testing alternative causes. The abuse of this analytical-subsystem is a result of lack of understanding of the nature of the spatial data contained in the subsystem. Beyond the problems of GIS abuse, there is a common belief that GIS is the panacea for all geographical problems. In many cases the user will be obliged to combine GIS tools with statistical analysis software, input-output modelling tools providing enhanced mathematical computations, geostatistical packages designed for advanced spatial analysis or sub-surface modelling called GIS applications development and customisation of softwares with respect to user's requirements.

Preprocessing procedures are used to convert a dataset into a form suitable for permanent storage within the GIS database for application development. Often, a large proportion of the data entered into a GIS requires some kind of processing and manipulation in order to make it conform to a data type, georeferencing system, and data structure that is compatible with the system. The end result of the preprocessing phase is a coordinated set of thematic data layers. The aim of this chapter is to introduce the range of data analysis methods available, and provide case studies of how they might be applied. The focus, in this chapter, is on the introduction of some analysis methods rather than a full explanation of the algorithms and concepts behind them.

The chapter begins by introducing methods for data format conversion and other methods of analysis. Some of the methods of analysis covered in this chapter are spatial measurement, reclassification, buffer analysis, map overlay analysis modelling surfaces (DTM), and network analysis models.

14.2 Format Conversion

Format conversion covers a number of different problems, but can be discussed in terms of two families: conversion between different digital data structures and conversion between different data media. The former is the problem of modifying one data structure into another. The latter typically involves converting source material, such as, paper maps, photographic prints, and printed tables into a computer based digital data either by means of manual digitisation or by means of scanning with automated digitisation. The conversion process is based on the digitiser. Each one of these devices has its advantages and disadvantages. The advantage of a manual digitiser is that, only the desired feature can be digitised and stored in the system and the remaining features may not be touched. The disadvantage of this type of digitiser is that if we want another feature, we have to repeat the entire process for this feature. The advantage of a scanner is that, whatever the feature you desire to digitise, the entire map can be scanned, stored, and used for subsequent automatic digitisation. The disadvantage of this device is that the storage space is unneccessarily filled with other than desired data.

14.2.1 Data Structure Conversion

The data produced by remote sensing systems is an array of brightness values for each wavelength band in the sensor. These systems generate datasets that are comparable to any other multivariate collection of raster data, including problems of geometric registration between the wavelength bands or between different dates of acquisition. There are, however, several ways to organise such datasets. The most important and commonly used methods of organising GIS datasets are band sequential, band interleaved by pixel and band interleaved by line. In band sequential each variable's data like elevation, annual rainfall, and the different spectral data from a multispectral sensor, can be arranged as a separate array and each array is kept as a separate file on the magnetic disk or tape. In this case, one data file would contain the elevation array, and a separate file would contain rainfall values. A common alternative, is band interleaved by pixel (BIP) (Mather, 1987). This organisation may be thought of as a single array containing multivariate pixels (Fig. 14.1). The first element in this second format would contain the elevation value for the pixel in the first row and column; the second element would contain the rainfall value for the same pixel.

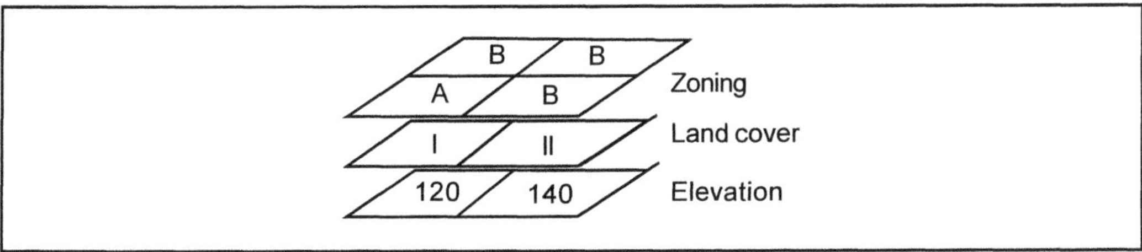

Fig. 14.1 Band interleaved by pixel format.

The **band interleaved by line (BIL)** raster organisation is a middle ground between the extremes of BSQ and BIP. In this form, adjacent ground locations (in the row direction) for a single theme are adjacent in the data file, and subsequent themes are then recorded in sequence for that same line. In this way, the different themes corresponding to a row in the file are relatively near each other in the file. Thus, one expects that its performance on specified tasks will fall between the pure sequential or pure pixel-interleaved forms. This intermediate type of multivariate raster is used in some commercial raster systems (Estes, 1995).

Fig. 14.1 shows the common raster data organisations. In band sequential data, each variable is stored in its own file. In band interleaved by pixel data, all the information about a pixel is kept together. In band interleaved by line data, all the values of a single line are stored before the values for another variable in the same line.

There are two common physical data organisations for BIL-structured data. The first is, the physical records hold all the themes from a single row in the array. Thus the number of physical records is the same as the number of rows in the array. In the other common BIL format, a physical record corresponds to a single thematic category. Thus in the second case, the number of physical records in the dataset is the product of the number of rows in the array times and the number of unique themes.

14.2.2 Conversion of Different Raster Formats

The problem of converting between the different raster data formats described above is relatively simple. Typically, a portion of the data in one format is read into a memory-based storage array, and the appropriate pointers created to extract the data values in whatever sequence is required for the new format. There are a wide variety of problems that develop when converting datasets between different vector data structures. As a guiding principle, it is expensive to generate topological information when it is not explicitly present in the vector data structure. To illustrate this point, converting data from an arc-node organisation to a relational one is very easy. In effect, these two data structures are really storing the same semantic information, with a slightly different syntax.

Consider the problem of converting data in a whole polygon structure to an arc-node structure. If a dataset is stored in a whole polygon structure, there is very little explicitly identified topology. The list of nodes that form the boundaries of each individual polygon is stored. Consider just the problem of extracting the arc-node list. We must go through the entire list of polygons, and create a list of the unique nodes. This might require a double-sort of all the points in the polygon file, and then a pass through the sorted list to identify the unique nodes. Creating the arc list requires another pass through the whole polygon file, this time cross-referencing edges of each polygon to the corresponding elements in the node list and generating the appropriate pointers. Furthermore, to identify all the polygons that border a given vector requires another complex sorting operation to identify all the shared edges.

Converting vector data into a raster data structure is conceptually straightforward, although practically difficult. For point data elements, the cell or pixel in the raster array whose center is the closest to the geographic coordinate of the point is coded with the attribute of the point. Thus, the elevation value from a surveyed benchmark is transferred to the raster cell whose location is closest to that of the original point. Of course, this operation usually changes the stored location of the point. It is unlikely that the original point location exactly coincides with the center of a raster cell. This approach also ignores the problem of different objects occupying the same cell. Because of these important limitations, the conversion from vector to a raster data structure is not normally reversible.

For linear data elements, the data structure conversion can be visualized by overlaying the vector or linear element on the raster array (Fig. 14.2(a)). The simplest conversion strategy would be to identify those raster elements that are crossed by the line, and then code these cells with the attribute or class value associated with the line. For lines that are not oriented along the rows or columns of the array, the raster representation shows a stair-step distortion (Fig. 14.2(b)). In this, first discussion, we have ignored the problem of specifying the thickness or width of rasterised line.

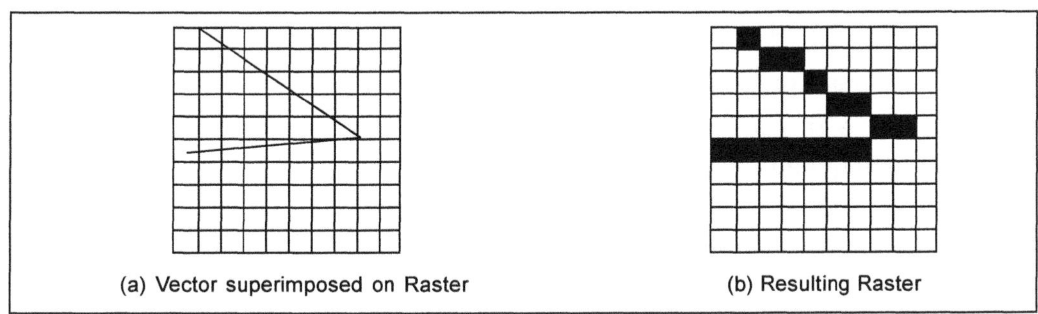

Fig. 14.2 Vector to raster conversion.

Polygons can be converted to a raster structure in two steps. First, the line segments that form the boundaries of the polygon can be converted, producing what is sometimes called the skeleton or hull of the polygon. Second, those raster elements contained by the polygonal boundaries are recoded to the appropriate attribute value. A simplistic approach for a trivial binary image (where the data are either in class 0 or class 1 and lines are only a single pixel wide) is illustrated in Fig. 14.3. Consider that each raster cell can be represented by a point in its centre. Further, this algorithm constrains all vectors to be of discrete lengths equal to an integer multiple of the raster spacing. Of course, vectors in real life do not have either of these restrictions.

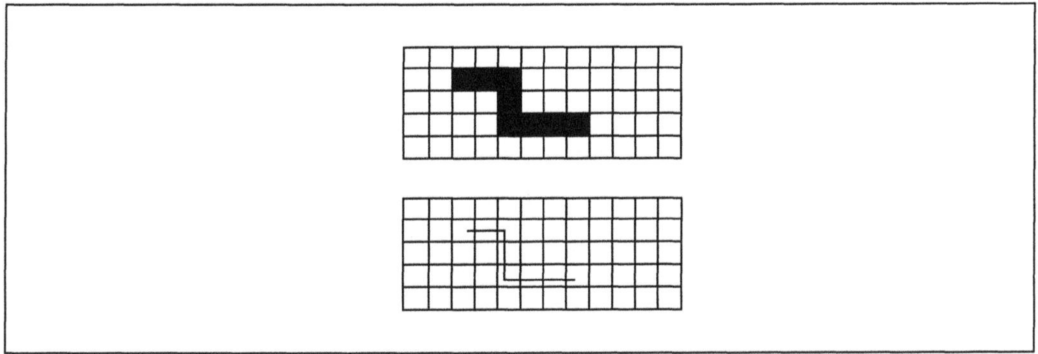

Fig. 14.3 Raster to vector conversion.

This approach will not be able to recover all vectors that have been converted to a raster. Consider the raster-coded vector in Fig.14.2(a). Since the straight-line nature of this data element has been lost in the process of conversion, we will not be able to recover the straight line without ancillary information. However, there are algorithms that can be used to extract straight lines from raster data sets under restrictive circumstances, at an increased computational cost.

14.2.3 Skeletonising

As Peuquet (1981) explains, operational systems require two general functions for converting a raster to a vector dataset. The first, skeletonising or thinning (Fig. 14.4), is required because the input data are not generally as simple as those we have discussed: single pixels for point data, and unit-width vectors for both vectors and polygon boundaries. Algorithms for determining the skeleton of an object are sometimes described as a peeling process, where the outside edges of thick lines are "peeled" away, ultimately leaving a unit-width vector. A symmetrical alternative approach is to expand the areas between lines, with the same ultimate goal. In either of these cases, the process is a sequence of passes through the data, with each pass producing narrower vector outlines. A third alternative, the medial axis approach, is designed to directly identify the centre of a line, by finding the set of interior pixels that are farthest from the outside edges of the original line.

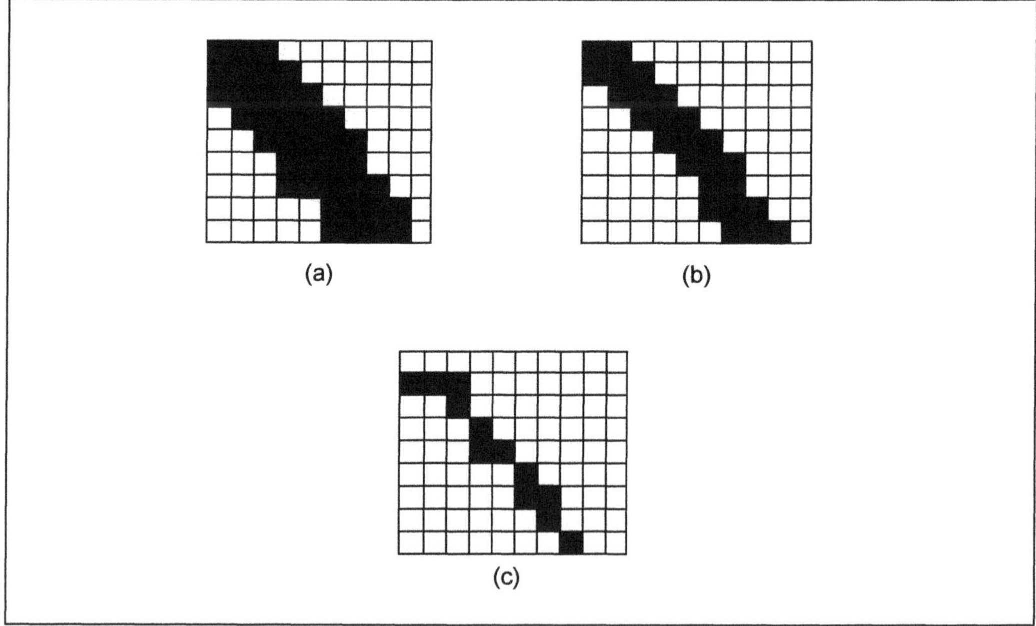

Fig. 14.4 Skeletonising.

After the raster data have had such thinning operations applied, the vectors implicitly stored in the raster are extracted. The extraction process may be based on the models discussed above. Finally, the topological structure of the lines is determined, by recognising line junctions and assembling the separate segments into connected vectors and polygons.

14.3 Data Medium Conversion

Most of the spatial data available for development of various applications are not in computer-compatible formats. These include maps on different scales, printed manuscripts, and imagery. Converting these materials into a format compatible with a digital geographical information system can be very expensive and time-consuming. The most common methods of converting maps and other graphic data to digital format is to use a technique called digitisation. This has already been discussed in the earlier chapters.

There are several technologies used in commercial digitising systems. Acoustic systems are relatively low in cost and often able to work with large-format materials. Such systems use acoustic transducers to triangulate positions on the map or graphic.

When it is necessary to digitise a map or other graphic with very high precision, the dimensional stability of the medium can become important. Photographic films are considered very stable with respect to changes in temperature and humidity, with distortions below 0.2 percent (Wolf, 1983). In contrast, paper shrinkage or expansion can range up to 3 percent, depending upon paper type and thickness, as well as processing methods.

14.3.1 Mode of Digitisation

When beginning a session with a digitising tablet, the user must specify the number of attributes of the map, as well as the map's location on the digitising tablet. Typically, the user will be prompted by the system for information about a map's scale and projection; menus with common choices can help the user to enter this information quickly and accurately. After entering this information, the tablet or stylus is used to specify both georeferencing information (for example, by placing the cursor at locations of known latitude/longitude) and a region of interest. In the process of converting the data compatible to any GIS, the most important function one should consider is mode of digitisation. In general there are three types of modes of digitisation (i) point mode, (ii) line mode and (iii) stream mode. In point mode, individual locations on the map, such as, elevation, benchmarks, road intersections, and water wells can be entered by placing the cursor over the relevant locations and pressing a button. In line mode, straight line segments, such as, short segments along political boundaries and straight road sections are entered by moving the cursor to one end of the line, pressing a button on the cursor, then moving to the other end and pressing a button again. The system automatically converts these two entered points to an appropriate vector. In stream mode, the location of the cursor on the map surface is determined automatically at equal intervals of time, or after a specified displacement of the cursor. Stream mode is particularly useful when digitising curved line segments, such as the boundaries of waterways. However, in stream mode it is often too easy to create very large data files, since data points are entered into the system very quickly.

14.3.2 Scan Digitising Systems

Scan digitising systems, generally called optical scanners or scanning densitometers, are typically larger and more expensive than digitising tablets. With many high-precision systems, the map (or graphic) to be digitised is attached to a drum. A light source and an optical detector are focused on the surface of the drum. The drum rotates at a constant speed around its major axis. Because of the rotation of the drum, the detector traces a line across the map, and the electronics in the associated computer system record numerical values that represent brightness on the map. This traced line across the map corresponds to a single row in a raster of data values. The detector then steps along the axis of the drum, in effect moving the detector to a new row in the raster, and the process repeats. In this way, the original map is converted to a raster of brightness values. Setting up a new session requires specifying the map coordinate system (UTM) scale and then indicating a control point coordinate location. Left, right and bottom reference coordinates are then indicated to calibrate the system for any relative rotation between the digitiser surface and the map's coordinate axes. Points, lines, and polygons are then digitised, with the user controlling the system from the cursor on the tablet.

14.3.3 Line Array of Detectors

An alternate mechanism uses a line array of photodetectors, which sweeps across the map in a direction perpendicular to the array axis. Such a device has a much simpler mechanical design than the drum systems mentioned above. In modern systems of either type, the map or graphic to be scanned can be on the order of one-meter square and the scanning step size as small as 20 micrometers. Based on these extreme values, the resulting raster dataset for a one-m-square map could contain 2.5 $X10^9$ pixels, before additional operations are begun. The resulting scanned raster of brightness may be used directly, or software can convert this initial raster dataset into other forms.

A normal sequence of steps in the use of a scanning digitiser would start with the actual scanning process. According to one set of figures, scanning a 24-by-36 in document at 20 lines per millimeter takes 90 minutes. The raster files are then interactively edited, to ensure that the skeletonizing process accurately extracts the graphic elements in the original dataset. Next, the preprocessed raster data are converted to a vector dataset. The vector data are then structured, to build whatever composite elements like chains as connected line segments, and topological relations like containment and adjacency, are required in the ultimate datasets. Finally, the data are interactively edited for quality assurance requirements.

14.4 Spatial Measurement Methods

Measurements allow to produce ratios of lengths to widths and of perimeters to areas. The GIS user need to describe not only what objects are, how many objects exist, and where they are, but also how large they are, how far apart and what the distance between them is like. Calculating length, perimeters, and areas is a common application of GIS. For example, measuring length of tankbund, perimeter of Hussain Sagar lake and its area are straightforward methods using GIS. The method of these spatial measurements depend upon the type of data, type of GIS (raster or vector) and the availability of software. It is important to remember that the measurements made using either raster GIS or vector GIS are approximating, since vector data are made up of straight line segments and all raster entities are approximated using a grid cell representation. In the remaining part of this section, simple methods of measurements of length, perimeter and areas of spatial patterns are presented. The reader can refer the cited references if he needs more than this.

In a raster GIS, the lengths are calculated using Pythagorean geometry. For example, let us consider raster image data in which the distance between 'A' and 'B' is to be determined (Fig.14.5). Calculating the line length AB in raster is a matter of adding the number of grid cells together to achieve a total. Let us begin by working with a straight line entity in grid format that occurs as a set of vertical or horizontal grid cells, knowing the resolution of each grid cell, generally assumed to be from side to side, we need only to add up the number of grid cells and multiply that value by the grid cell resolution value. If a line is composed (vertical/horizontal) of 20 cells with the grid cell resolution value of 50 m^2, then the total length will be 1000 meters.

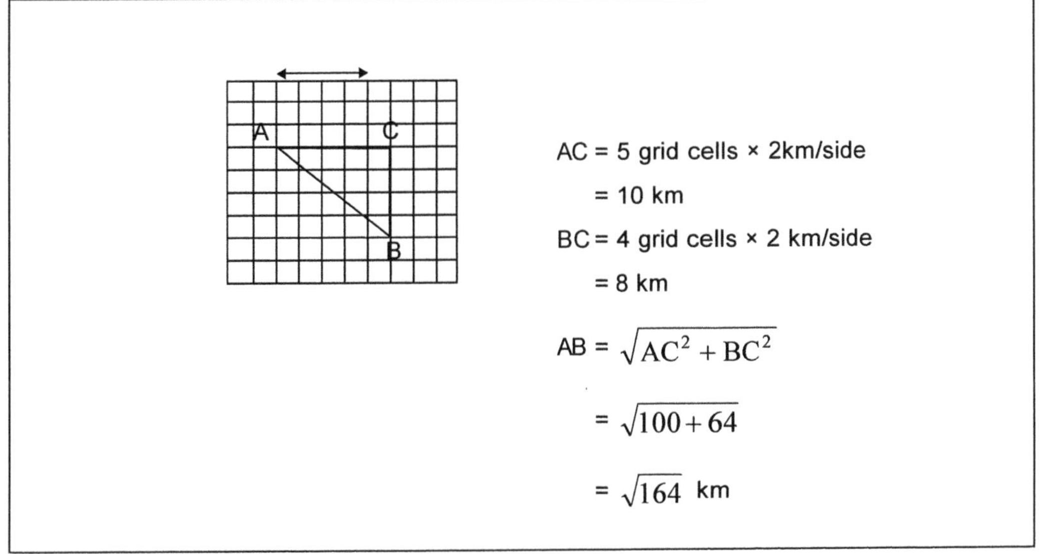

Fig. 14.5 Pythagorean geometry.

Suppose the line is perfectly diagonal. Most sophisticated raster systems are able to calculate the diagonal distance between each cell in a diagonal line of grid cells through the use of simple trigonometry (ESRI, 1993). Because a diagonal line through a square grid cell produces a right triangle, the right angle sides which are identical to the grid cell resolution, the hypotenuse is calculated with the Pythagorean theorem. To obtain a perimeter measurement in a raster GIS, the number of grid cell sides that make up the boundary of a feature is multiplied by the known resolution of the raster grid (Fig. 14.6). For area calculations, the number of cells a feature occupies is multiplied by the known area of an individual grid cell (Fig. 14.7). The area under MCH jurisdiction, discussed in chapter 16, is 180 sq.km, computed using this method.

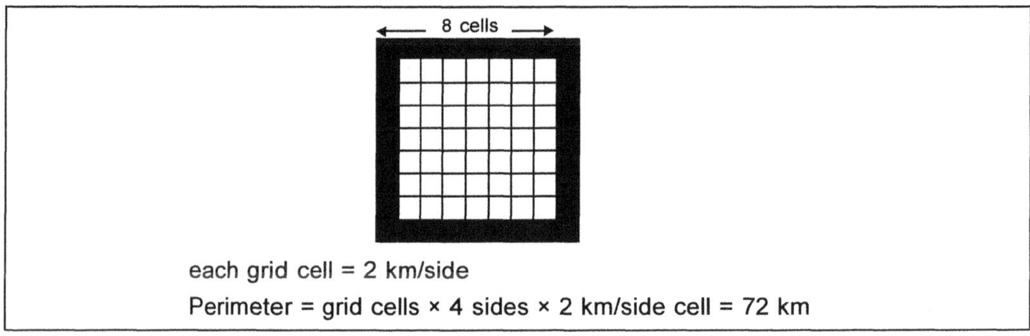

Fig. 14.6 Perimeter calculation.

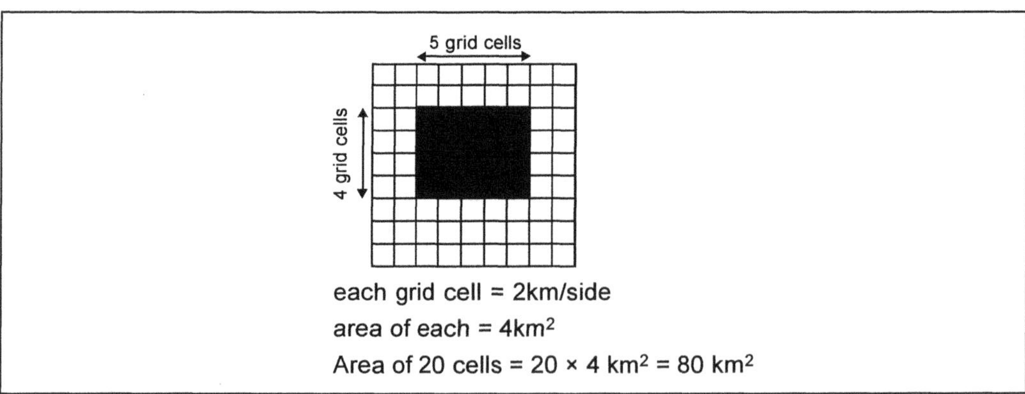

Fig. 14.7 The area calculations.

In a vector GIS, distances are measured using Pythagorean's theorem to obtain the Euclidean distance (Fig.14.8). The vector coordinates describe points, lines or area available for calculating distances. A series of modifications of the standard Pythagorean theorem can be used to calculate the distances. The original formula for Euclidean distance between two points (x_i, y_i) and (x_j, y_j) is

$$SQRT\ [(x_i - x_j)^2 + (y_i - y_i)^2]$$

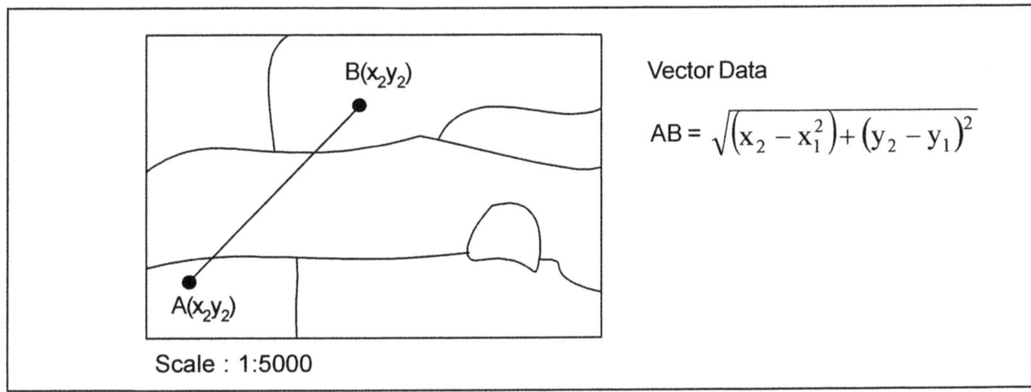

Fig. 14.8 Distance (vector).

The geometrical formula can be used to calculate perimeters and areas. Perimeters are built up of the sum of straight line lengths, and areas are calculated by totalling the areas of simple geometrical shapes formed by subdividing the feature of interest (Fig.14.9). For example, the area of the feature shown in Fig. 14.9 can be calculated. The area of ABC can be computed by using mathematical formulae.

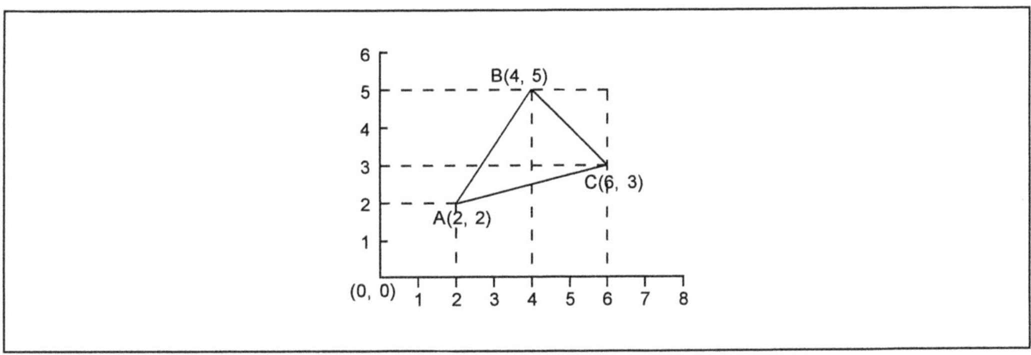

Fig. 14.9 Geometrical shapes

14.5 Reclassification

Reclassification is an important variation of the query idea in GIS and is used in place of a query in raster GIS. Performing queries on a GIS database to retrieve data is an essential part of most GIS projects. Queries offer a method of data retrieval which can be performed on data that are part of the GIS database and on new data produced as a result of data analysis. Queries are useful at all stages of GIS analysis for checking the quality of data and results obtained. For example, a query may be used to obtain the information on the name and address of the hotel in any city from the hotel database. There are two general types of queries that can be performed with GIS, namely, spatial and aspatial. Aspatial queries are those about the attributes of features. How many luxury hotels are there in Hyderabad city is an aspatial query.

Boolean operators need to be used with care in creating the queries in GIS environment.

In raster GIS, the method of reclassification can be used in place of query to obtain information under search from GIS database. Consider the raster land use image referred to plate 8. For instance, we wish to extract the information on areas of parks. Based on query based GIS, ask what is the area under parks. But to answer this query the answer could be obtained by classifying the image. Reclassifications would result in a new image. For example, in a raster image, if cells representing parks in the original image had a value of 20, a set of rules needed for the reclassification could be:

(i) Cells with values 20 (parks) should take the new value of 1

(ii) Cells with values other than 20 (other features) should take the new value 0.

As a result of this reclassification, a new image will be generated with all park areas coded with value 1 and all areas that are not parks coded with value 0. This is called a Boolean image. The similar concept has already been discussed in MAP model of raster GIS in chapter 10. Reclassification of this nature or conversion of raw image into Boolean image, produces a two-code image which is very much useful in further analysis since it contains only value 1 and 0. The resulting reclassified image is very much useful in application like forestry and agriculture, land use/ land cover and environmental studies.

Other reclassification methods, namely, neighbourhood functions, filtering techniques (roving windows), polygonal neighbourhoods, and extended neighbourhood method can also perform this analysis method.

14.6 Buffering Techniques

Buffering is the creation of polygons that surround other points, lines or polygons. The user may wish to create buffers to exclude a certain amount of area around point, line or polygon, or to include only the buffer area in a study.

A buffer is a polygon created as a zone of influence around an entity or around individual objects or multiple objects. The creation of buffer is based on the location, shape, characteristics of influential parameters, and orientation of an existing object. However, a buffer can be more than just a measured distance from any other two dimensional object and is controlled to some degree by the presence of friction surfaces, topography, barriers, and so on. The creation of buffer also influences its neighbours or the character of an entity. This leads to the recalculation of cells in raster image based on characteristic of neighbours. A buffer may be termed as point buffer, line buffer, and area buffer based on the type of the entity; point, line, and area/polygon and the measure of the influence of an entity and its corresponding application. Fig.14.10(a) shows point buffer. The creation of point buffer is very simple conceptually but poses complex computational operations. Creating buffer zones around point features or entity is a circle simply drawn around each point as centre and the area of influence under study as the radius of the circle. Line buffers can be created by measuring a specified distance in all directions from the line target object (Fig. 14.10(b)). Plate shows the example of line buffer for the alignment of pipline.

An area buffer (Fig.14.10(c)) can be created by measuring the distance with an area measured from its outer perimeter. Sometimes, there may even be a requirement for a buffer around a second buffer. Such buffers are called doughnut buffer.

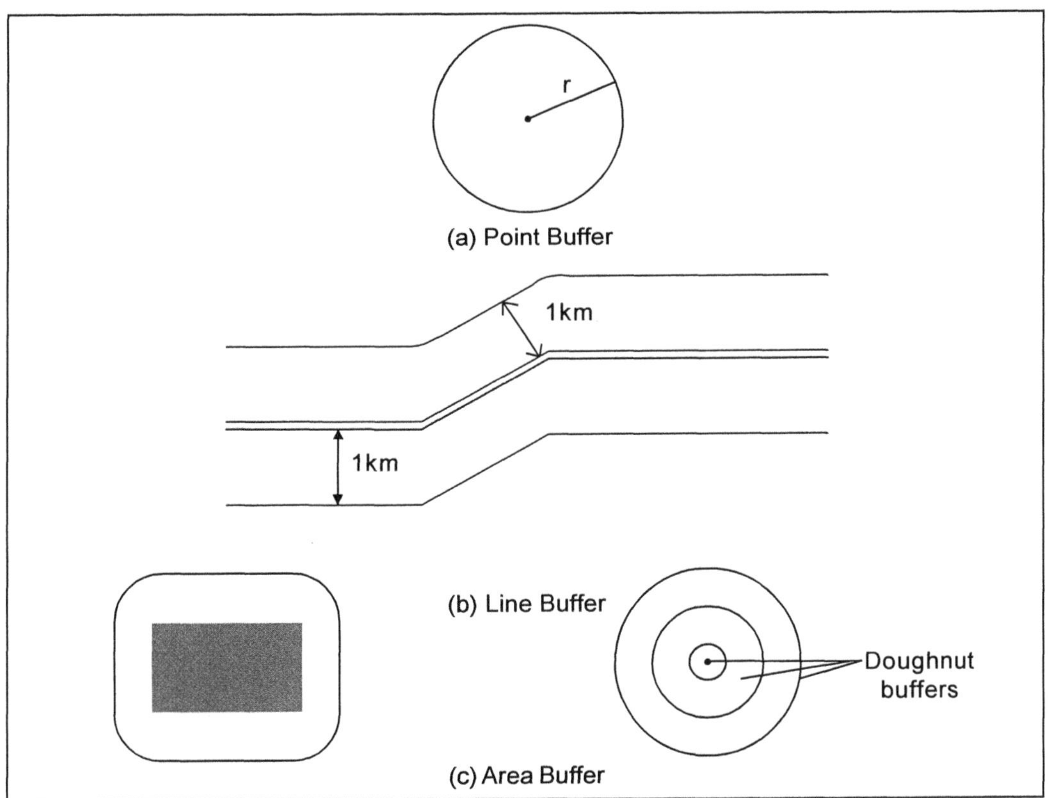

Fig. 14.10 (a) Point Buffer (b) Line Buffer (c) Area Buffer.

Buffers are useful methods for analysing the landscapes, environmental problem solving, water quality studies, road highways studies, pipeline alignment studies and so on. For example, in using GIS, to help in identifying the possible sites for a new water well, the location of chemical factories may have a 8-10 km buffer drawn around them so that these buffered areas are excluded in the list of possible sites.

14.7 Overlay Analysis

Map overlay is an important technique for integrating data derived from various sources and perhaps is the basic key function in GIS data analysis and modelling surfaces. Map overlay is a process by which it is possible to take two or more different thematic map layers of the same area and overlay them on top of the other to form a composite new layer. This technique is used for the overlay of vector data (for example, pipelines) on a raster background image (often a scanned topographic map) overlays where new spatial data sets are created involving the merger of data from two or more input data layers to create a new output data layer. There are some fundamental differences in operations and analyses in the way map overlays are performed between the raster and vector worlds. In vector-based systems map overlay is time-consuming, complex and computationally expensive. In raster-based systems it is just quick, straightforward and efficient.

One of the most important benefits of an overlay analysis of GIS data is the ability to spatially interrelate multiple types of information stemming from a range of sources. This concept is illustrated in Fig. 14.11, where we have assumed that a hydrologist wishes to use GIS to study soil erosion in a watershed. As shown, the system contains data from a range of source maps (a) that have been geocoded on a cell-by-cell basis to form a series of land data files or layers, (b) all in geographic

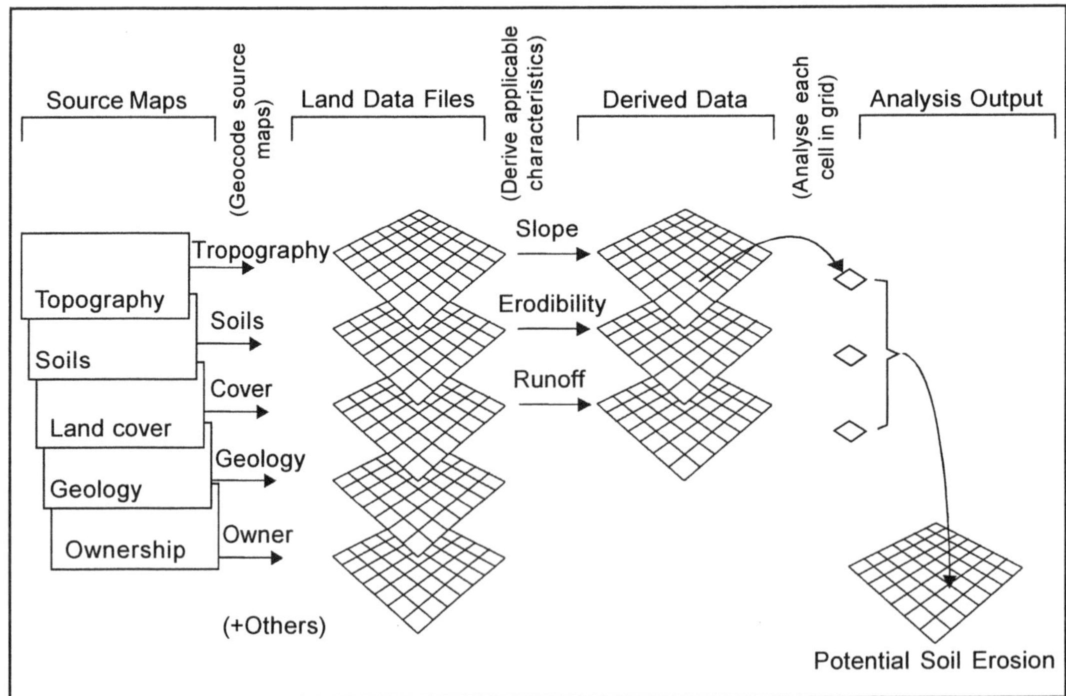

Fig. 14.11 GIS analysis procedure used for soil conservation studies.

registration. The analyst can then manipulate and overlay the information contained in, or derived form, the various data files. In this example, we assume the potential for soil erosion throughout the watershed involves the simultaneous cell-by-cell consideration of three types of data derived from the original data files: slope, soil erodability, and surface runoff potential. The slope information can be computed from the elevations in the topography file. The erodability, which is an attribute associated with each soil type, can be extracted from a relational database management system incorporated in the GIS. Similarly, the runoff potential is an attribute associated with each land cover type (land cover data can be obtained through interpretation of aerial photographs or satellite images). The analyst can use the system to interrelate these three sources of derived data where combinations of site characteristics indicate high soil erosion potential, that is steep slopes and highly erodible soil-cover combinations.

The above example illustrates the GIS analysis function commonly referred to as overlay analysis. The number, form, and complexity of other data analyses possible with a GIS are virtually limitless. Such procedures can operate on the system's spatial data, the attribute data, or both. For example, aggregation is an operation that permits combining detailed map categories to create new, less detailed categories, like combining "dense forest" and "sparse forest" categories into a single "forest" category.

Because each cell in a two-dimensional array can hold only one number, different geographical attributes must be represented by separated sets of cartesian arrays, known as 'overlays'. The overlay idea for separating data is not restricted to computer cartography, having been used by cartographers for a very long time. Fig. 14.12 illustrates the raster overlay concept in which each attribute is described and mapped separately.

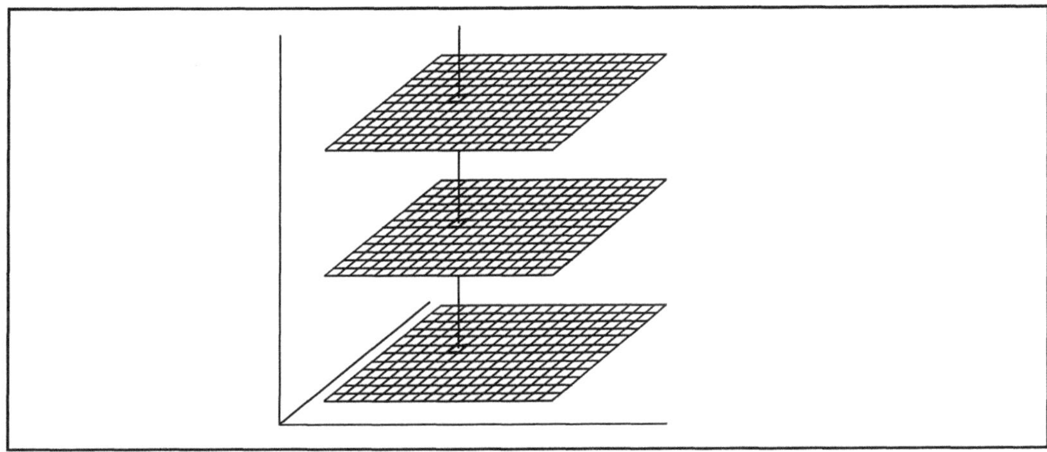

Fig. 14.12 Three dimensional arrays used for coding map overlays in raster database structure.

14.7.1 Vector Overlay Capabilities

Vector GIS displays the locations or all objects stored using points and arcs. Attributes and entity types can be displayed by varying colours, line patterns and point symbols (Fig.14.13). Using vector GIS, one may display only a subset of the data. For example, one may select all political boundaries and highways, but only areas that had urban land uses.

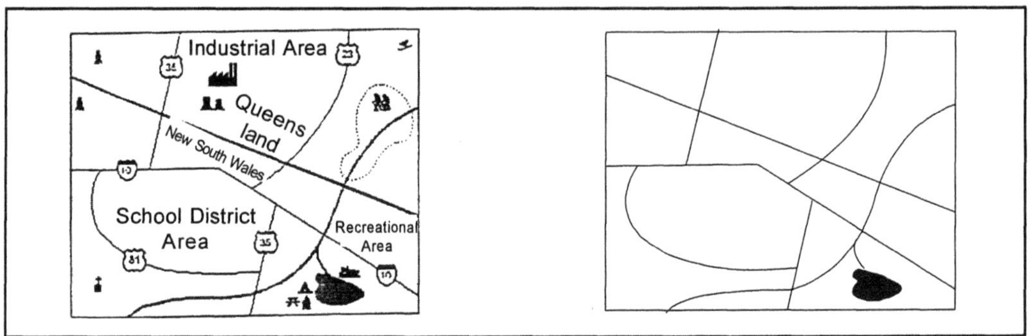

Fig. 14.13 Vector GIS display.

Relational query is an important concept in vector overlay analysis. Different systems use different ways of formulating queries. Structured Query Language (SQL) is used by many systems. It provides a "standard" way in querying spatial data bases. Using relational queries, the user can select objects interested in producing map output using colours, symbols, text annotations and so on.

Reclassify, dissolve, and merge operations are used frequently in working with area objects. They are used to aggregate areas based on attributes. Consider a soil map (Fig. 14.14). We wish to produce a map of major soil types from a layer that has polygons based on much more finely defined classification scheme. To do this, we process the data using three steps: (i) Reclassify areas by a single attribute or some combination; for instance reclassify soil areas by soil type only (ii) Dissolve boundaries

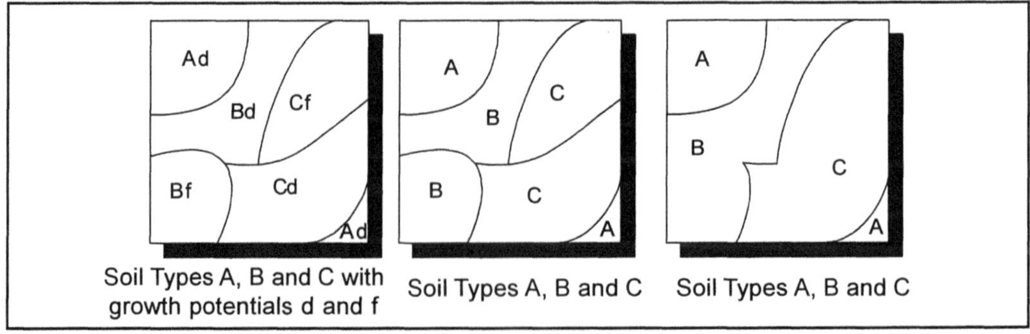

Fig. 14.14 Reclassify, dissolve and merge.

between areas of the same type by deleting the arc between two polygons if the relevant attributes are the same in both polygons, (iii) Merge polygons into large objects by recording the sequence of line segments that connect to form the boundary, that is, to rebuild topology and assign new ID numbers to each new object.

14.7.2 Topological Overlay

Suppose individual layers have planar enforcement when two layers are combined, or overlaid, or surperimposed (Fig.14.15), the result must have planar enforcement as well. In this case, new intersection must be calculated and created wherever two lines cross, and a line across an area object will create two new area objects. Topological overlay is the general name for such an overlay followed by planar enforcement. When topological overlay occurs, spatial relationships between objects are updated for a new and combined map. However, in some circumstances, the result may be information about relationships (new attributes) for the old maps rather than the creation of new objects. Examples of topological overlay are demonstrated by a number of ARC/INFO operations.

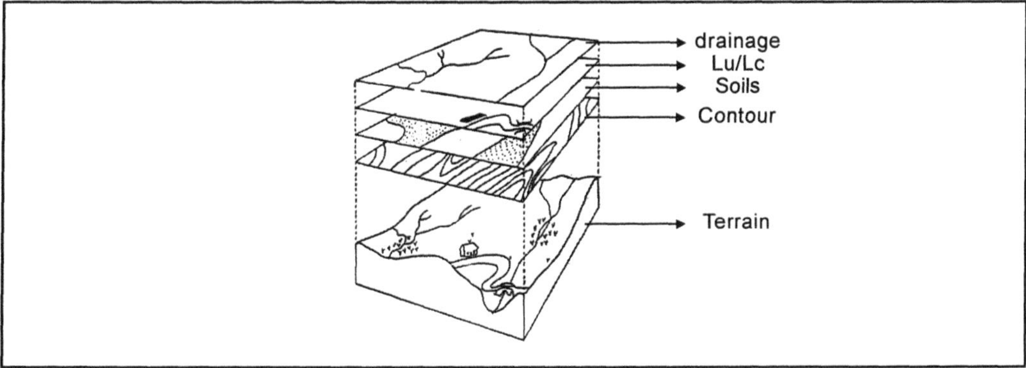

Fig. 14.15 Vector overlay concept.

Vector map overlay relies heavily on the two associated disciplines of geometry and topology. The data layers being overlaid need to be topologically correct (Fig. 14.15), so that lines meet at nodes and all polygon boundaries are closed. To create topology for a new data layer produced as a result of the overlay process, the intersections of lines and polygons from the input layers need to be calculated using geometry. The three main types of vector overlay are (i) point-in polygon, (ii) line-in-polygon, and (iii) polygon-in-polygon. The overlay of two or more data layers representing simple features results in a more intersections and more line segments on either of the input layers. Point-in-polygon overlay is used to findout the polygon in which a point falls. Line-in-polygon overlay is more complicated. Imagine that we need to know where roads pass through a very dense urban conglomeration to plan the traffic control strategies. To do this, we need to overlay the road data on a data layer containing very dense urban land use.

Lines are broken at each area object boundary to form new line segments and new attributes created for each output line specifying the area it belongs to Fig.14.16.

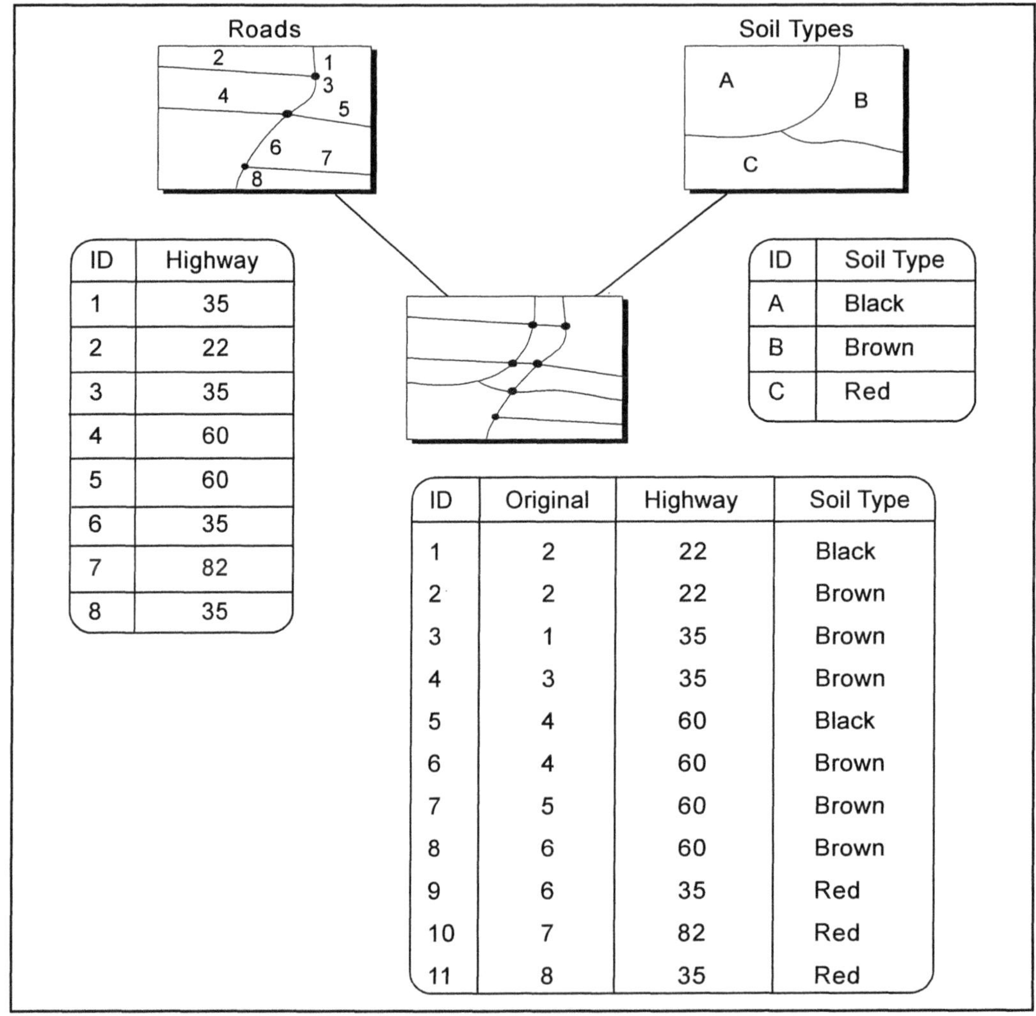

ID	Highway
1	35
2	22
3	35
4	60
5	60
6	35
7	82
8	35

ID	Soil Type
A	Black
B	Brown
C	Red

ID	Original	Highway	Soil Type
1	2	22	Black
2	2	22	Brown
3	1	35	Brown
4	3	35	Brown
5	4	60	Black
6	4	60	Brown
7	5	60	Brown
8	6	60	Brown
9	6	35	Red
10	7	82	Red
11	8	35	Red

Fig. 14.16 Line-in-polygon.

Polygon-on-polygon algorithm performs overlay of two layers of area objects. Boundaries of polygons are broken at each intersection and new areas are created (Fig. 14.17). During polygon overlay, many new and smaller polygons may be created, some of which may not represent true spatial variations. These small un-wanted polygons are called sliver or spurious polygons. The problem cannot be removed by more careful digitising since more points in digitising simply lead to more sliver. To solve the problem, a tolerance value may be set and the spurious polygons may be deleted during overlay operations.

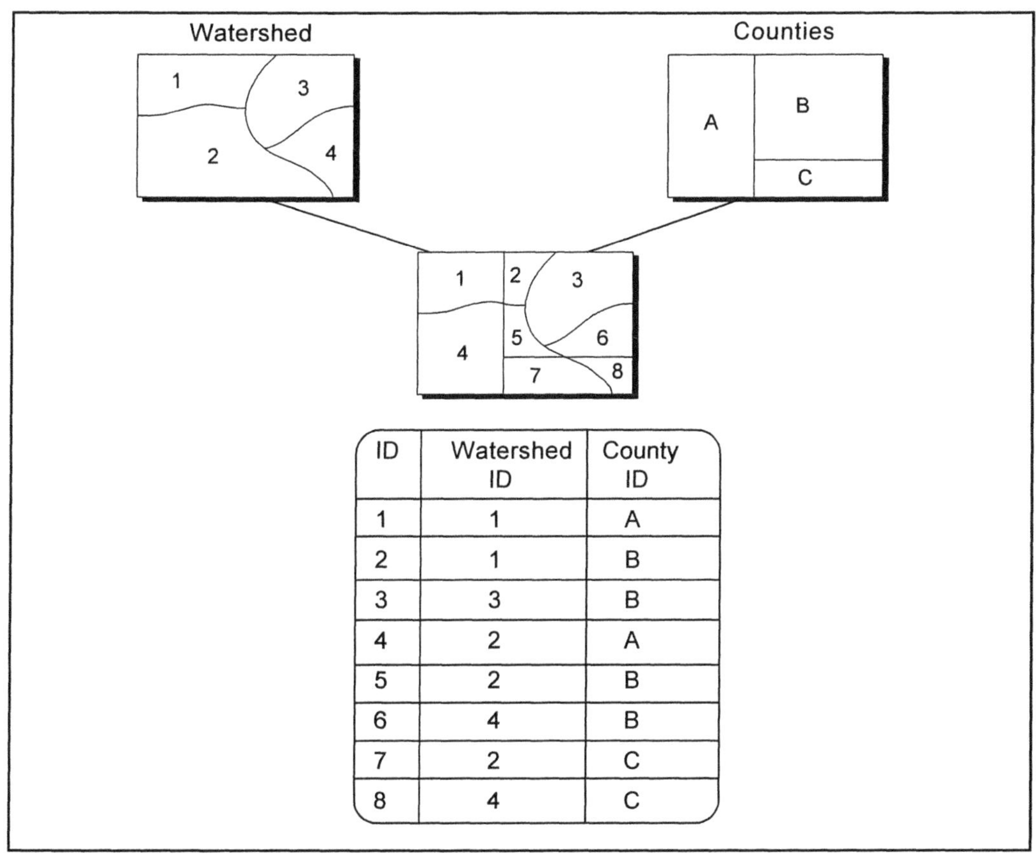

ID	Watershed ID	County ID
1	1	A
2	1	B
3	3	B
4	2	A
5	2	B
6	4	B
7	2	C
8	4	C

Fig. 14.17 Polygon-on-polygon.

14.7.3 Raster Overlay

In the raster data structure everything is represented by gird cells. A point is represented by a single cell, a line by a string of cells and an area by a group of cells. Therefore the method of overlaying various thematic layers are different from vector overlay. Raster overlay can be performed by using map algebra or mathematics (Berry, 1993). Using map algebra input layers may be added, subtracted, multiplied or divided to produce an output value. This most important function in raster overlay, is basically an operation of entities like appropriate coding point, line and area features in the input data layers.

To understand the concept of raster overlay and map algebra, consider a terrain which consists of four thematic layers, namely, well stations, the road network, the land use, and boundaries of waterbodies. Let the well stations be represented in a raster data layer for which the value '1' has been given. Similarly road networks are coded '2' in the road network layer, the land use patterns are coded '3' in the land use layers and '4' for the boundaries of water bodies. On all data layers '0' is the value given to cells that do not contain any information content. The raster overlay using map algebra is illustrated in Fig. 14.18.

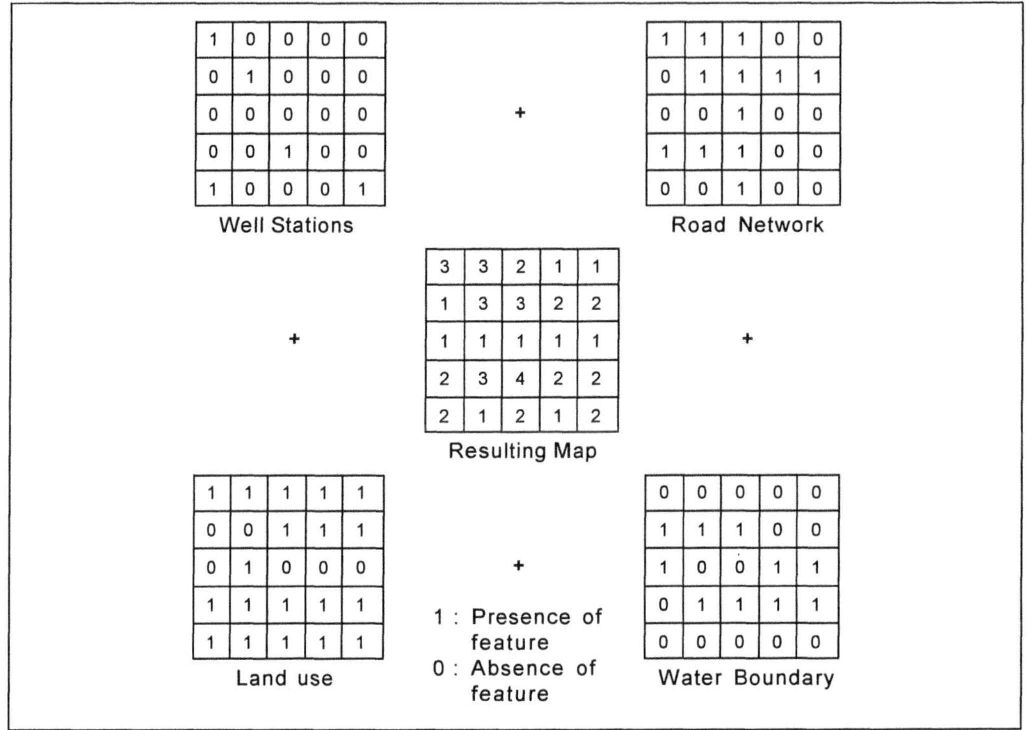

Fig. 14.18 Map algebra.

By using the concept of point-in-polygon, line-in-polygon and polygon-on polygon, the output integrated layers with the various combinations of these raster data layers can be produced by the method of map algebra. The algebraic manipulation of images in raster GIS is a powerful and flexible way of combining, integrating, and organising the data analysis models. Equations can be developed with variables to allow the creation of spatial models.

Two issues are specifically considered in performing raster overlay; resolution and scales/levels of measurement. These two parameters of digital data effect the results of raster over lay modelling. Consideration of these two issues are very much useful in reducing the degree of uncertainty and improving the accuracy and precision of GIS data analysis.

14.8 Modelling Surfaces

Surface such as the surface of the earth are continuous phenomena rather than discrete objects (chapter 1). In other words, to fully model the surface, one would need an infinite number of points which in turn require and infinite data storage. Thus, our purpose is to represent the continuous surfaces in a digital form using a finite amount of storage. Since the earth is three-dimensional, it would seem that all GIS applications include some element of three-dimensional analysis. To meet this demand, the digital terrain modelling techniques have emerged.

The term digital elevation model or DEM is frequently used to refer to any digital representation of a topographic surface. It is, however, most often used to refer specifically to a raster or regular grid of spot heights. The term digital terrain model or DTM may actually be a more generic term for any digital representation of a topographic surface. A DTM may be understood as a simplest digital representation of a portion of the earth's surface, but it is perhaps the most commonly used. As the overhanging cliffs and faults are relatively rare in nature, topographic surfaces are most often represented as 'fields', connected surface models, having unique z-values over x and y. In this sense DTM is '2.5 -D' rather than a 3-D model. In a more general sense, a DTM may be used as digital model of any single-valued surface like geological horizons and even air temperature or population density. Here, however, attention is focused on digital models of terrain since many aspects of the digital modelling of the other surface phenomena are functionally related to terrain modelling.

The input data, models and algorithms required by digital models of terrain or other surfaces are quite different from those used in representing planimetric or two-dimensional data. The activity of modelling and processing digital terrain data may thus be regarded as a system component of GIS that is functionally disparate from modelling 2-D data, yet needs to be closely linked to other processing functions of GIS like polygon, network, and raster processing. Digital terrain modelling encompasses the following general tasks:

1. DTM generation : sampling of original terrain data, formation of relations among the diverse observations (model construction);
2. DTM manipulation : modification and refinement of DTMs, derivation of intermediate models;
3. DTM interpretation : DTM analysis and information extraction from DTMs;
4. DTM visualisation : graphical rendering of DTMs and derived information; and
5. DTM application : development of appropriate application models for specific disciplines.

A DTM application forms the context for digital terrain modelling: each particular application has its specific functional requirements relative to the other terrain modelling tasks. Flexibility and adaptability to given problems are fundamental objectives of a digital terrain modelling system. Thus, as shown in Fig. 14.19, deriving products from a DTM should not be viewed as a one-way process, but rather as the result of various interrelated stages in modelling. For example, a DTM may be modified by model manipulation procedures. It might then be displayed by visualisation procedures, or through interpretation functions. Visualisation and interpretation in turn may require or support further modification or adaptation of the original DTM. Thus, results of individual modelling steps may feed back into previously run procedures.

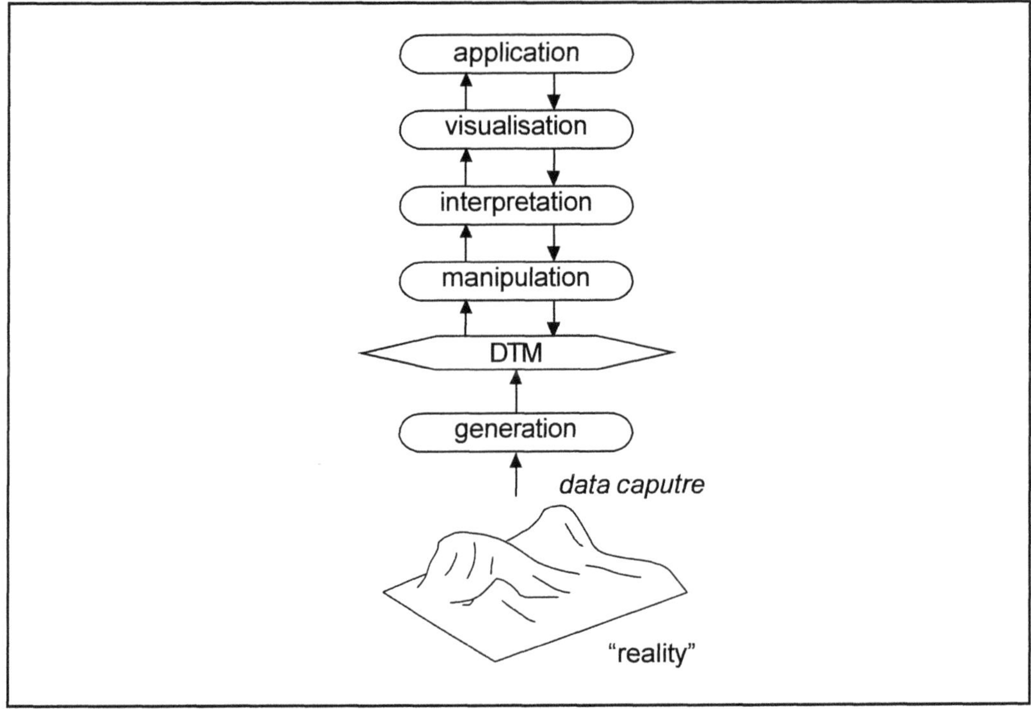

Fig. 14.19 The main tasks of a DTM system

14.8.1 DTM Generation

The choice of data sources and terrain data sampling techniques is critical for the quality of the resulting DTM. Data for a DTM should consist of observations about terrain elevations and shape of the terrain surface, that is, structural features, such as, drainage channels, ridges and other surface discontinuities. There may be other criteria apart from DTM quality requirements, which will guide the selection of a particular sampling technique and scheme for any given application like efficiency, cost and technological maturity. At present, most DTM data are derived from three alternative sources: ground surveys, photogrammetric data capture in manual, semi-automated, or from digitised cartographic methods. Other methods occasionally used include radar or laser altimetry, and sonar (for subaquatic terrain). Data for geological models are obtained from either borehole records or seismic surveys.

The process of terrain data capture generates a set of relatively unordered data elements. In order to construct a comprehensive DTM, it is necessary to establish the topological relations between the data elements, as well as an interpolation model to approximate the surface behavior. The original data must be structured to enable handling by subsequent terrain modelling operations. At the present, the majority of DTMs conform to one or other of two data structures rectangular grid (elevation matrix), and TIN (Triangulated Irregular Network).

Grids present a matrix structure that records topological elevations between data points implicitly. Since this data structure reflects the storage structure of digital computers, the handling of elevation matrices is simple, and grid-based terrain modelling algorithms therefore tend to be relatively straightforward. On the other hand, the point density of regular grids cannot be adapted to the complexity of the relief, so that an excessive number of data points is needed to represent the terrain to a required level of accuracy. Some research, however, reported the use of quadtree method to handle digital elevation data in the grid form so that the data redundancy can be reduced.

To construct the DTM, we need the estimation of elevation for each point of the grid. To do this, we need to know first whether the point is exactly at a point where the sampling data is available, or between the sampling points. In the first case, some method for estimating elevation need to be used. One such methods is called interpolation.

14.8.2 Triangulated Irregular Network (TIN)

The Triangulated Irregular Network (TIN) model is a significant alternative to the regular raster of a DEM and has been adopted in numerous GIS, automated mapping and computer packages. The TIN model was developed in the early 1970's as a simple way to build a surface from a set of irregularly spaced points. Several prototype systems were developed in the 1970's. Commercial systems using TIN began to appear in the 1980's as contouring packages embedded in GIS. Irregularly spaced sample points can be adapted to the terrain, with more points in areas of rough terrain and fewer in smooth terrain. An irregularly spaced sample is therefore more efficient at representing a surface than a regular spaced sample such as grid-based DEM. In a TIN model, the sample points are connected by lines to form triangles and within each triangle the surface is usually represented by a plane. By using triangles we ensure that each piece of the mosaic surface will fit with its neighbouring pieces. Thus the surface will be continuous as each triangle's surface would be defined by the elevations of the three corner points (Fig.14.20).

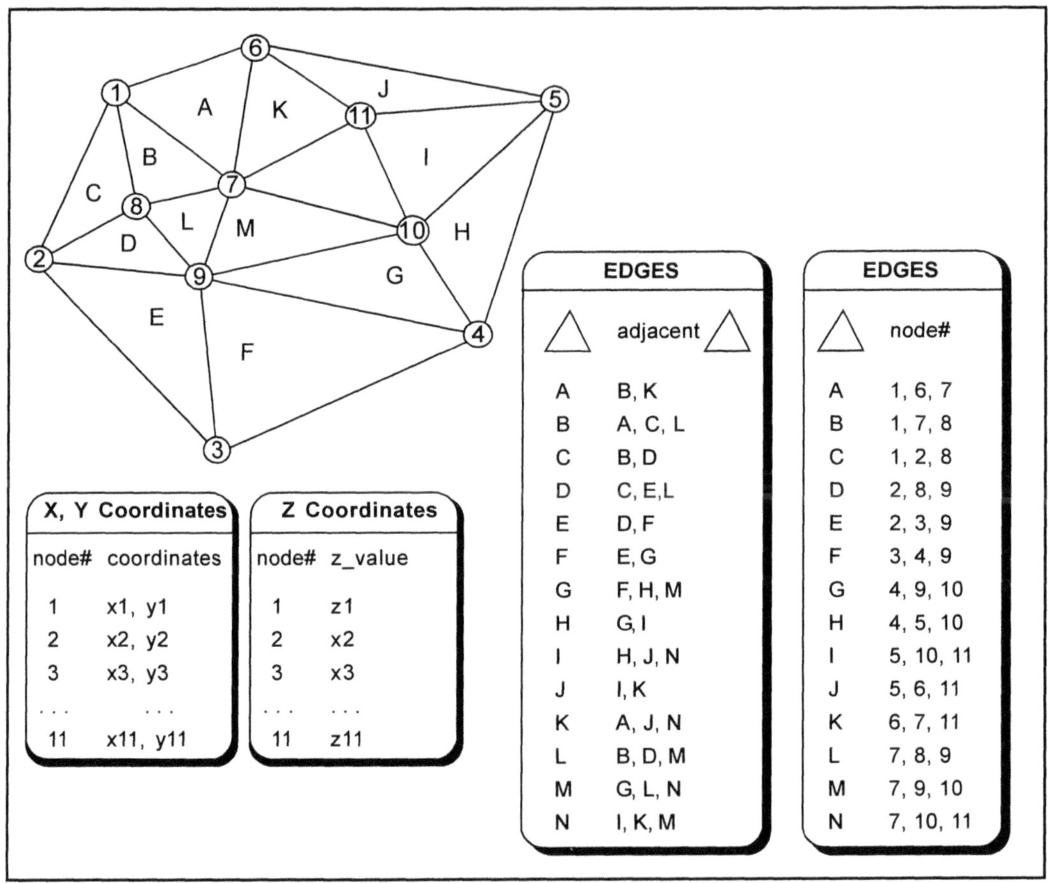

X, Y Coordinates

node#	coordinates
1	x1, y1
2	x2, y2
3	x3, y3
. . .	. . .
11	x11, y11

Z Coordinates

node#	z_value
1	z1
2	x2
3	x3
. . .	. . .
11	z11

EDGES

△	adjacent △
A	B, K
B	A, C, L
C	B, D
D	C, E,L
E	D, F
F	E, G
G	F, H, M
H	G, I
I	H, J, N
J	I, K
K	A, J, N
L	B, D, M
M	G, L, N
N	I, K, M

EDGES

△	node#
A	1, 6, 7
B	1, 7, 8
C	1, 2, 8
D	2, 8, 9
E	2, 3, 9
F	3, 4, 9
G	4, 9, 10
H	4, 5, 10
I	5, 10, 11
J	5, 6, 11
K	6, 7, 11
L	7, 8, 9
M	7, 9, 10
N	7, 10, 11

Fig. 14.20 TIN Model : The structure of a TIN as topological model.

It might make sense to use more complex polygons as mosaic tiles in some cases, but they can always be broken into triangles. For vector GIS, TIN can be seen as polygons having attributes of slope, aspect and area with three vertices having elevation attributes and three edges with slope, aspect and area with three vertices having elevation attributes, and three edges with slope and direction attributes. The TIN model is attractive because of its simplicity and economy. Certain types of terrain are very effectively divided into triangles with plane facets, like fluvially-eroded landscapes. Other landscapes, however, are not well represented by flat triangles like glaciated landscapes. Triangles work best in areas with sharp breaks in slope, where TIN edges can be aligned with breaks, along ridges or channels. Some of the applications of TIN are slope and aspect, contouring, finding drainage networks, creating cross sections, visualisation of the terrain and volume classification and transportations network studies.

14.8.3 DTM Manipulation

Along with DTM generation procedures, the DTM manipulation processes are of crucial importance for the performance and flexibility of a DTM system. They are needed for the modification and refinement of existing models. DTM manipulation consists of processes for DTM editing, filtering, and merging, as well as for conversion between different data structures.

DTM editing involves updating and error correction. An editor is required for interactive and selective modification of the properties of individual elements of a DTM. Edit operations for DTM elements should include query, delete, add, move, change height and change attribute. For gridded DTMs, however, editing is restricted to modifying elevations at grid points.

DTM filtering serves two purposes, smoothing or enhancement of DTMs, and data reduction. Smoothing and enhancement filters for DTMs are equivalent to lowpass and highpass filters as they are known in image processing. The effect of smoothing, that is applying low pass filters is to remove details and make the DTM surface smoother. Enhancement, that is, highpass filtering has the opposite effect by which surface discontinuities are emphasised, while smooth shapes are suppressed. DTM filtering procedures are also used to reduce the data volume. Processes of this kind may be desirable to eliminate redundant data points, to save storage space and processing time, or to reduce a DTM's resolution.

In joining and merging phase of manipulation, DTMs may be combined either by joining adjacent models or by merging overlapping models. For gridded DTMs, joining is only straightforward if the grids correspond in grid resolution, orientation, and so on. Otherwise, a resampling process has to be used to establish continuity.

14.8.4 DTM Interpretation

Within GIS, DTMs are most valuable as a basis for the extraction of terrain-related attributes and features. Information may be extracted in two ways: by visual analysis and graphic representations, through interpretation. A first objective of DTM interpretation is the derivation of geomorphometric parameters. Geomorphometric analysis may take the forms of *general geomorphometry*, "the measurement and analysis of those characteristics of landform which are applicable to any continuous rough surface"; and *specific geomorphometry*, "the measurement and analysis of specific landforms such as cirques, drumlins and stream channels, which can be separated from adjacent parts of the land surface according to clear criteria of delimitation". The most frequent use of general geomorphometry is the derivation of slope values from DTMs. Other uses of DTM include interpretation of landform features, that is terrain positioning, and derivation of drainage networks.

14.8.5 Slope Models

Perhaps the most straightforward usage of a DEM is the generation of slope information including slope angle and aspect which are frequently used in DTM applications. Slope is a vector, that is it has a direction and length. A common algorithm to calculate slope is the one generating slope and aspect values from its normal vector (Fig. 14.21).

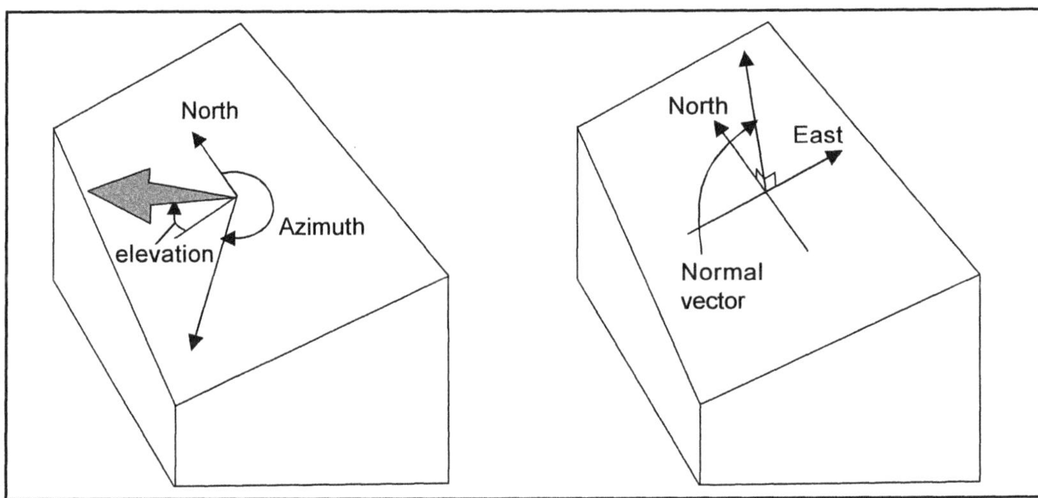

Fig. 14.21 Slope and aspect models.

Slope and aspect are calculated in two ways according to the type of DTM being used. In raster DTMs slope and aspect are calculated using a 3 × 3 windows over a database to determine a best fitted tilted plane for the cell at the center for the equation Z = a + bx + Cy where Z is the height at the point of interest. The center of the window, (x, y), is the co-ordinate of the point at the centre of the window and a, b, c are constants. Slope and aspect for the center cell can be calculated using the formulae,

$$S = b^2 + c^2$$
$$A = \mathrm{Tan}^{-1}(c/b) \qquad \text{..........(14.1)}$$

where, S = Slope and A = aspect.

In the vectors TIN model, slope and aspect variables are usually calculated using a series of linear equations when the TIN is generated. The equations calculate the slope and aspect of the individual triangles formed by the TIN model.

14.8.6 DTM Visualisation

Results of DTM modelling operations are most often communicated to user in graphical forms. Visualisation thus plays a vital role in a DTM system. Results of interpretation need to be displayed and interpretation operations may in turn lead to improvements in visualisation. Moreover, graphics themselves may directly support decision making without involving any quantitative analysis. Visualisation commonly pursues two goals: interactive visualisation, which helps the researcher to explore models and refine hypotheses, and static visualisation, which is used to communicate results and concepts. The usefulness of visualisation products mainly depends upon their communicational effectiveness and their ability to support interpretation. Visualisation in GIS can be performed by means of contouring, relief shading, and perspective display.

Contour lines (isolines) are probably still the most widely used technique for displaying relief. Contours represent a method for quantitative visualisation of the third dimension. They are used to satisfy the requirement for extracting quantitative information from relief displays. The major drawback of contours is that they give no immediate impression of the topographic forms. Basic contouring may be refined by index contours and highlighted by special symbolisation. Contour labels added to index contours and contour drawing may be suppressed in areas of steep gradient. Where contours are too densely clusttered, only index contours are drawn.

The method of relief shading, also called hill shading has been developed by cartographers as an important technique to improve the visual qualities of maps, particularly with respect to portraying relief difference in hilly and mountainous regions. However, old manual methods of relief shading relied on hand shading and airbrush techniques to produce the desired effect. They were expensive and very dependent upon the skills of the cartographer. With the development of computer technology, when DTM became generally available, the techniques of automatic relief shading emerged.

The relief shading model (Fig. 14.22) use of the optical theory (Lambert's Cosine Law) states that the brightness of any small area of a perfectly diffuse undulating surface varies as the cosine of the angle of incident parallel light.

(a)

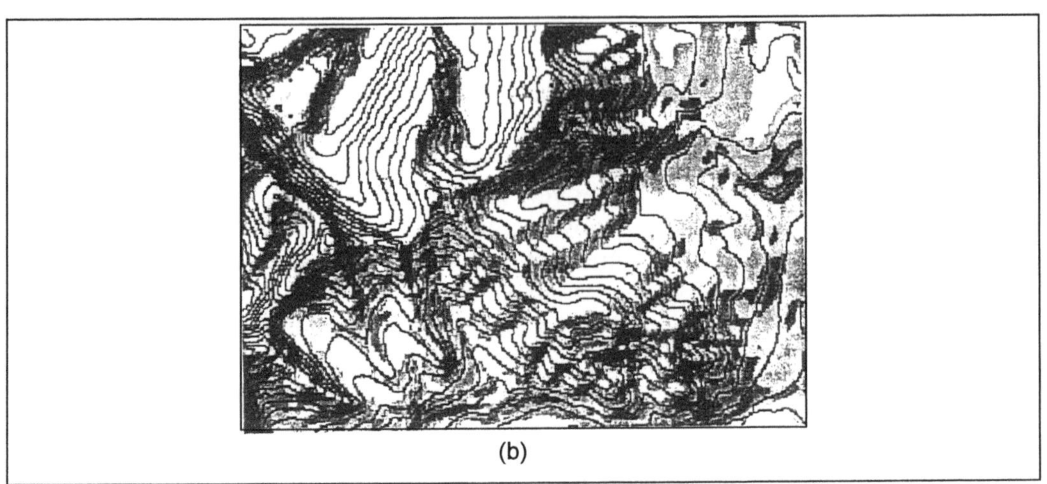

(b)

Fig. 14.22 Relief and Hillshade model (a) Relief (b) Hillshade

The advantage of orthographic displays is that all parts of the terrain surface are visible and relatively undistorted. Perspective displays, on the other hand, provide much more convincing visualisation results. It is common to display a DTM in a perspective view in GIS applications and it is often an important criterion in selecting and implementing a GIS into an environment-related project. A classical example of DTM is shown in plate 6.

14.8.7 DTM Applications

Due to the recent technological advances, terrain modelling systems of increasing complexity are being implemented and these systems offer powerful solutions to applications. It is becoming feasible to create specific applications based on this core functionality. Practical use will then provide feedback for further enhancement of DTM concepts and techniques. Some examples of common applications of DTM are addressed to the following disciplines : surveying and photogrammetry, civil engineering, planning and resource management, earth sciences, and military applications.

14.9 Modelling Networks

A network is a set of interconnected linear features through which materials, goods, and people are transported or along which communication of information is achieved. Network models in GIS are abstract representations of the components and characteristics of their real-world counterparts. They are essentially adaptations of the vector data model made up of the same arc (line segments) and node elements as any other vector data model but with the addition of special attributes as illustrated in Fig. 14.23.

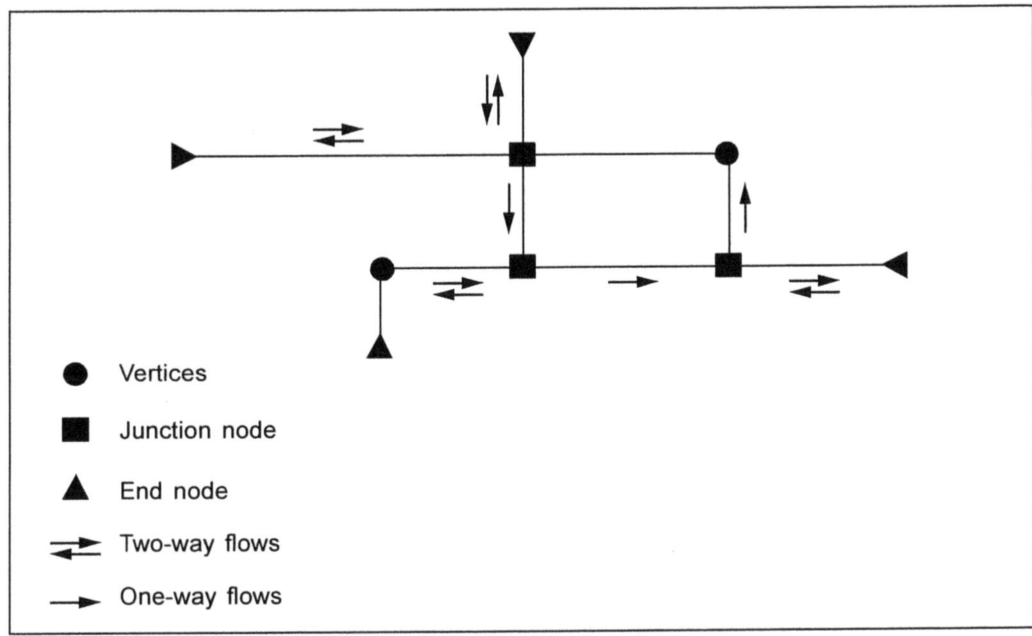

Fig. 14.23 Network data model

In the network model the arcs become network links representing the roads, railways and air routes of transport networks; the power lines, cables and pipelines of the utilities networks, and the rivers and streams of hydrological systems. The nodes in turn become network nodes, stops, and centers. Network nodes are simply the endpoints of network links and as such represent junctions in transport networks, confluences in stream networks, and switches and valves in utilities networks. Stops are locations on the network that may be visited during a journey. They may be stops on a bus route, pick-up and drop-off points on a delivery system, or sediment sources in a stream network. They are points where goods, people or resources are transferred to and from some form of transport system. Centres are discrete locations on a network at which there exists a resource supply or some form of attraction. Examples include shopping centres, airports, schools, and hospitals. At a larger scale centres may be a whole city if the transport, resource, or information networks for an entire country are being considered. Turns represent the transition from network links and greatly affect movement through the network systems. For example, turns across oncoming traffic on a road network take longer than turns down slipways, whereas turns that go against the flow of traffic on one-way streets are prohibited altogether. All the data regarding the characteristics of network links, nodes, stops, centers, and turns are stored as attribute information in the vector model database. Two key characteristics of network features are impedance, and supply and demand.

Correct topology and connectivity are extremely important for network analysis. Digital networks should be good topological representations of the real-world network they mimic. Correct geographical representation in network analysis is not very important, so long as key attributes, such as, impedance and distance, are preserved. A classic example of this is the famous map of the Culcutta underground system. This bears little resemblance to the real-world map of the underground systems, which would be far too complex for underground users to follow. Instead, the map was redrawn to simplify the network making it easily understood whilst at the same time maintaining relative distance and connectivity between all the stations on the network.

There are several basic network type problems, including identifying shortest paths, the travelling salesperson problem, allocation modelling and route tracing. Creation of drainage network model as a classic example of network shown in Fig. 14.24, is called dendritic drainage pattern which is an indication of the presence of hardrock terrain in the watershed. This also indicates the occurrence of more runoff of rainfall water. In this case, a watershed is defined here as an attribute of each point on the network which identifies the region upstream of that point. To find a watershed, we begin at the specified cell and label all cells which drain to it, and so on until the upstream limits of the basin are defined. The watershed is then the network, in which the polygon is formed by the labelled cells.

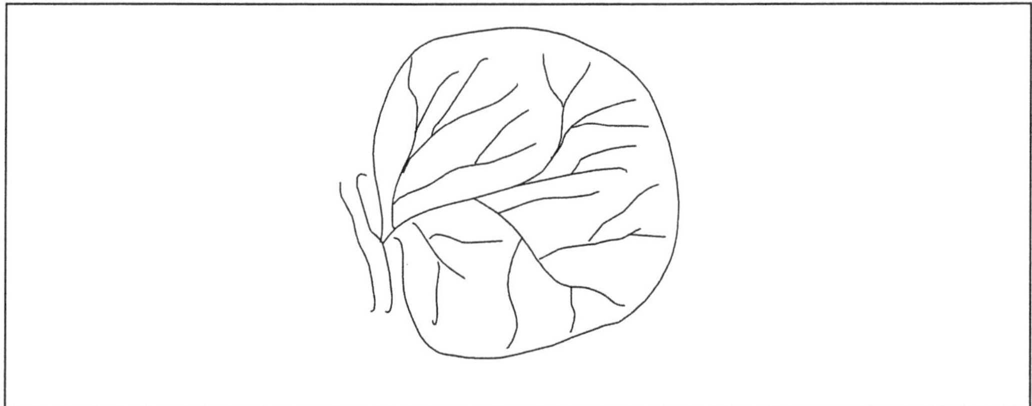

Fig. 14.24 Dendritic drainage pattern as an example of netwrok model

Another example of GIS data network analysis and modelling is the creation of pipeline distribution system for water supply of a very fast growing city like Hyderabad. The existing pipeline network of water distribution system is modified or strengthened by developing another pipeline network (Fig. 14.25) using ARC-INFO GIS. The parameters used in strengthening the existing pipeline network for water distribution system are attribute database including population, density of dwelling units, and quantity of water, and the spatial database including location of open spaces, location of dwelling units and land use/ land cover derived from satellite imagery.

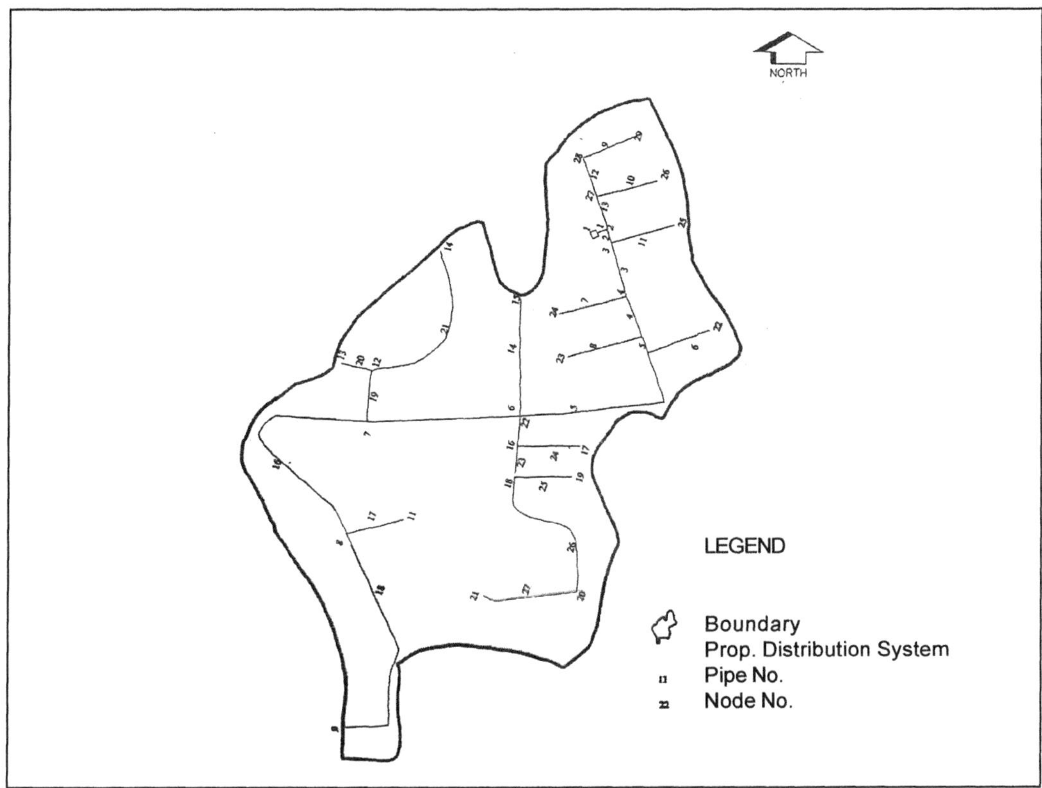

Fig. 14.25 Water distribution system of a part of Hyderabad city.

14.10 GIS Output

Informed decision making and problem solving rely on the effective communication and exchange of ideas and information. If GIS assists these activities the information it produces must be meaningful to users. Who are aware of the different forms of information. Hence a geographical Information system must include software for displaying maps, graphs, and tabular information on a variety of output media. Cartographic functions should permit production of the kinds of maps that clearly depict the spatial distribution of various kinds of phenomena. Several non-map graphic products can also be of value when communicating the results of a spatial analysis to the users. The choice of the type of display to use depends upon a number of factors including the nature of the data itself, required scale and resolution, hardware and software limitations, and of course the ultimate user for the output products. User community should see GIS output not as the final goal of a GIS project but rather as the starting point of informed decision making or problem solving.

The most common form of output from GIS is a map. In many cases the map will be thematic in nature and illustrates the spatial distribution of any feature type of interest for any specific application. For example, a map produced for the integrated watershed management shows the land use/land cover and shows the groundwater potential zones of Shivannagudem watershed. Alternatively GIS output maybe a 3-D model of the terrain produced by drapping the satellite data over a digital contour data or terrain surface. In addition to the major output, a map of a different kind, other output products like bar charts, pie charts, graphical representation of the results of GIS analysis are also presented in this chapter.

14.10.1 Maps as Output

The most common output produced by Geographic Information Systems are maps of various kinds. We use a very general definition of a map a two dimensional scale model of a part of the surface of the earth (chapter 1). This model is a systematic depiction of the earth, generally using symbols to represent certain objects and phenomena. Maps are an effective way of presenting a great deal of information about objects and spatial relationships of objects. A map depicts the terrain and achieves its objectives, it is necessary to consider a number of key map design elements. These include the frame of reference, the projection used, the features to be mapped, the level of generalisation, annotation used and symbolism employed. All these design elements constitute the basic map components. In this section, some (Plate 1) kind of commonly used maps are described in the following paragraphs.

Thematic maps (Plate 1) concentrate on the spatial variations of a single phenomenon or the relationship between phenomena like, different classes of land cover. Thematic maps portray the structure of a given distribution, that is, the character of the whole as consisting of the interrelation of the parts. Thematic maps may be used to characterise a wide variety of terrain characteristics.

Chloropleth maps are typically used to communicate the relative magnitudes of continuous variables as they occur within the boundaries of unit areas. These kinds of maps are common for census data, such as, population density choropleth map showing geology of in area is given in (Fig.14.26), as it varies by area, or average annual per capita income as it varies from area to area. In these maps different tones, colors, and shading patterns are used to convey the different values assigned to each predefined polygonal area. The areas are predefined, and thus the potential boundaries between the displayed regions are predefined in that they represent existing political or other boundaries. When constructing choropleth maps, the tones (or colors or patterns) for indicating the data values within regions must be chosen with care. Choropleth maps are extremely common geographic information system products. In a vector-based GIS, an analytic process can provide us with an attribute value for each predefined polygon, and we can then instruct the system to shade or colour the interior of each polygon appropriately. In a raster-based system, this operation is typically more complex, generally using presentation. Raster-based systems often use colour to code the different contourline values. This necessiates the use of a legend to explain the correspondence between colour and the corresponding attribute value.

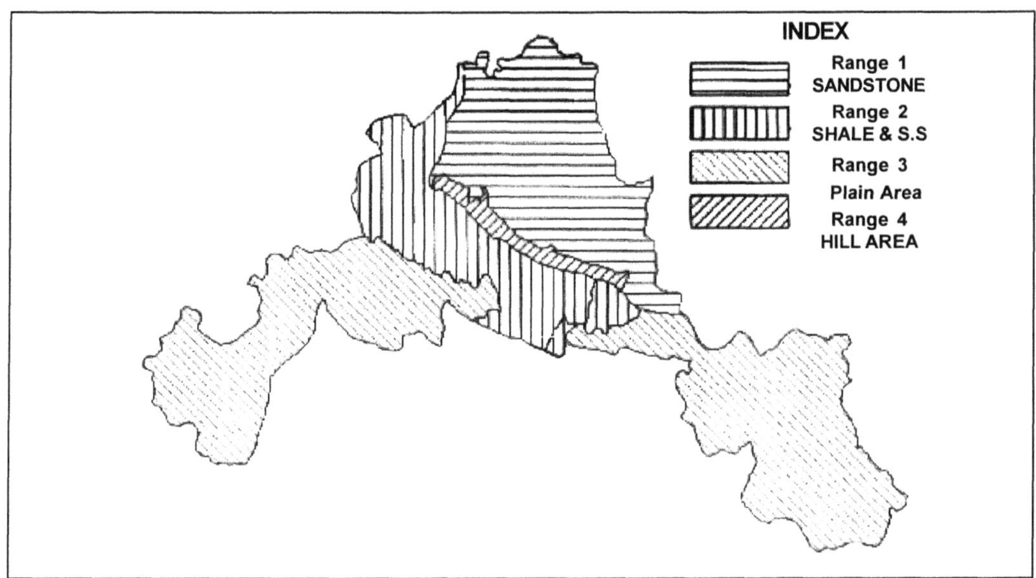

Fig. 14.26 Choropleth map showing the geology of the area.

Developing a contour map on a raster-based GIS is often done by simply creating an appropriate look-up table. The look-up table is generally a portion of the graphic display system in which the system stores the cross-reference between a cell's attribute value and its displayed representation. Contour map is verymuch useful in creating digital elevation models of the terrain.

A cartogram is another unusual form of map. It is based on the idea that the data display for a homogeneous subdivision of the earth's surface is not a function of the object's area, but is a function of the value of some other characteristic or attribute of the object. While there are computer cartographic systems which have functions for the creation of cartograms, these are relatively uncommon in geographic information systems.

Animated maps have become more common, as computer graphics techniques make them easy to create. Animation makes it possible to easily display sequences through time. These methods of data display are sometimes called movie loops. An example of the use of an animated map would be to show the growth of a city as its population and area increase through time. In this case, the animated map is an excellent summary tool, since patterns occurring through years of change may be efficiently presented to the audience in a minute or two.

14.10.2 Graphical Output

The output of GIS data manipulation and analysis subsystem may be presented in the form of a bar chart, pie chart and scatter plots. The selection of any one of these output formats is based on the application, data used for the study, availability of hardware and software, and more specifically the user's needs.

A bar chart can be used to illustrate differences in an attribute between categories. Bar charts can be organised either vertically or horizontally. In either case, the length of the bar is used to indicate the value of the attribute. Bar chart (Fig. 14.27) portrays the time-varying distribution of land use in an area. Three classes of land use, each of two years, are displayed. The heights of the bars indicate the total number of hectares covered by each of the land use categories, and the shading pattern indicates the year of the data. Note the

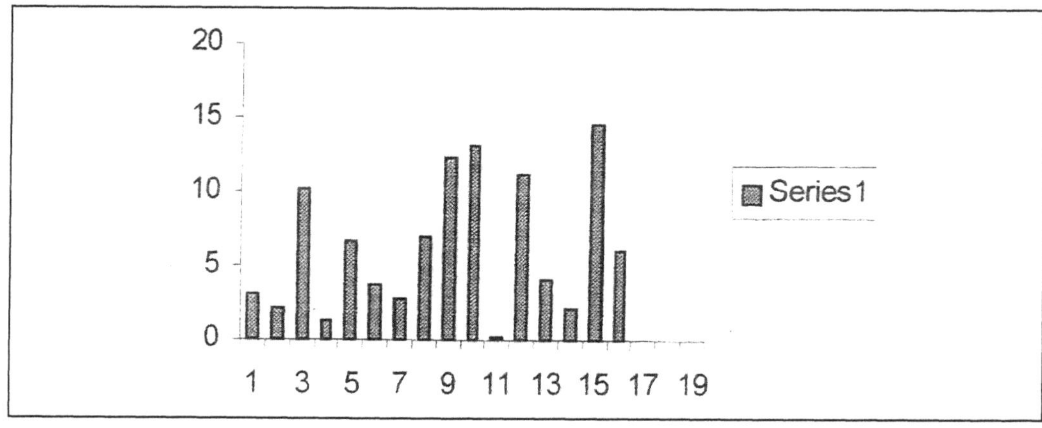

Fig. 14.27 Bar chart.

legend in the figure which explains the correspondence between shading and year. This kind of display can present changes through time in a very clear and effective manner. This graphic output might have come from two data layers in GIS, each recording the land use in a different year. To generate these bar charts, the system must be able to accommodate the total area occupied by each of the land use categories, and then pass the derived data plus the annotation to a display module.

A pie chart presents information by dividing a circle into sectors, and in this way, illustrates proportions of the whole. In Fig. (14.5) the size of the area in percentage corresponds to the amount of each kind of land use. Note that it is difficult to determine the actual amount of land use in each of the categories from a pie chart, when compared to the bar chart. Furthermore, one sometimes "explodes" one or more categories from the pie, as we have done here, to highlight a particular category. Pie charts can be very confusing when they must portray a great many categories, since the slices of the pie become small.

Scatter plots are extremely valuable for displaying the behavior of one attribute versus another. In Fig.(14.28), we show a scatter plot of reflectance values of different bands. In this case, we have asked the system to compare reflectance of band 1 and band 2. The scatter plot is a common capability of many statistics software packages; in more sophisticated systems, various regression lines may be superimposed on the data.

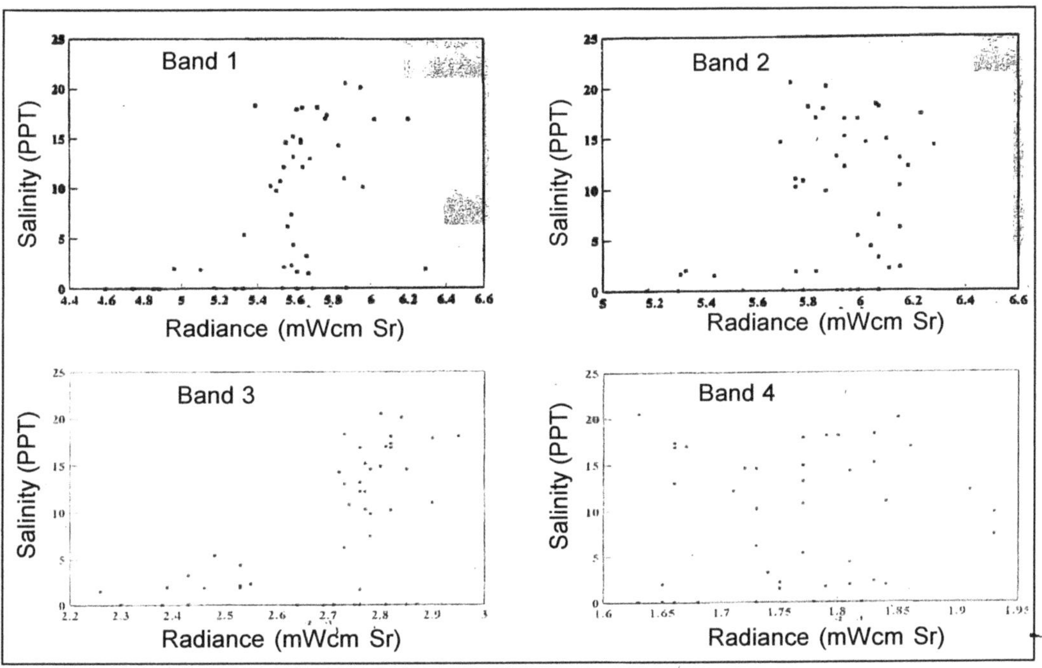

Fig. 14.28 Scatter plot of reflectances values of different IRS bands.

15 Integration of Remote Sensing and GIS

15.1 Introduction

GIS and Remote Sensing are linked both historically and functionally. In a historic context, some of the early work leading to the development of GIS revolved around methods to provide better access to aerial photographic coverage of specific areas. GIS facilitates the storage of and the access to many diverse data types. GIS, correctly employed, also permits the data held within a database to be readily updated. The synergy that exists between remote sensing and GIS technologies is built on the foundation that, for many applications, remote sensing can be employed effectively and efficiently to update GIS data layers. Such data layers in a GIS can, appropriately employed, improve the interpretability and information extraction potential of remotely sensed data. Earlier studies on the application of GIS technique for creating decision support systems and other computer based spatial information extraction have revealed that 75% to 85% of the spatial data layers have been derived from the analysis of aerial photographs and satellite image data. As a result of this, remote sensing and GIS have become very important and have been associated with each other.

This chapter focuses upon the importance of remotely sensed data as an input to GIS, and describes some of the ways in which they may be managed for operational use. Key characteristics and source of remotely sensed data commonly employed as input to GIS are described, as key minimum set of GIS facilities required to support the use and effective management of remote sensing data. In short, this chapter is designed to inform managers of the sources and potential of remotely sensed data so that they can discharge their responsibilities for monitoring or operations with maximum effectiveness but at minimum cost.

Remote sensing data can be readily merged with other sources of geo-coded information in a GIS. This permits the overlapping of several layers of information with the remotely sensed data, and the application of a virtually unlimited number of forms of data analysis. On the one hand, the data in a GIS might be used to aid in image classification; on the other hand, the land cover data generated by a classification (through VIP and DIP) might be used in subsequent queries and manipulations of the GIS database. Remotely sensed data is almost always processed and stored in raster data structures. When working simultaneously with an image processing system and raster geographic information system, it is usually easy to move data between the two. Typically, a single theme of information is extracted from the remotely sensed data.

The combination of image processing and GIS technologies is astounding. As indicated earlier, they both are coming closer and it is now becoming extremely difficult to differentiate each other. The hardware was almost similar, and now software is being integrated. The problem of transfer of data is also being circumvented. Private companies are combining their efforts in this direction.

15.2 Remote Sensing and GIS Synergy

The relationship between remote sensing and GIS has received considerable attention in literature and , indeed, remains the subject of a continuing discussion. Much of this discussion revolves around the scientific and technical issues relating to integration of the two technologies. Remotely-sensed images can be used both as a source of spatial data within GIS and to exploit the functionality of GIS in processing remotely sensed data. Despite this, the actual progress towards the goal of full integration is surprisingly slow. While this is undoubtedly due to the technical challenge of accessing, manipulating, and visualising vector, raster, and tabular data simultaneously, it seems unlikely that technical constraints have been the sole barrier to achieving full integration. It might be argued that competing imperatives in both remote sensing and GIS have tended to draw attention away from the issue of integration. The recent focus on monitoring global environmental change using coarse spatial resolution sensors and the assimilation of the data that they produce in various environmental simulation models has deflected some of the attention away from the traditional issues of large-scale mapping, which are more closely allied to the concerns and use of GIS. Similarly, one can see a number of other developments in the application of multimedia technology within GIS that have consumed much of the research and development effort in the field of GIS.

There are at least two reasons why the issue of integration is likely to receive fresh impetus in the near future. The first is the increasing availability of data from the very high spatial resolution as well as commercial satellites that are scheduled for launch over the next few years. These will produce data appropriate to many of the large-scale mapping projects in which GIS has often been used, and is likely to compete directly with the traditional aerial photographs market. The second is that these high resolution images require the development of new data processing algorithms, such as, syntactic (or structural) pattern-recognition techniques, to extract the maximum amount of information about the observed scene. There is a considerable overlap between the objectives and functionality of these techniques and those used in mainstream GIS, at least in terms of their potential for spatial analysis. This may also bring the two communities closer together.

15.3 Raster Data for GIS

GIS users routinely require timely input data in order to optimise their systems for analysis and decision making. In many instances, raster data are preferred. Typically in the past, only a single type of photography or imagery was acquired for a given application. While this is still true, users are increasingly addressing more sophisticated applications which require multiple images from different regions of the electromagnetic spectrum and/or different dates, at scales ranging from local studies to global investigations. These more sophisticated investigations are employing data with a variety of spatial, spectral, and temporal resolutions. More importantly, applications that require the integration of both raster and vector data have been evolved. Processing and analysis of these data are improved when an analyst has access to collateral information, often in vector format, for an area covered by the imagery being analysed. In addition, raster data from satellite sensors have inherent information content beyond their use as an image backdrop. Indeed, it is important to note that most operational base cartographic products produced by mapping organisations around the world are compiled using photogrammetric techniques to process remotely-sensed data. It is maps such as these that supply the base upon which GIS applications are accomplished.

While remotely sensed data are used in making maps, they are also being employed to measure a variety of environmental parameters. Surface and cloud top reflectance, albedo, soil and snow water content, fraction of photosynthetically active radiation, areas and potential yield of given crop types, and the height and density of forest stands are but a few of the measurements that can be made with the aid of remotely sensed data. Such data measured and/or mapped over a time, constitute the basis for monitoring. As we measure map, and monitor environmental features and phenomena we can then employ the data generated as input to models. Remote sensing data are being employed today in a wide variety of modeling activities. Finally, environmental planners, resources managers and public policy decision makers are employing remotely sensed data within the context of GIS to improve the management skills and efficiency.

Thus there is a wide variety of satellite-based sensors currently providing operational raster remotely sensed data for GIS developers and users. Our emphasis here is upon information on the sources of those satellite remote sensed data being operationally acquired for use within GIS today. The spatial, spectral, and temporal characteristics of these systems vary according to specific design goals and engineering trade-offs which were explained in the first part of this book. Once the remotely sensed data has been converted to a desired data type, transferring this data to a raster GIS is relatively simple. Header and trailer records or files may need to be modified during the conversion process. In our experience, most operational image processing and raster Geographic Information Systems provide mechanisms to read and write pixel raster arrays.

15.4 Vector Data for GIS

Much work is involved when transferring raster data derived from remote sensing systems to a vector-based GIS. In the following paragraphs we present only the outline of the process. One possible sequence, based on derived continuous data, such as, vegetation abundance, involves extracting the contours of abundance (often called isolines) on the image processing system, and converting these raster representations of contour lines to vectors. The vectors may then be passed to the GIS, along with labels to indicate values associated with the contour lines. When working with information which has been classified into discrete categories, like land use or rock type, appropriate transformations are made to convert the continuous multivariate brightness data into discrete categories. Then, to isolate the implicit homogeneous polygons in the derived image, the pixels that form the boundaries of the areas are detected. This is exactly like the skeletonizing algorithms used in conversion process. These boundary pixels may be used to develop the vectors surrounding the areas, and attribute values and class names are assigned to the bounded areas.

It is important to understand that the conversion processes are limited by the underlying data, in terms of precision and accuracy. A 512-by-512 array of 24-bit pixels, displayed on a high-quality colour monitor, may be very satisfying to the analyst. However, the vectors that are developed from this discrete array may look poor on a high-quality plotter, since the plotter has many more addressable positions than the image display screen. Often, the analyst will smooth the plotted lines on the final product to make a more pleasing presentation.

15.5 Need for Integration

Recently, it has been proved that Digital Terrain Models (DTMs) can be developed using GIS technology. The widespread distribution of Landsat and SPOT imagery in the case of Canadian GIS, and the availability of digital elevation models and street files in many other countries have certainly led to applications well beyond those used to justify the data's compilation. In another example, the digital database was created for the wastelands and their aerial extent using ARC/INFO GIS. The spatial data used for this database was derived from the analysis of IRS LISS satellite data. This data is then integrated with GIS along with other collateral data. With better resolution and improving software, the topographical mapping requirements are being met by GIS and remote sensing combination. Remote sensing and GIS have almost become an unavoidable source for cross-checking or updating in digital surveying. Furthermore, the GIS software can now accept the Global Positioning System (GPS) information in their program, as an additional advantage. For example, the ARC/INFO GIS software now accepts GPS data through its geolink module. Experiments have been carried out for various applications through such efforts. It is not always transitional land based information, which is generated, but unconventional themes have also been tackled. Urban and regional planning are obvious offshoots. Environmental issues related to ground cover such as forests, water bodies, and wastelands also serve as examples.

Along with the above application areas, more research work should be oriented towards the development of application oriented remote sensing software that can be easily integrated with GIS. A recent development in this field has been integration of raster based data in conventional vector based GIS software like ARC/INFO. Apart from raster/vector conversion facilities in the GIS packages, vector overlays on remote sensing or other raster data is now possible. This has, to certain extent, facilitated the remote sensing data integration with GIS. Any software should support the functions like live link, direct usage of compressed data in GIS, automatic registration of GIS and remote sensing datam, projection support, preparation of ortho-images and image maps, and solution based functions for specific application.

The future GIS and image processing softwares are likely to include at least some of the above functions. However, integration of GIS and remote sensing data are being carried out in different institutions and laboratories with different rates of manual/digital involvement. The remote sensing discipline has developed a number of different techniques for transforming general purpose data sets into thematic information for many related fields for subsequent analysis using GIS.

In other instances, discrete categories of surface cover may be distinguished through a classification algorithm. Many different decision rules may be developed to isolate or characterise components of the earth's surface. Often, a sequence of discriminating steps is required to provide an acceptably accurate or precise result. For example, simple binary decision rule may be able to exclude water bodies from consideration. Finally, an unsupervised classification can then permit us to focus on cultivated vegetation. This example is over simplified, but illustrates a general principle that is parallel to the use of interpretative keys in photo interpretation.

15.6 Facilities for Integration

A wide variety of photogrammetric, image processing and statistical analyses are utilised to extract information from raster data. The systems to accomplish such processing range from the relatively simple PC-based desktop mapping, softcopy photogrammetry, and combined image processing/GIS system to complex analytical stereoplotters, orthophotoscopes, mainframe and super computer-based image processing/GIS systems. The choice of the hardware and software to be employed in any remote sensing/GIS application depends upon a wide range of factors. These include: cost, type of application, timeliness required in data production, level of understanding/training of the staff involved; and the appropriateness, accessibility, and availability of the input data; hardware, and software. In addition to these facilities, a digitiser or/and a scanner must be available to convert the paper based data to digital base through the process of digitisation.

To accomplish the integration of remotely sensed data into vector based GIS requires the additions of relatively sophisticated image processing package to these systems. For instance, ESRI's ARC/INFO, Arc/View, and ARC/GRID software accomplish much of this integration. Correspondingly, some raster based image processing systems have considerable GIS functionality. ERDAS/Imagine, ER Mapper, EASI/PACE, and Intergraph's MGE Base Imager (MAI) version are examples of the systems that are commercially available today. An important consideration in linking any of these systems to a vector based GIS is the grid interface between the two systems and the file structures of the systems. Care should be taken by the managers involved, and advice must be sought from users, before going forward with an attempt to link raster and vector systems for a particular applications area.

15.7 General View on Applications

This section deals with the applications of GIS, remote sensing, and both in combination. In many respects applications are the most important aspects of GIS since the only real point of working with geographical information systems is to solve substantive real-world problems. GIS is perhaps best considered a methodology or collection of tools which when applied can bring great benefit. Remote sensing and GIS can contribute a great deal to our study of patterns and processes on the surface of the Earth and to create decision support systems. Even though the innovation of GIS is itself quite recent, it is possible to classify GIS applications as traditional, developing, and new. Traditional GIS application fields include military, government, education, and utilities. The developing GIS application fields of the mid 1990's include a whole raft of general business uses like, banking and financial services, transportation logistics, real estate, and market analysis. New application areas, which are probably due for takeoff in the next decade include small office/home office and personal or consumer applications. This simple classification, although useful in itself, hides a complexity of approaches to applying GIS. More specifically, a different way of examining trends in GIS application is to look at the diffusion of GIS use. Fig 15.1 shows the classic model of GIS diffusion developed by Rogers (1986).

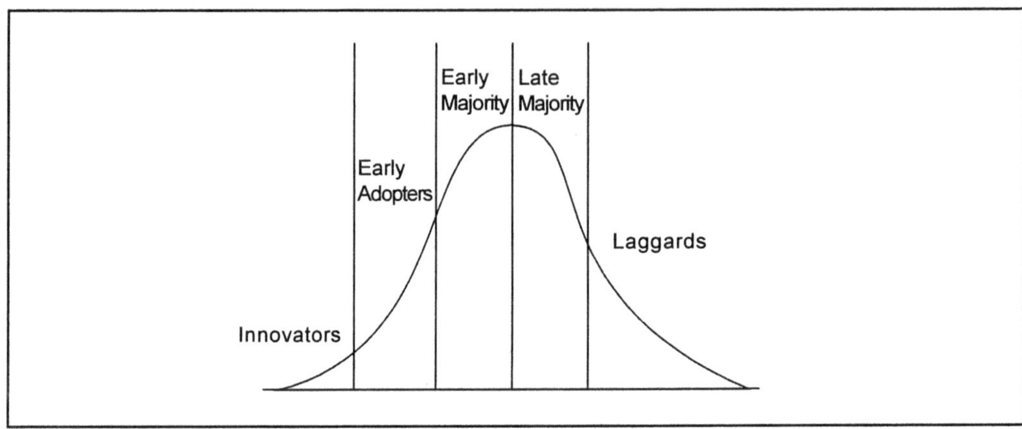

Fig. 15.1 The classic model of GIS diffusion

Grimshaw (1994) has provided different levels at which GIS can be used within organisations. The levels are operational, tactical, and strategic as shown in Fig. 15.2.

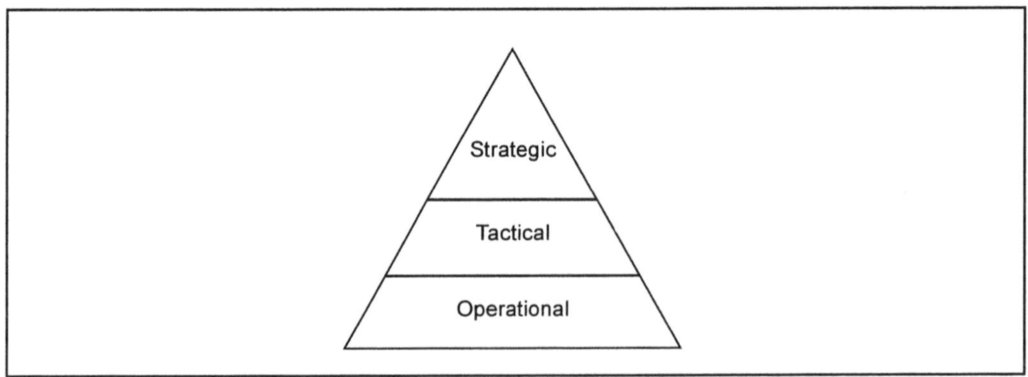

Fig. 15.2 Different levels at which GIS can be used within organisations

Operational activities are the basic day-to-day activities of many organisations. The tactical activities are typically the domain of middle managers and strategic activities involve senior management.

There are many books which are themselves devoted to the description of remote sensing and GIS applications individually. While there are now plenty of examples of routine and oppurtunistic GIS and remote sensing applications, truly pioneering applications are rare. A list of various application areas are given in Table 15.1 which provides the broad view of applications. A very upcoming application area "Urban and regional planning with a case study of Hyderabad city", is provided in chapter 16.

15.8 Software Scenario

***Remote Sensing Software* :**

ENVI : The Environment for Visualizing Images (ENVI) provides the most complete set of image processing tools and functions available for our land cover analysis using Landsat TM and SPOT data. It runs exactly the same way on windows, Unix and Macintosh systems. Its orthorectification features help correct aircraft or satellite position, topography and other camera effects.

The functionalities are as follows :

Input formats : ARC/Info Images, Arcview Shape, AutoCAD DXF, ERDAS 7.5 & IMAGINE 8X,GEOTIFF, MapInfo, Microstation.DGN, PCI (. pix) Files RADARSAT, SPOT (1A, 1B, 2A,CAP);

Preprocessing & Calibration : Preprocessing of AVHRR and Calibration, Calculate SeaWIFS Image Geometry, MSS & TM-Specific Preprocessing, IAR Reflectance Calibration, Landsat, Flat Field Calibration;

Registration & Rectification : AVHRR Georeferencing, Image to Map Registration, Image to Image Registration, Interactive Ground-control Point Collection, Resampling methods, SeaWIFS Automatic Georeferencing;

Ortho-rectification : Orhto-rectification of Air photos, Interactively Select GCPs, Import GCPs from File, SPOT (1Aor2B);

Transforms : Band Ratios, Decorrelation Stretch, HSV to RGB/RGB to HSV, HSL to RGB/RGB to HSL, IHS Sharpening (data fusion), Minimum Noise Fraction (MNF),NDVI, Saturation stretch (automatic);

3D Surface View : Interactive 3D surface modeling, Drape Image Over Surface,Output to VRML 2.0,Set vertical Exaggeration, Smooth Image;

Filters : Adaptive, Convolution, Interactive Fourier Filtering, Morphology, Texture, User-Defined Filter Kernels;

Table 15.1 Application Areas of Remote Sensing and GIS

Agriculture and Soils
- Crop acreage estimation / condition / yield prediction
- Soil categorization/resources/erosion assessment/watershed management

Geology
- lithological discrimination / structural mapping /drainage analysis
- Mineral exploration / coal fire mapping
- Oil field detection / explorations / geo chemical analysis / seismic studies-analysis

Ocean resources
- Wealth of oceans / explorations / productivity / SST / potential fishing zone, coral reef mapping
- low tide / high tide marking

Forestry and Environment
- Loss of biological diversity / biosphere reserves / ecologically hot spot areas / wet land environment
- Forest cover mapping & surveillance / forest stock mapping / deforestation-afforestation grassland mapping.

Major river valley projects

Coral reef mapping

Disaster warning mitigation

Climatic change and greenhouse gases-atmospheric aerosols / atmospheric winds

Wild life habitat assessment

zoom cultivation-desertification.

Street network-based applications
- Vehicle routing and scheduling
- Location analysis-site selection- evacuation plans

Natural resource-based applications
- Management of wild and scenic rivers, recreation resources, flood plains, wet lands, agricultural lands, aquifers, forests, wild life etc,.
- Environmental Impact Analysis (EIA)
- Viewshed analysis
- Hazardous or toxic facility siting
- Ground water modelling and contamination tracking
- Wild life habitat analysis, migration routes planning

Land parcel- based applications
- Zoning, sub division plan review
- Land acquisition
- Environmental Impact statements
- Water quality management
- Maintenance of ownership

Facilities management
- Locating underground pipes, cables
- Balancing loads in electrical net works
- Planning facility maintenance
- Tracking energy use.

Mathematics & Statistics : Average Min, Max, Standard Deviation, Band Math, Trigonometric Functions;

Classification : Accuracy Assessment, Binary Encoding, Image Threshold to ROI, Raster to Vector Conversion, Supervised Classification, Unsupervised Classification, ROI Statistics; Interactive Contrast Enhancement, that – is, Interactive Histograms, Lookup Tables (LUTs); Vector Processing, Mapping Functions, Spectral Analysis Tools, and Radar Functionality.

Operating Systems : Digital UNIX (OSF/I), Linux (Slackware Pro), Windows 95/98,Windows NTTM (Alpha, Intel); Interface & Operation, that is, 8-and24-Bit Color, Geo-Browser Image Selection, GUI (Graphical User Interface), Interactive Multiple Displays, Interactive 2D Scatterploting, Virtual Mosaic Capability, Vector Overlays & GIS Capabilities; Batch Operation Output Formats, that is, ARC/INFO Images, BMP, DXF, PCI, TIFF; Tape Utilities, that is, Generic BSQ, BIL, BIP, Landsat MSS & TM, Support for Windows NT, Windows 95,UNIX, SPOT, AVIRIS, RADARSAT; Documentation, that is, Complete Printed Documentation, Programmer's Guide, Tutorials & Sample Data, User's Guide.

EASI/PACE

It is a remote sensing software developed by PCI Geomatics, Ontario, Canada.It is designed for Remote Sensing, Image Processing, Data Visualization and GIS Support.The functional components are Image Works; Xpace; GCP works, EASI; PCI Modeler and ACE.

Image Works : Image classification and image enhancements.

GCP works : Georeferencing and Image Registration, Projections and Mosaicing,

PCI Modeler : Analysis, Data inter change, Image correction, Image processing, Radar analysis PACE Packages.

ACE : Advanced Cartographic Environment is for Cartographic Functions.

PCI AUTHOR : It is a custom development environment that enables users to create graphical interfaces for EASI scripts. It can be used quickly and easily to develop applications.

ERDAS

ERDAS (Earth Resources Data Analysis System) is the mapping software company specializing in Geographic Imaging Solutions since 1978. The company is world leader in highly customizable, easy–to-learn and-use Geographic Imaging software. The software is precisely for Visualization. The essentials of this software are vector module, virtual GIS, NIFT, radar module.

The functional components are :

IMAGINE ESSENTIALS – for image enhancements, geo-correction, visualization and mapping.

Work with a variety of vector data sources, including ARC coverage and Shape files.

Index, view, manage and archive image with a map-based graphical tool.

IMAGINE ADVANTAGE- for orthocorrection, mosaicing, image processing, and GIS analysis.

Orthorectify SPOT, TM, MSS and aerial Photography using advanced sensor modeling techniques

Use an existing DEM or create one from contours, GPS surveys or SPOT heights with the surfacing tool.

IMAGINE PROFESSIONAL – for radar analysis, advanced classification and graphical spatial modeling.

Supervised, unsupervised and hybrid land cover classification

Fuzzy classification

"Iterative" classification processing, including cluster-busting capabilities

Automatic seed/ region- growing tools for signature extraction

Sophisticated feature –space signature editing, extraction, and evaluation.

Accuracy assessment.

GIS SOFTWARE

Arc info

Arc info was developed by Environmental Systems Research Institute (ESRI), Redlands, California, USA.

Arc Info data structure

Arc Info is a vector- based GIS package, capable of handling both spatial and non-spatial data. It organizes geographical data using vector topological models and non-spatial data using relational models in a DBMS. The arc node and polygon topology are organized to identify points, lines, and polygon relations. The cartographic data are then linked to the attribute data through a link item.

Arc Info functionalities : Arc Info has a wide range of functions which have been developed based on a tool-box concept –where each function can be visualized as a tool and having a specific utility. The major modules of Arc Info functionalities are :

(a) **ADS and ARCEDIT :** data base creation in Arc Info is possible through the process of digitisation using the Arc Digitising System(ADS) and the ARCEDIT module. ARCEDIT is a powerful editing utility having capabilities for feature –based editing. These modules include the functions for coordinate entry using different devices – digitisers, screen cursors and so on;

(b) **INFO :** INFO is the manager for tabular data associated with geographic features in map coverage of Arc Info. INFO provides facilities for data definition of data files, use of existing data files, data entry and update and, sorting and querying;

(c) **Analysis Modules :** Arc Info offers spatial overlay capabilities based on topological overlay concepts. Union/ intersect overlays, buffer generation, proximity analysis, feature aggregation , feature extraction, transformation, nearness functions and other integration utilities are available;

(d) **ARCPOLT :** This module has capabilities for generating cartographic quality outputs from the database. This includes utilities for interactive map composition, editing map compositions, functionality, the incorporation of coverage features to the required scale, generalisation, symbolisation, transformation, and so on. Placement of non-coverage features, include legends, free text, and logos, graphic elements.

(e) **TIN :** The TIN module of Arc Info can be used to create, store, manage, and perform analysis pertaining to the third dimension data. The modeling capabilities include calculation of slope, aspect, isolines or contouring range estimation, perspectives, and volumes. Additional functions for determining spatial visibility zones and line of sight are also provided;

(f) **NETWORK :** The NETWORK module of Arc Info performs two general categories of functions : net work analysis and address geocoding. Network analysis is possible for optimal path determination and resource allocation analysis. The geocoding module allows for associating addresses to line networks and determining the spatial framework of addresses in an application;

(g) **COGO :** It is the coordinate geometry module of Arc Info; supports the functions performed by land surveyors and civil engineers for the design and layout of sub-divisions, roads and related facilities, as well as the special plotting requirements. COGO allows definition, adjustment and close traverse including adding curves on a traverse; it computes area, bearing and azimuths;

(h) **GRID** is a raster–based module of Arc Info. GRID has an interface to Arc Info, so coverage can be converted to GRID and from GRID to Arc Info. GRID supports powerful modeling tools of raster integration, potential mapping, spread/grow operations and so on.

Arc Info also supports ERDAS system, DEM data, Autocad –DXF format, IGES format and a flat file format. ARCVIEW module is a desktop mapping package oriented towards viewing and querying Arc Info databases.

Arc Info Platforms : Arc Info is available on a wide range of platforms : PCs and workstations on DOS ,UNIX and NT platforms.

PAMAP GIS

The PAMAP GIS is a product of PAMAP Graphics Limited, Canada, integrated group of software designed for open system environment . This package is modular and designed to address the wide range of mapping and analysis requirements of the natural resource sector.

PAMAP data structure

PAMAP adopts an integrated raster as well as vector representation of the spatial elements. It uses vectors for data capture and storage and rasters for analysis purposes.

PAMAP functionalities : PAMAP GIS has seven major modules;

GIS MAPPER : It is the basic module for data entry to create a data base of maps by generation and editing of the vector database, which forms the base for subsequent raster –based analysis. It includes Planner, for quick interactive report generator. GIS MAPPER also supports pen plotter output to several plotters;

ANALYSER : This module allows the user to perform data conversion for polygonisation: raster creation from the vector boundaries of the polygonal areas , overlay operations for two or more polygonal overlays to generate another level of output, proximity analysis, and corridor analysis around specified map features

TOPOGRAPHER : This is for processing of three dimensional data and Dem. Different products like slope, aspect, perspective views, and volume calculations can be derived from this module;

INTERPRETER : This is for importing remotely sensed images from digital image analysis system in to the PAMAP GIS as surface covers.

MODELLER : This module integrates multiple –surface rasters or multiple data base attributes to make planning decisions quickly and accurately. It has three main functions – combination of modeling, regression analysis and correlation, and covariance analysis;

NETWORKER : This module is used to create, analyse, and manage networks.

FILE TRANSLATOR : This is for importing and processing map files created in various data formats like IGDS, SIF (Intergraph), DLG and DXF (Autocad).

PAMAP platforms : This is available on variety of platforms —on pentium 486 PCs; UNIX workstations and VAX systems and also on MS Windows with multitasking capability.

SPANS

Spatial Analysis System (SPANS) is a GIS package developed by TYDAC Technologies, Canada has a powerful modeling function for applications.

SPANS data Structure : It adopts a mixed vector tessellation approach to the GIS and has developed the region –quadtree data structure. It has the ability to read and process vector and raster formats used by other GISs.

***SPANS functionalities* :** It has the following modules :

CORE GIS MODULE : Includes digitisation, editing , raster and vector to quadtree conversion and vice versa, projection coordinates to latitude/longitude transformation and vice versa. Polygon analysis, logical overlays, matrix overlay, indexing/weighted overlay and spatial modeling. It has corridor analysis, point query, nearest point and point to area conversion

TYDIG : It is a digitising system with extensive editing features for transfer of map data to digital format.

CONTOURING/DEM MODULE : This is a module for converting georeferenced point observation into an interpolated plane, to generate contours, slope aspect and angle of incidence maps.

POTENTIAL MAPPING MODULE(POTMAP) : It is a point interpretation programme. Features include user specified distance, decay function, weighing schemes and various interpolation options such as moving average, slope aspects, density, and statistical analysis

RASTER INTERFACE MODULE : This is used to import and export a wide range of raster–based GIS/ imagery data sources. And the imported data can be converted to SPANS quadtree structure or standard raster format.

SPANS PLATFORMS : This is available on PCs and workstations on DOS, MS-Windows, NT and UNIX paltforms

GENAMAP

The Genamap package is marketed by Genasys, an international developer.

Genamap Structure : Genamap handles spatial data in both vector and raster form.

Genamap functionalities : Complementary modules to the GIS provide for “cell” modeling, TIN modeling, enhanced analysis, net working, links to an external RDBMS in a true client-server environment, image visualization, document storage and retrieval, and many more extension services.

Genasys provides a module, Genavive; has a wide range of functions for handling and viewing pre-classified remotely sensed images or scanned maps, ortho-photos and so on.

Genius, the graphical user interface to all products, takes full advantage of the client server architecture.

Genamap Platforms : Available on a wide range of PC and workstation platforms.

INTERGRAPH MGE

Intergraph provides MGE (Modular GIS Environment) as a solution for mapping/ GIS applications for infrastructure, environmental and natural resources management, and digital cartography.

MGE data structure : MGE adopts a vector –cum-raster data structure for spatial data handling and includes object oriented and relational saptial analysis and a unique raster analysis capability which links raster files to a RDBMS.

MGE functionalities : Application modules which are based on the MGE platform include:

MGE geo data manager, which is a feature level data management environment for a seamless map base;

MGE parcel manager is client application for maintenance of cadastral parcel fabric

MGE projection Manager, is a complete suite of geometric transformation tools to support common map projections/ coordinate systems

MGE analyst, for spatial analysis and display of topologically combined and integrated geographic data

MGE dynamic analysis, providing dynamic query interface for object based analysis of MGE data sets.

MGE grid analyst, for manipulating, displaying, and analyzing multiple layers of grid data.

MGE terrain modeler, for surface analysis of contours and drainage data using triangle, or grid based terrain models

MGE network modeler for modeling and management of utility networks in a GIS environment

MGE voxel analyst, a general purpose visualization tool for dynamic modelling of geotechnical data in the context of their geographic location.

MGE PLATFORMS : it runs on PC, Windows NT, Macintosh and UNIX Platforms.

ISROGIS

Developed by Indian Space Research Organization.

ISRO data structure : it adopts PM quadtree data structure, which is edge based structure that decomposes the vectors in map into quads, and is then organized using the vector structure.

ISROGIS functionalities : it has CREATE (for creation of maps and themes), EDIT (systematic editing of spatial features), MAKE (to provide symbolisation, annotations), ANALYSE (overlay analysis), QUERY (for obtaining information related to spatial & attribute data), LAYOUT (cartographic work), 3-D MODULE (for handling Z-axis of spatial data).

ISROGIS platforms : It is available on PC platforms on MS- Windows and UNIX and on a wide range of workstations.

Some other GIS packages

Idrisi GIS: has been developed by Clark University, USA. It has advanced features like import /export capabilities, a new digitizing module, and image processing tools. It runs on PC's. Manipulation and import of vector data is supported.

GRASS : A public domain UNIX package. It has several image processing tools and good support for spatial statistics and analysis. It has strong support for raster/vector integration. GRASS is mainly used for hydrologic / watershed modeling applications.

MapInfo : has very good spatial data handling capabilities. It supports dBase and also has its own data manager.

Small world : The functionalities include tools for network tracing, polygon operations, query system, object editor, topology editor, geometric constructions, graphical styles, real-world data modeler, report generator, and on -screen digitisation.

16

Urban and Muncipal Applications

16.1 Introduction

The adiministrators of city or a town adopt two approaches to evaluate the urban places and the impact of unrbanisation. The first approach undertakes the study of size, function, growth rate, and tributary area of cities in the general fabric of settlements. The second approach deals with the internal structure of cities and the related factors, which control the layout and buildings, the character and intensity of land use, the movement of persons and goods between various functional areas. The land use planning is a part of larger process of city planning. It is basically concerned with location, intensity, and amount of land development required from various space using functions of city life, such as, industry, wholesaling business, housing, recreation, education, and religious and cultural activities of the people. The study of urban land use and its planning is of considerable significance in the overall planning of urban places like Hyderabad city.

Decline and decay of our cities calls for all policy makers, administrators and politicians to stop destructive forces of the urban development and to forge alliances among all local and national forces which are concerned with the social, political, economic, ecological, physical and cultural development of our cities and towns. Despite the claims that developments in communications imply that the cities are going to lose their function and will change dramatically, it cannot be predicted that the new paradigm of the 'information city' will only represent another device of those who do not have access to information technology.

Today nearly half of the world's population lives in cities. In developing countries people are deserting rural areas while population is rising rapidly. In less than 20 years from now, these two factors will combine to drive over two billion people into urban areas, which in some cases are already overcrowded. Most urban growth falls outside formal planning controls, thus increasing economic and social pressures and health and hygiene problems. "Urban growth and rural depopulation are going to be the main issues of this millennium and their impact will be felt in many areas, like water resource management. To resolve this problem, we must identify and distinguish zones that can or cannot prevent urban densification.

To keep track of urban growth, objective data are necessary on population influx to assess the impact of growth and to plan development actions. It is also necessary to analyse urban encroachment on arable land, so that we can plan and upgrade infrastructures, to update land use and land cover maps, and much more. The rate of urban growth in the modern era is making it increasingly difficult to keep track of populations, and traditional survey and census methodologies are proving inadequate,. Not only are censuses time-consuming and costly, but the sheer pace of growth in the world's cities means that information collected is virtually obsolete even before it reaches our desks. As censuses are generally carried out every ten years, they are of limited use for monitoring populations.

It is in this context that urban planners require nearly continuous acquisition of data to formulate government policies and programs. These policies and programs might range from the social, economic and cultural domain to the context of environmental and natural resource planning. The role of urban planners and municipal administrators is becoming increasingly more complex. Consequently there is an increased need to have timely, accurate, and cost effective sources of data in different forms. This can be well served by the new and upcoming technology, namely, Remote Sensing and GIS.

16.2 The Role of Satellite Imagery and other Data Sets

The need for recent data to conduct detailed spatial analysis means that urban planning departments are increasingly turning to satellite for a broad picture of urban areas untrammeled by boundaries on the ground. Satellite image resolution is currently good enough to produce maps at scales of 1;1,00,000,1:

50,000, 1:25,000, and 1:10,000 in certain cases. Satellite imagery is a vital source of global, frequently update and reliable information for producing and updating maps. Satellite data are easily integrated in Geographic Information System (GIS) and are an ideal tool for change detection in urban areas. The utility of satellite imagery is further enhanced when combined with other data. For example. Photo-interpreters can analyse high resolution satellite imagery to identify homogeneous habitat zones, and then survey them using aerial photography to refine their analysis in conjunction with field surveys.

Multidate satellite imagery is also well suited for detecting change in land use and land cover at intervals of 5, 10 and 15 years. Keeping track of change over time is vital to provide accurate input for planning guidelines, and here again the IRS, SPOT, IKONOS satellite systems, have much to offer. For example, in Africa, on average 30% of the territory is covered by zones like towns and farmlands generating economic value. Our global archive contains over seven million scenes. That means, we can provide images of Johannesburg acquired weekly for over ten years. If the archive doesn't meet requirements, we program the satellites to acquire fresh imagery.

Local government agencies need accurate and regularly update geographic information for immediate use. It must be preinterpreted and be ready for retrospective analysis and landuse change detection. A number of local authorities have already adopted the remote sensing data products to help planning departments make the right decisions concerning large areas experiencing rapid change, because it lets them compare observed patterns of change at different dates. Thus high resolution satellite data provides a clear picture of urban and agricultural land use. With very high resolution (VHR) imagery, we will be able to move inwards from the semi-urban fringes and zoom in on urban centres, where data will have to be refreshed very frequently.

16.3 The Indicator Function of Urban Land Uses

There is a casually quoted statistic that roughly half of all GIS implementations fail. Most failures are related to institutional issues, resistance to change, lack of political support, insufficient funding, and the fact that GIS innovation results in a radical change in information flow within an organization. The spatial information component of a project needs to be an integral part of overall project design from the beginning. Insertion of a GIS can be well executed and products generated will be effectively used. The user needs assessment, training, data collection, pilot project, and a complete project success. An awareness of spatial data products and analysis capabilities typically needs to be cultivated in end-users early-on, in order for these products to be used to their fullest.

The user needs assessment is a vital component of GIS implementation within a municipality. Thoroughly exploring potential data sources, integrating the GIS with more traditional information management within the municipality, and promoting an understanding of spatial information and analysis capabilities early-on are crucial to project success. It is also important to explore whether there have been previous GIS initiatives within a city, and whether there are any project remnants (equipment, trained staff, or digital data sets) which can be utilised.

Analysis of the land use pattern with regard to the major functional areas of land use in a city, and how they change over the years, is the basis of urban ecology. Urban ecology is a term, that the sociologist has adopted from the biological sciences to describe the physical change process in the city. It concerns the physical, spatial, and material aspects of urban life. It is also connected with the interrelation of living things and their environment. These ecological factors of land use thus can be explained in terms of "ecological processes" with their physical context and "organisational process" with their social structure context. In other words, this approach considers the existing land use pattern as an outcome of the interacting determinants. The main idea is that in analysing the pattern by considering related contexts, the determinants and their way of action can be identified. It is then possible to view the changes through the time and the evolution of context. In a further development some land uses or combination of a set of land uses could be recognised as indicators for analysing the land use pattern.

This indicator function has considerable appeal in land use analysis and planning procedure to cope with the scale and rationale of the ever-expanding horizons of urban planning. It also produces vast and important factual references which are necessary to these professionals interested in physical aspects of current urbanisation.

The sets of indicators could outline the formation of hierarchical land use pattern reflected by social and physical context. Also associated with the social process, land use indicators could breakthrough the identification of distinct cultural areas, islands of ethnic groups, areas of high religious practice, and so on. The extent and rapidity of the sorting could cover all aspects of land use pattern as the function of all attitudes, decisions, and actions at stake. They frequently involve such matters as deed restrictions, zoning, tightness of the housing market, cultural ties, location of place of work, school environment, and so on.

Some sets of land use could replace the insufficiency of economic or social indicators relevant to land use analysis. For example, where the detailed statistical data about the distribution of income groups in urban areas are not available, the gradient concentration of pastry and flower shops could indicate the average income of surrounding residents. One of the major achievements in the preparation of such an effort could be the producing of the most appropriate standards in an emerging field of study, among the vast wealth of material available. In order to maintain a manageable form it is therefore, necessary to experiment and select the facts that would represent a real criterion rather than to attempt the inclusion of imagined information.

16.4 Appropriate Methodologies

In general, there is a lack of accurate, current data for urban areas in developing countries. Projects often rely on catch-as-catch-can basis, where the best available data of reasonable scale from a variety of sources is integrated. This is essentially a make-do approach. Satellite imagery is a significant alternative source of data for development and maintenance of data layers for a municipal GIS. Digital satellite data is often of overwhelming size and format for installations.

The costs are also often prohibitive. Manual interpretation from hardcopy images is a valuable alternative, which requires as little image processing as possible. This technique is useful for developing land use classifications, for interpreting additional data sets, and for the development of point-based property information systems. High resolution sources exist, such as SPOT 10 and 20 meter resolution data and Russian KVR 2-3 meter resolution photography, IRS PAN 5.8 m resolution, IKONOS 1 m resolution and so on. These are valuable data sources for many areas. Cloud cover remains a problem for space-borne sensors in the tropics. Beyond frequent revisits and dry-season collection, there is no real solution. The cost of both digital and hardcopy high resolution satellite data products is still prohibitive for 86 most developing nation users. Simplified and relatively low-tech approaches to data generation, automation and consolidation are often effective in developing country environments. Three upcoming technologies which are effectively used for urban and muncipal planning and providing alternative urban landuse practices, are discussed in the following sections.

16.4.1 Rapid Land-Use Assessment

A streamlined approach to land use classification called the Rapid Land use Assessment (RLA) methodology assesses and quantifies land use through a combination of satellite imagery, local field knowledge/expertise and resources, and simple GIS techniques. High resolution data is valuable for identifying property locations. This can be explained in the following case.

In Honduras the RLA methodology allowed the municipality of San Pedro Sula to determine and quantify by area 18 categories of land use for a 900 square kilometer area in three weeks using SPOT 10 meter satellite data and local staff. Image maps of the satellite data were processed in the US and taken to Honduras to be used as black and white ortho maps for interpretation. Utilising simple aerial photographic techniques, field surveys, and local knowledge of the area, the land use database was constructed. Once the data had been interpreted it was then transferred digitally by trained local municipal staff into a simple, PC-based GIS. This resulted in an accurate and up-to-date land use classification for the municipality. Road networks and surface hydrology data layers can serve as a base for development of a rich database for environmental and land use planning.

16.4.2 Rapid Land Information System Development

A streamlined approach is developed for Land Information System to LIS that incorporates key data into a land registration structure, subsequently transferring them into a fully automated information system. Under this approach, a land parcel is not stored as a polygon or area in the LIS and hence is not used as the base framework for the related database. Instead, a single point feature representing each property is used as the identifier and geographic locator and are usually termed as, "lots by dots." This is the critical difference from the polygon-based approach which attempts to reconcile geometry and compile all land parcels together into a contiguous map of polygons. This is practically impossible to achieve even in the United States since individual surveys of land parcels or 'lots' are often, not entirely accurate and do not actually reconcile with one another.

This point-based property database can be developed from hardcopy very high resolution satellite imagery or aerial photographs, or through ground-based collection using Global Positioning System (GPS) techniques. As the field person collects the geographic (point) position of each land parcel the property identifier number as well as physical characteristics of the land, such as, land cover, soil condition, and number of structures can also be logged. This effectively allows an LIS database to be built in the field during survey. Additional complementary data can also be integrated into the LIS, such as, scanned property documents.

The benefits of this point-based LIS versus a polygon-based system include reduced time for input and reduced processing overhead as the point databases are smallest and faster to manage, analyse, update and use than polygon databases. All relevant cadastral information is stored within the database just as it would be with a polygon-based system, no resources are expanded on resolving land geometry problems during database and system development.

16.4.3 GIS as an Emerging tool

GIS based method involves, basically, two approaches in urban planning procedure. The first method consists of the application of GIS in preparation of urban planning projects. In recent years CAD and more recently GIS software have been introduced by urban planning agencies. The initial utilisation of such software is concentrated in mapping procedures. In particular, the GIS software is used to prepare the thematic maps of urban related projects. The land use map and some related maps such as types and states of structures or the land-use intensity site ratings, road network maps, land suitability maps and dynamic changes of urban patterns have been prepared by the using of GIS software products.

Where previously there was a lack of appropriate information, now the planning staff had another problem; "How to deal with so much of information ?". Accumulation of data within the GIS for analytical processing for land use planning. Also in this field the GIS computational capabilities have provided oppurtunities for modelling the prototypes of urban land use planning procedures, and even possibilities of observing the results through virtual simulations. These oppurtunities have advanced the exploration of new ideas and the elaboration of new applicable theories.

The second and the most major proceeding involves the land use planning field. Some theoretical studies concerning the process of more appropriate methods for land use planning are in course of materialisation. A notable field in these studies involves land use determinants. For example, about 2000 urban land uses have been identified during recent statistical survey of establishments in Iranian cities. In a parallel study using the field notes of land use surveys of seventeen Iranian cities the distribution of these 2000 land uses in the cities are studied and considered with city size and rank, city's regional function, population densities, and so on. Also these land uses have been sorted in different ways, applying the criteria relevant to land use planning. These findings showed four major types of land use patterns for cities. In this study each land use is considered in terms of its urban activity. The second is contrary to the first, that is the outlying drift of certain land uses. The third and most important criterion included two main factors : (i) land value effective in attracting or repelling a land use pattern, and (ii) specific regulation's factor which surmounts the rent or cost factor. The fourth criterion is the function area of each land use.

It is also included that, in Iranian urban agglomerations a grocery store indicates the existence of a neighbourhood unit, and its function area indicates the boundaries of that neighbourhood unit. The periodical investigations had proved that when there were two grocery stores in one neighbourhood unit, the second was obliged to close due to lack of trade. On the other hand, in a neighbourhood unit where even due to specific regulations the establishment of stores is forbidden, investigations show that an unofficial grocery store could exist. Subsequent studies confirm this dynamic relation to the functional area. At the same time the space requirements and localisation factors could be entrenched in relation mainly to the population density of the area, and some other relevant criteria. So a map of grocery store distribution could show the formation of neighbourhood units in an urban area.

16.5 An Analysis System

The municipal applications of Remote Sensing and GIS, involves determining of homogeneous zones, trend analysis of land use, integration of new parcels, area changes or change detection, inventory of municipal real estate appraisals, parcel mapping and cartographic adjustments, production of cadastral maps and advertisement zones. In very general terms, this urban planning involves the redistribution of urban land uses due to social, economic and environmental needs. The analysis system's methodology for this model of urban/municipal planning process consists of two principal phases. The first phase essentially includes field surveying and data gathering. In this phase the major problem lies in the working procedure of data gathering. This working procedure is not able to recognise the determinant development factors and give a clear image of the city's present state, while it is useful to give a clear image of the city's present stage and describe broad and general tendencies at work in the patterning of urban land uses. In the developed countries, specialised institutions produce the necessary information and its related data in a systematic way. These can be added directly to a land use planning procedure to understand the present state and its determinant factors in physical formation of the cities. Generally, this does not hold true in developing countries. Consequently the identifying of affecting factors is usually estimated by some samplings, and so, as a result, the analysis of urban land use structure is imperfect and in many respects it is an oversimplification. So the result of this procedure, especially in physical aspects has no major relevance to the land use planning process.

The second phase is data processing. In this phase, space requirements and appropriate standards are needed to be set down for an appropriate plan. Appropriate standards can only be provided where locally prevailing culture/nature relationships are thoroughly studied and have been adopted by a rational decision making procedure. Once again, this is not so in developing countries. Alternatively, the utilisation of the standards of developed countries are preferred, which again are not necessarily appropriate. Further uncertainties follow this procedure of land use conceptualisation. One of the most important factors is the uncertainties concerning the testing out of various locations for size for arriving at a procedure of land use conceptualisation and land utilisation. This must be done in assent to analytical procedure throughout the whole sequence of study, but the procedure described does not provide an appropriate base for a rational conclusion. The importance of this inappropriate state is stressed considering the statutory character of these land use plans.

Based on the land use planning for a new development plan, functional area of existing patterns and potency of localisation land uses are sorted (Mohammed Barar, 2000). The results of this sorting are classified in dynamic, semi-dynamic and passive land use patterns.

16.5.1 Dynamic Urban Land-Use

Dynamic urban land use locates itself in an urban context according to a direct relation to its area of function and potency of its functionality. The spatial localisation of these land uses are usually the central areas of their functionality boundaries. The periodical investigation of dynamic land use locations has proved that any change in their zone of function or their potency of functionality has contiguously affected their localisation in urban context. In a gradient manner due to the extent of functionality area, mainly the commercial, private office building and some governmental or public land uses could be categorised as dynamic.

16.5.2 Semi-Dynamic Land-Use

The semi-dynamic land use locates itself in an urban context with an acceptable relation to its zone of utility. The potencies of functionality of semi-dynamic land uses are regularly determined due to governmental or some public interventions. The spatial localisations are usually in a range of their zone of utility but are far from the central areas of dynamic land use area of functionalities. Public schools, general hospitals and other similar public or private services are usually categorised as semi-dynamic.

16.5.3 Passive Land-Use

Passive land uses have a poor potency of localisation due to their area of function. These usually are the quasi-public utilities which are not supported by any specific regulations or due to restrictions, the private sector is not interested. Cultural, entertainment or recreational centres are among the main passive land uses.

In urban land use planning procedure this sorting gives the planner the criteria to intervene to support the appropriate land use redistribution. Its usefulness rests on its approach in giving the planner, the basic pattern of information required to aid in solving the many varied and complex problems of our cities. Also, the material which is comprehensive in scope, will be highly valuable to other disciplines related to urban planning.

Dynamic, semi-dynamic, and passive land uses are viewed as a group. These three modes of processes offer a means of understanding the ongoing aspects of the patterning of the city. A land use table is produced by interactively contemplating these modes in relation to land use functional categories, and jurisdictional considerations as well. This table which mainly deals with space requirement, also indicates where and in which parts of categorised land uses special consideration or planning type interventions are needed.

The land use table referred to earlier is also extended for hierarchical urban divisions. For each division a specific land use table is assembled containing the land uses with a similar area of function. These tables indicate how to deal with every category of land use by divisions (classified functional areas). Describing the present situation of land use distribution in each stage of urban hierarchical division is also possible by using these tables.

16.6 Land use/Land cover System in India

The land use/land cover system adopted by almost all concerned organisations and scientists, engineers and remote sensing community who are involved in mapping of earth surface features, is a system derived from the United States Geological Survey (USGS) land use/land cover classification system. This system was designed on the basis of the following criteria (Lillesand and Kiefer 1999) : (i) the minimum level of interpretation accuracy using remotely sensed data should be at least 85 percent, (ii) the accuracy of interpretation for the several categories should be about equal, (iii) repeatable results should be obtainable from one interpreter to another and from one time of sensing to another, (iv) the classification system should be applicable over extensive areas, (v) the categorization should permit land use to be inferred from the land cover types, (vi) the classification system should be suitable for use with remote sensor data obtained at different times of the year, (vii) categories should be divisible into more detailed subcategories that can be obtained from large-scale imagery or ground surveys, (viii) aggregation of categories must be possible, (ix) comparison with future land use and land cover data should be possible, and (x) multiple uses of land should be recognised when possible.

The basic USGS LU/LC classification system for use with remote sensor data is shown in Table 16.1. On the basis of this system a multi-level system has been devised because different degrees of detail can be obtained from aerial and space images, which depend upon the resolution. Fig. 16.1 illustrates a sample aggregation of classifications for levels III, II and I. One more level that is level IV is also devised for local users. In principle, levels IV and III are designed for local level or very large scale mapping whereas levels II and I are meant for small scale mapping. Table 16.2 lists representative interpretation formats for various land use/land cover classification levels.

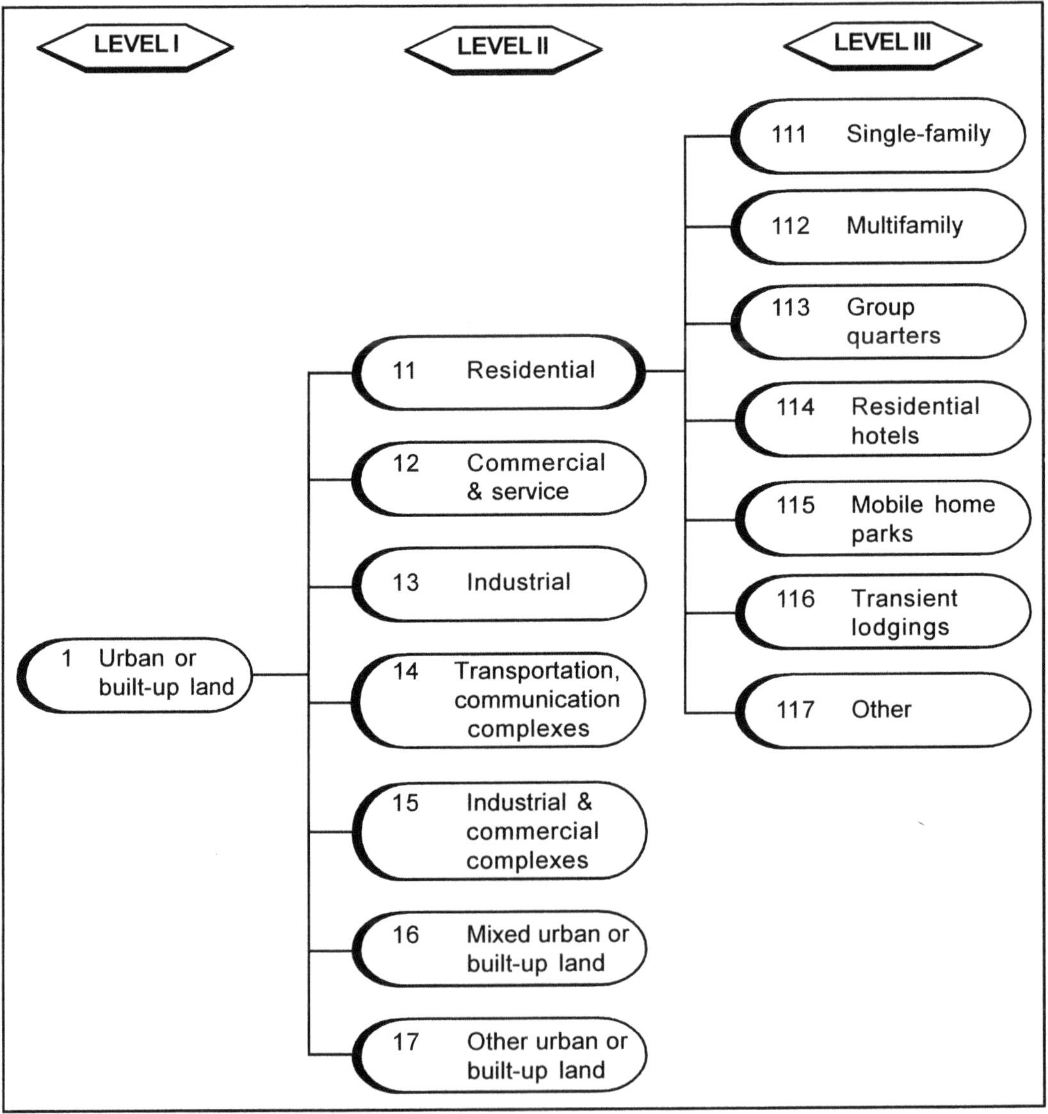

Fig. 16.1 An example of aggregation of land use/land cover types (Lillesand & Kiefer, 2000).

Table 16.1 USGS Land Use/Land Cover Classification System for use with Remote Sensor Data

Level II	Level II
1. Urban or built-up land	11 Residential 12 Commercial and service 13 Industrial 14 Transportation, communications, and utilities 15 Industrial and commercial complexes 16 Mixed urban or built-up land 17 Other urban or built-up land
2. Agriculture land	21 Cropland and pasture 22 Orchards, groves, vineyards, nurseries, and ornamental horticultural areas 23 Other agricultural land
3. Rangeland	31 Herbaceous rangeland 32 Shrub and brush rangeland 33 Mixed rangeland
4. Forest land	41 Deciduous forest land 42 Evergreen forest land 43 Mixed forest land
5. Water	51 Streams and canals 52 Lakes 53 Reservoirs 54 Bays and estuaries
6. Wetland	61 Forested wetland 62 Nonforested wetland
7. Barren land	71 Dry salt flats 72 Beaches 73 Sandy areas other than beaches 74 Bare exposed rock 75 Strip mines, quarries, and gravel pits 76 Transitional areas 77 Mixed barren land
8. Tundra	81 Scrub and bush tundra 82 Herbaceous tundra 83 Bare ground tundra 84 Wet tundra 85 Mixed tundra
9. Perennial snow or ice	91 Perennial snowfields 92 Glaciers

National Remote Sensing Agency (NRSA), Government of India, has devised a generalised land use/land cover classification system with respect to the Indian conditions based on the various categories of Earth surface features, resolution of available satellite data, capabilities of sensors, and present and future applications. Table 16.3 shows the general legend adapted for land use/land cover categories. This system is used for the development of land use/land cover map for the project area, namely, MCH area of Hyderabad.

Table 16.2 Representative Image Interpretation Formats for Various Land Use/Land Cover Classification Levels(Liilesand & Kiefer, 1999)

Land Use/Land Cover Classification Level	Representative Format for Image Interpretation
(i)	Low to moderate resolution satellite data like Landsat MSS data.
(ii)	Small scale aerial photographs; moderate resolution. Satellite data like Lnadsat TM data.
(iii)	Medium scale aerial photographs; high resolution. Satellite data as acquired by commercial high resolution systems.
(iv)	Large scale aerial photographs

16.7 Case Study of Hyderabad City

Hyderabad registered a decadal growth rate of 57.48%. All this rapid and haphazard growth of urban sprawl and increasing population pressure is resulting in deterioration of infrastructure facilities, loss of productive agricultural lands and green open spaces, loss of surface water bodies and depletion of groundwater aquifers zones, besides causing air pollution, contamination of water, health hazards, and micro-climatic changes. To address these issues effectively, it requires up-to-date and accurate data at regular intervals of time on the changing urban sprawl, urban land use, urban environment. and urban resources. Realising the need of such environmental Baseline Information System the present study has been made. The information system will be organised on municipal zone basis and will address issues/ information contents related to all grids relevant to its land use/land cover and major road network.

The satellite remote sensing with its ability to provide reliable and accurate data offers excellent possibilities to map, monitor, and measure the various facets of

Table 16.3 Legend for Land use/Land cover Categories

Level - I	Level - II	Level - III	Level - IV	Code
1. Built-up-Land	1.1 Town/Cities Villages	1.1.1 Residential		01
		1.1.2 Industrial		
			1.1.2.1 Salt Pans	02
			1.1.2.2 Others	03
2. Agricultural Land	2.1 Crop Land	2.1.1 Kharif		04
		2.1.2 Rabi		05
		2.1.3 Kharif+Rabi		06
		2.1.4 Summer		07
		2.1.5 Kharif+Summer		08
		2.1.6 Rabi + Summer		09
		2.1.7 Kharif + Rabi + Summer		10
	2.2 Fallows			11
	2.3 Plantations	2.3.1 Agro Horticulture		12
		2.3.2 Horticulture		13
3. Forest	3.1 Evergreen/ Semi Evergreen	3.1.1 Dense		14
		3.1.2 Open		15
		3.1.3 Scrub Forest		16
		3.1.4 Forest Blanks		20
	3.2 Deciduous	3.2.1 Dense/Closed		18
		3.2.2 Open		19
		3.2.3 Forest		20
		3.2.4 Forest Blanks		21
	3.3 Forest Plantations			22
	3.4 Mangrove	3.4.1 Dense		23
		3.4.2 Sparse		24
	3.5 Evergreen/ Semi-Evergreen (Unnotified)	3.5.1 Dense (Unnotified)		25
		3.5.2 Open (Unnotified)		26
		3.5.3 Forest Blanks (Unnotified)		27

Level - I	Level - II	Level - III	Level - IV	Code
	3.6 Deciduous (Unnotified)	3.6.1 Dense (Unnotified)		28
		3.6.2 Open (Unnotified)		29
		3.6.3 Forest Blanks (Unnotified)		30
	3.7 Forest Plantation (Unnotified)			31
	3.8 Mangrove (Unnotified)	3.8.1 Dense (Unnotified)		32
		3.8.2 Sparse (Unnotified)		33
4. Wastelands	4.1 Salt Affected Land			34
	4.2 Gullied/Ravinous Land			35
	4.3 Land with Scrub			36
	4.4 Land without Scrub			37
	4.5 Sandy Area (Coastal/Desertic)			38
	4.6 Mining/Industrial Wasteland			39
	4.7 Barren Rocky/ Stony waste/Sheet Rock Area			40
5. Water Bodies	5.1 River			41
	5.2 Canals			42
	5.3 Lake/Pond			43
	5.4 Reservoir			44
	5.5 Tank			45
	5.6 Cooling Pond/ Cooling Reservoir			46
	5.7 Abandoned Quarries With water			47

Level - I	Level - II	Level - III	Level - IV	Code
	5.8 Bay	5.8.1 Back waters		48
		5.8.2 Estuary/Kayal		49
		5.8.3 Creek		50
		5.8.4 Lagoon		51
	5.9 Cut-off Meander			52
6. Wetlands	6.1 Inland/ Wetlands	6.1.1 Water Logged		53
		6.1.2 Marshy/Swampy		54
		6.1.3 Oxbow Lake		55
	6.2 Coastal	6.2.1 Marsh Veg.		56
		6.2.2 Algae		57
		6.2.3 Mud Flats	6.2.3.1 High Tidal Flats	58
			6.2.3.2 Inter Tidal	59
			6.2.3.3 Sub-Tidal	60
			6.2.3.4 High Tidal With Salt Encrustations	61
		6.2.4 Sand	6.2.4.1 Spit	62
			6.2.4.2 Bar	63
			6.2.4.3 Shoals	64
			6.2.4.4 Beach Ridges	65
			6.2.4.5 Plantations On Sand	66
		6.2.6 Rocky Coast		67
7. Grass Land/ Grazing Land	7.1 Dense			68
	7.2 Degraded			69

urban planning and development. Considering the high dwelling density and low floor space area (FSA) here due to compact parcel size and lack of physical spacing and homogeneity in the surface built-up features in Hyderabad, remotely sensed data of high spatial resolution along with multispectral resolution data were used for detailed urban surveys for the area under municipal corporation of Hyderabad. The availability of space data from IRS-ID becomes more relevant and important for urban and municipal planning and for making proper decisions for the overall development of Hyderabad city. Keeping the importance of the existing land use/land cover and widening of road network within city, MCH has assigned a consultancy work to prepare a map showing the various land use/land cover patterns and an overlay of major road network to the Centre for Environment, Jawarharlal Nehru Technological University, Hyderabad.

16.7.1 Growth and Development of Hyderabad

According to 1991 census, Hyderabad is the fifth largest metropolis of India with a population of 4,334,437. The population of Hyderabad has increased from 0.448 millions in 1901 to 0.502 millions in 1911 but came down to 0.406 million in 1921. It again went up to 0.447 million in 1931, 0.739 million in 1941 (an increase of 65.3%), 1.28 million in 1951 (52.5% increase), 1.429 million in 1961 (10.71% increase) and 1.796 million in 1971 (43 .8% increase). Between 1971 and 1981 the population went up to 2.759 million, and the rate of increase in the urban agglomeration was 42.65%. Between 1981 and 1991 the population went up to 4.34 million and the rate of growth the highest so far, is 67.04%.

Hyderabad city took its birth from the southern side of Musi river, which area is now referred to as the old city. The area is densely populated with very little vacant space left. As the city grew, parts of the northern side of river Musi started filling up. The city's development picked up rapidly with the establishment of Secunderabad contonment. Huge vacant areas around Hussain Sagar lake, are being occupied by various types of residents.

The land use of this metropolitan city shows a pattern of diversified functions, which are well distributed all over the city and its region. Historical, economic, and cultural factors have direct influence over the land use pattern, beside the topographical and transportation factors. From the chronology chart Fig. 16.2, it is clear that the city's growth in the last few decades has been very much towards the northern, eastern, and south eastern sides. Substandard and haphazard development is taking place at a rapid rate especially on the major arterial roads and highways of the region. The area under each existing land use along with percentage and the total development area are shown in Table 16.4. Table 16.5 shows the land use changes that have taken place during the period 1927 to 1991. The land use/land cover for the year 2000 has been developed in this project for the planning and management of this fast growing city for the benefit of muncipal administrators. The boundary of this area is based on the maps obtained from MCH, Hyderabad.

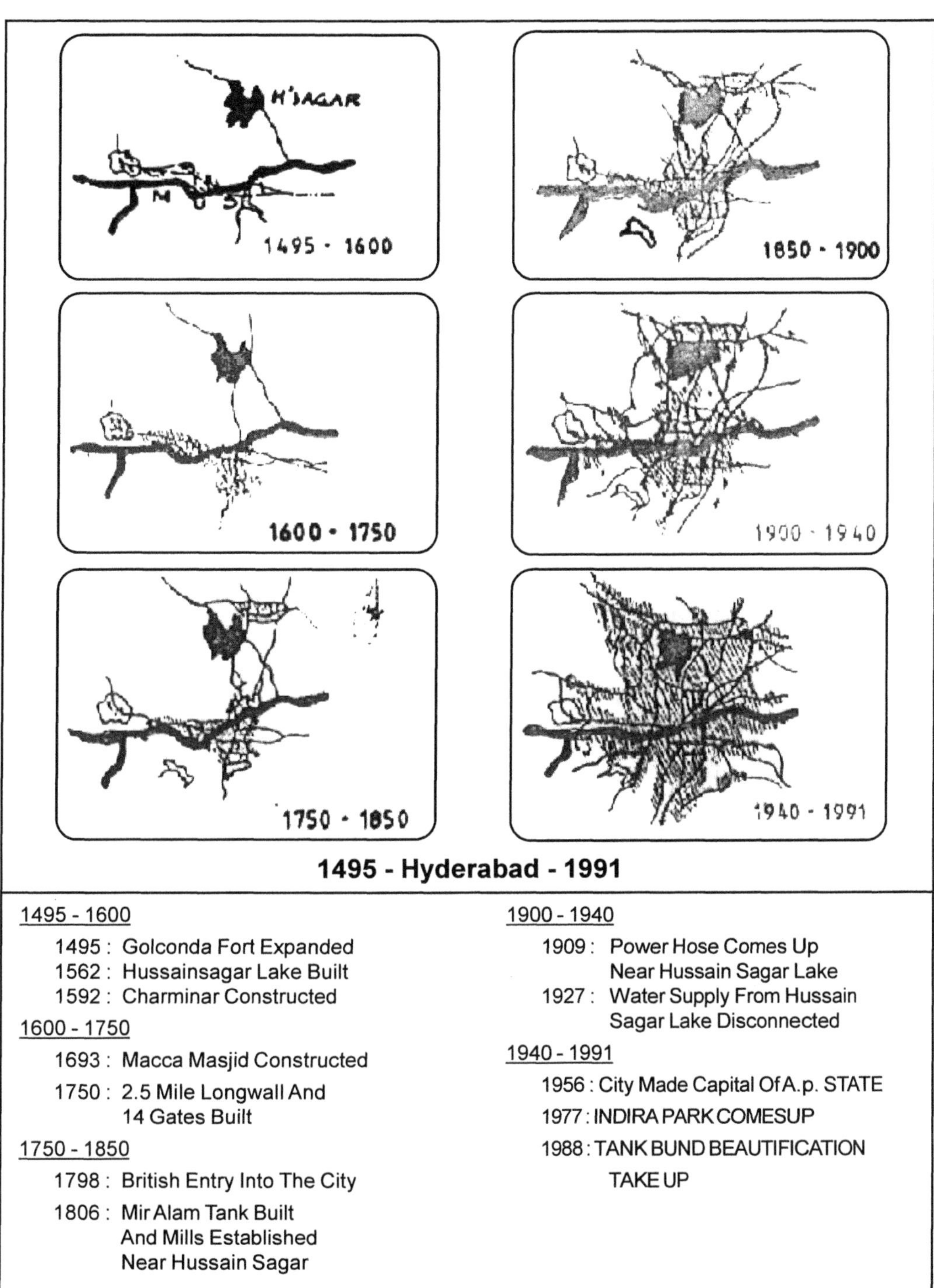

Fig. 16.2 Hyderabad 1495 - 1991.

Table 16.4 Existing land use - Hyderabad Development Area

Land use	Area (Hectares)	% Area to the Total Development Area
Residential	12824.24	8.36
Commercial	1043.12	0.68
Industrial	4341.22	2.83
Public & Semi-public	10507.90	6.85
Agriculture	45667.18	29.77
Circulation	2301.00	1.50
Parks and Play grounds	184.08	0.12
Vacant Land	50161.80	32.70
Water Bodies	7762.04	5.06
Hillocks and Rocks	12072.58	7.87
Forest	6534.84	4.48
Total	**153400.00**	**100.0**

Table 16.5 Estimation of loss of Agricultural Lands and Water bodies due to urban spread in Hyderabad Metropolitan area

Year	Built up area (Sq.km)	Agricultural Lands (Sq.km)	Waterbodies (Sq.km)
1927	99.75	–	–
1973	245.13	785.14	117.98
1983	354.98	740.01	114.63
1991	522.49	684.71	112.01
Growth/loss	+ 277.36	-100.43	-5.97
(1973-1991)	(+113.20%)	(-12.8%)	(-5.11)

16.7.2 Division of Planning Zones

According to existing records of Municipal Corporation of Hyderabad (MCH), the total municipal area is divided into 11 planning zones (Fig. 16.3). Some of the planning zones are fully developed like divisions 1,3,7,9 and 10. With increase in population and built-up area the pressure has increased on the civic amenities affecting the quality of environment. Most of the lakes and rivers are forced to serve as drainage channels, which in turn are affecting the groundwater quality of the surrounding area.

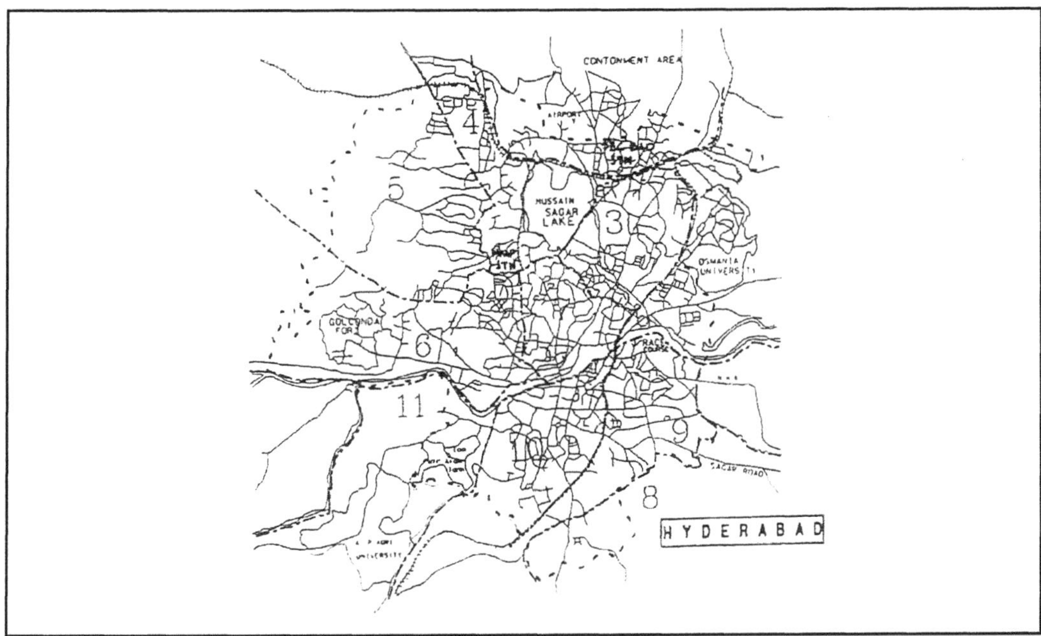

Fig. 16.3 Map showing 11 zones of MCH, Hyderabad for planning purpose.

16.8 Methodology

The broad objective of the project of MCH discussed in this chapter is to prepare municipal zone-wise land use/land cover and major road network using PAN+LISS III merged digital data of IRS ID satellite on 1:25000 scale. This can be performed by image processing of digital data with the help of EASI/PACE software and ARC/INFO GIS software. The land use/land cover map will be used as baseline information for taking necessary decisions for the improvement of area under Municipal Corporation of Hyderabad.

Remote sensing and Geographic Information System (GIS) addresses the problems related to urban planning and management. As discussed earlier, GIS can be defined as a set of tools for collecting, storing, retrieving, transforming, and displaying geographically referenced spatial data with its corresponding attribute information. There are essentially two kinds of data bases: one, the specific characteristic of a location called spatial data, and two, the attribute data (Statistics of written text, tables and so on). New maps can be generated precisely by easily integrating innumerable layers of data. Thus a GIS has a database of multiple information layers that can be manipulated to evaluate relationships among the chosen elements in the different layers under consideration called topology, in order to develop Hyderabad Environmental Information System (HEIS) in general, and land use/land cover information system of Hyderabad using GIS technique in particular. Basically, two types of data are to be generated along with other collateral data from different sources for desired accuracy parameters. The two types of data base are : (i) spatial database creation and (ii) non-spatial database creation (attribute data).

Apart from these data products, other data related to environmental impact assessment (EIA), have been obtained from different State and Central Government Organisations. The various sources of data products needed for the creation of land use/land cover information system are given in the following Table, 16.6. The spatial database and non-spatial database are created and used as GIS input to ARC-INFO software.

Table 16.6 GIS data collection and sources

S. No.	Data Product	Source
(i)	Satellite Data IRS ID-PAN, LISS III of IRS 1D	National Remote Sensing Agency Government of India,
(ii)	Toposheets of 1:50,000, and 1:25,000 scale.	Survey of India. Government of India.
(iii)	Maps showing existing information of Hyderabad and its boundary of MCH area	Municipal Corporation of Hyderabad (MCH) Centre for Environment, JNT University, Hyderabad. A. P. State Pollution Control Board, A.P. State Remote Sensing Application Centre. Central Pollution Control Board
(iv)	Field Data	through intensive field work (Groundtruth).

Since the main objective is to prepare the land use/land cover information system with major road network (the sites for the development of green vegetation), a large number of data layers like land use/land cover, topological information, such as, roads, parks, other protected areas, and so on are to be mapped as the project is multidisciplinary in nature, and hence the output of this project is the analysis of all types of data. Some of these data products are obtained from various organisations and the other data products are derived from the analysis of satellite digital data obtained from the NRSA (Table 16.6).

16.8.1 Data Source and Collection

The Current exercise envisaged the use of three types of data, namely,

(i) Satellite digital data which corresponds to IRS 1C PAN and LISS III of IRS 1D in four bands.

(ii) Survey of India topo maps on 1:25,000 scale for preparing forest boundaries, base map and identification of GCPs.

(iii) Sample ground data for verification of doubtful areas arising during the pre-classification stage and for creation of ground control points.

The second generation operational Indian Remote Sensing Satellites IRSIC and IRS-1D were launched during 1994-95 and 1998 respectively by the indigenously developed Polar Satellite Launch Vehicle (PSLV) from Sriharikota, India. The satellites were placed in a near circular, sun-synchronous, near polar orbit with nominal inclination of 98.53^0 are a mean altitude of 870 km. There are three sensors, namely (i) Panchromatic Camera (PAN), (ii) Linear Imaging and Self Scanning Sensor (LISS III), and (iii) Wide Field Sensor (WIFS). More details on these satellite sensing systems are given in chapter 5. In this project, the remote sensing data in the digital mode is used.

Two maps of 1:50,000 scale and 1:25,000 scale obtained from Survey of India covering the entire project area are used to extract the Ground Control Points (GCPs) and to demarcate the boundary of Municipal Corporation of Hyderabad. This information is then used for image registration of LISS III and PAN digitally using EASI/ PACE software. The boundaries of MCH drawn from these maps are then superimposed on classified land use/land cover image in order to develop the action plan for identification of sites for green vegetation and other decision making processes.

16.8.2 Data Processing

The project (case study) is executed through the following steps (Fig. 16.4).

(i) Acquisition of satellite data from NRSA, Balanagar, Hyderabad and toposheets from survey of India (SOI), Hyderabad.

(ii) Geo coding and geo-referencing of LISS III and PAN digital data by extracting the Ground Control Points (GCPs) from SOI toposheets.

(iii) Digital image enhancement and application of correction models for making the digital data free from error and distortions in terms of radiometry and geometry of the satellite data.

(iv) Fusion of Pan and LISS III for merged product preparation of a mosaic which shows the continuous imagery of Hyderabad under Municipal Corporation. This is FCC mode and is used for visual interpretation to extract the land use/land cover information by applying both previsual interpretation, ground truthing and post visual interpretation of this Image mosaic.

(v) Preparation of cartographic output for making the data layer ready for scanning for further GIS analysis.

(vi) Scanning of cartographic output using A0 colour Abakos scanner, digitisation of this hard copy using AUTOCAD software and editing the digitised data compatible to ARC/INFO GIS software.

(vii) GIS data manipulation and analysis, linking the spatial data file and attribute data file for the creation of topology.

(viii) GIS output in the form of land use/land cover map showing various land use/land cover patterns of MCH jurisdiction.

(ix) Overlay of major road network on the land use/land cover map for the final product as required by MCH administration.

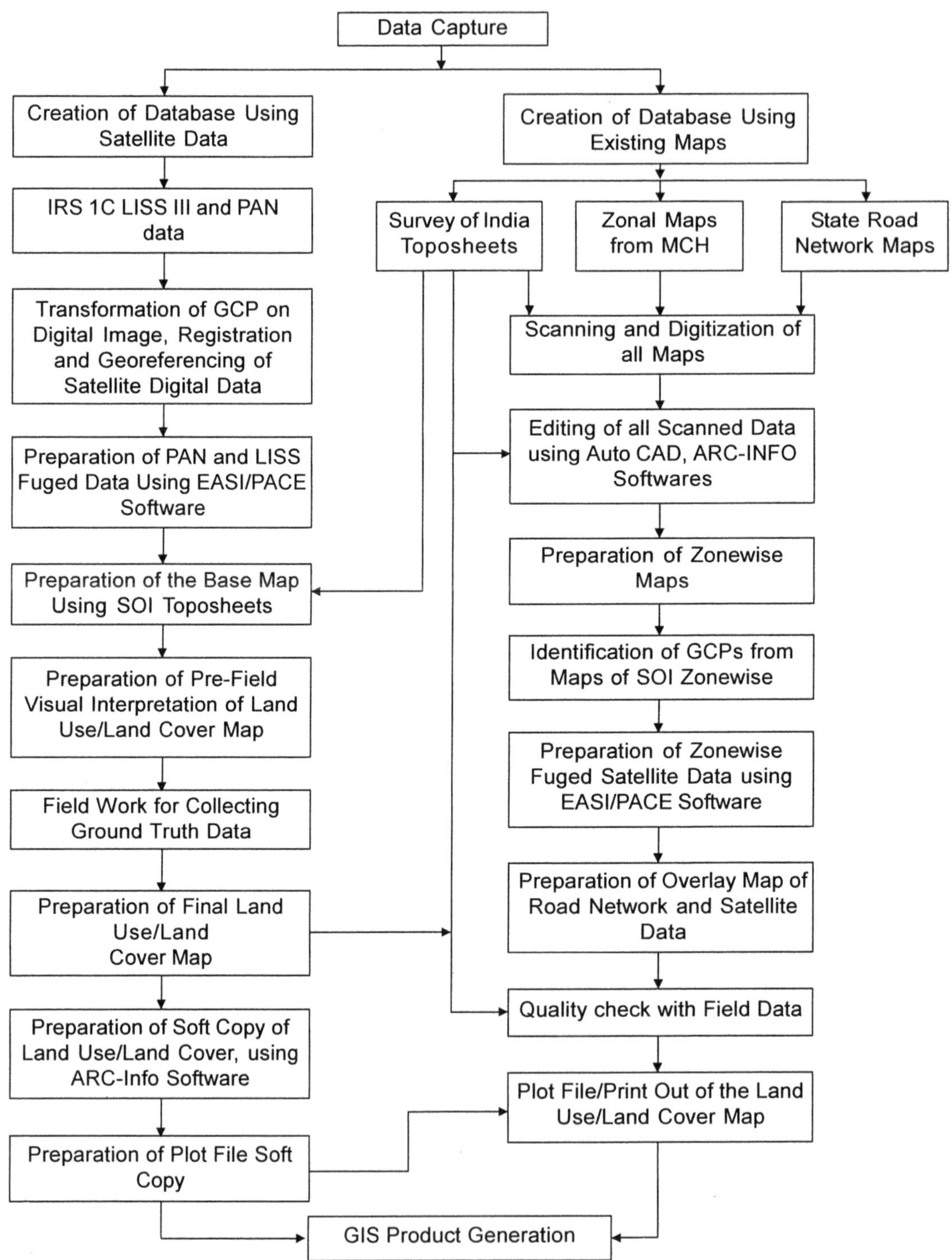

Fig. 16.3 Flow chart showing the methodology for GIS data processing.

16.8.3 Geocoding and Georeferencing

The standard techniques have been adopted for Georeferencing of LISS III and PAN data. EASI/PACE (Image Processing Software) has been used for this purpose. 1:25,000 scale toposheets are scanned and raster files for project area are created. These are georeference based on the longitudinal and latitudinal coordinates. After georeferencing, all the maps are edge-matched and a digital mosaic is prepared which depicts the continuity of the project area.

The LISS III data is processed for initial corrections like drop outs, stripping and earth rotations. Sufficient number of well distributed ground control points are selected both on the maps and corresponding imagery. Care is taken to satisfy the condition on density of GCPs for image registration. Georeferencing was carried out based on the module "GCP Works" of EASI/PACE software. The georeferenced image was further mosaiced and then feature matching was carried out. At the end of this process the digital data which is free from all distortions is available for digital image enhancement, and classification for land use/land cover. (Anji Reddy and K. M. Reddy, 1996)

16.8.4 Digital Image Enhancement of LISS III Data

Image Enhancement deals with the individual values of the pixels in the image. The goal of spectral enhancement is to make certain features more visible in an image by bringing out more contrast. Initial display of LISS III data through EASI/PACE software revealed that the features like minor roads, streams and Musi river are not clear/visible as the contrast of the imageries is very dull because the raw data values fall within a narrow range. Therefore, an attempt is made to apply linear contrast stretch technique in order to improve the contrast of the image, which can be capable of expanding the dynamic range of radiometric resolution. To perform this technique, Lookup Table (LUT) is created that converts the range of data values to the maximum range of the display device. Based on these LUTs an enhanced image is produced.

16.9 Land Use/Land Cover Map Generation

The primary data on which the classification procedure is based is land use/ land cover, such as, forests, built-up agricultural lands, and wastelands in the project area. All these earth surface features have different spectral reflectance values. Based on this a thematic map of land use/land cover is prepared. The preparation of this map is mainly based on image classification of LISS III digital data. The two methods of digital image classification are Supervised Classification and Unsupervised Classification. The supervised classification technique is considered for the analysis of land use/land cover keeping the limitation of LISS III data in view. LISS III data have been founded to be adequate to bring out all the level 1 classes derived on 1:50,000 scale from United States of Geological Survey (USGS) and adopted by NRSA. Apart from level 1 classification of land use/land cover, a few other information classes are also derived keeping their importance on the present project objectives.

The preparation of land use/land cover map is mainly based on the satellite digital image which contains a detailed record of spectral reflectance values in four bands, of different earth surface features. An image analyst systematically examines digital image converted to pictorial product for generating the final land use/land cover classification system using visual interpretation techniques.

16.9.1 Image Interpretation Process

This process consists of a set of image elements or characteristics like color/ tone, texture, pattern, size, and shape which help in the recognition of various land use/land cover classes systematically on the enhanced fused satellite imagery during interpretation process. A preliminary image classification key is prepared for classifications of surface features. This is then finalised after the ground truth.

Using the image interpretation key, preliminary interpretation is carried on SOI toposheets available on 1:50,000 scale and a base map is prepared. Based on this map, the fused product (Plate 7) is used and interpretation key for classification of land use/land cover patterns and 16 different patterns/classe categories are identified and marked on the image and validated during the subsequent field work conducted during the project. The doubtful areas (due to similar spectral response and spectral signature) identified during the preliminary image classification are listed out before ground verification. After finalising the ground traverse plan, the doubtful areas are noted.

Based on the ground information collected, corrections and modifications of misclassified land use/land cover details and doubtful areas were carried out on enhanced imageries for final land use/land cover classification. The final land use/ land cover classes were prepared by assigning standard color with respect to land use/land cover classes. This rapid land use assessment methodology allowed us to quantify the entire area of MCH into 16 land use/land cover categories for 173.812 sq km area. Each one of these 16 categories or classes or patterns represents one type of feature on the MCH area. These categories are shown in Table 16.7.

16.10 Production of GIS Output

The land use/land cover map developed using visual interpretation techniques is transformed into a cartographic output. This output is scanned by using AO Abokos colour scanner and digitised using AUTO-CAD software. This software has converted the hard copy of land use/land cover into a digital database through digitisation. This digital database is then transformed into a database compatible to ARC/INFO GIS software. In simple terms, utilising the satellite data, field surveys, existing data products, and local knowledge of the area, a land use/land cover digital database is constructed. This database is processed resulting in an accurate and up-to-date land use/land cover map of the area under MCH jurisdiction Plate 8. The total area has been divided into 4 zones, in accordance with the administrative boundaries of each zone supplied by MCH. These zones are (i) North zone, (ii) South zone, (iii) East zone, and (iv) West zone. The area statistics of each zone is given in Table 16.8.

16.11 Area Statistics

Plate 8 shows, 16 land use/land cover categories. This study revealed that these 16 categories of land together occupied an area of 173.812 Sq. km. Areawise distribution of the land under each one of the 16 categories and the percentage occupancy of each class or category are given in Table 16.8 and is also shown in a pie chart (Fig. 16.5).

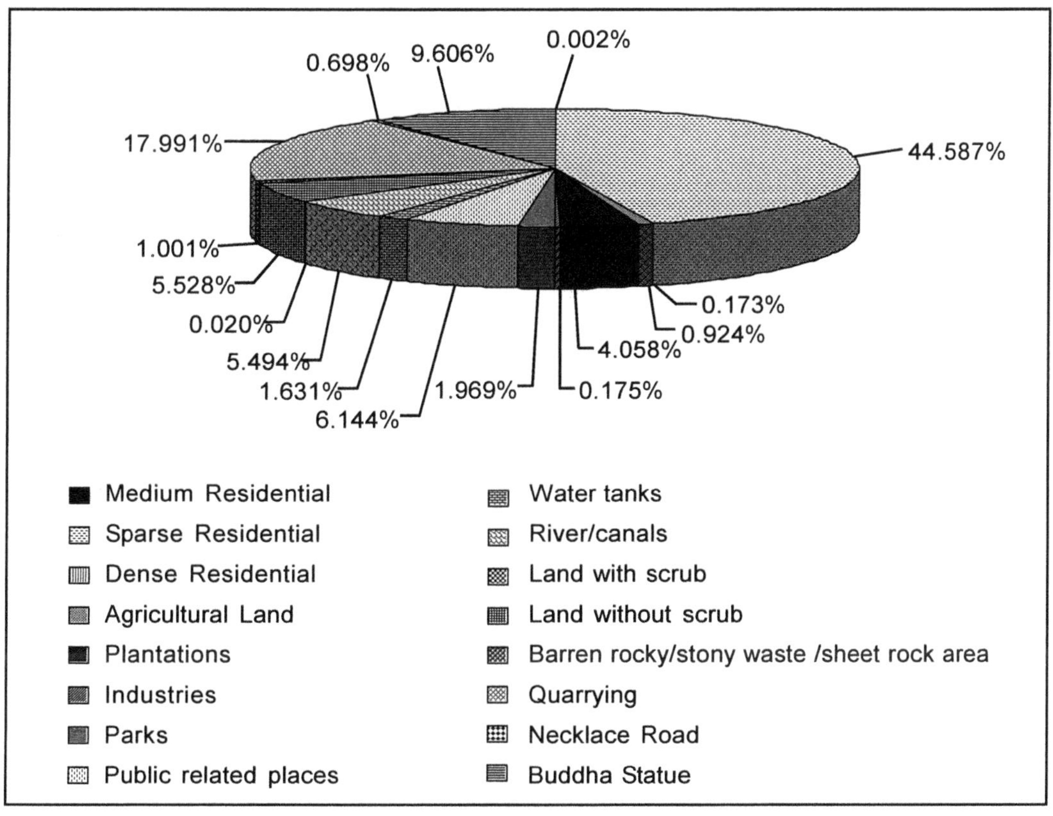

Fig. 16.5 Area Distribution under various classes of Land use/ land cover (MCH).

Table 16.7 Area statistics of Land use/land cover of Hyderabad city under MCH. (Area in Sq.Kms.)

S. No.	Description of the land use/land cover pattern	Area	% of area
1.	Medium Residential	31.270	17.995
2.	Sparse Residential	16.696	9.605
3.	Dense Residential	77.497	44.586
4.	Agricultural Land	1.606	0.923
5.	Plantations	0.304	0.174
6.	Industries	1.214	0.697
7.	Parks	2.835	1.632
8.	Public related places	1.740	1.001
9.	Water tanks	7.053	4.058
10.	River/canals	3.422	1.968
11.	Land with scrub	10.679	6.144
12.	Land without scrub	9.550	5.495
13.	Barren rocky/stony waste/sheet rock area	9.609	5.528
14.	Quarries	0.300	0.172
15.	Necklace Road	0.034	0.019
16.	Buddha Statue	0.003	0.002
	Total Area	173.812	100.00

Table 16.8 Area Statistics of 4 zones (Area in Sq. Kms)

Sl. No.	Description of land use/ land cover pattern	East Zone		West Zone		North Zone		South Zone	
		Area	%	Area	%	Area	%	Area	%
1.	Medium Residential	6.885	23.591	10.992	17.349	4.937	21.335	8.456	14.548
2.	Saprse Residential	0.803	2.751	9.369	14.787	3.469	14.991	3.055	5.256
3.	Dense Residential	18.894	64.739	15.892	25.082	10.088	43.596	32.624	56.124
4.	Agricultural Land	--		1.520	2.399	--		0.080	0.138
5.	Plantations	--		0.092	0.145	--		0.276	0.475
6.	Industries	--		--		0.387	1.672	0.737	1.268
7.	Parks	0.394	1.350	1.971	3.112	--		0.469	0.807
8.	Public related places	0.540	1.850	--		0.414	1.789	0.786	1.352
9.	Water tanks	0.228	0.781	4.928	7.778	0.333	1.439	1.585	2.727
10.	River/canals	0.335	1.148	1.538	2.427	--		1.559	2.682
11.	Land with scrub	0.662	2.268	4.884	7.708	1.030	4.451	4.105	7.062
12.	Land without scrub	0.444	1.521	4.525	7.142	2.482	9.726	2.100	3.613
13.	Barren rock/stony waste/ sheet rock area	--		7.336	11.578	--		2.274	3.912
14.	Quarries	--		0.278	0.439	--		0.022	0.038
15.	Necklace Road	--		0.035	0.055	--		--	
16.	Buddha Statue	--		0.003	0.045	--		--	
	Total	29.185	100.00	63.359	100.00	23.140	100.00	58.128	100.00

17

Forest Resources Management

17.1 Introduction

We live in a landscape and fulfil our most of the needs from its diversity. The increasing human population satisfies their demand from various kinds of economic activities carried out on the landscape, which ultimately, along with various geomorphic and biological processes, give the birth of spatial and structural changes. Renewable natural resources, especially forests, are conserved for three major sets of purposes: the maintenance of ecological processes and life support systems; the sustained use of the resources for consumptive and social benefits (AKCA., A., 1971). and the conservation of biodiversity units own right for ethical, moral, aesthetic and evolutionary reasons. Only if all three of these can be met and integrated into the socio-political process will true sustainable development achieved, to meet them requires quantification of the current status and of changes on the resources.

Forests, 'all lands bearing a vegetation association dominated by trees of any size, exploited or not, capable of producing wood or other products, of exerting an influence on climate or on the water regime or providing shelter for livestock and wildlife", are the most vulnerable renewable resource of the 'biodiversity' as the plants basically support the survival of each and every member of the biome. The forest resource is large in volume and diverse in character, on the other hand, the excess population pressure intensifies the demand of forest and forest products side by side. But unfortunately, this natural resource has been overlooked by the human kind regarding the aspect of degrading trend of forest resources all over the world.

The forest resources and forest lands need to be managed in such a way that they can meet the needs of not only this generation but for future too. Stringent strategies are to be evolved to conserve the existing forests afforest the state owned forest land as well as degraded private lands. To develop such strategies quick, repetitive and accurate information about forest cover is required at the local, regional, state and national levels. The temporal and spatial distribution of forest cover can be derived using remotely sensed data and the strategies can be developed using GIS technology by integrated studies.

The key factors involved in developing forest resources management strategies are - large economic development programs involving resettlement, agriculture and infrastructure, pressure for increased food production conversion of forest land to agriculture, increase in the demand and decrease in forest area, and future demand for forest products.

Vegetation is an important dynamic characteristic and very prominent feature of earth's surface and forms a major element of the landscape and is either aquatic or terrestrial. The natural vegetation on land is generally grouped as forest, grassland, scrub etc. Forest provides two benefits: direct and indirect. The direct benefits are harvesting timber, fuel and other products whereas the indirect benefits include conservation of soils, water and air, controlling and regulating the global cycling of oxygen, nitrogen and carbon.

17.2 Geomatics in Forestry

Mapping and monitoring of forest cover is one of the foremost requirements for its planning, management and conservation. Remote Sensing, GIS and GPS play vital role in mapping of existing forest resources information at a particular period or time. The multi-spectral data obtained from remote sensing satellites like Landsat, IRS, SPOT, IKONOS have been used to study forest features either by visual interpretation or by processing digital data using digital computers. The spectacular development in space technology, particularly, the repetitive satellite remote sensing across various spatial and temporal scale, offers the most economic means of assessing, planning, managing and monitoring the forest resources. It provides excellent capability to monitor composition of forest ecosystem and impact of management and degradation processes. Sometimes the Remote Sensing provides unique data that cannot be obtained by any other source, but much more often, the collected data facilitates fieldwork and enables tasks to be completed at lower cost and much more quickly. It is unlikely, nowadays, forest inventory would be undertaken without remotely sensed data.

Remote sensing is a useful tool for assessment of environmental conditions, either in an existing forest or prior to planting a new one in accordance with the climatological conditions. Data for climate analysis can be obtained through microwave or radar sensing, which is ideal for gathering information about the atmosphere. The microwave data provides an uninterrupted flow of accurate data for interpretation, and such data can be used to track rainfall and winds, to measure water and ozone content in the air to determine the compatibility of the climate with forest growth, and to predict potential problems in forest growth. It also gives the regeneration status using SAR imagery (C band-VV Polarization). Remote sensing contributes to fire-fighting efforts, as well. Radar and thermal sensing allow for constant observation of fires, unaffected by clouds, smoke, or other conditions that hinder aerial observation. After a fire, damage can be quickly and in expensively assessed by using the data obtained by thermal sensors like AVHRR, HCMM. This accurate spatial information of the burn scar can be used to know the amount of biomass destroyed, the amount of smoke produced and the air pollution, by intensity of forest managers. The Geomatics can be used to predict the fire risk areas, which will be useful in taking effective management decisions (Agarwal, N. K. and Sharma, M. K. 1983).

During every stage of forest management, foresters can use remote sensing data to estimate future urban spread and population growth. Then, forest management can be planned taking into account the future needs of settlements. Urban planning data can also be applied to the management of urban forestry, to create inventories of trees in parks and on streets.

IRS-IC/ID LISS-III multispectral scanner with 23.5m resolution enables discrimination of different forest types, grasslands, plantations etc. The discrimination of forests into deciduous and evergreen depends on the use of optimal season satellite data based on phenology and different band combinations. The dry period (Feb-May) satellite data is used for delineation of evergreen & deciduous forests. The use of middle IR data facilitates delineating vegetation types such as evergreen, semi-evergreen, moist deciduous & dry deciduous. The LISS-III data of March help in delineating the deciduous forest being in leaf fall condition with other vegetation areas. Pan data with 5.8m resolution is useful in delineating the vegetation into various canopy density classes and also provides micro level details for stock mapping. Hence the fused Pan and LISS III data will be useful in revision and updation of stock maps. Wide Imaging Field Sensor (WiFS) payload with repeativity of 5 days permits rapid forest cover monitoring and mapping. It provides forest cover details at global/regional level, used for annual forest cover mapping which saves adequate time and money. WiFS data is used for forest fire monitoring and to account damages and suggest control measures for large scale forest fires because the WiFS data can be acquired for every 5 days. NOAA Satellite AVHRR sensor data is extensively used for large area vegetation monitoring, the spectral bands used for this purpose are visible band (0.58-0.68um) and NIR (0.73-1.l, um). Various mathematical combinations of multi-spectral data have been found to be sensitive indicators of the presence and condition of green vegetation. These mathematical quantities are called as vegetation indices.

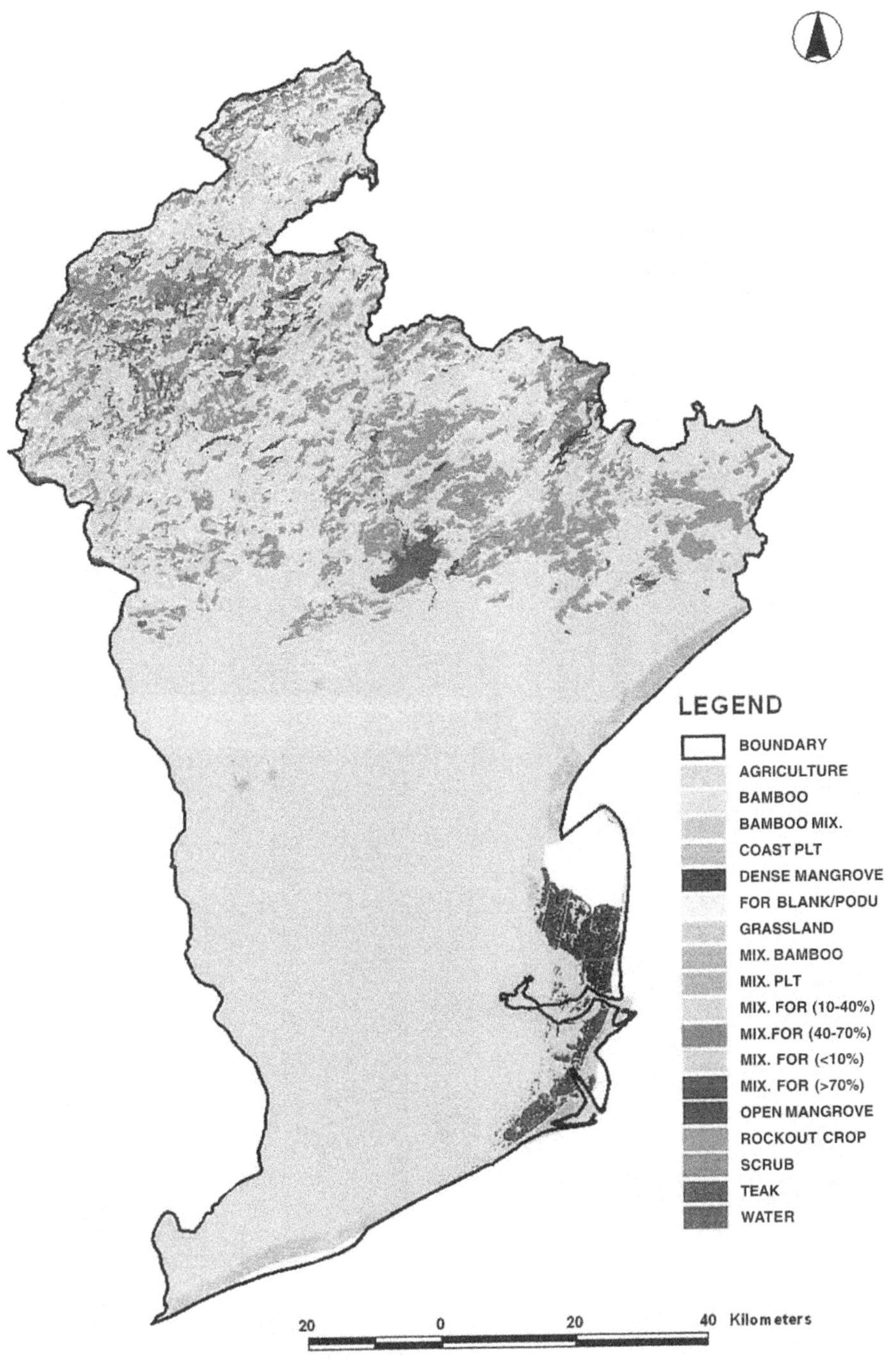

Plate 9 Stock Map of East Godavari District of Andhra Pradesh State, India.

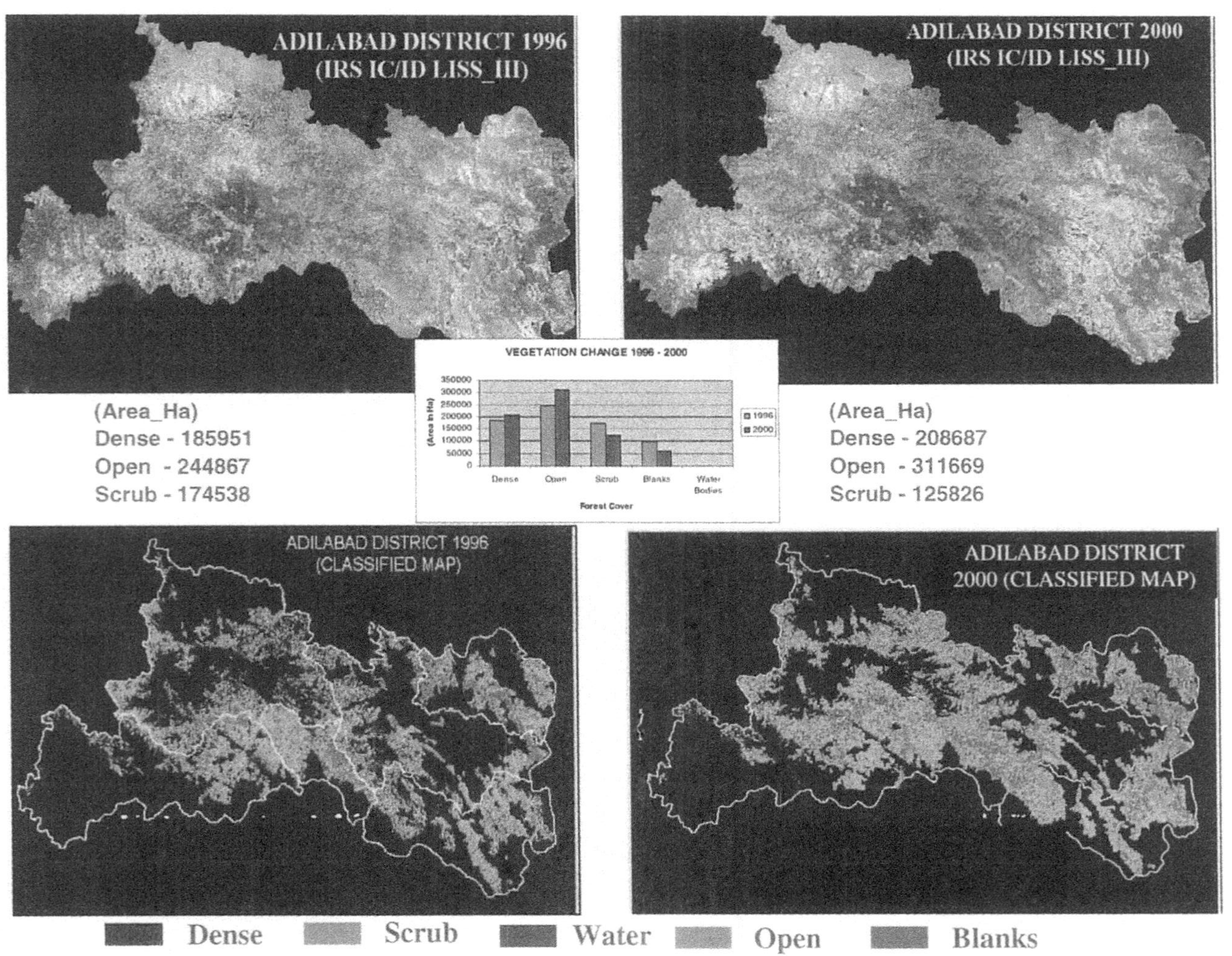

Plate 10 Forest Density and Vegetation change map of Adilabad District of Andhra Pradesh State

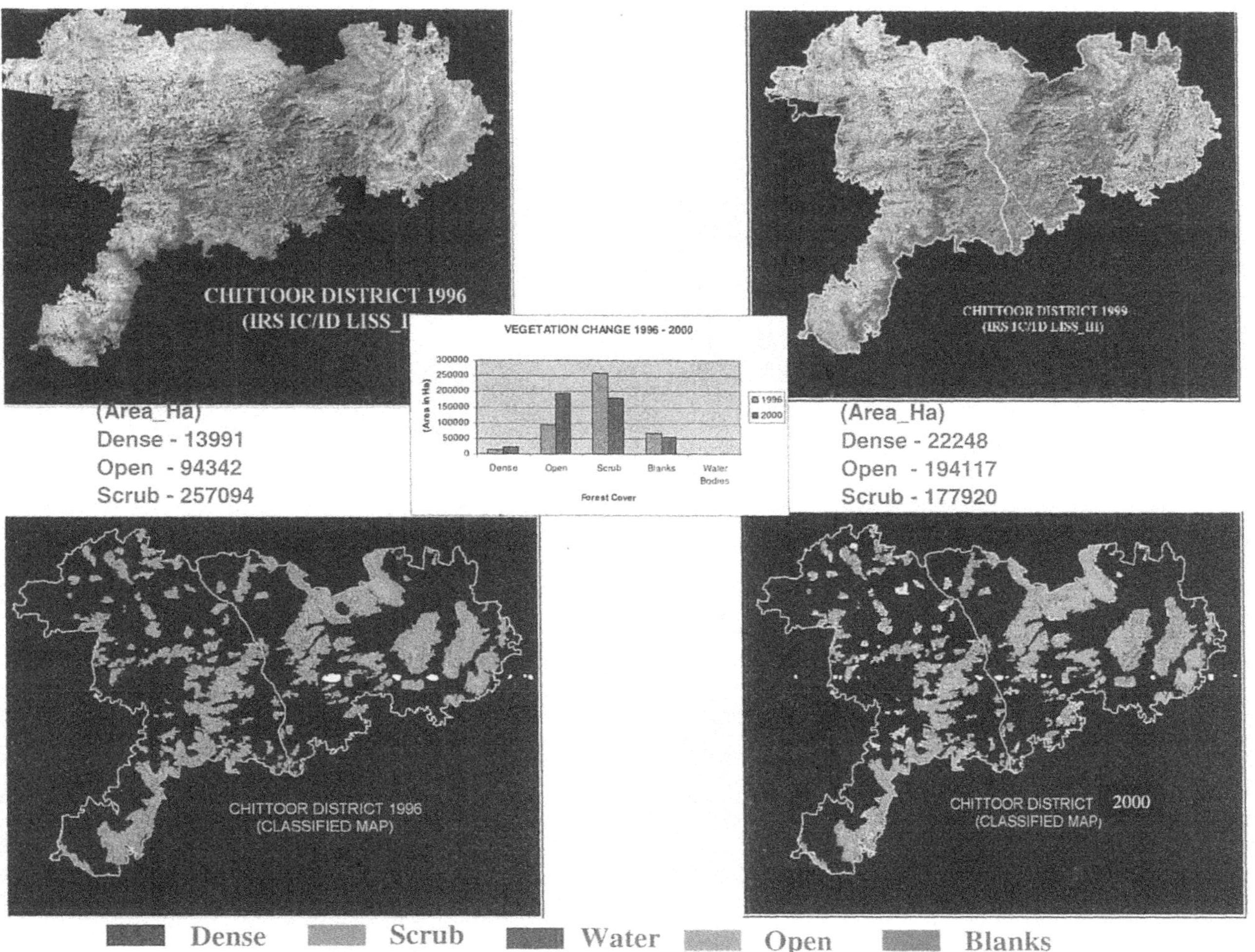

Plate 11 Forest Density and Vegetative change map of Chittor District of Andhra Pradesh State

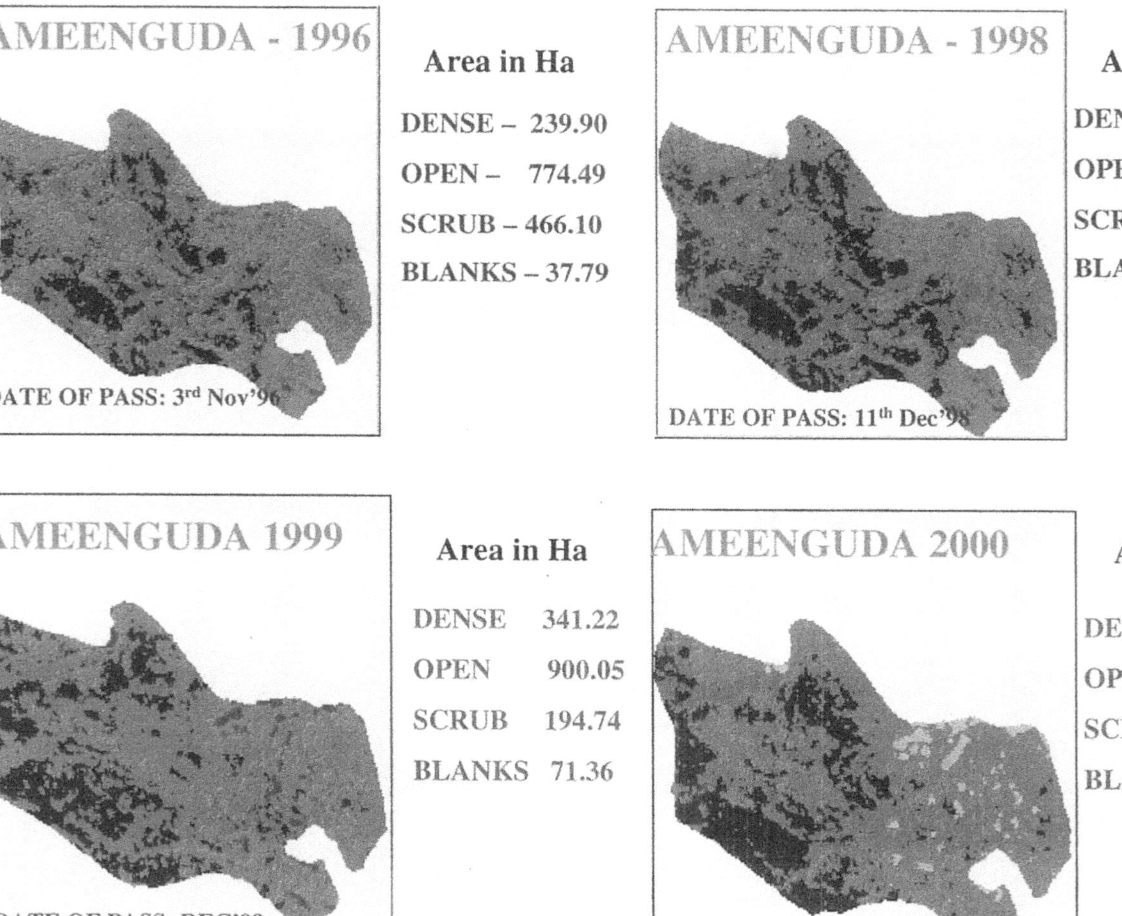

Plate 12 Vegetation change in single Beat

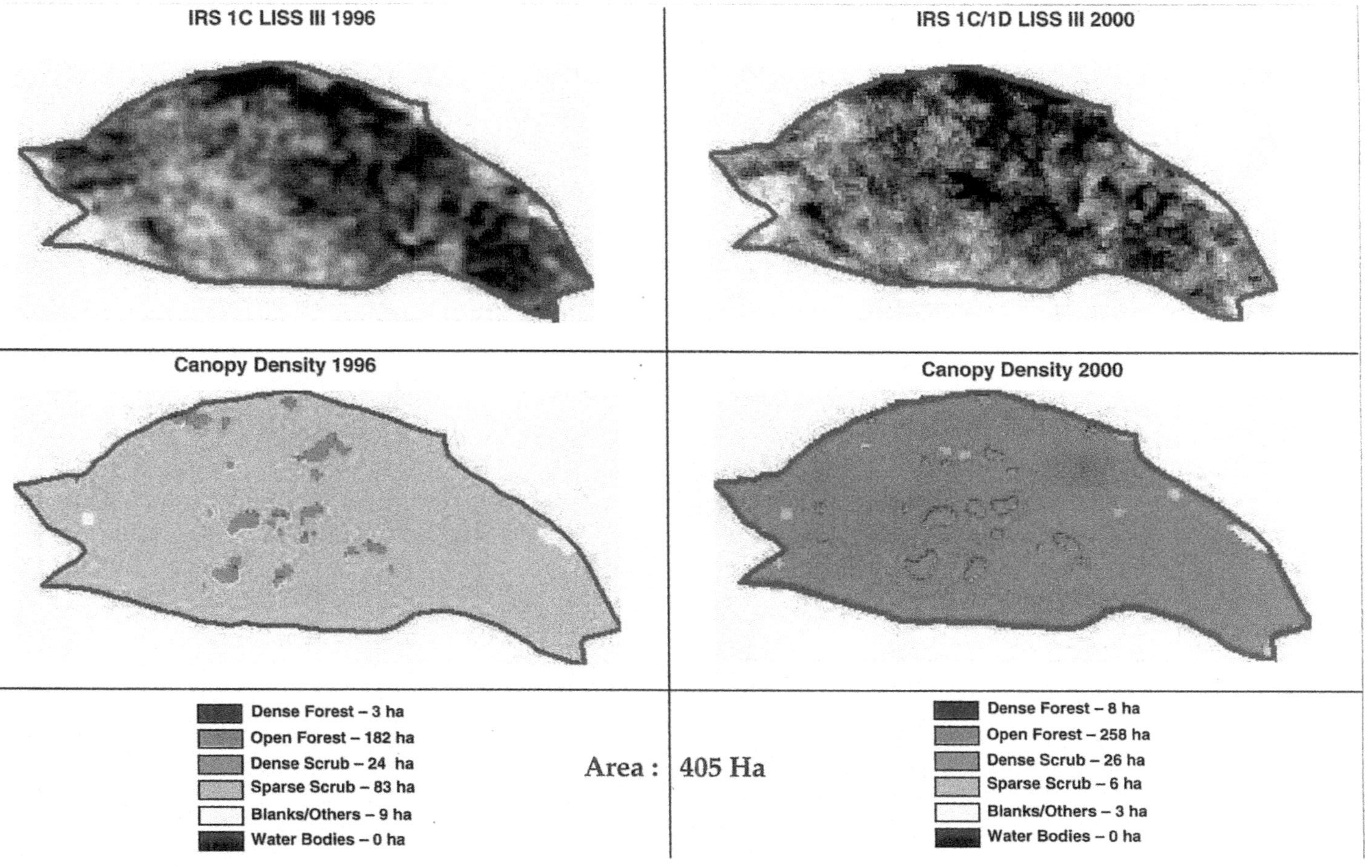

Plate 13 Vegetation Status Kotha thanda VSS

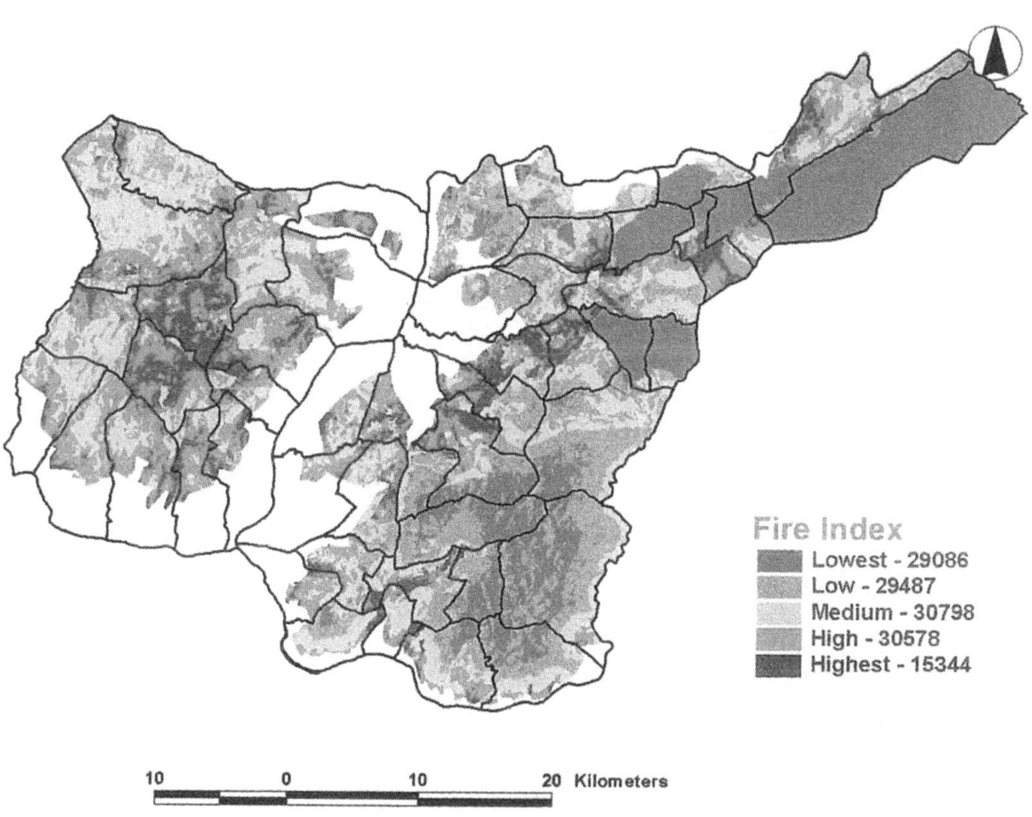

Plate 14 Fire Risk Zonation Map of Bhadrachalam South Division

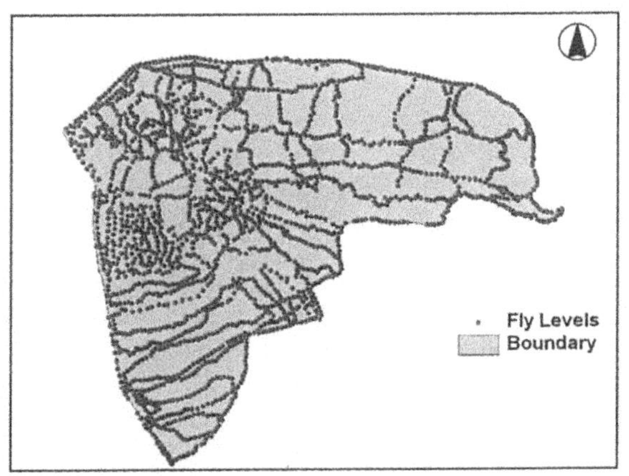

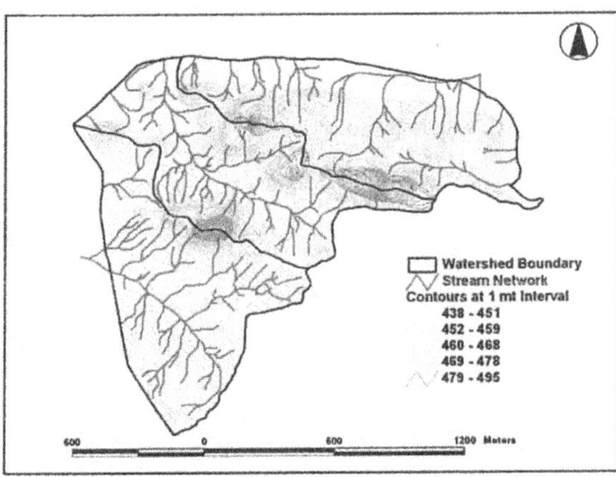

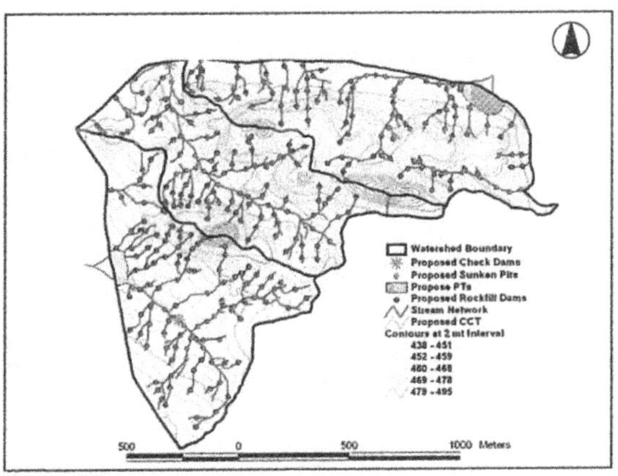

Plate 15 Spatial data for forest resource management (a) DGPS data (b) Contour of stream network and (c) Watershed management plan

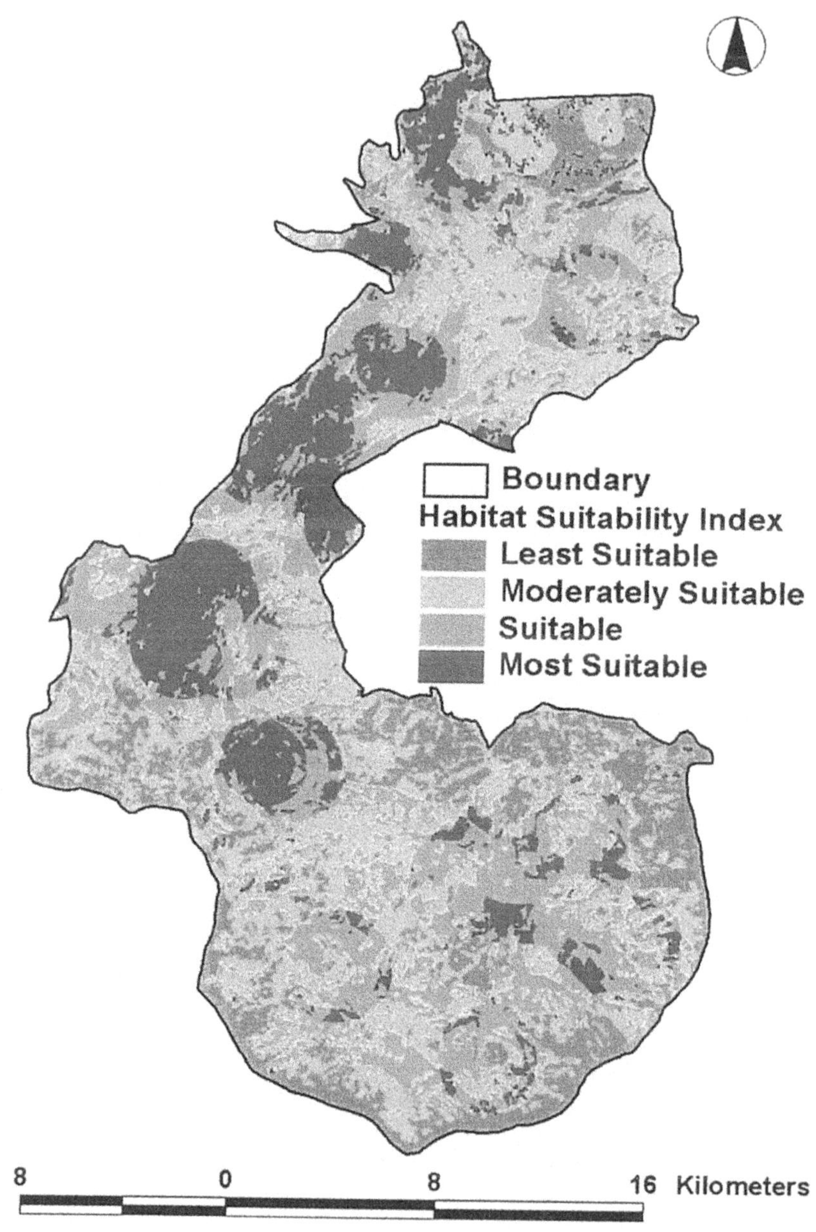

Plate 16 Habitat Map of Mannaur Range

GIS technology has been widely accepted by public as well as private forestry agencies at local, regional and global level. In large part, this has been a result of the benefit of using GIS technology over current forest maps. It is used to assess the existing forest resource and develop harvest schedules and treatment programs to project future timber supplies and for other operational planning activities. Forest inventory data can be collected using remote sensing techniques. The conventional forest inventory is in general, done progressively with a small portion of the forest being inventoried each year. To update a forest cover map could take 20 years or more with expensive manual drafting. With GIS, the forest cover maps can be updated on a constant basis and it provides the forest managers more current data than what was previously available. The inventory can be more effectively designed by stratifying the satellite imagery into homogeneous units and the sampling intensity and number of sampling points in each unit can be estimated based on their probability of existence. With GIS technology, the average age of the information in the forest database could be reduced from 20 years to only a few months. The time factor alone has led to a wide acceptance and large demand for GIS applications in forestry.

In itself, the use of GIS to update the forest inventory maps is not much more than automated cartography but it is the analytical power of GIS that sets it apart from cartography. The GIS can store and analyze the forest information in ways that could not be previously done. It can be used to calculate the harvestable timber in the forest or model the spread of a forest fire.

GIS technology provides the means to collect and use geographic data to assist in decision making for natural resources management. A digital map is generally of much greater value than the same map printed on paper as the digital version can be combined with other sources of data for analyzing information. GIS makes it possible to synthesize large amounts of different data, and to manage and retrieve the data in a useful manner. With GIS, different layers of information can be combined and analyzed to better study and understand the complex relationships between the Earth's ecosystems. So, the integration of all these new technologies will give integrated information and help in effective planning with accurate and valid information. The integrated information can assist in evolution of appropriate action plans for initiating sustainable forest management.

17.3 Forest Cover Mapping and Change Detection

Deforestation, desertification, pollution and natural resources depletion are now serious problems in the world, which threaten the survival of half of the living things in the world. These environmental change processes, are to be understood to conserve renewable and non-renewable resources for future generations. The forested landscape is not only that part of the earth's surface dedicated to the production of timber, but also extends to areas of woody vegetation sustaining other resources and providing protective benefits. Based on the landscape that has been threatened or influenced by man's activities for thousands of years and supports the current development functions, the forests may be categorized as industrial forestry, environmental forestry, rural development forestry and urban forestry. The impact of the increasing human population pressure on the forested landscape and deforestation accelerates the indirect and direct threats of pollution. This causes the major changes in the patterns of land use, soil erosion, declining soil fertility, decreasing productivity, climate change and in some regions desertification.

Forest change detection and monitoring has been confined mainly based on comparing two or more temporal sets of data, comparing old aerial photographs and present satellite imagery and comparing satellite imagery with existing old maps. Monitoring from satellite sensing system includes the detection and evaluation of sudden forest damage and continuous forest inventory. The temporal satellite sensor data is used to monitor changes which are sudden fire damage, wind blow, gradual over a short period like bark or beetle damage and gradual over a long period of time like forest decline. These change detection studies can be done by using visual interpretation and digital analysis techniques with the help of satellite data acquired on two different dates for the same ground area taken on an identical scale. The digital analysis of the satellite data has the advantage of being fast and able to handle large quantity of data whereas the visual interpretation is based on the tonal variation of different features observed on imagery. By allocating a single color to each multidate black and white transparency which is called as overlay analysis, the imagery can be compared which has been taken on several different dates. The major limitation in using satellite imagery for change detection of forest cover is the spatial and temporal resolution of the sensors. The high spatial resolution obtains low temporal resolution like thematic mapper sensor of Landsat, which cannot detect rapid change, and low spatial resolution obtains high temporal resolution like AVHRR of NOAA satellite and WiFS of IRS ID which cannot detect detailed spatial change. To understand causes of changes and deforestation process, the relationships between deforestation and landscape attribution has to be analyzed using GIS by generating a deforestation risk map.

GIS can be associated with satellite based forest cover monitoring since it is not only relatively easy to store and recall earlier satellite data sets for comparison with future datasets and to mask out those areas that are not relevant, but also the spatial changes can be associated quickly with other categories of field data such as climatic factors, soils, transportation routes environmental parameters which in turn help in the generation of a final change detection map of any project area. A flowchart showing the step-by-step process of preparing a final forest change detection map is given in Fig. 17.1.

Observations of temporal changes through the forest change detection map can be helpful for forest development activities like forest damage assessment, afforestation working plans, identification of areas of new plantations, separation of deciduous and evergreen forest cover and forest canopy studies (Heller, R.C. and Ulliman, J. J. 1983).

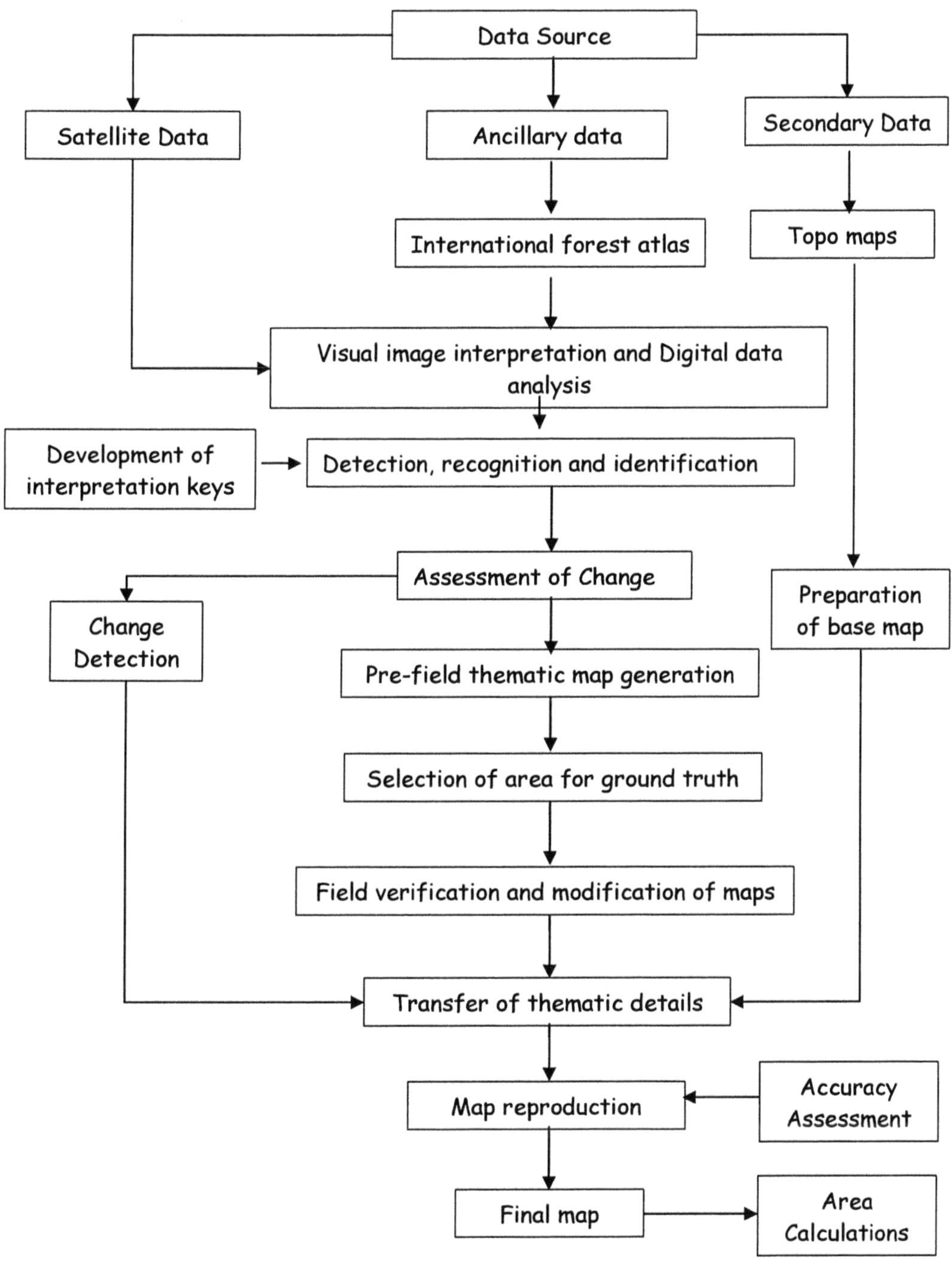

Fig. 17.1 Flow-chart showing the methodology for change detection.

17.4 Forest Inventory (Stock Mapping)

Forest stock maps are defined as the maps which contain detailed spatial information on extent of recorded forest lands including the administrative jurisdiction at various levels, infrastructure and communication facilities, water resources and the status of forest vegetation. The status of forest vegetation include categorization of forest into density classes, species distribution, assessment of growing stock, growth data for various species and age distribution. The forest stock maps depict forest type, density, encroachments, cultivation patches, human habitats, regeneration status and provide an idea about available resources. The purpose of forest inventory is to assess and monitor to the extent, state and sustainable development of nations forests in a timely and accurate manner by considering all the relevant parameters.

The disadvantages of conventional method of enumeration of the forest inventory are uncertainty in working plan staff, and lack of generation of real time information highly time consuming unrealistic. By the time enumeration of the forest stock of one division is completed and the working plan is prepared for the next 10 years or so, there may be a drastic change on the ground reality and the forest working plan may have become obsolete. The other disadvantage of using conventional method for forest inventory is that the forest stock maps are not geo-referenced so that, not possible to analyze the data on GIS platform even after converting the data into digital mode under computer-based GIS.

These disadvantages can overcome with the use of Geomatics technology. Remote sensing with its multi-spectral, multi-temporal and synoptic view has the potential to provide both spatial and temporal information on forests. This is possible because of the characteristic differences in structure, phoenological state, internal biochemical and biophysical properties of different vegetation types which result in reflection properties unique to each vegetation type. The spatial and spectral resolution of satellite data is of great advantage in precise spatial mapping of forest type distribution.

Recent studies conducted by Forest Department of Government of Andhra Pradesh and National Remote Sensing Agency (NRSA), Government of India have shown that the interpretation of high-resolution satellite data like IRS-ID PAN and LISS III can be used to prepare forest stock mapings and the possibility of achieving more informative stock maps than conventional ones. Forest density is one of the most important components in revising stock maps. With the advent of high-resolution satellite data it is now possible to obtain better information on stand density as shown in Plate 9. The stock maps and the forest density maps derived form satellite remote sensing and GIS provide the data for global changes, information on episoci events like conservation strategies, land use planning, availability of resources, master plan for afforestation programs, environmental impact assessment, planning for developmental activities, presenting the management options, reforestation of degraded forests and management of protected areas for biodiversity conservation. A classical example of forest density map of Adilabad District for the year 2000 is shown in Plate 10 and 11.

Plate 12 and plate 13 show the vegetation change in single and vegetation status of Kotha thanda which can be used to protect the forest environment.

17.5 Dynamics of Forest Ecosystem and Forest Canopy

Forests cover approximately 30% of the earth's land surface and represent important sites for evaporation, transpiration, photosynthesis and the cycling of carbon and nitrogen. These processes control the exchange of energy and matter in vegetation and they are directly linked to net primary production. Forests represent a major source for terrestrial carbon and change in forest area, particularly the effects of forest clearance in the tropics, may be partly responsible for the increase in global atmospheric Change in the magnitude of the flows of energy or matter in forest ecosystems may have widespread implications for the earth. Clearly, there is a need for information on these forest ecosystem processes, and models have been developed to describe the dynamics of such systems. The application of these models, however, requires the input of quantitative information on a range of vegetation canopy variables. The multi-spectral reflectance properties of forest canopies are known to be affected by a wide range of biophysical, biochemical and structural variables. The most important factors affecting canopy reflectance are :

- Leaf area index
- Leaf optical properties reflectance, transmittance and absorption
- Leaf location in three dimensional space: vertical and horizontal distribution of leaves and degree of clumping
- Leaf orientation in 3 dimensional space: leaf inclination angle, distribution and leaf azimuth angle distribution
- Reflectance of the under storey vegetation, soil or other background
- Geometry of illumination and view

17.5.1 Leaf Reflectance

The spectral reflectance of individual leaves or needles is controlled by a number of factors, the most important of which are absorption by photosynthetic pigments and water. The diffuse spectral reflectance caused by internal scattering and specular reflectance caused by leaf surface. The shape of the reflectance invisible region (0.4 p - 0.7 μm) is based on the composition and concentration of the photosynthetic pigments, chlorophyll and the cartotenoids Leaf reflectance in near infrared region (0.7 μm - 1.3 μm) is mainly due to the internal structure of the leaf, the number of air spaces and their arrangements. The reflectance in the near infrared and middle infrared (1.3 - 3.0μm) region are the effects of absorption by leaf water content, and biochemical constituents of leaves. It can be observed that the spectral signatures or spectral signatures or spectral reflectance curves of dry and deciduous tree leaves are similar in nature and hence the wavelength region of (1.0 μm - 2.5μm) electro-magnetic spectrum is more appropriate for estimating canopy biochemsitry of forest cover. The reflectance of leaf varies with the variation of stress factors like air pollution, the presence of heavy metals, the drought conditions, Environmental and physiological factors, the concentration of pigments, water, and chemicals in leaves and maturity of the leaf.

Variation in leaf or needle spectral reflectance will also occur from species to species as the spectral reflectance of broad leave species is different to those of conifer needles.

17.5.2 Leaf Area Index (LAI)

LAI is a quantitative measure of the surface area available for the interpretation of photosynthetically active radiation (PAR) and for transpiration and is the single most useful characteristic of vegetation for quantitative analysis of interest in global ecology. Vegetation canopy reflectance depends on the number and size of the foliage elements and their arrangement on individual plants. The relationship between canopy LAI and reflectance are well known for many agricultural crops and grassland areas but they are less well known for forest canopies. In the former, there is generally a negative relationship between red reflectance and LAI up to an asymptote which, for agricultural crops is normally reached at an LAI of 1 to 2. Interaction of near infrared with successive leaf layers results in a positive relationship between with LAI reaching an asymptote at an leaf area index of 3 to 4. The relationship between middle infrared reflectance and LAI is similar to that for the visible wavelengths. In unstressed canopy water content is normally correlated with LAI and the form of the relationship between LAI and middle infrared reflectance is likely to be negative and asymptotic. Negative relationship between LAI and red and infrared reflectance, and positive relationship with NIR were found early in the season when there was full canopy over.

17.5.3 Effect of Canopy cover

Forest canopy cover plays a key role in determining the spectral response of forest canopies because it controls the amount of under storey vegetation, which is visible to the sensor. In agricultural crops there is normally a simple functional relationship between LAI and canopy cover but in forest canopies this covariance cannot be assumed as forest canopy cover and forest canopy LAI may be independent variables and changes in forest LAI may be confounded by change in canopy cover.

17.5.4 Effect of Leaf Location and Orientation

The directional reflectance properties of the canopies are similar between those of agricultural crops, natural grass land and forest canopy. In the more sparse canopies the clumping of vegetation gave rise to increased forward scatter and decreased back scatter. The directional reflectance of forest canopies is highly dependent on the structure of the canopy elements.

17.5.5 Effect of Illumination and viewing geometry

The target illumination and viewing conditions are also important. These are the solar zenith angle, the solar azimuth angle, the sensor zenith view angle and sensor azimuth view angle.

17.5.6 Forest Stand Measurements

The term forest stand is often loosely used as a synonym for forest crop and precisely it may be defined as descriptive of an aggregation of trees sufficiently uniform in composition to be distinguishable from adjacent crops with different species composition, age classes and size classes.

17.5.7 Stand Measurements using Aerial Photographs

There are three types of Stand measurements can be possible using aerial photographs, namely, (i) Stand height (ii) Crown closure/Crown cover and (iii) Stocking.

Stand Height: the stand heights of a small sample of the trees are measured using the photographs, which is sufficient to provide a reasonable estimate of the mean height of the forest stand. Determination of stand height of forest cover can also be done using stereo-photogrammetry, which deals with the 3-dimensional measurements from a stereo-pair. But under many circumstances, this overcomes the problem that the physiognomy of the forest does not permit the stereoscopic measurement of the height of all trees recorded on stereo model permits the determination of stand height stereo-photography of the individual dominant and co-dominant trees using image parallax method discussed in chapter 9. Thus the mean stand height' determined using the parallax formula is that of the co-dominants and dominants and not the mean stand height of all trees, as could be measured on the ground. The application of image parallax method may not be possible to measure stand height in rain forest, since the ground is seldom seen and the crowns coalesce and are entwined with lianas.

Crown closure / Crown cover

This is defined as the proportion of unit area of the ground covered by the vertical projection onto it of the overall tree crowns. Based on parallax methods in combination with stand height, it is used to stratify the forest into timber volume classes which is measured on aerial photographs using either stereo gram or dot grids or crown cover percentage cards.

Stocking

Stocking is defined as the number of trees per unit area. The stocking and crown closure are the two direct measurements of stand density obtainable from aerial photographs. Counting of trees from stereo pairs of photographs depend on the skill of the interpreter scale of photograph, physiognomy of forest and the age of the stand.

17.5.8 Stand Measurements using Satellite Remote Sensing

The role of Satellite remote sensing will remain mainly contemporary and its attractiveness will be the comparative low cost per unit land area of data acquisition, and the lower cost per unit land area of image analysis. Satellite remote sensing can be used particularly in large area surveys and for monitoring frequently at national, continental and global level changes in forest cover, density and landuse.

Visual image analysis of satellite imagery can be applied to forest stands. The advantage of visual analysis is that it can be intimately associated with field work and local knowledge. These studies will be qualitative in approach as compared with aerial photographic interpretation. The smallest classifiable areas are likely to be a group of nine pixels or probably 25 pixels.

Most forest activities, involving the computer processing of remotely sensed digital image data have been carried out in many countries like China, India, Thailand, Latin America etc. This digital data analysis is used to provide informative image classification to improve geometric accuracy, to enhance the imagery for field use, to update maps and to provide layered information as input to GIS. That means, the output of visual as digital image analysis is the input data to GIS for subsequent spatial mapping. Various steps involved in the process are :

- selection and review of input data
- decision on classification level and type of desired classes
- radiometric and geometric corrections of digital imagery
- image enhancement and unsupervised classification scene-wise
- probability proportionate stratified random sampling to derive the number of sample points to be ground verified in each class
- ground truth data collection and error correction
- supervised classification
- post classification filtering of classes and seamless mosaicking
- ground checking and error correction
- classification accuracy analysis
- planimetry of classes & cartographic completion and production of hard copy outputs.

17.6 Forest Damage Assessment

Forest damage assessment can be performed using low altitude remote sensing data obtained from a light winged aircraft or a helicopter or a light aircraft using small format cameras. The damaged areas in Forest lands can be mapped using photo interpretation techniques qualitatively and metric photogrammetric techniques can be used to make the quantitative measurements with the help of these low altitude remote sensing data products. Generally, the count dead trees will be done on 1:4,000 scale and the spatial information about the subtle differences in stress on 1:1000 scale using these photographic products, more specifically, the colour infrared photographs. So far satellite remote sensing has been modest in mapping damaged areas as the coverage of each scene is very large but very much useful for the aerial reconnaissance survey. The usage of GIS techniques is in primitive stage and it attracts the positive momentum in using these technologies because of the availability of very high resolution satellite data like IKONOS and PAN and LISS III of IRS.

The fundamental principle involved in using the satellite data for assessing the forest damage is occurrence of major differences in the spectral signatures between the healthy and unhealthy vegetation covering extensive areas at various wavelength regions of the spectrum. It is possible not only to discriminate the damaged forest from healthy forest cover, but also to recognize the host forest type. But the forest species identification using satellite sensor data needs research to establish the methodology based on a number of environmental considerations (Townshend, J.R. and Justice, C., 1981). Since damage is associated with individual tree species and differences in stand density, the classification of the forests using satellite data needs to be better than level II and that the major constraint to the wider use of satellite data in disease monitoring and insect infestation monitoring is the failure to be able to identify the host tree species at an acceptable level of accuracy.

17.6.1 Disease and Insect Damage

The assessment of the forest damage due to disease and insects is performed using colour infrared film with a yellow filter. In general, in the ensuing years until 1950s, panchromatic black and white aerial photography is little used and confined to severe physiognomic damage including fire and wind storms. Heavily defoliated and dying trees appear conspicuously in dark tone on black and white IR aerial photographs, but severe defoliation could be interpreted on black and white panchromatic photographs. Variation in the spectral reflectance in the visible and near infrared between healthy and diseased foliage in color aerial photographs is now being extensively used to study the classes of damage caused by insects and disease. In general, color IR film has the advantage of providing a record of spectral reflectance in both the visible and near IR. Although on processed imagery, healthy foliage often appears red/reddish and dead or dying foliage appears blue/bluish-green, there is no exact hue, value and chroma indicator of disease.

17.6.2 Pollution Damage

Increasing consideration is being paid to using color aerial photography and high resolution satellite data in assessing water, air and soil pollution, which leads to foliage damage. This can result in the death of the trees along roadsides in urban areas and in wide-scale forest damage in the countryside. Infrared film and the satellite sensor data in infrared region of electromagnetic spectrum is used to identify the forest damage due to air oxidant. Air oxidant injured trees can be easily identifiable on colour photographs and false colour composites of satellite data. This is possible because of the absence of lush foliage on injured trees and the high spectral resolution data at an appropriate wavelength permit the identification and discrimination of positively affected trees. According to the earlier investigations, it is revealed that healthy conifers usually retain their needles for upto five years, appearing green/ green yellowish on normal color prints, whereas smog-affected foliage may have only a current growth of short needles at low density, appearing yellow/yellow-reddish on the prints. One of the most extensive and severe occurrences of pollution damage of forest stands is caused by sulfur dioxide. This damage can be readily seen and monitored through satellite imageries like IRS IC/ID LISS and Land Sat TM. Industrial pollution in urban areas causes the decline of forest health and a gradual process of tree deterioration in condition and viqour and the interaction of a number of hidden parameters like climate changes, changes eventually lead to the death of the trees.

17.7 Parameters of Forest Inventory

Forest inventories are the main source of data for the national forest inventory. Management agencies are continually updating and upgrading their forest inventories. The age of their inventories is always changing. As a result, the system of forest inventory is a snapshot of different aged information, collected to a number of different standards.

The current approach to national inventory is cost effective in that it is based on existing data. The process of developing forest inventory system is a well established and accepted by the contributing agencies internationally. The approach provides detailed information about forests that are consistent with forest management information. The inventory also contains location-specific information on the characteristics and quantity of the forest resource, providing mapping and spatial analysis capabilities. The forest inventory system should accomplish the information on the nature and rate of changes to resources, current status of the forest, vegetation attributes, data related to complete and consistent area coverage, stratification of sampling points, species diversity, wood volumes, attributes related to land use, ownership, protection status, human influence conversion of forest lands, origin of exotic trees and spatial maps of all these parameters.

17.8 Development of Working Plan

Management of forest Resources in the country primarily require growing stock maps which is generally used in the states for working plan preparation. Working Plan is the document of the forest department, written for each forest division, which incorporates the spatial, quantitative and qualitative information of the forests under the control of the division. The Working Plan is an extremely vital document since it presents the various strategies for the systematic management of forest, to be followed for the next ten years or so. Working Plan prescriptions are prepared on the basis of forest stock map, which contain information on extent, density, forest blanks, topography, soil etc. Generally for preparation of working plan maps, information is required at forest block/compartment levels. Using remote sensing techniques it is possible to identify density classes at the compartment level by overlaying the forest compartment boundaries. The study of spatial distribution of forest types would help greatly in forest management for accounting the changes that have resulted in the changing quality and composition of forests. This can be best achieved through use of multi date satellite data since conventional ground surveys are time consuming and strenuous. The use of satellite data during the optimal green season and deciduous season would help in understanding the distribution of various types on a timely and cost effective manner. The two important components of working plan are (i) description of forest boundary (Forest Map) and (ii) description of forest vegetation (Forest inventory). Survey of the forest boundary to check about their correctness and encroachment etc. is done mainly by the staff of Working Plan Divisions. The conventional plain table survey method adopted in this case is not only time consuming but also perennial in nature, since detection of any change of boundary or encroachment can be confirmed only through another detailed field survey like DGPS.

GIS has tremendous potential in storing the above voluminous spatial and non-spatial data. The socio-economic data can be combined with this data on village basis for better understanding of the human dependencies. Using GIS analysis the forest areas can be categorized into various regions defining the ecological and social role as Production forest, Protection forest, Natural reserves, Wildlife sanctuaries, Village forestry, Tree lands etc. that needs different approaches in the forest management for maximizing the benefits from forest and maintaining ecological balance.

The WP suggests the working / management of the forest as per their structure and resource availability and divides the forests into various working circles by handling complex data sets through GIS techniques by taking into consideration of locality factors and Socio-Economic conditions of an area as Production WC, Protection WC, Afforestation and Reforestation WC, Bamboo WC, Village wood WC, NTFP WC, Grass WC, Rehabilitation of Degraded Forest WC -JFM WC, Agro Forestry WC etc.

17.9 Forest Management Information System (FMIS)

The FMIS is envisaged as an integrated system which will be used to support the planning, implementation and monitoring of multi-objective forest management activities. FMIS has the ability to maintain current forest inventories and generate maps of spatially oriented data such as attributes of entities depicted on a map such as population of a village, whose location can be fixed on a map. FMIS requires the integration of information from various sources, information technology, and in particular the integration of database management system, remote sensing, image processing and GIS provides a broad range of tools for forest area mapping, monitoring and management to achieve optimization in their conservation. The elements of FMIS are shown in Flow chart (Fig. 17.2). The FMIS can be used for strategic, tactical and operational planning and implementation, and operational control in and across administrative units and levels of the organisational hierarchy. Functional areas addressed by FMIS are

(i) Administration
(ii) Research and extension
(iii) Protection
(iv) Land management
(v) Forestry project
(vi) Production
(vii) Plantation
(viii) Wild life
(ix) Development and planning
(x) Working plan
(xi) Accounts
(xii) GIS

The purpose of creation of FMIS is to :

- Monitor the status and evaluate forest resources
- Coordinate in case of forest fire
- Make effective decision and for everyday planning
- Collect information from various sources and map with integrated information
- Communicate results and findings to a broad audience in effective way

The critical issues that need to be addressed while implementing FMIS are :

- Linking the planning at all levels (strategic, tactical and operational), and providing facilities to reflect the changes in the actual situations and their implications for the plan.
- Providing a framework for the integration of GIS, remote sensing and GPS to resolve the problem of non-availability of the latest map or spatial information.
- Helping the decision makers to effectively achieve the social, environmental and economic objectives of the forest department's plan in a multi-objective decision making environment.
- Planning for and ensuring the availability of trained people during the entire life cycle of FMIS.

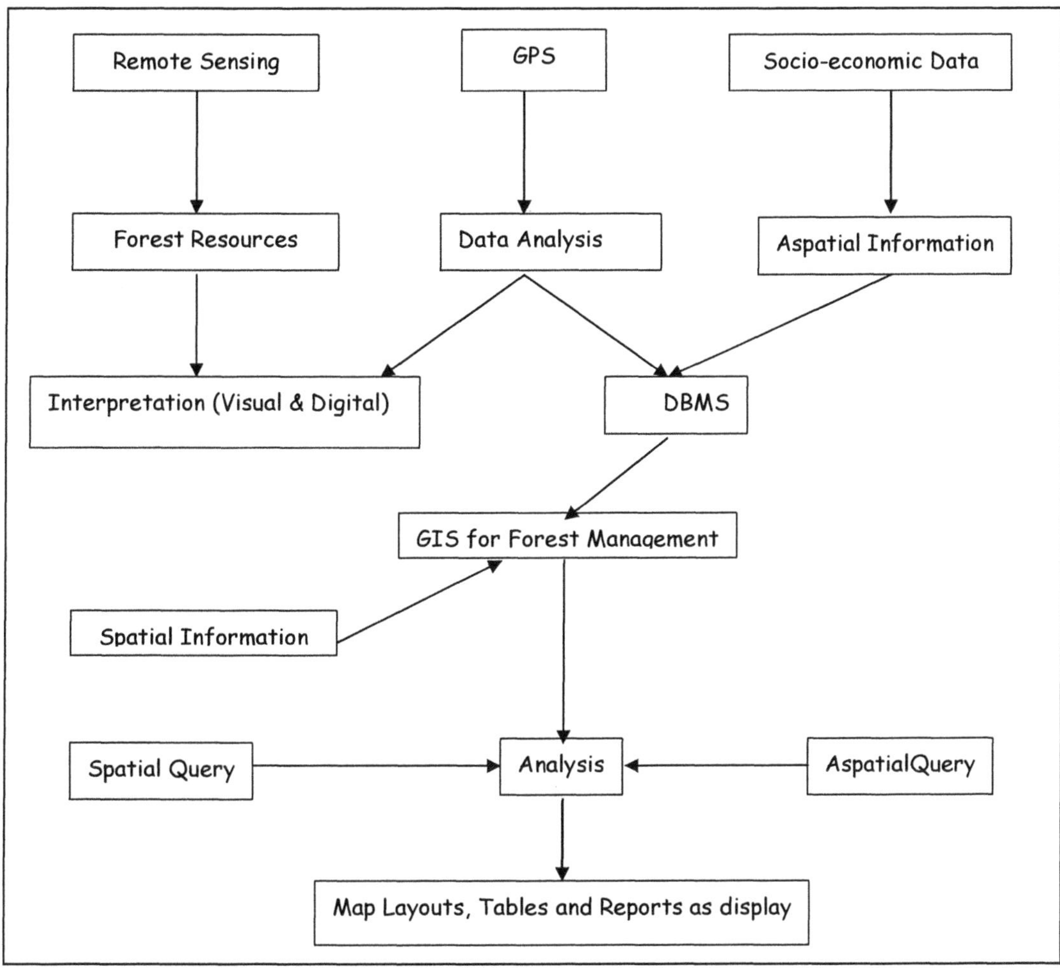

Fig. 17.2 Elements of FMIS.

17.10 Forest Fire Forecasting and Risk Area Mapping

Forest fires have an important influence on the vegetation cover, animals, plants, soil, stream flow, air quality, microclimate, and even general climate. The loss of timber is obvious and so is the damage to life and property. The loss of recreation value of the forest and the destruction of wildlife habitat are also consequences of forest fires.

The causes of the forest fires can be classified in three main categories (i) Natural, (ii) Intentional/deliberate due to man and (iii)Unintentional/accidental due to man.

Natural fires occur mainly due to lightning or sometimes due to rolling stones and rubbing of dry bamboos with each other in the strong wind. Nothing can be done to avert such fires. Intentional/deliberate forest fires are caused by the villagers, who set fire to drive away the game destroying their crops and to catch the wild animals the enmity with forest staff, neighboring rural population, villagers for collecting minor forest produce (MFP) like honey, mahua, flowers etc. that fall on the ground, and sparks from coal-fired railway locomotives. Unintentional/accidental fires occur in a forest due to careless throwing of match- sticks and burning ends of cigarettes/bidis, carrying of naked fire by people passing through forests, spread of fire from labor camps and recreation or picnic sites. These are the factors governed by parameters like vicinity to settlements and distances from roads. For any area to be prone to fire depends on many factors such as vegetation type/density, soil type, humidity of the area, vicinity to settlements, distances from roads and the host of others. While it is not possible to account for the natural occurrences or the deliberate attempts, the areas prone to accidental fires can be mapped.

17.10.1 Forest Fire Model

Geomatics can be used to detect fire spots, assessment of forest fires & burnt areas, fire hazard rating system. The layers used for such fire hazard zonation are the Forest cover map, Forest density map, Contour map, Road network map which give the required indices. The inflammability index or fuel type index is generated from the forest type and density map; elevation, slope, aspect indices from the contour map; accessibility index from the settlements and the road & communication network map. The layers and other ancillary information like wind, temperature, humidity etc., can be integrated in GIS environment to generate forest fire index map. The forest department, government of Andhra Pradesh, with the aid of Ministry of Environment and Forests (MoEF), Government of India GOI has completed the fire risk modeling for 15 districts in-house and remaining districts are to be completed. A sample map showing the zones of five risk can be seen in Plate 13. The objectives of the department are

- forest fire risk modeling
- mapping forest fire burnt areas
- designing of preventive measures for alleviating the forest fires
- detection of fire spots if equipment for recording NOAA AVHRR data
- rational utilization of funds.

17.10.2 Fire Behaviour Prediction Model

Researchers and scientists have long been trying to predict the behavior of a forest fire. Computer modeling has been the effort of many scientists using high resolution remote-sensing satellite imagery, powerful software, and GIS. In order to model a forest fire, the techniques for obtaining, analyzing and displaying spatial information in a timely and cost-effective manner are needed. As forest fires are spatial, GIS is used as a tool for modeling. A fire simulation program called FIRE! has been developed using ARC/INFO. "The model puts the power of comprehensive fire behavior prediction into the hands of qualified ground resource managers where it can be most effectively applied." (Weinstein, Green, Campbell, and Finney). Through the integration of fire behavior models with GIS models, new insights in the fire danger situation can be gained. One example is the damage potential that arises from fires starting at a certain point to sensitive objects and areas like buildings, railway lines, fire-sensitive ecosystems. Proximity is a concept which is used in a great many GIS-related models. The need for accurate and dependable tools for forest fire management has led to the demand for GIS in fighting forest fires.

In the past, the fire behavior prediction tool used was BEHAVE. BEHAVE is a non-spatial tool that inputs fire fuel types, topography data, weather data, and initial fuel moisture data. While BEHAVE is useful for predicting fire characteristics in a given area, it cannot give spatial output. It is unable to predict spread rates, flame lengths, fire line intensities, and heat calculations when these parameters change and they do constantly during a fire. Whenever there is a change, the entire BEHAVE model must be recalculated. Obviously, this is a very inefficient way to model the behavior of a forest fire. However, with the use of GIS technology, the prediction of fire behavior has not only become possible, but incredibly efficient and effective.

The GIS-based forest fire model puts the power of sound, accurate and efficient information into the hands of forest fire management personnel charged with coordinating the containment and extinguishing of wildfires. The model can now become one of the most effective tools for managing personnel, financial and equipment resources for battling one of our most destructive and dangerous forces of nature." (Weinstein, Green, Campbell and Finney).

17.10.3 Fire Group

To improve local decision support to integrated fire management through improved information and knowledge, we think that the documentation of events, in combination with local information and insight, must be customized to local needs to support improved knowledge, and hence, efficient decision making. Customized information improves local understanding of fire issues, allowing better definition of strategies, better management of resources, and better assessment of their implementation.

The Fire Group has expertise in fire monitoring, forestry, socio-economics, remote sensing and GIS, and decision support. It is also part of the Natural Resources Institute (NRI), which is an internationally recognized centre of expertise in research and consultancy in the environment and natural resources sector.

Most of NRI's work supports development assistance programs and contributes to environmentally sound management of physical and biological resources ('natural capital') for use by future generations. We can offer experience in low-cost technologies for the Local Application of Remote Sensing Techniques (LARST) and state-of-the-art expertise in the extraction of locally relevant fire management information.

17.11 Bio-Diversity Characterization

We live in a landscape and fulfil our most of the needs from its diversity. The increasing human population satisfies their demand from various kinds of economic activities carried out on the landscape, which ultimately along with various geomorphic and biological processes give the birth of spatial and structural changes.

Renewable natural resources, especially forests, are conserved for three major sets of purposes: the maintenance of ecological processes and life support systems; the sustained use of the resources for consumptive and social benefits; and the conservation of biodiversity units own right for ethical, moral, aesthetic and evolutionarily reasons. Only if all three of these can be met and integrated into the socio-political process, true sustainable development will be achieved, to meet them requires quantification of the current status and of changes on the resources. To ensure political and public support for conservation and wise use of such resources may require economic value.

Sustainable development of natural resources has become a key issue for survival of planet earth. In recent times, conservation of biodiversity has been put to the highest priority through Global diversity convention. There has been severe threat to the ecosystem by human interference and unscientific exploitation of natural resources. Therefore, there is an urgent need to conserve the gene pool in situ before it lost forever. The most appropriate method to that would be, to assess the ecological susceptibility by way of understanding the ecosystem/landscape complicities and their uniqueness. The landscape elements will be useful to generate scientific basis for further decision-making.

17.11.1 Scientific rationale

The landscape change spatially and temporally due to the tectonic, climate, biotic interaction with the environment. Of these the influences brought by the human beings through indiscriminate utilization of the natural resources and action, without due regards to the environmental impacts and sustainable development, can be prevented and controlled by scientific and technical know how, if proper data base is built. The landscape characterization has been proved to be effective for evaluation of landscape ecologically, and for the sustainable utilization of natural resources, to maintain the essential ecological processes and the life support system, to conserve the biological diversity and to prioritize the conservation areas and creation of network of protected areas.

- Study of structure, function and change in landscape elements.
- Study of patch characterization for habitat suitability analysis.
- Prioritization of natural and biological conservation areas by identifying "hotspot" regions and creation network of protected areas for present and future generations, to follow the WCS, and NCS.

- Study of fragmentation and disturbance indices for the attention of policy makers and the managers of the ecosystems.
- Study of landscape patterns for better management practices.
- Acquisition and preparation of long term sets for perception of global climatic change, deforestation, forest dieback, desertification, acid precipitation etc., for management intervention.
- To meet the objectives of sustainable developments.

17.11.2 Landscape ecology

Landscape can be defined as heterogeneous land surface composed of a cluster of interacting ecosystems that is repeated in similar from throughout. Therefore its study focuses on -

Structure : The relationships among the distinctive ecosystems or elements present i.e. distribution of energy, materials, and species in relation to their sizes, shapes, numbers, kind and configuration of the ecosystem.

Function : The interactions among spatial elements

Change : The alternation in the structure

The above three characteristics are closely related to each other, by studying one of them it is also possible to obtain information and insight concerning the others. Landscapes differ functionally according to their structure. Also changes in the structure of the landscape are related to changes in this functioning, and vice versa. Finally, the overall structure of the landscape influences the biodiversity, the species flow and the energy flow.

Landscape ecology focuses on the relationship between the various spatial units. Each landscape is formed by several landscape elements, which appear as patches and vary markedly in size, type, heterogeneity and boundary characteristics. Each of these characteristics has its own significance in the ecosystem and is important in evaluating landscape structure.

In addition, landscape ecology involves the application of these principles in the formulation and solving of real-world problems. Landscape ecology considers the development and dynamics of spatial heterogeneity and its affects on ecological processes, and the management of spatial heterogeneity.

There have been recent advances in computer hardware, remote sensing, GIS information theory, hierarchy theory, fractal geometry, and mode developing. Geographic information systems (GIS) are computer-based systems for the manipulation and analysis of spatially distributed data (Plate 15).

17.11.3 Remote Sensing and Landscape Ecology

Remote sensing plays an important role in analysis of environmental, cultural and natural resource management characteristics and because of its multi-spectral capabilities, it presents a unique perspective for observation and management of biophysical characteristics. Remotely sensed data can be collected at multiple scales and at multiple times, thereby offering the opportunity for analysis of various phenomenons synoptically from local to global scales throughout the time. These attributes make remote sensing data appealing to landscape ecology. Moreover, remote sensing data can be used as an important source of data for the development and refinement of models and can be used to validate models that predict landscape change.

Remote sensing promises to bridge the gap between the ecological research and the better planning and management of landscape and the distinct advantages are :

- Observe interference is prevented because of the large distance between the sensor and the object.
- Regional or global measurements can be done repeatedly, and
- There are a wide variety of spectral ranges and sensors available to provide remotely sensed data.

The remote sensing techniques are effectively useful on landscape ecology viz., inventory and mapping of resources, qualification of environmental characteristics, describing the flows of matter and the energy in the ecosystem, and evaluating change and alternative solution for ecosystems management etc. It is also useful in identifying community characteristics from a synoptic perspective; the ability to detect and observer temporal changes in habitat conditions.

17.11.4 Landscape Ecology and GIS

The GIS as a tool can be used to :

- Analyze temporal change,
- Determine spatial coincidence of physical and biological features,
- Determine spatial characteristics such as proximity, contiguity, patch size and shape
- Analyze the direction and magnitude of fluxes of energy; organisms or materials produce graphic outputs
- Interfaces with simulation models to generate new spatial data.

GIS capabilities with respect to landscape analysis are :

- Provide a database structure for efficient storing and managing ecosystems data over large regions.
- Enable aggregation and desegregation of data between regional, landscape plot scales,
- Assist in location of study plots and/or ecologically sensitive areas,
- Support spatial analysis of ecological distributions.
- Improve remote sensing information extraction capabilities, and
- Provide input data/parameters for ecosystem modeling

Geomatics can be used to evaluate landscape parameters from the analyzed raster data that is generated from the remote sensing data and also to generate many metrics useful in determining landscape pattern and structure like **Basic measures - Area statistics** -Patch statistics, Perimeter statistics, Information theoretic measures, Contagion, Diversity, Dominance, Evenness; **Probabilistic measures** – Electivity, Adjacency; **Shape measures** - Average perimeter to area ratios, Fractal dimension; **Textural measures** - Angular second moment, Inverse difference moment, Lacunarity; **Structural measures** – Connectivity etc., which will be useful in evaluating the biodiversity value, the disturbance index, the ecosystem uniqueness and finally the biological richness of the area.

17.12 Wildlife Habitat Mapping

Habitat analysis is considered most important for planning and management of protected areas. Wildlife Habitat includes wide variety of factors viz., vegetation, cover characterization including human influence on all these, soil, and topography and water availability. Since each species required a particular habitat to meet the requirement of space, food, cover and other needs of survival, the database with respect to above parameter can be used to identify the species needs for particular wildlife for Habitat suitability analysis.

Remote sensing has great potential in wildlife management because of its role in providing information about three most important parameters viz., cover, and food value and water resources and the fourth physiography indirectly required for Habitat suitability analysis. Besides other analysis like home range, tourist route planning and carrying capacity analysis can also be made. The change in the habitat that is mainly due to change in cover characteristics, degradation due to reduction in canopy density, fragmentation of forested areas, degradation due to fire, over grazing and cultivation can also be analyzed. A hapitat map of Mannanur Range of Andhra Pradesh State in seen in Plate 16.

The Geomatics will be helpful in :

- Monitoring of Physical setting of the protected area and land cover changes in the vicinity.
- Wildlife habitat suitability mapping for carrying capacity analysis and creation of corridors.
- Monitoring of the corridor of the protected area, important population migration and genetic exchange
- Monitoring of Positive and negative impacts on the habitat conditions.
- Management and monitoring of various changes in flora and fauna.

18
Watershed Management

18.1 Introduction

Soil, water, and vegetation are the most vital natural resources for the survival of man and his animals. To obtain the maximum and optimum production of vegetation, all the three resources have to be managed efficiently. For their efficient management, one has to look for suitable unit(s) of management so that these three resources are handled and managed effectively, collectively and simultaneously. Soils can be managed on the basis of a soil series or type or any other convenient unit of land; vegetation can be managed on forest type/forest sub-type or some similar classification for trees/grasses etc., water can be managed if a watershed is taken as a unit. Since soil and vegetation can also be conveniently and efficiently managed in this unit, the watershed is considered the ideal unit for managing these three vital resources of soil, water and vegetation.

Watershed conditions affect and influence water quality, quantity and regime. These three water parameters are vital for human and animal consumption, irrigation, industry, power generation, recreation, transportation and in many other ways. By manipulating different watershed characteristics the quality, quantity and regime can be regulated according to requirement.

Watersheds should be properly managed through scientific planning to prevent loss of top soil and water that results in low productivity, to prevent land from deterioration, to mitigate flood hazards in the downstream reaches, to prevent rill/ gully formations even at low rainfall events etc., which can be achieved by adopting suitable soil and water conservation measures to intercept and reduce runoff, and induce larger and extensive adoptions of rain water, to trap eroded material and thus reduce sediment transport into streams and in turn into reservoirs, to restore/reclaim degraded lands into productive lands, to increase total productivity and provide a stabilized environment.

Many countries all over globe are facing acute water problem due to population explosion, urbanization, conversion of forests to agriculture, indiscriminate use of surface and ground water. Water on earth goes through a hydrologic cycle. Rain water is the major source for agricultural production for major parts of the world and especially the tropical countries. Major part of the total cropped area in these countries is under rain fed/unirrigated. Hence effective rain water management is vital to increase and stabilize their yields. Realizing the fact the Andhra Pradesh State Government, India is implementing massive Neeru-Meeru (Water and We) (Anji Reddy., M. 2002) program for harvesting the rain water (Fig. 18.1).

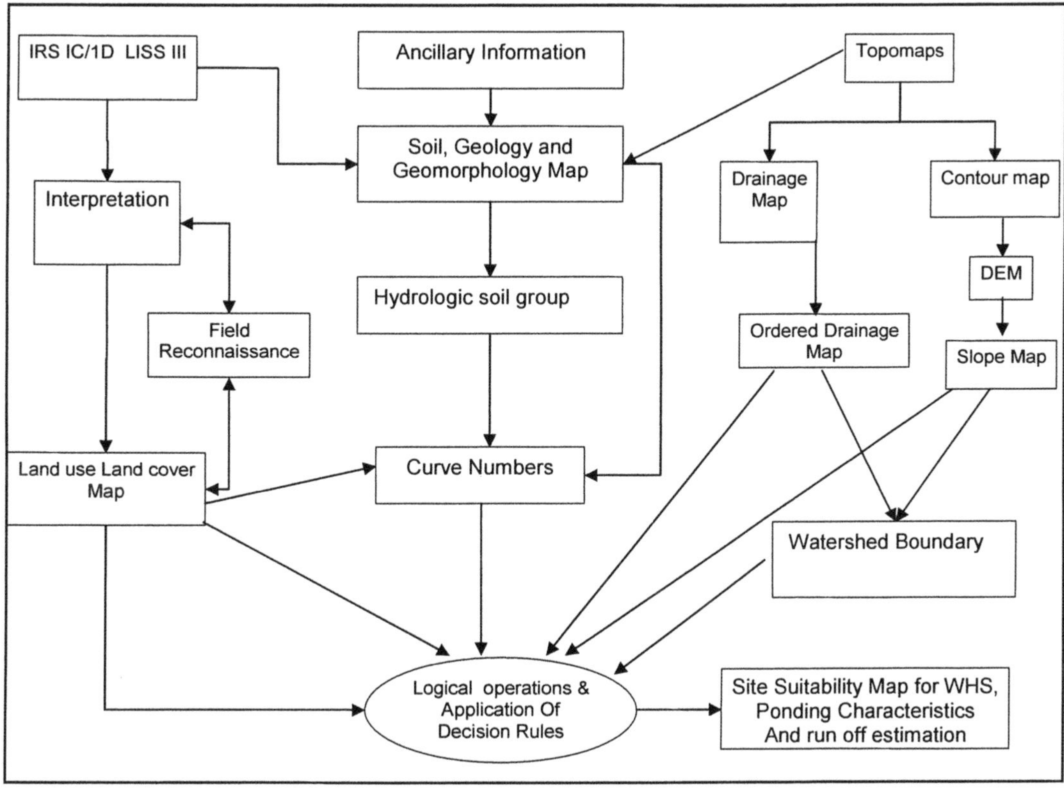

Fig. 18.1 Schematic representation of preparation of site suitability map of water harvesting structures.

The traditional rain water management practices have not yielded satisfactory results. GIS, RS, GPS and various analytical tools associated with them, are proven to be capable, providing sources of data and means of manipulating data appropriate for large scale management of natural resources i.e., forests, water, soil etc., and appropriate planning on sustainable basis. The DGPS technique in conjunction with the GIS and RS can be used for site suitability analysis of water harvesting structures.

Water is the elixir of life. The first civilisations, Mesopotamia and Harappa (Indus valley Civilisation) flourished on the banks of rivers. Thus the importance of the water was known to the first civilisations of our world. Even though groundwater was discovered long before Christ, no technology could replace the necessity of ground water. Thus the exploration for water is still continuing even in the twenty-first century. Science and technology are growing at such a rapid pace that we can communicate with a person who is on the other side of the earth or in space, but the scarcity of even drinking water is still persisting.

The growing pressure on land for food, fibre and fodder in addition to industrial expansion and consequent need for infrastructure facilities due to ever-increasing population, has given rise to conflicting demands on finite land and water resources. India is a land which has been blessed with a number of perennial rivers like the Ganga, and the Bramhaputra in the north; the Godavari and the Krishna in the south, and so on. The water from these rivers can be put to many uses, such as, drinking, agriculture, domestic and industrial uses, and navigation and power generation.

Rainfall is the source of water in our country. About 175 Mha of land in India, consisting about 53% of her total geographical area, suffers from deleterious effect of soil erosion and other forms of land degradation. A country like India that supports 16% of world population on 2% of the global land area, the problem is serious enough. Keeping in view of the ever increasing population and need for food security for future generation, it is realised that the water and land resources need to be developed, used and managed in an integrated and comprehensive manner.

It has already been realised that the soil and water conservation measures carried out on a watershed basis play a prominent role in the strategy of comprehensive land and water management. A watershed is an area from which runoff resulting from precipitation flows past a single point into a large stream, river, lake, or ocean. Thus watershed is the surface area drained by a part of the totality of one or several given water courses and can be taken as a basic erosional landscape element where land and water resources interact in a perceptible manner. Watershed management is the process of formulation and carrying out a course of action involving modification of the natural system of watershed to achieve specified objectives. It implies the proper use of land and water of watershed for optimum production with minimum hazard to natural resources environment. Remote Sensing and GIS techniques are being effectively used in recent times as tools for watershed management.

Watershed management is an integrated approach to the management of water, land, and vegetative resources for multiple objectives. Watershed being a natural hydrological entity responds effectively to various engineering, biological, and cultural treatments designed to maximum production. The quantitative analysis of drainage system is an important aspect for characterisation of watershed. Earlier studies indicate a relationship between cumulative stream length, stream order, bifurcation ratio, drainage density, stream frequency, and different topo elements, such as, texture ratio, form factor, circulatory ratio, and elongation ratio, and the prioritisation depending upon the Sediment Yield Index (SYI).

India is well endowed with rainfall and is one of the wettest countries in the world with annual precipitation estimated at around 400 millions hectare-meters; however, a vast area of the country is drought prone. The arid and semi-arid regions of India covering 53.9% of the geographical area are drought areas. Precipitation is confined to three or four months a year, and most of the rainfall is lost as runoff and evapo-transpiration (Seshagiri Rao, K. V., 2000).

In developing countries like India, the major thrust of any Government is the sustainable development of rural areas. This is possible only with proper planning of watershed, which is essential for the conservation of land and water resources management. Characterisation and analysis of watershed are prerequisite for watershed management. Watershed characterisation and its productivity can be achieved by using watershed concept. Watershed characterisation requires generation and gathering of information on a number of parameters both static and dynamic in nature comprising geology, geomorphology, hydrology, soils, landuse/landcover, soil erosion, and climate. Sustainable development aims at maintaining an equilibrium between human needs and economic development within the parameters of environmental conservation through efficient use of natural resources to ensure trade-offs between desired production-consumption levels within the natural ecosystem.

Watershed characteristics

- Physiography: size, shape, land slope, drainage density and pattern
- Edaphic conditions: soil and geology, land use, vegetative cover, rainfall
- Socio-economic factors: Demographic profile, social stratification, farm structure, attitudes and behavior, land tenure systems
- Organizational structure: people's participation
- Peak runoff rate

18.1.1 Concept of Sustainability

The concept of sustainable development drew worldwide attention after the publication of Bruntland report. Sustainable development is defined as a development that meets the needs of the present without compromising the ability of the future generation to meet their own needs. Turner (1992) emphasised that sustainable development should be based on economic development, and depends upon interactions among three systems : the biological system, the economic system, and the social system. The goal of biological system is maintenance of genetic diversity and biological productivity. For the economic system, it is the satisfaction of basic needs (reduction of poverty), equity enhancement, and an increase in useful goods and services. The aim of the social system is to ensure cultural diversity, institutional sustainability, social justice, and people's participation.

In order to make use of the existing technologies for the sustainable development of any part of the developing countries, the management of local institutions should adopt a three tier system. This system consists of (a) group level (b) community level and (c) locality level (Rajora, 2000). Group level consist of persons in a small residential area based on occupation, age, and gender. The user groups of Andhra Pradesh at micro watershed level are an instance in this regard. The community level consists of persons from self-contained socio-economic residential unit as watershed development committee of Andhra Pradesh. The third is the locality level with a multi-disciplinary team drawn from local institutions, district level institution and project implementing agencies.

18.2 Philosophy and Concept of Watershed

The significance of watershed strategy is traced back to two independent historic processess. The attempt to restore Alpine in the last quarter of nineteenth century and the conservation movement of the 'U. S' have pushed the watershed to the surface. However, it is FAO, which has been playing the catalytic role for the last five decades at the international level in campaigning for the watershed approaches. The land restoration and torrent control techniques developed in Europe and the vegetation manipulation, and soil and water conservation techniques of U.S along with conceptual and theoretical inputs have formed the basis for the FAO's watershed strategy.

Watershed is a spatial unit with complex relations among its entities and is defined by stream pattern where the rainwater is collected through surface and sub-surface runoff and culminated at the common point down the stream. The hydrological regime of watershed has inalienable relationship with flora and fauna, soil, climate, and population, specific to local eco-system. Human interface with nature on spatial and temporal scale in terms of the resource extraction/exploitation is a right strategy to understand the forces and relations of development. The indiscriminate exploitation of the resources during the phase of modernisation clearly demonstrated the historical negation of resources.

The multi-dimensional nature of the watershed impinges upon the strategies to be multiple in nature and will be invariably tackled by interdisciplinary approaches. It is increasingly recognised that watershed management not only tackles soil erosion, soil percolation, soil moisture, surface and sub-surface runoff, conservation of water and soil, deforestation, contour trenching checked on development but also poverty, unemployment, housing, illiteracy, soil inequalities, economics and social justice, cells for people's mobilisation towards environmental awareness, and right to access for unpolluted resources.

The concept behind planning and management of any watershed involve in developmental activities using Natural Resources available in accordance with socio-economic conditions of that area. The two general natural resources for the development of watershed are : land and its related resources and both surface and subsurface water resources and the details of cultivation and other infrastructure facilities. Some of the developmental activities in any watershed in general comprises forestry along ridge lines and steep slopes; pasture development in the forest areas along ridges; soil conservation needed in most of the area spreading from forest boundary up to common areas; dry farming practices in the rainfed cultivated areas; water conservation measures including construction of percolation tanks in the rainfed cultivated areas; construction of irrigation tanks across minor streams; groundwater exploration works where groundwater use is available; sericulture under well ayacut; and making people aware of the importance of watershed management.

If all these developmental activities are implemented on watershed basis instead of administrative boundaries, in a planned manner they have a greater impact on the economic benefits since most of the activities support each other.

18.2.1 Classification of Watersheds

Classification of watersheds becomes important for the development of countries at village level for its sustainability on a systematic, scientific, and rational approach to watersheds. In order to apply scientific and technological results to planning and developmental activities, it is essential to have the hierarchical system of delineating bigger hydrologic units into watersheds of different sizes. The size of the watershed may vary from a few square meters to thousands of square meters. On the otherhand for the creation of an action plan for land and water resource development or for the construction of a percolation tank, a few hectares of watershed are sufficient.

Therefore for all practical applications on land and water resources development, all the hydrological units have to be classified as having a workable size of the watershed. The size of the watershed is also affected by afforestation, grass land development, cultivation, and so on. Many physiographic features like valley, undulating hillocks, and rugged hilly terrain also influence the size of the watershed. Based on the developmental activities, physiographic parameters, geomorphometric characteristics, aims and objectives of a particular developmental programme, and planning and implementation of an action plan, Bali (1980) has identified the hydrologic unit taking their dimensions into consideration. (Table 18.1)

Table 18.1 Terminology against the size of the watershed

Terminology	Size
Sub-watershed	100-500 sq. kms.
Milli-watershed	10-100 sq. kms.
Micro-watershed	1-10 sq. kms.
Mini-watershed	<1 sq. kms.

The Central Water and Power Commission (CWPC), delineated India into 6 distinct Water Resource Regions (WRR). They are :

Region (i) Rivers flowing into the Arabian Sea, excluding Indus system.

Region (ii) The Indus basin in India.

Region (iii) Rivers flowing into the Bay of Bengal, other than the Ganga and the Brahmaputra systems.

Region (iv) The Ganga system.

Region (v) The Brahmaputra system.

Region (vi) Rajasthan.

The number of hydrologic units at different stages of delineation of our Country are shown below :

Water Resources Regions	- 6
Basins	- 35
Catchments	- 112+
Sub catchments	- 500+
Watersheds	- 3237+

As shown in Table 18.2 watersheds in India are delineated on the basis of size Table 18.3 shows the terminology used for the watershed, symbolism for the boundary configuration and respective code.

Table 18.2 Terminology, size and digital code adopted for watershed delineation

Delineation order	Suggested Terminology	Mean size (+50%) sq. km.	Example of Digital code
(i)	Water resources Region	5,00,000	4000000
(ii)	River Basin	50,000	4060000
(iii)	River Sub-Basin	5,000	4060400
(iv)	Watershed	500	4060480
(v)	Subwatershed	50	4060486
(vi)	Mini-watershed	5	Open
(vii)	Micro-watershed	Below. 2.5	Open

Table 18.3 Hydrologic Units, Boundaries and Code Numbers

Terminology	Symbology	Code
Region		2
Basin		D
Catchment		4
Sub-catchment		D
Watershed		8

Water resources regions are assigned Arabic numbers 1, 2, 3,....
Basins are assigned letters as A, B, C,..
Catchments are assigned Arabic numbers 1, 2, 3,....
Subcatchments are assigned letters as A, B, C,...
Watersheds are assigned Arabic numbers 1, 2, 3,....

Thus watersheds will have the codes like 1A1A1, 2B2A3, 3A5C4,...

18.3 Technology Vectors and Social Dynamics

Deforestation, poor management of top soil, water resources, and total neglect of environment have intensified poverty in this area. The extinction of biodiversity caused by anthropogenic impact on the environment is destroying the rich natural heritage built up through decades of evolutionary process. The long-term vision of a sustainable and equitable world requires harmonisation of cultural diversity with human solidarity to achieve a social revolution with global planetary consciousness.

Science and technology which alone can usher in a quantitative transformation of society and ensure its very survival, is helpless unless the society as a whole is able to imbibe the scientific temper and make science and technology an integral part of its cultural fabric.

Long-term solutions for solving the root causes are suggested to meet basic human requirements within the constraints imposed by limited land resources and sustainability of environmental integrity (Venkateswarlu., P. 2000).

The Ideal Integrated Watershed Management Model (IIWMM) is presented in this book by examining a case study. This study is concentrated to meet the micro level priorities and needs by incorporating the ridge-to-valley concept keeping in mind the dictum of thinking globally and acting locally.

18.4 Role of Remote Sensing and GIS

The concept of watershed as a planning unit for development of land and water resources, has gained importance since 1974, when the Ministry of Agriculture, Government of India, initiated various developmental programmes like Drought Prone Area Programme (DPAP) and Desert Development Programme (DDP). Therefore it is necessary to delineate watershed boundaries at various levels of hierarchy to identify developmental potentials under various activities in each watershed. The plans for watershed development should bear in mind the fragile economic system of a region and ensure user's participation in developmental activities.

Remote Sensing through air borne/space borne sensors have been gainfully used for watershed characterisation, assessment of watershed (priority) evaluation problems and potentials, management requirements, and periodic monitoring. Satellite image interpretation and analysis of digital raster/vector data has brought out substantial improvement in watershed survey and mapping techniques during the last two decades. Remote Sensing invites integrated satellite inputs of various land information units for micro level watershed management. GIS technique can be used to integrate all the themes obtained from remote sensing data interpretation for integrated multi-discipline watershed approach for its proper management for sustainable development of rural areas.

Space technology in the form of remote sensing can play useful role in watershed management. Modern technologies employed in watershed management are remote sensing and GIS. The advent of Remote Sensing and GIS techniques has opened new possibilities for the study of various components of watershed management including land resources, water resources and evaluation of greenery status. The spatial data obtained by remote sensing satellite along with the attribute data about the watershed can be integrated on GIS platform for the overall management of any given watershed. On this basis a plan on micro level watershed management for sustainable rural development of drought prone areas using Remote Sensing and GIS techniques, is presented in this chapter for understanding the concept of watershed approach by examining a model watershed.

The Geomatics can be used :

- To create Digital Elevation Model / Digital Terrain Model and to estimate slope percentage (DGPS will be employed to collect the fly levels for specific micro watershed areas to generate contours at 1 mt interval) and to generate the stream network (upto streamlets) and to delineate the watershed boundaries
- To derive the various watershed parameters from the satellite remote sensing and other ancillary data.
- To integrate the watershed parameters in GIS environment by framing various decision rules to identify suitable sites for water harvesting to adopt appropriate soil and water conservation measures and their prioritization as per feasibility and local importance and to prepare a comprehensive watershed management plan.
- To characterize the geometry of flow and ponding at a selected site, to estimate the timing and magnitude of surface water flow and accumulation, to establish the actual content and extent of surface water ponding at the selected site at two or more specific times, to compare the results of simulated estimates of water distribution with the measured data on actual ponding.
- To assess the impact of the watershed management practices.

Remote Sensing, GIS and GPS technology will provide much information and can be used to estimate the runoff or used as an aid for estimation of equation coefficients and model parameters and also in selecting suitable sites for water harvesting structures and prioritization according to criteria and budget availability. Satellite data can be interpreted to derive a number of thematic information in land use, soil, vegetation, surface water, snow cover, stream network, land form, erosion intensity etc., which combined with conventionally measured climatic parameters (temperature, precipitation, evaporation etc.,) and topographic parameters (height contour, slope) should provide the necessary inputs to the models, both empirical and conceptional, are in practice reliable and timely estimation of runoff (Fig. 18.2).

The DGPS system will be employed for sensitive micro watersheds to collect the fly levels and the information will be utilized for generating Digital terrain model which in turn helpful in generating the contours at 1 mt interval, slope and stream network. Basing on various criteria rules framed with the help of contours, slope, soil, drainage pattern etc., and other collateral information, the site suitability analysis for WHS will be carried out. The runoff will be estimated using other parameters governing the watershed development.

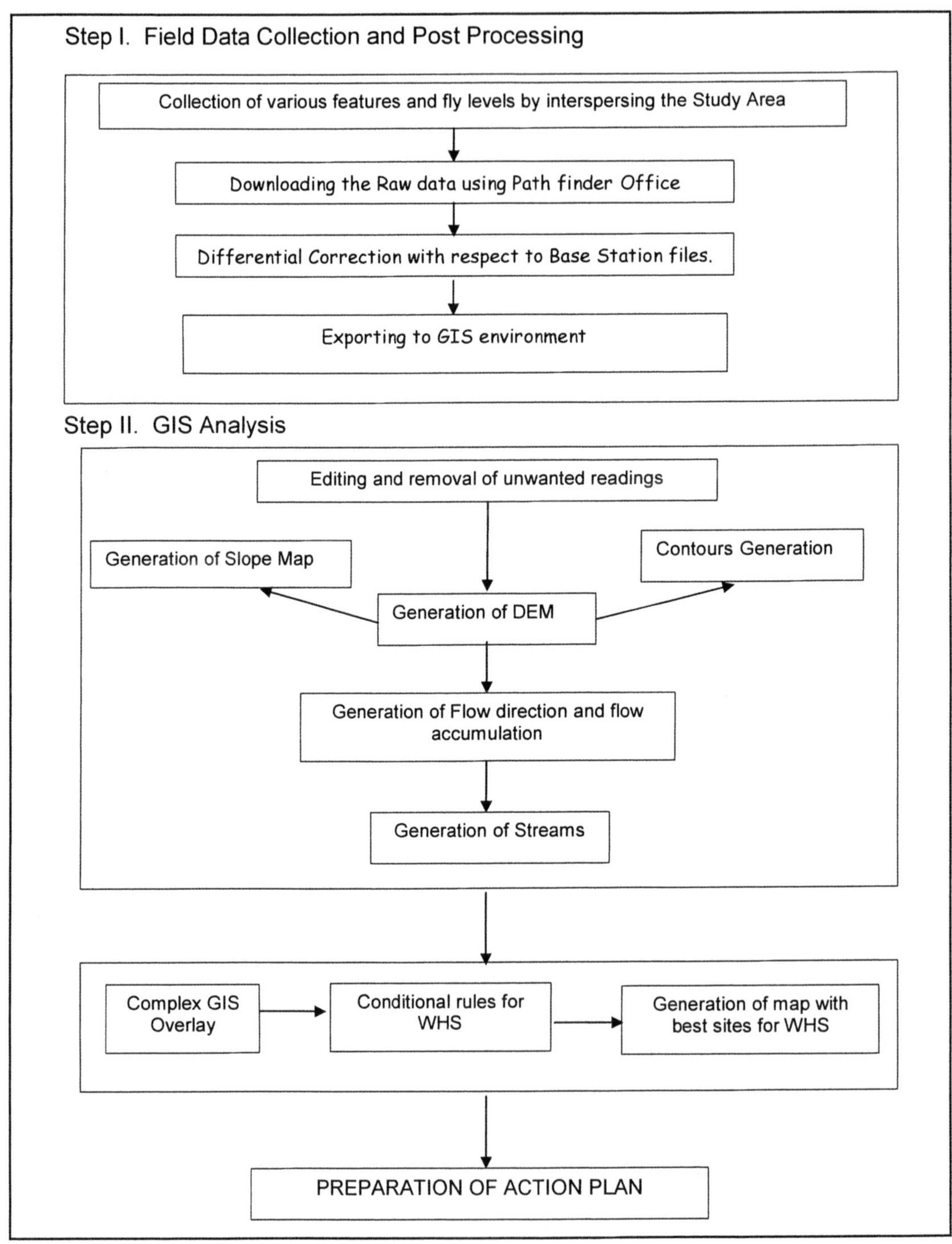

Fig. 18.2 DGPS techniques for preparation of action plan for water harvesting structures.

18.5 GIS Database for Watershed Management

The database required for the management of watershed is furnished below.

(a) Natural Resource data bases :

Soils, geology, forest, land resources
Rainfall
Surface water sources (canals, tanks, rivulets, lakes, and rivers)
Ground water (open wells, tube wells, depth to water table - & water quality) and
Minerals / other natural resources

(b) Population :

Occupation
Male, Female
Age-wise distribution
Employment status
Per capita income
Crops grown, kharif / rabi / summer

(c) Infrastructure :

Education :

Primary, Secondary, Higher and Technical Institutions

Health :

Primary health centres
Local fund hospitals
General hospitals

(d) Communication :

Roads
Post & Telegraphs
Railways
Waterways

(c) Marketing :

Banking facilities
Markets and
Administrative units

18.5.1 Suggested Developmental Plans

Development plans include :

- Scale of operation
- Extent of peoples participation
- Expenditure involved
- HRD benefits anticipated
  - Self-employment
  - Employment generation
- Value of benefits
- Built-in checks for ensuring sustainability
- Maintaining & following mechanism

18.5.2 Technical Inputs

- Aerial photographs
- Remote sensing data analysis
- Database
- GIS representation of basic data
- DEM/DTM analysis

Hydrologic analysis: soil erosion/loss, water conservation measures, and associated structural measures - sizes, costs involved and benefits likely to occur.

The problems encountered in the watershed management study are : (i) drinking water problem, (ii) problems related to dry land agriculture, (iii) degradation of forest land due to commercial activity, human interference and so one (iv) soil leaching due to erosion of soils, (v) problem of salinity, and (vi) problem of fluoride and so on.

18.6 General Objectives of Watershed Management Program

Earlier investigators have revealed that soil and land degradation in the catchment areas pose ecological problems like sedimentation of reservoirs causing reduction in the irrigation potential of the command areas, low hydro-electric power output, floods, and drought. Realising the seriousness of the problem of sedimentation following soil erosion, a theme of micro-level watershed management is initiated to prevent soil erosion, and develop watershed. Keeping this in view, the following objectives have to be designed for the development of any area on the basis of watershed for its sustainability:

(i) To evaluate the natural resources of the watershed using Remote Sensing satellite data for the extraction of feasibility condition for its development.

(ii) To derive required attributes of each sub-water drainage basin with the help of morphometric analysis.

(iii) To estimate sediment yield for the watershed and prioritisation of sub watersheds within the micro-level catchment area. The requirements of prioritising the sub-watershed are mainly based on the sediment yield, drainage characteristics, intensity of eroda bility, and topography of the area.

(iv) To integrate the information obtained from the analysis of the satellite data, results from various characteristics and the quantity of sediment of sub-watershed within the catchment area and socio economic factors into GIS for providing indicators of developmental activities and to prepare a master action plan for its sustainable development.

18.6.1 Operational Objectives

(i) To study the existing natural resources, both land and water, for their assessment, management, proper utilisation, and feasibility condition for their development. The natural resources mapping package consists of hydrogeomorphology, geology, groundwater potential zoning, soils, land use/ land cover, drainage network, slope, and surface water bodies.

(ii) To promote the economic development of the project area which is directly or indirectly dependent on the watershed through (a) optimum utilisation of the watershed's natural resources to mitigate the adverse effects of drought and prevent further ecological degradation. (b) employment generation and development of the human and other economic resources at village level in order to promote savings and other income generation activities.

(iii) To create a digital database information system for the watersheds under study consisting of spatial and non-spatial information of natural resources and socioeconomic data.

(iv) To introduce a physical drainage network model based on basin geo-morphometry for the development of action plans for land and water resources and application of this model for prioritising to sub-watersheds in the study area for sustainable development.

(v) To formulate an ecofriendly and economically viable physical model for integrated watershed management, which includes the generation of action plans for land and water resources development. This model is an ideal integrated watershed management model. The formulation of this model is based on (a) existing natural resources database, (b) drainage basin geomorphometric parameters, and (c) socioeconomic status of the study area.

18.7 Research Approach

The approach to research based on the use of remote sensing and GIS begins with acquiring the satellite imagery and toposheet of the required study area. The following steps are adopted for the proper management of watershed using remote sensing and GIS techniques. The conceptual frame work of integrated mission for sustainable development is presented in Fig.18.3.

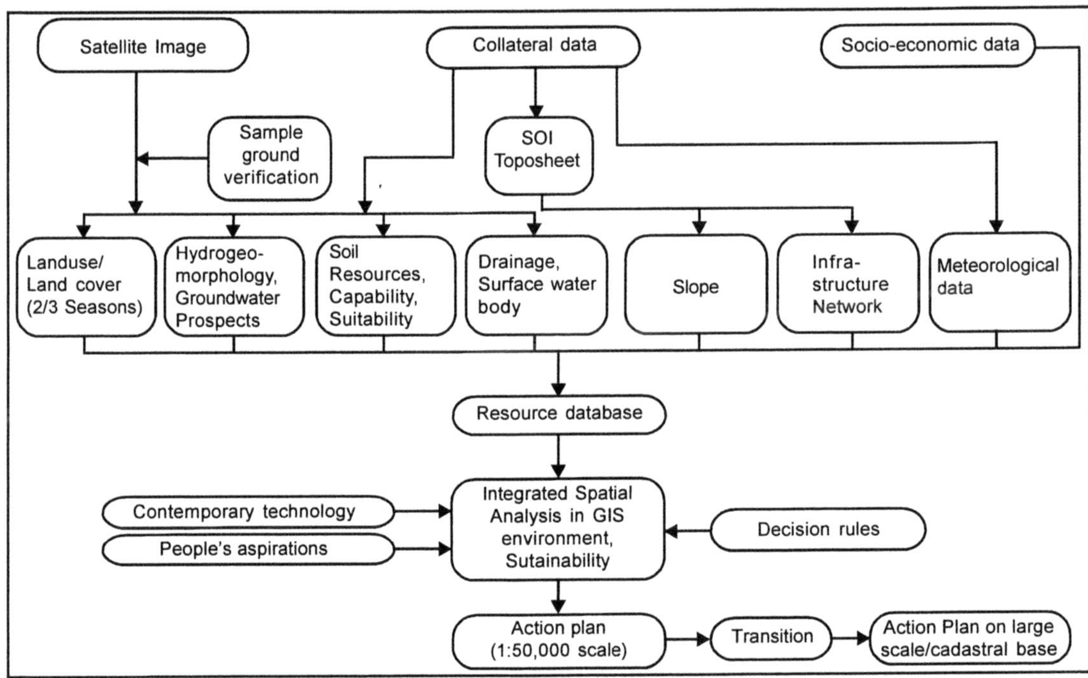

Fig. 18.3 Flow chart showing the steps involved in integrated mission for sustainable development (IMSD).

Step (i) : Land use/Land cover map is prepared to know the various drainage characteristics like drainage density, stream frequency, and basin slope for each micro-water shed.

Step (ii) : Land use/Land cover map is prepared to know the various uses of the land, vegetation, water bodies, rock/soil, artificial cover and others which are due to land transformation. This map is also used to find the areas of the crop land, forest, and watershed for each sub-watershed for calculating Sediment Yield Index and vegetative cover factor.

Step (iii) : Hydrogeomorphology map is used for finding out ground prospects to suggest water harvesting structures, such as, percolation tanks.

Step (iv) : Slope map is used to know the terrain properties like steep slope, and plains.

Step (v) : A GIS digital system like ARC/INFO is used for input, manipulation, and creation of error-free digital database for all the natural resources.

Step (vi) : Precipitation data of the watershed is used in finding Sediment Yield Index.

Step (vii) : Depending on the combination of the above-mentioned resource themes, action plan for land and water resources and treatment plan for catchment area are generated for the development of water resources.

18.8 Model Watershed

This section of the book highlights the fundamental procedure for the study that deals with a comprehensive evaluation and integrated analysis of Shivannagudem watershed. The watershed is situated in the south-west corner of Nalgonda district of Andhra Pradesh state at a distance of 25 km. from Chautuppal. The watershed under study forms a part of Musi catchment which inturn forms a part of Krishna river basin. It lies between North latitudes 16° 58' 20" and 17^0 I' 40" and East longitudes 78^0 and 78° 57' 48". Chautuppal and Devarakonda are the nearest towns to the watershed (Fig 18.4). It is connected by a convenient network of both metalled and unmetalled roads to the mandal head quarters. This watershed is covered by parts of Indurti, Shivannagudem, Kottala, Namapuram, Saran and Lankalapalli villages. Agriculture is the principal avocations of the people living in the area. The total extent of the watershed is 3589.00 hectares.

The area has been categorized into 10 land forms/geomorphic units. Pediplain with moderate weathering and pediplain with shallow weathering are the most prominent units in the model watershed area. Groundwater occurs in these geomorphic units under both unconfined and semi-confined conditions from the weathered and fractured/jointed rocks. Rainfall is the principal source of recharge in the study area. The concentrations of flouride vary from 3.5 to 3.8 mg/l, and so the water in this area is not potable.

18.8.1 Objectives of Model Study

(i) To study the existing cropping patterns on the available water resources of the study area using Remote sensing data.

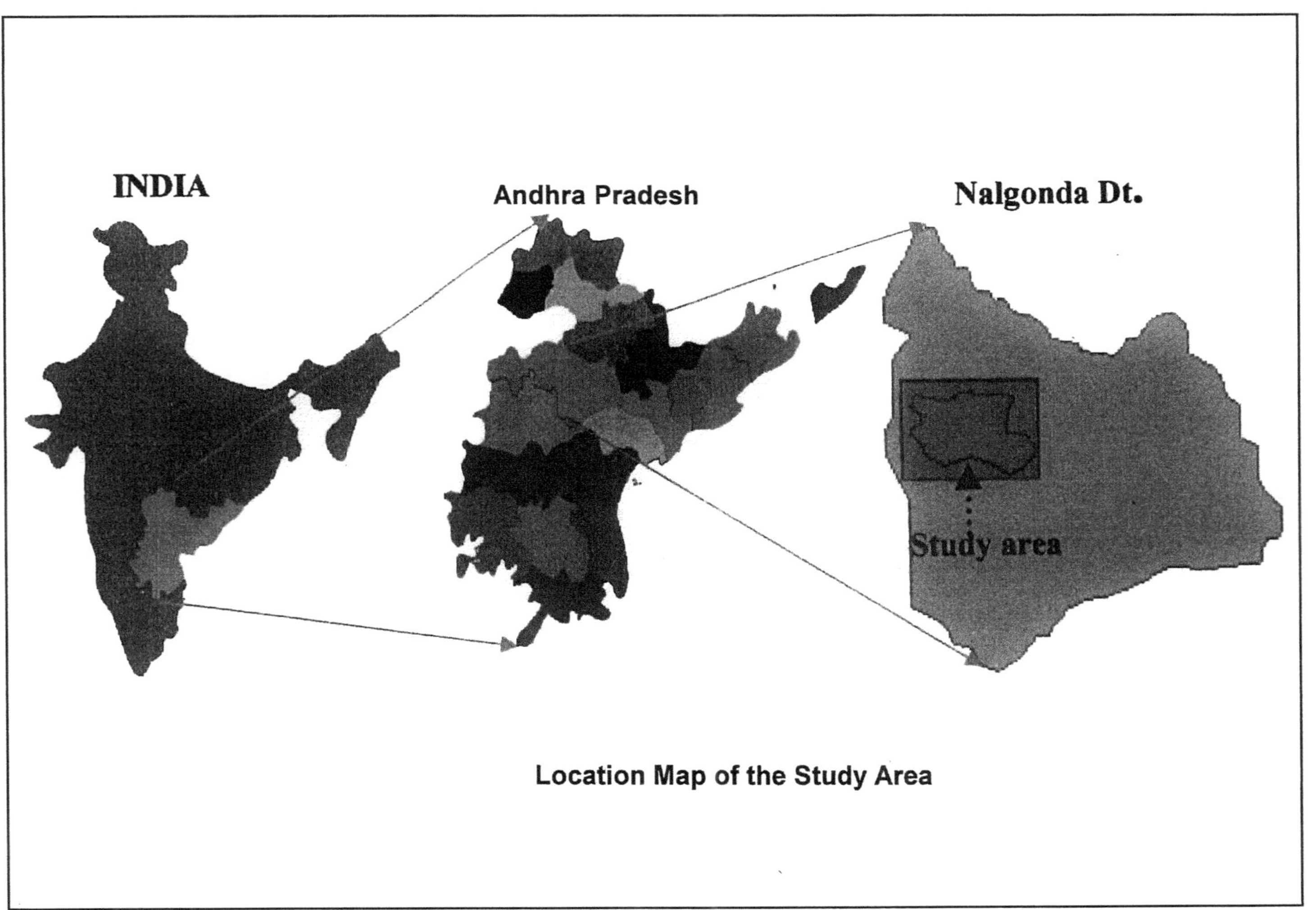

Fig. 18.4 Location map of model watershed.

(ii) To map all the thematic information by interpreting the satellite imageries and SOI toposheets and production of maps through ARC INFO GIS software.

The Thematic information includes:

(a) Land Use/land Cover

(b) Slope analysis

(c) Soils and their classification

(d) Hydrogeomorphology

(e) Drainage

(f) Groundwater prospects

(iii) To select a suitable cropping pattern for the study area, an action plan for land resources is prepared by the integration of the above themes.

18.8.2 Method used in Model Study

Watershed is made up of its natural resources, both physiographical and hydrological, and human resources. A comprehensive development of watershed requires the analysis of the present status or conditions of its natural resources, such as, soils, topography, climate, drainage, land use, water resources, and the present trend of land degradation. So, the basic prerequisite is the availability of timely, accurate, and detailed information on the watershed resources. With the advent of modern information systems, such as, satellite remote sensing, and owing to its repetitive coverage in time and space, accuracy, and cost effective nature, it has become possible to apply watershed models efficiently. The required database combines remote sensing data and other ancillary data in the form of toposheets, reports, maps, and groundtruth data. The important watershed resources considered in the present study include climate, slope, drainage and surface water bodies, Hydrogeomorphology, land use/land cover, and soils.

Realising the importance of comprehensive and integrated approach to the problem, watershed resources were evaluated from different sources using new ideas, techniques, and tools. The thematic mapping of various watershed resources is done on 1 :25,000 scale. A flow chart showing the steps involved in the development of action plan for watershed management is given in Fig. 18.5.

The watershed area of Shivannagudem is delineated and a base map, slope map, and drainage map are prepared from the SOI toposheets of 1:25,00 scale. The land use and land cover, soil and groundwater prospect maps are prepared from 2,3, and 4. False Colour Composite imageries on 1:25,000 scale. The maps are then scanned using Logitech Scanner, and digitised using Autocad 14. Then the drawing output is given, using Arcview GIS software.

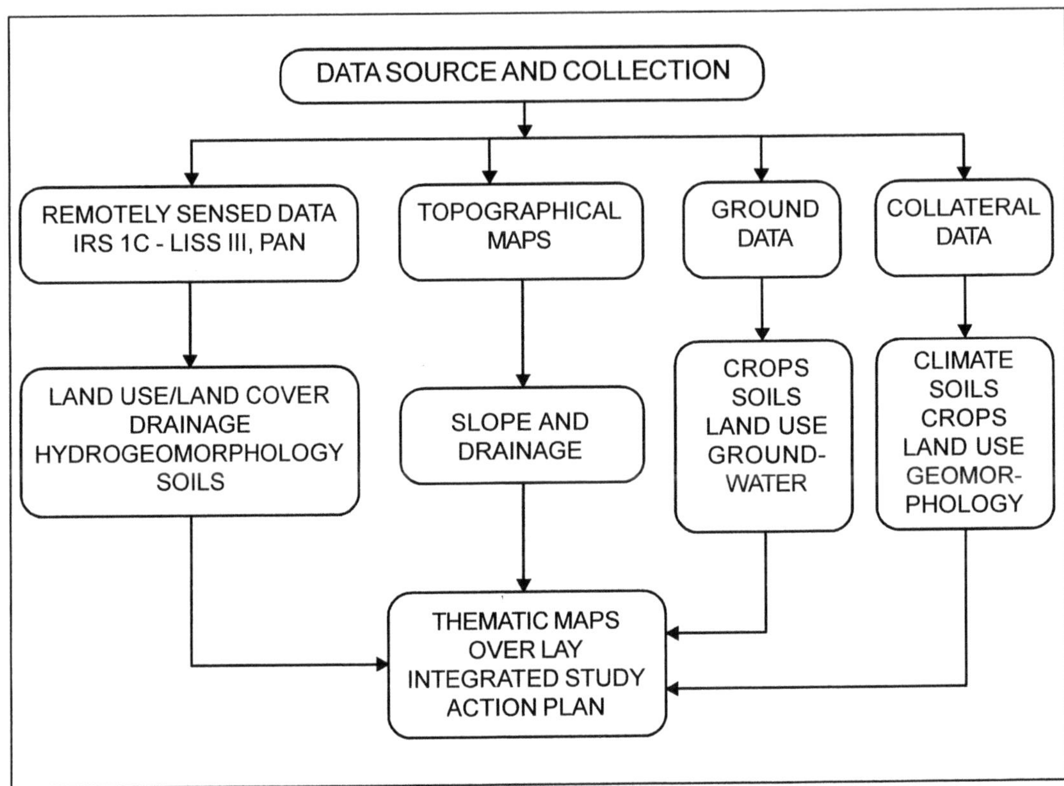

Fig. 18.5 Flowchart showing the methodology adopted in the model study.

18. 9 Land Use and Land Cover

A comprehensive information on land use and land cover is the basic prerequisite for land resource evaluation, assessment, utilisation, and management. Today, with increasing population pressure on land and the resulting changes in the land use pattern and processes, a considerable degree of land transformation and environmental deterioration is being witnessed. Therefore, it is important to understand the cause and effect of the change through a scientific study.

Preliminary interpretation of satellite data has been carried out on 1:25,000 scale using visual interpretation techniques for delineating various land use/land cover classes in the study area. The satellite data related to kharif and rabi seasons are used for mapping the land use/land cover classes. The doubtful areas identified during the interpretation are physically verified on the ground during the field work, and necessary modifications and corrections of the misinterpreted land use/land cover boundaries are made. The final thematic details are transferred on to the base map prepared on 1:25,000 scale. The various land use/land cover classes (Plate 1) delineated in the watershed include (i) Built up land, (ii) Agricultural land, (iii) Wasteland, (iv) Water bodies, and (v) Others like quarrying. The area-wise estimates of land use/ land cover classes of the study area are given in Table 18.4.

Table 18.4 Area wise estimates of land use/land cover of Shivannagudem watershed

S. No.	Class	Lu/Lc	Area	% area
1.	Built-up land	Town/village	60.60	1.69
2.	Agricultural land	Cropland		
		Kharif irrigated	58.40	1.63
		Kharif Unirrigated	2188.00	60.97
		Double Crop	1000.00	27.86
3.	Wastelands	Land with Scrub	2.80	0.08
		Land without Scrub	133.40	3.72
		Barren rocky/stony waste/sheet rock area	23.40	0.65
4.	Water Bodies	River/Tank	110.20	3.07
5.	Others	Quarrying	1.50	0.04
		Total	3589.00	100.00

18.10 Slope Analysis

Slope is one of the most important terrain characteristics and plays a vital role in geomorphologic and runoff processes, soil erosion and land use planning, and watershed management. It is very important to have an understanding of the spatial distribution of slopes for the development, planning, and management of both land and water resources.

The slope map of Shivannagudem watershed has been prepared on 1: 25,000 scale using Survey of India Topographical maps. The contour lines drawn on toposheets have been used in the preparation of the slope map. The nature of the slope of a terrain is explained by the horizontal spacing of the contours. The closely spaced contours correspond to steeper slopes and the spare contours reveal gentle slopes. The slope categories are expressed in percentages. The slope map (Plate 17) has been prepared by following the guidelines of All India Soil and Land Use Survey on slope categories. The details of slope classes that correspond to the categories and percentages are presented in Table 18.5.

Table 18.5 Details of Slope classes corresponding to the categories and percentages

S.No	Slope Class	Slope Category	Slope %
1.	1	Nearly level	0-1
2.	2	Very gently sloping	1-3
3.	3	Gently sloping	3-5
4.	4	Moderately sloping	5-10
5.	5	Strongly sloping	10-15
6.	6	Moderately steep to steep sloping	15-35
7.	7	Very steep sloping	> 35

The study area has been categorized into six slope classes as 1, 2, 3, 4, 5, and 6. Slope class 1 is the most predominantly occurring class in the study area. This Slope class which corresponds to nearly level ground is generally associated with pediplain of granite gneiss with shallow weathering and pediplain of granite gneiss with moderate weathering.

Slope class 2 which represents nearly very gently sloping land corresponds to pediments of granite gneiss and dolerite dykes. This is generally associated with barren rocky/stony waste/sheet rock area land with scrub and land without scrub. This slope class is observed in the study area in the northern and north-eastern parts of the study area.

Slope class 3 which represents gently sloping land is observed in the north-western part of the study area. Slope classes 4, 5, and 6 generally correspond to hilly terrain of residual hills and inselberg of granite gneisses. These are generally observed in the study area southern side of Namapuram, north-west and south-west parts of the watershed. In general, the slope of the watershed is towards the east. The nearly level ground corresponding to the pediplain of granite gneiss with shallow and moderate weathering is the most predominant unit in the watershed.

18.11 Soil Mapping

Geological forms and geomorphological variations, influenced by the varied physiographical setups, have their own say in soil formation. In fact, the soil survey conducted envisaged assessment of the transformation of basic rock to solum and the influences of climate, microbial actions and the conditioning by the gradient over a period of time, all collectively accelerating the pedogenic process. The area under investigation is located in a semi-arid environment and is under the influence of hyperthermic soil temperature regime.

The study area comprises moderate to high hills ranging from 700-800 mm. Dykes have 400-500 m elevations and valleys are located in 300 m-MSL range in the existing topographical system. Thus the soil catena starts from about 700 m and extends up to 300 m with increases in textures and depths gradually from hill to valley. The runoff passing through streamlets and rivulets has been harnessed by small to medium tanks of various sizes and capacities under which irrigation is practised. Except for very small areas, no hydromorphic soils are encountered. Rest of the area located between hills and valleys is an undulating terrain with the elevational variations ranging between 300-500 m.

The soil legend has been initially compiled during the visual interpretation of satellite data duly combining with the information derived from Survey of India topographical maps. This constituted the initial legend, and the validity of boundaries has been checked in the field to ascertain the associated soil composition. The methodology followed in case of soil mapping is briefly discussed in the following sections.

18.11.1 Soil Mapping Procedure

As the source is remote sensing data and the bases are on 1:25,000 scale, detailed reconnaissance type of soil survey technique is deployed for mapping soils in the survey area. Under this system of survey specific areas are covered by detailed survey and information extrapolated to other areas where reconnaissance type of survey is carried out.

Discriminating features, as manifested by varying tones, textures, shape, size, composition, location aspect, and other relevant components available in the imagery have been identified and segregated. Differences are connected to topographic features available in the toposheets and integrated with local knowledge of the terrain. Thus, an initial legend was formulated for all the themes like soils, landuse/landcover, and Hydrogeomorphology. This legend formed the basis for further operations in all the surveys. Depending upon the variations arrived at during preliminary interpretation of satellite imagery suitable sample strips and sample study areas have been selected in the watershed.

Semi-detailed soil survey of low intensity is carried out in selected areas and extrapolation done to the rest of the area. Profile pits were excavated in almost all the mapping units of all sample strips, up to murrum or 100 cm. Profile descriptions are made following the FAO guidelines. Stipulations laid down in the Soil Survey manual of All India Soil and Land Use Survey have been closely followed throughout the period of field work. Apart from the profile study, a number of shallow pit observations are made to confirm the percentage of soils at the series association level.

18.11.2 Soils of the Area

Eight mapping units have been identified in the survey area. In total, 11 soil series have been recognised and mapped. Description of each mapping unit along with the series and taxonomic classification is provided in the following sections (Plate 19).

The details of mapping units identified in various physiographic positions along with the soilscapes are compiled and presented in the form of a Final Soil Legend in Table 18.6.

18.11.3 Suitability for Sustainable Agriculture

Due to inherent characteristics of soils, surface features and conditions related to water holding and nutrient supply, differences in adaptability to various crops exist under rainfed conditions. In the total area 2188.00 ha. is the area under rainfed agriculture. The rest of the area is either barren, hilly or under irrigated agriculture. The rainfall is received in two spells, most of it occurring in rabi seasons. Failure of rains is frequent. Soils are leached and moderately deep in the rainfed zone. Slopes are moderate and surface textures are coarse in most cases. Keeping in view all these factors, an attempt is made in this section to asses the suitability of each mapping unit to some principal crops that are in vogue in the survey under the classification for suitability to rainfed agriculture. In all, five classes have been envisaged.

(i) Soils with very few limitations for growing crops under rainfed conditions indicated as Very good (VG)

(ii) Soils with few limitations for growing crops under rainfed conditions indicated as Good (G)

(iii) Soils with moderate limitations for growing crops under rainfed conditions indicated as Fair (F)

(iv) Soils with severe limitations for growing crops under rainfed conditions indicated as Poor (P)

(v) Soils with severe limitations for growing crops under rainfed conditions indicated as Unsuitable (US).

The irrigability classification aims at assessing the inherent properties of the soils that express their degree of limitations for development under irrigation agriculture. If water is available with varied inputs and costs, any soil can be brought under irrigation.

Table 18.6 Soil taxonomic classification in Shivannagudem watershed

MU	PHYSIOGRAPHY	ASSOCIATION OF SOIL SERIES	TAXONOMIC CLASS
1	Hills	Rocky	
		Cherlagudem	Loamy skeletal, mi, hyp, Typic Ustorthents
2	Dykes	Rocky	
		Indurti	Loamy skeletal, mi, hyp, Lithic Ustorthents
3	Upper pediplains	Paluvayi	Coarse loamy, mi hyp, Typic Ustorthents
		Kottala	Fine loamy, mi, hyp, Typic Ustochrepts
4	Middle pediplains	Kottala	Fine loamy, mi, hyp, Typic Ustorthents
		Parlapalli	Fine loamy, mi, hyp, Typic Ustochrepts
5	Lower pediplains	Gudalakunta	Fine, loamy, mi, hyp, Typic Ustochrepts
6	Buried pediplains	Vinjamur	Fine, montomorillonitic, hyp, Leptic Haplusterts
		Battugudem	Fine, mont, hyp, sodic Haplusterts
7	Narrow valleys	Godkondala	Fine loamy, mi, hyp Typic Ustorthents
8	Broad valleys	Shivannagudem	Fine, mixed, hyp, Aeric Halaquepts
		Kudasapalli	Fine, mixed, hyp, Typic Ustochrepts

But the core idea of classification is to indicate the best soils suitable for irrigation with limited management and sustained use of the soils under irrigation. The following soil properties are generally considered in classifying the soils under varied degrees of suitability for sustained use under irrigation. A generalised picture of these various parameters are to be considered before deciding any soil on its suitability for irrigation. Since the consideration is for sustained use under irrigation, considering the present status of the soils or the present condition is not simply sufficient.

The long-term behavior of the soils under irrigation should also be considered by evaluating the following soil properties :

Depth of soil
Texture at surface
Water holding capacity
If salts present, the amount and nature
pH of the soil
Cation exchange capacity
Depth of groundwater

Major soils are moderately eroded except the areas under irrigation, where lands are levelled and erosion is moderate in rainfed areas and severe in Gully. Taming with loose boulder structures, contour trenching, gaps filing with afforestation, and grassland development as under cover to plantations, are some of the measures to be taken up in the model study area.

18.12 Hydrogeomorphological Mapping

The delineation of various geomorphic units and the knowledge about their groundwater prospects play an important role in the optimum utilisation of land resources. Besides, the information on land forms and geological information like lithology and structures also help in identifying the groundwater potential zones. The Hydrogeomorphological map has been prepared on 1: 25,000 scale by sensing techniques (Plate 18).

Visual interpretation techniques have been adopted in the delineation of various geomorphic features. Besides, the linear features like dykes and structures like lineaments/fractures are also delineated. The interpreted geomorphic units/land forms and structures have been transferred onto the base map which was already prepared using Survey of India topographical map. A detailed programme of field work has been planned to verify the doubtful areas and arrive at the information regarding the groundwater prospects of various land forms/geomorphic units.

Based on the groundtruth information, corrections and modifications of the misinterpreted units were made and the final thematic details were transferred onto the base maps. The final drawings were prepared and each Hydrogeomorphic unit/ structure was assigned a particular symbol. The various geomorphic units occurring in the study area were described in the legend and presented in Table. 18.7.

Table 18.7 Legend of Hydrogeomorphology of Shivannagudem watershed

Mapping symbol Prospects	Geomorphic unit	Litho Stratigraphy	Structure	Description	Ground Water
PPM	Pediplain With Moderate Weathering	Archean gneissic	Fractures/ Lineaments	Flat and smooth surface of pediplain with 8-20 m. Thick weathered material and covered with red soil	Moderate to good. Very good yields along fracture/lineament.
PPMA	Pediplain With Moderate Weathering covered shallow weathering	- do -	-do-	Flat and smooth surface of Pediplain with 8-20 m. Thick Weathered Material and Covered with Alkaline Alkaline soils	Moderate to good. Quality-wise not suitable for drinking and agricultural Soils purposes.
PPSA	Pediplain With Shallow Weathering Covered With Alkaline Soils	- do -	-do-	Flat and smooth surface of pediplain with 0-8 m. Thick weathered material and covered with black soil	Poor to moderate. Moderate yields are expected along fractures/lineaments. Quality-wise not suitable for drinking and agritural purposes.
BPPS	Buried Pediplain With Shallow Weathering	- do -	-do-	Flat and smooth surface of pediplain with 0-8 m. Thick weathered material, covered with black soil	Poor to moderate. Moderate yields are expected along fractures/ lineaments
P	Rocky Pediment	- do -	-do-	Gently sloping smooth surface of erosional bed rock between hill and plain with never of detritus.	Negligible to poor. Poor yields are expected along the fractures.
I	Inselberg	- do -	-do-	Isolated hill occupying small area	Poor
RH	RESIDUAL Hill	- do -	-do-	Isolated hill occupying considerable area	Poor
	Dyke/Dyke Ridge	DOLERITE	-----	Massive, linear feature cutting across various geomorphic units.	Poor
----	Fracture	----	-----	Linear vertical/subvertical features generally associated with drainage pattern	Moderate to good

18.13 Groundwater Prospects Map

In recent years, the increasing use of satellite remote sensing has made it rather easy to define the spatial distribution of different groundwater prospect classes on the basis of geomorphology and other associated features. Visual interpretation of image is in hydrogeomorphological mapping, and each landform was studied and mapped (Plate 20). Depending upon the geology, structure, and tectonic phenomenon that have caused development of secondary porosity have been coupled with elevation and drainage information from the SOI toposheets and potential zones for groundwater development are delineated as shown in Table 18.8.

Table 18.8 Legend of groundwater prospects

Mapping symbol	Geomorphic Unit	Groundwater Prospects
PPM	Pediplain With Moderate Weathering	Very Good
PPMA	Pediplain With Moderate Weathering Covered With Shallow Weathering	Moderate - Poor
PPS	Pediplain With Shallow Weathering	Moderate - Good
PPSA	Pediplain With Shallow Weathering Covered With Alkaline Soils	Moderate - Good
BPPS	Burried Pediplain With Shallow Weathering	Moderate - Poor
P	Rocky Pediment	Poor
I	Inselberg	Nil
RH	Residual Hill	Nil

The ground water potential zones were deduced after the integration of hydrogeomorphological and lineament maps through GIS. Various hydromorphic units are grouped into potential zones, moderately potential zones and poor zones. Generally there will be more recharge in alluvial plains, flood plain, buried pediplain and Infield valleys. Other parameters are also taken into consideration like lineaments, particularly joints, fractures and their intersection to enhance the potential of these units. Intersecting lineaments are also the potential zones for ground water exploration and development in the study area.

18.14 Drainage Mapping

Drainage pattern is a plan which the individual stream courses collectively form. It refers to both the spatial relations of individual streams and the overall pattern made by the individual drainage lines. In the present study, the drainage map is prepared using satellite imagery and SOI toposheets on 1:25000 scale. The drainage map is scanned using Logitech deskscanner and digitised by AutoCAD 14. Then it is exported to Arc/Info 3.5.1 in which the editing is done. The output is given using the Arcview

3.0 software. This process is the same for the preparation of all the thematic maps. This drainage map is used as an action plan for water resources and is shown in (Plate 21).

The drainage pattern in the Shivannagudem watershed is found to be of a dendritic pattern. As it is a hard-rock granitic terrain, drainage follows the weak zones in which the erosion is, much, like joints and fractures The dendritic pattern is the most common type of pattern formed by a main stream with tributaries branching and rebranching freely in all possible directions. It generally develops upon homogeneous rocks of uniform resistance, such as, horizontally bedded sedimentary rocks and, massive igneous terrain offering no variation in the resistance to the flow of water. The resulting stream runs in all directions with no definite preference to any particular direction.

18.15 Action Plan Generation

The base map, slope map, drainage and watershed map have been prepared from SOI toposheets. The land use/ land cover, hydrogeomorphology and soil maps are prepared by the visual interpretation of IRS-IC satellite image. These maps are digitised using AutoCAD after scanning the above cartographic maps. The digitised maps are then converted into a form compatible to Arc/INFO GIS software. These digital data are transformed through editing and analysing these layers.

A total of six tanks were constructed across the drainage in the watershed. 8 soil mapping units and 12 soil series were identified in the watershed. The land use/land cover analysis indicates that 90% of the watershed area is covered by crop land of which 61% of the area corresponds to Kharif unirrigated land. The wastelands in the form of land with scrub, land without scrub and barren rocky/stony waste/sheet rock area amounts to about 4% of the total watershed area. Based on the technical guidelines given by the Department of Agriculture, Govt. of India, various water resource development and management techniques, soil and water conservation measures, optimal land use patterns were suggested for the overall sustainable economic development of the watershed.

Action plan for Water Resources is developed from the integrated study in which three check dams are recommended based on the slope, soil, land use and land cover, and hydrogeomorphological criteria, to obstruct the surface water flows and thereby increase its influence over the command area and the groundwater levels. A percolation tank is also suggested on a major stream in the watershed with the stream overflow and area criteria under consideration. In the remaining part of the project area, the slope is gentle and the major streams have more area of influence. Action plan for Land Resources is made from the integration of the prepared thematic maps and suitable crops have been recommended. (Plate 22) Through the implementation of the recommendations made, the study area can be well developed.

19

Natural Disaster Management : Landslides

19.1 Introduction

The term landslides has been used to designate the movement of surface and near surface earth material down slopes and towards streams valleys and coastlines. Landslides are natural phenomena such as mudflows, earth flows, snow or debris avalanches and subsidence or natural events would occur with or without human activity. However, human activity has increased some of these events.

The same 'Landslides' is used both for the geomorphic process, which involves rapid gravity movements and for the resulting landform that is created by the displaced material. Furthermore, the term 'landslide' is also a misnomer because the process of landsliding has come to include a broad range of different types of motion whereby earth material dislodged by falling, sliding and flowing.

Natural disasters happen every year and their impact and frequency seem to have greatly increased in recent decades, mostly because of environmental degradation, such as deforestation, intensified land use, and the increasing population (Vincent, 1997). The urban growth is extending into hazard prone regions such as low elevation areas more prone to flooding. The result is that the magnitude of the impact of a disaster increases since it is dependent on the susceptibility of the land the vulnerability of the society (Verstappen, 1995). The vulnerability is also compounded with the increased complexity of interdependent systems (Comfort, 1999). Given that it is the transportation domain which interconnects those systems, interest has risen in determining methods which can be used to (1) measure transportation's vulnerability before such a disastrous event, (2) monitoring its availability during an event, and (3) assess the damages after an event. The occurrence of Natural Hazards is a serious constraint on economic development, particularly in developing countries, where the economic loss due to the impact of natural hazards often makes the difference between economic growth and stagnation. On the other hand practice has shown that adequate hazard mitigation is possible. Slope instability hazard zonation is defined as the mapping of areas with an equal probability of occurrence of landslides within a specified period of time. Slope instability hazard assessment is based on the analysis of the terrain conditions at sites where slope failures occurred in the past. Many landslides are natural phenomena that occur independently of human actions and many of these have enacted large human costs. There are also many landslides that have inflicted losses that have been used by the action taken to make the land suitable for some human purpose.

Population pressures and increasing urbanization along with local zoning laws have been forcing man flat lying areas and flood plains into the adjacent hill slopes. The hill slope as an environmental setting can help us to avoid the severe flooding population of the low lying areas, but can be equally hazardous because of potentially unstable earth materials. Therefore, a recognition of landslide - prone topography is becoming increasingly important in land use decisions.

Landslides can affect most hillslopes. The three principal causes for landsliding are excessive rainfall, human activity and earthquakes. Even nearly flat terrain, when undeclained by fine grained unconsolidated sediments is vulnerable. Even prolonged precipitation produce disaster on the leveled ground.

Thus, there are few areas that are completely safe from landslides although the probability for their occurrence is greatly enhanced by such factors as steep slopes, weak bedrock, unstable regolith, absence of vegetation and high moisture content and forces seeking to change hill side equilibrium.

19.2 Major Types of Landslides

The names of the major types of landslides are assigned on the basis of the type of displacement. Based on the mass movement, landslides are divided into four major groups.

(i) Slow flowage : Rock Creep, talus creep and soil creep
(ii) Rapid flowage : Earth flows/movements, mudflows, debris avalanche
(iii) Sliding : Slumps, rock slides, rock falls, and landslips
(iv) Subsidence : Sinking of mass

Fall : A fall may be defined as a free fall or rock fragments of various sizes. They take place very rapidly and consist of materials of which cliffs and bluffs are made.

Slide : Downward movements of intially intact masses of earth materials on well defined surfaces are called slide. There are two major types of slides such as rotational slides and transitional slides. Rotational slides are characterized by movement along curved surfaces that are concave upwards. The rupture surface may resemble the bowl of a spoon. The transitional slides entail movement on planer surfaces of repture that may be nearly horizontal or steeply inclined.

Lateral spreads represent a further development of nearly horizontal transitional slides, differing from them in the extent to which they consist of separate blocks that are broken and divided or of material that collapses and moves uniformly as it moves laterally.

Flows : Comprise a variety of slides that are distinguished by the chaotic movement of the material within the moving mass. This material especially sand, silt and clay, tends to appear during flow as if it were a viscous liquid. The surface of the flow has an irregular character, retaining no vestige of the original land surface.

Creep : It is a form of slope movement that takes place in a blanket of material that lies relatively close to the land surface. Creep is imperceptible except through observation of its long - term cumulative effects such as the formation of curved tree trunks. The average rate of downslope movement in creep is greatest near the surface and decreases with depth, so no sharp discontinuities separate the creep material from the material remaining in place.

Slump : A slump is a rotational slide in which the materials commonly retain coherence and move along a shear plane, which is concave upward.

Subsidence : Subsidence may be defined as the downward movement of the natural ground surface. Subsidence may be instantaneous and violent or gradual and may be caused by natural forces or by the activities of man.

19.3 Common Features of Landslides

The following features are to be considered in characterizing particular landslide

(i) The material of which the landslide consists
(ii) The shape and location of the rupture or slip surface
(iii) The displacement or distance travelled during the slide event.

In interpreting the processes of landslide movement, it is known the rate of movement, the behavior of the materials within the landslide during movement, and the record of any pre-landslide phenomena include earthquake precipitation and human modification of the terrain. Keeping in view of its importance, causes of hazards and controlling measures of landslides are discussed in the following.

19.4 Causes of Landslides and Related Phenomena

In many cases the real cause of landslides is the increase in driving force or decrease in resisting force are marked by immediate causes such as earthquake shock vibrations, or a sudden increase in the amount of moisture in slope material.

Landslide causes may be grouped into two, as external and internal. External causes produce an increase in the shear-stress (driving force per unit area) at relatively constant shear strength (resisting force per unit area). External causes include loading of a slope steepening of a slope by erosion or excavation, and earthquake shocks. A slight vibration or jerk to the mass would greatly add up against the frictional resistance and the mass would become unstable. The heavy traffic on hill roads is of great contributing factor towards causing the imbalance of the masses. Internal causes include changes due to increase in water pressure or decrease in cohesion of the slope materials. The steeping of the slope, water content of the stratum and mineralogical composition and structural features, which are tending to reduce the shearing strength of the rocks. (S.S.Ramakrishnan, et al.).

In addition to the above some other causes of landslides are intermediate having both external and internal causes. For example, rapid drawdown involves an increase in the shear stress and accompanied by decrease in shear strength other intermediate causes include spontaneous liquefaction and sub-surface weathering and erosion. Landslides occur when there is a loss of support under or infront of the shear surface, or when there is a change in the physiochemical constitution of the earth's materials. The three factors that influence stability of hill slopes are :

(i) Internal properties of the earth materials

(ii) The geomorphic setting and environment

(iii) Independant external factors.

(i) Internal properties of the earth materials : These include the type of material and its structural character such as degree of compaction and cementation thickness and arrangement of different rock types, strike and dip of zone of discontinuity, size and distribution of fault joints and other features and bedding surfaces.

(ii) The geomorphic setting and environment : Hill slope characteristics include the amount of relief, steepness of slope, shape of the land surface, slope orientation and aspect, nearness to rivers, vegetation, antecedent moisture conditions etc.

(iii) **Independant external factors :** These are termed the triggering machanisms because they provide the immediate stress that initiates the movement of mass. Prior to displacement, there is an accommodation equilibrium that exists the movement of mass. Prior to displacement, there is an accommodation equilibrium that exists between the earth materials and the geomorphic and environmental setting. The three most common disturbances that cause the critical threshold of cohesion to be exceeded are excessive precipitation, human activities and earthquakes.

19.5 Landslide Analysis

Analysis of a landslide is important to determine what factors have controlled the formation of the rupture surface and the movement on that surface. This requires an engineering analysis of the stability of the landslide mass and a geologic analysis of the changes in geologic and meterologic conditions that are correlated with landslide activity. Earth materials on slopes may fail and move or deform in several ways that includes flowage, sliding, falling and subsidence.

19.5.1 Slope Stability

The nature and extent of slope stability for the various types of downslope movements generally explained by examining the forces on slopes and their interrelated, variable types of earth materials, slope, topography, climate, vegetation, water and time.

19.5.2 Forces on Slopes

The causes of many landslides are examined by studying relations between driving forces, those which tend to move earth materials down a slope and resting forces those which tend to oppose such movement.

Slope stability is evaluated by computing a safety factor. Safety factor may be defined as the ratio of the resisting forces to the driving forces. If the safety factor is greater than one, the resisting forces must exceed the driving force and the slope is considered stable. If on the other hand the safety factor is less than one, then the driving forces exceed the resisting force and a slope failure is expected. Anything that reduces the resisting forces or increases the driving forces will lower the safety factor and thus increase the chances of a landslide.

19.5.3 Role of Earth Material

Depending upon the type of the earth material the frequency and magnitude of the landslides may be more or less for instance when a permeable rock such as sandstone overlies a weak rock such as shale rotational slides are most common. However, a resistant rock overlies a rock of very low resistance, than rapid undercutting of the resistant rock may cause a slab failure.

19.5.4 Role of Slope and Topography

Angle greatly influence the driving force on slopes as the angle of potential slip plane increases the driving force also increases. Thus everything else being equal landslide should be more frequent on steep slopes.

19.5.5 Role of Climate and Vegetation

Climate and vegetation may influence the type of landslide. Climate is significant in influencing the downslope movement because it controls the nature and extent of the precipitation and thus the moisture content of slope materials. For example, both earthflows and mudflows involve downslope movement of water saturated earth materials, but earthflows are common and mudflows relatively rare in humid regions. This results due to the infiltration of good amount of water into slopes in the humid areas facilitating earth flows.

The role of vegetation in landslides and related phenomena is complex because the vegetation in an area is a function of several factors including climate, soil type and topography. Vegetation is a significant factor in slope stability. Vegetation provides a cover that cushions the impact of rain falling on slopes thus facilitating infiltration of water into the slope while retarding grain by grain erosion on the surface. Vegetation has root systems that tend to provide an apparent cohesion to the slope materials. Lastly vegetation adds weight to the slope.

Problem concerned with slope stability and vegetation result from disturbance or removal of vegetation from slopes. In some cases however, vegetation increases the probability of landslides. Disturbance or removal of vegetation by logging has also been associated with an increase in landslides. Clear cutting or removal of all trees has caused several problems. First the rate of transpiration is greatly reduced thus soil moisture increases, sending to reduce the stability of the scopes leading to landslides.

19.5.6 Role of Water

Water is always directly or indirectly involved with landslides and its role is thus particularly important. Water is the universal solvent, and much weathering of rocks takes place. The weathering contributes to the instability of rocks which inturn leads to landslides.

The effects of water on slopes and landslides are quite variable, saturation of earth material causes a rise in more water pressure as the water pressure increases, the shear strength (resisting force) of the slope decreases, and the weight (driving force) increases, thus, the net effect is to lower the safety factor.

19.5.7 Role of Time

The forces on slopes often change with time. For example both driving and resisting forces may change seasonally as the moisture content or water table position alters.

Furthermore, these changes are greater in especially wet years, as reflected by the increased frequency fo landslides during wet years. A slope safety factor may also decrease with time because of progressive wetting which causes disarrangement of soil particles in the slope, lowering internal friction and thus the strength of the slope material.

19.6 Human Causes of Landslides

There has been a sharp increase in the number of man - induced landslides during the twentieth century. Because of expanding population, urbanization and construction, we have been rapidly modifying the landscape on a massive scale and have triggered through our own carelessness.

New engineering techniques, heavy land moving equipment have permitted the metamorphosis of slopes throughout the world on a scale of extraordinary dimensions. The nearly universal construction of dams and buildings of immense highway systems have added their toll to increase landslides. Most man made landslides are caused by changes in the slope configuration or by activities that permit additional moisture to enter the substrate material. Thus, any activity that increases hillside gradient, undercuts earth materials, adds weight to the slopes, or produces more water content can lead to instability and get the stage for landsliding.

19.6.1 Highway Construction

Road building is one of the most common ways that have directly caused landsliding. The majority of the landslides occur where a hill side cut is made and the design of slope stabilization structures is inadequate. When the underground water flow is change and the shear resistant of earth materials lessen, such artificial excavations become prime candidates for earth's movement. The construction of the roads in areas to be logged is an especially a serious problem because roads may interrupt surface drainage, alter sub surface movement of water, and adversely chage the distribution of mass on a slope by cut and fill operations.

19.6.2 Hillside Development

Building on hill side is a risk to structures, especially when there are improperly design or poorly located. Although single buildings rarely cause significant or large scale landsliding, house developments on this slopes lead to landslide. The most probable caused would be the increase in pore water pressure from lawn watering and septic tank effluent or cess pools.

19.6.3 Dams and Reservoirs

The construction of dams and development of artificial lakes and reservoirs create another type of conditions that can provide fertile settings for landslides. The primary cause of this new instabilities is the artificially induced changes in contiguous water tables.

19.6.4 Mining

Mining operation also provide triggering mechanism for landslide. In mining opertaions the over steepening of a cut into a terrace may produce failure in slope stability and leads to landslides. Other causes for landslides in mining area is due to the deforestation, transportation of ore materials and use of heavy machinery for mining operation in open cast and underground mines. Apart from this, landslides are also trigger by blasting involved in mining operations.

19.6.5 Lumbering

Timber operations especially clear cuttings may accelerate mass wasting process and produce a variety of different types of landslides. Landslides produce by multiple events such as removal of forest cover and excavation and dumping of the excavated materials in and around accompanied by rains produced landslides. Thus the shear charge of addition load coupled with diminishing of shear resistant of trees and the increased water content exceeds the strength of the earth materials, and failure is the result.

Earthquakes are the major immediate cause for some landsliding and produces disastrous effects.

19.7 Remote Sensing for Landslide Mapping

The purposes for using remore sensing include (Verstappen, 1995) : "to investigate the susceptibility of the land and the vulnerability of the society, to construct hazard zoning maps and potential damage maps, to monitor potential hazards, and to deal with emergency situations after a disaster". Before discussing remote sensing application in predicting monitoring, and assessing hazards and disasters, we should first define these terms. A disaster is "an event, concentrated in time and space, in which a society (or a community) undergoes severe danger and incurs such losses to its members and physical possessions (or appurtenances) that the social structure is disrupted and the fulfillment of all or some of the essential functions of the society is prevented" (Fritz 1961). A hazard is " the danger that a disaster event may develop or ensue" or "a threat to humans and what they value" (Kates, et al. 1985).

The use of remote sensing depends highly on the type of disaster or hazard of concern. Disasters or hazards can be divided into two groups, slow and rapid. In slow, "creeping" ones (e.g., desertification), remote sensing is genrally used for the purpose of monitoring and low temporal resolution images are often sufficient. In rapid, "instantaneous" disasters, the required capability of remote sensing varies largely. Due to the temporal requirements of the technology, it is usually not possible to monitor rapid disasters, e.g., landslide, earthquake, etc. Sometimes there is no clear distinction between slow and rapid when labeling a disaster. For example, a flood could either be classified as slow or rapid disaster depending on the conditions under which the event occurs. No matter what type of disaster occurs, rapid or slow, there are always some clues that remote sensing system can capture and use to predict a disaster (asses the hazard), monitor its progress, and assess the impact afterwards.

Satellite imagery has been used in the analysis of landslide occurrence primarily through the analysis of colour composites. Several studies have experimented with the use of a true colour composite (TCC) (Sauchyn and Trench 1978; Greenbaum et al. 1996). In most cases the primary restriction has proven to be spatial resolution, with only landslides of approx. 50 m × 50 m or larger being easily resolved. Rather better results were achieved using the 5.8 m spatial resolution of IRS-1 instrument however (Nagarajan et al. 1998).

False colour composite (FCC) images have proven useful in some cases, especially where a landslide scar provides a clear change in the surface properties, such as the removal of forest to expose bare soil. Greenbaum et al. (1995) successfully used this technique for the examination of landslides in Pupua New Guinea for example, similar results were also achieved by Rothery (1987) to identify rock avalanche deposits.

Mixed success has been achieved with the use of more complex analysis techniques. For example, Huang and Chen (1991) reported considerable success using a principal components analysis. Kusaka et al. (1996) on the other hand used the LANDSAT thermal bands to identify areas of perennially wet soil, which were linked with potential landslides. In all cases, spatial resolution has proven to be a key limitation, and shadows cause considerable problems.

Remote sensing has been used to study characteristic properties of ground surface due to the advantages of its broad area observation and periodicity. The integrataion of GIS with remote sensing data and thematic map data may facilitate greatly the assessment and estimation of regional landslide hazards.

19.8 Landslide Analysis in Geographic Information Systems

Geographic Information Systems (GIS) have become the promising tool for an effective analysis associated with the study of geologic hazards. The increasing computer - based tools are found to be useful in the hazard mapping of landslides. One of such significant tools for hazard mapping of landslides is Geographic Information Systems (GIS). GIS is an indeal tool for landslide modeling owing to its versatility in handling a large set of data, providing an efficient environment for analysis and display of results with its powerful set of tools for collecting, storing, retrieving, transforming and displaying spatial data from the real world. The key factor in landslide hazard mapping is the assessment of risk associated with the failures.

Landslides are one of the complex analyses, involving multitude of factors and need to be studied systematically in order to evaluate the hazard. One of the main advantages of the use of this technology is the possibility of improving hazard occurrence models, by evaluating their results and adjusting the input variables. An important aspect of landslide investigations is the possibilities to store, treat, and analyze spatiotemporal data that are available. Better interpretation and understanding of the failures could be derived from them in a GIS environment.

The unique capability of GIS to work with data referenced by vast spatial or geographic coordinates and its ability to incorporate appropriate engineeing models, have caused its proliferating application across the wide sections of engieering, especially in civil engineering where management of spatial-data is pivotal for the analysis. Any spatially-distributed data with a geo-reference to real world could be stored as points, lines and polygons (vector model) or as continuous fields (raster data model). Beyond GIS being used as a spatial database, it assists in modelling applications through handling a special form of data that would otherwise be compromised in conventional analysis (Miles et al. 1999). Also, GIS does not only serve as a database for parameter data, rather qualitative data can be integrated through spatial relationships rather than through relationships between attributes that may not exist (Frost et al. 1997). The other facilities such as : Query languages and user interfaces permitting rapid modification of parameter values : Convenient and quick updating of model parameters; Overlay function, where multiple maps are either visually or topologically combined and its potential in visualization of data using the graphic features, that assists the engineer in verifying data and information pertaining to the model and its application; and developing elevation maps and subsequent slope, aspect and hillshade themes (which are useful in the landslide analysis) are worth mentioning, as they are common to many applications of engineering models (Miles et al. 1999). In particular, the ability of GIS to present the data and analysis results in map forms play a key role in identifying the critical areas (where more rigorous analysis and improved solution is required) by its interactive visualization in a spatially optimized mode.

For the analysis of the causative factors the application of geographic information systems (GIS) is an essential tool in the data analysis and the subsequent hazard assessment. When making use of GIS techniques, the following methodological approaches can be differentiated.

- **Heuristic qualitative approach,** particularly suited for small - scale regional surveys. The scale of such surveys is in the order of 1:100.000 to 1.250.000. They are used mostly by regional planning agencies;

- **Statistical quantitative approaches** for medium scale surveys. These are in the range of 1:25.000 - 1:50.000, and are used by consulting firms or planning agencies for the preliminary planning of infrastructural works, such as the definition of road corridors. The methods at this scale can be sub-differentiated as follows :
- data driven multivariate statistical analysis;
- experience driven bivariate statistical analysis;
- predictive modelling through the application of probability and favourability functions.
- **deterministic approach** for detailed studies at large scale (1:2.000 - 1:10,000), without entering at the level of the engineering geological site investigation. Such small scale studies are used by consulting firms or local planning agencies for the detailed planning of infrastructural works.

All three methodologies require a good idea on the spatial distribution of landslides as the essential element for the analysis. The professional expertise, leading to a classification, is essential. Historic databases are most useful for this work. However, most of the landslide distribution mapping are to be obtained by aerial photo interpretation, and image processing.

19.9 Hazard mapping of landslides

Landslide hazard maps can be constructed by different methods, such as, from inventories; by consideration of site conditions including geology, hydrology, topography, and/or geomorphology; by statistical correlation of landslide frequency with geologic and geomorphic factors; using safety factors from stability analysis. Mechanics based models have also been used to estimate failure probability based on uncertainties about infiltration of rainfall, or pore pressure, and soil strength (Wu et al. 2000). A prior study of causal factors of landslides indicate, the ground conditions, geomorphological processes, physical processes and man made processes as the significant contributors. The pertaining data needed for landslide analysis for a particular slope are :

- Geologic and Geomorphologic features/setting
- Types and quantification of soil and/or rock properties
- Details of vegetation cover, folds, faults, etc.
- Past records of rainfall and earthquake incidences
- Appropriate hydrologic and stability models

Natural hazard is defined as the probability of occurrence of potentially damaging phenomena within a specified period of time and within a given area. Zonation also refers to the division of the land in homogenous areas or domains according to the degree of actual or potential hazard (Varnes, 1984). The analysis of hydrologic and stability models could be readily incorporated in the Arc View either using map calculator or avenue scripts. The results give the factors of safety in the form of maps or contours. The task of engineer becomes easier to identify, visually which areas are critical that deserve further attention. Hazard rating could be assigned based on suitable criterion. Table 19.1 gives such a rating criterion, based on a quantitative estimation of landslide occurrence over a given region, without mentioning the time period.

Table 19.1 Quantitative hazard rating of landslides (Miles et at., 1999)

Stability Criterion	Hazard rating
F > 1.5	Stable slope
1.25 > F < 1.5	Low Hazard
1.00 > F < 1.25	Medium Hazard
0 > F < 1.00	High Hazard

F is the factor of safety

19.9.1 Infiltration and Drainage Model :

The rainfall-induced failures are the common occurrences in soil slopes. Depending upon the intensity and duration of rainfall and permeability characteristics of the slope, infiltration causes development of pore pressures. Rise in piezometric level in hill side slopes, decreases the effective stress, eventually leading to failure of soil mass. Hence prediction of piezometric levels constitutes an important element in the evaluation of landslide hazards. Suitable hydrologic model predicting the level of pore pressures induced for a particular intensity of rainfall and for the given soil, hydraulic and slope conditions should be coupled with the stability model. The lumped parameter model given by Wu et al (1996), gives the recharge of the groundwater as a function of rainfall characteristics and soil properties. Then a finite-difference solution be used to calculate the pressure head at different points in a slope.

19.9.2 Stability Model

Slope failures triggered by any causal factors, involves movement of mass or slide. The exact configuration of the movement and its mass is difficult to predict. Though, depending upon the material characteristics, slope conditions, suitable theoretical models could be used for analysis. Limit equilibrium solutions are available for stability analysis, depending on the assumption of the failure plane, whether it is plane failure of finite/infinite slope or circular failures. However, in actual slope failures, the assumption of single slope failure is not realistic. It may include a slide of mass with circular failure surface and also with plane surface failure, as multiple slope failures are possible. Hence, the stability model should capture both the types of failures according to the conditions of its occurrences. Appropriate criterion could be established for identifying which type of failure would occur for the given conditions.

19.10 Case Study : Kohima Area

Natural disasters are brought about by forces of nature that cause a great damage to human civilisation. The Himalayas are one of the world's most tectonically active areas. Landslides occur frequently and represent a very real hazard to life and property. Since study of landuse/landcover is a part of management of natural hazards, study of landuse/landcover has been carried out in Kohima area, Nagaland, using multidated remote sensing data of 1989 and 1999. The changes in landuse/landcover classes are classified using maximum likelihood classifier applied on IRS IC LISS III digital data through image processing software, namely, EASI/PACE. In the study area, it is mainly found below the settlement areas, rolling Jhumland, slopes and road construction sites. Lack of proper drainage accounts for forty percent. Jhumland areas and deforestation are also responsible for around twenty percent of the landsliding while road construction site is responsible for another 40% occurrence. Kohima, the state capital of Nagaland, is located in the North-Eastern region of India between the latitudes of 25^0 30' N to 25^0 45' N and longitude 94^0 0' E to 94^0 15' E. It is bounded in the north by Wokha district, in the south by Manipur, in the east by Zunheboto and Phek district, in the west and north-west by Assam. The area shows a rugged topography of high hills, narrow valleys, steep cliffs, deep gorges are common features. The maximum and minimum heights encountered are over 3700m and below 360m respectively above mean sea level. The slope in the study area ranges between 15^0 and 36^0. Areas close to drainage and streams are comparatively steep. It has a variety of climatic conditions ranging from warm sub-tropical over the foothills and the plains to cool and temperate in the high hills. The average annual rainfall is in between 200 and 250 cm. Two groups of rocks are found, namely, the disang and barails. The causes observed are increase in population, illegal felling of trees, and wasteland management. Based on the analysis of the result at various stages of work, it can be said that remote sensing method has been able to detect the changes in landuse/landcover classes during the selected time period. Using Remote Sensing and Geographical Information System (GIS), natural disasters can be identified and preventive measures undertaken to minimise the losses. Certain types of disasters cannot be predicted directly through Remote Sensing but potential zones can be identified fairly accurately. Landslides are an inherent part of the environment that required control and management strategies. Hence, GIS and Remote Sensing tools can be administered for disaster management and preventive measures for the study area.

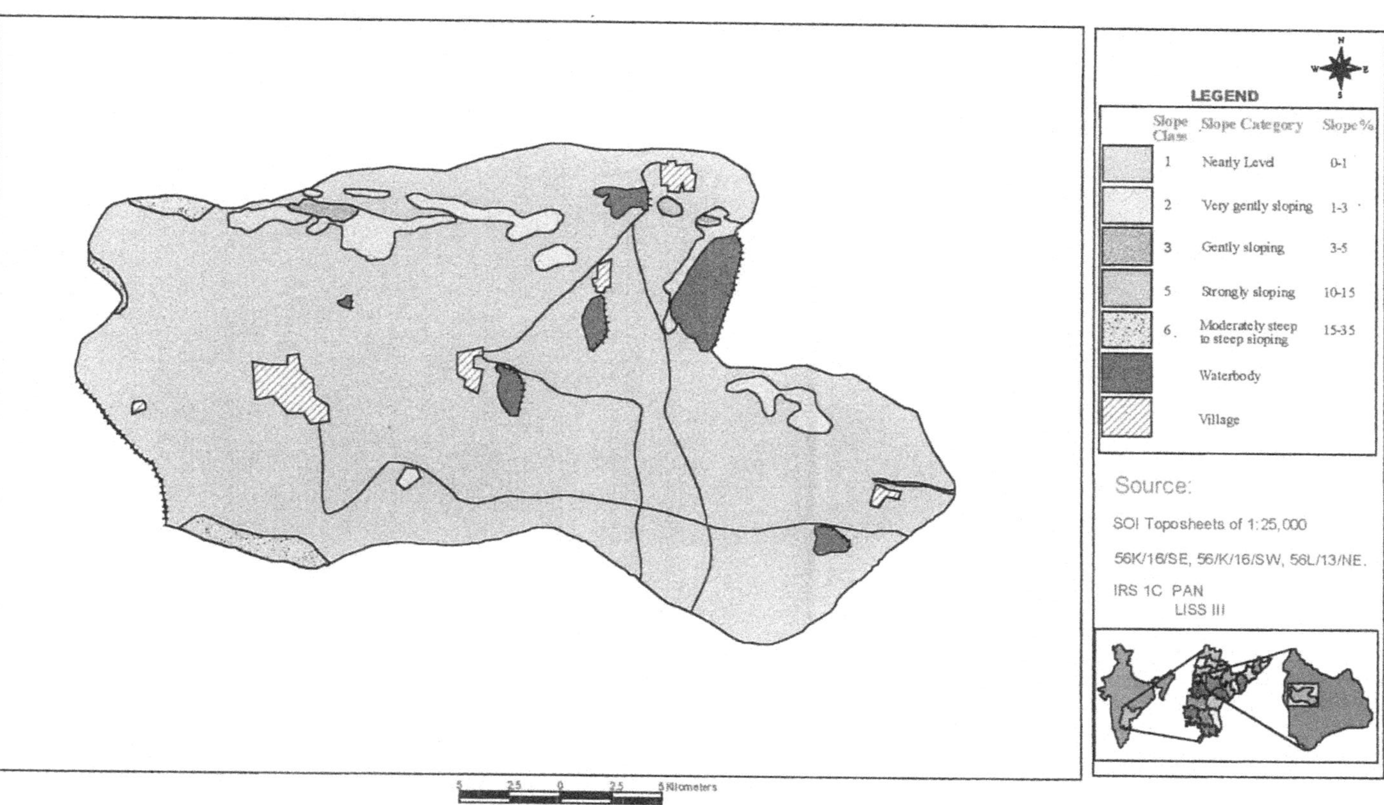

Plate 17 Slope map of Shivannagudem watershed, Nalgonda Dt., (A.P) INDIA

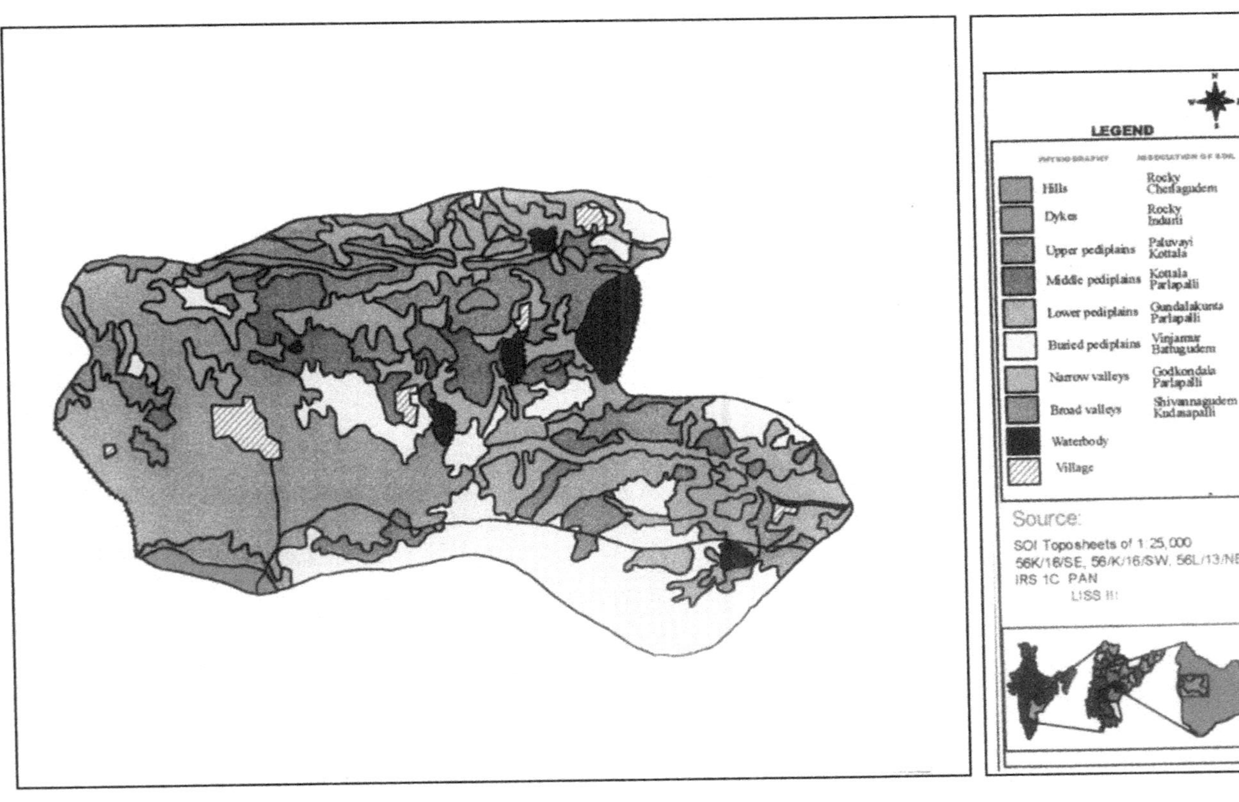

Plate 18 Landuse and Landcover map of Shivannagudem watershed, Nalgonda Dt., (A.P.) INDIA.

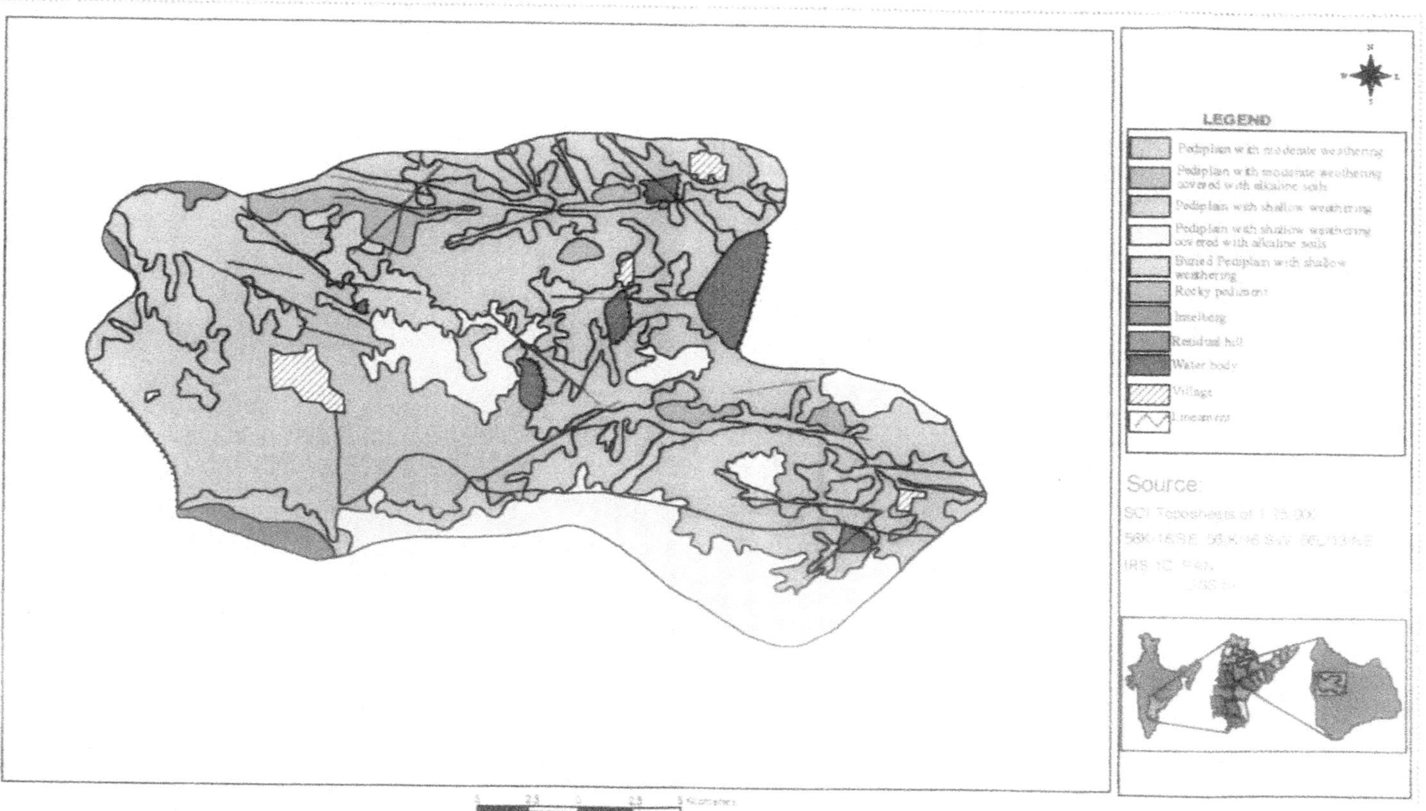

Plate 19 Hydrogeomorphological map of Shivannagudem watershed, Nalgonda Dt., (A.P.) INDIA.

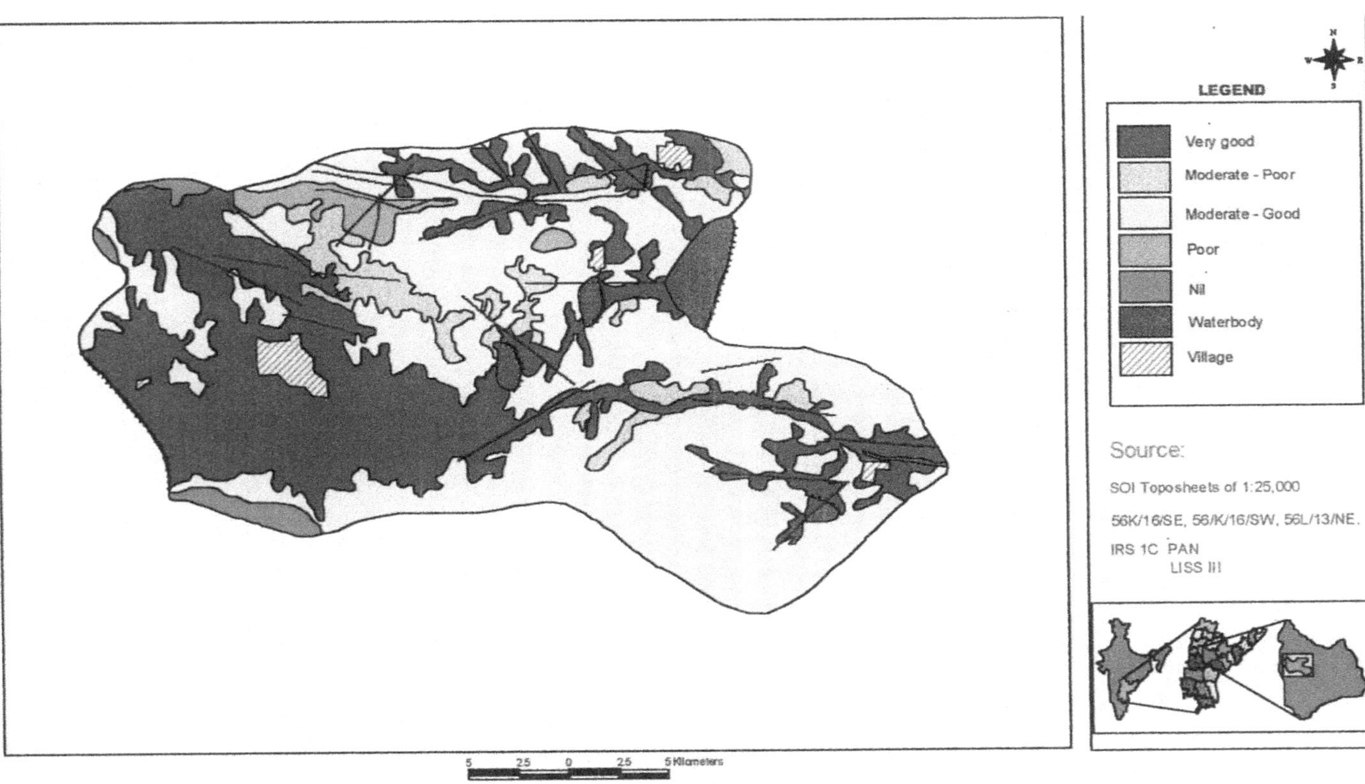

Plate 20 Groundwater prospectus map of Shivannagudem watershed, Nalgonda Dt., (A.P.) INDIA.

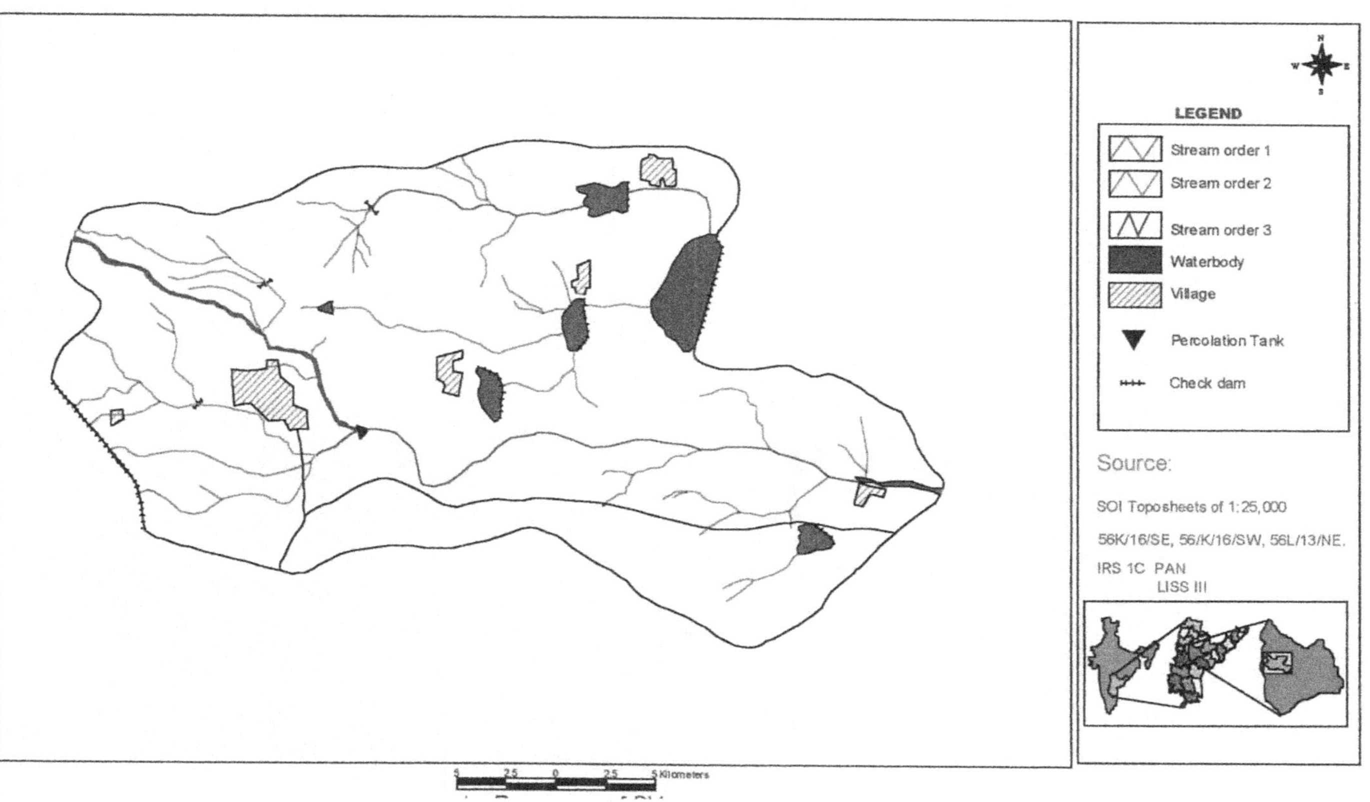

Plate 21 Action plan for Water Resources of Shivannagudem watershed, Nalgonda Dt., (A.P.) INDIA.

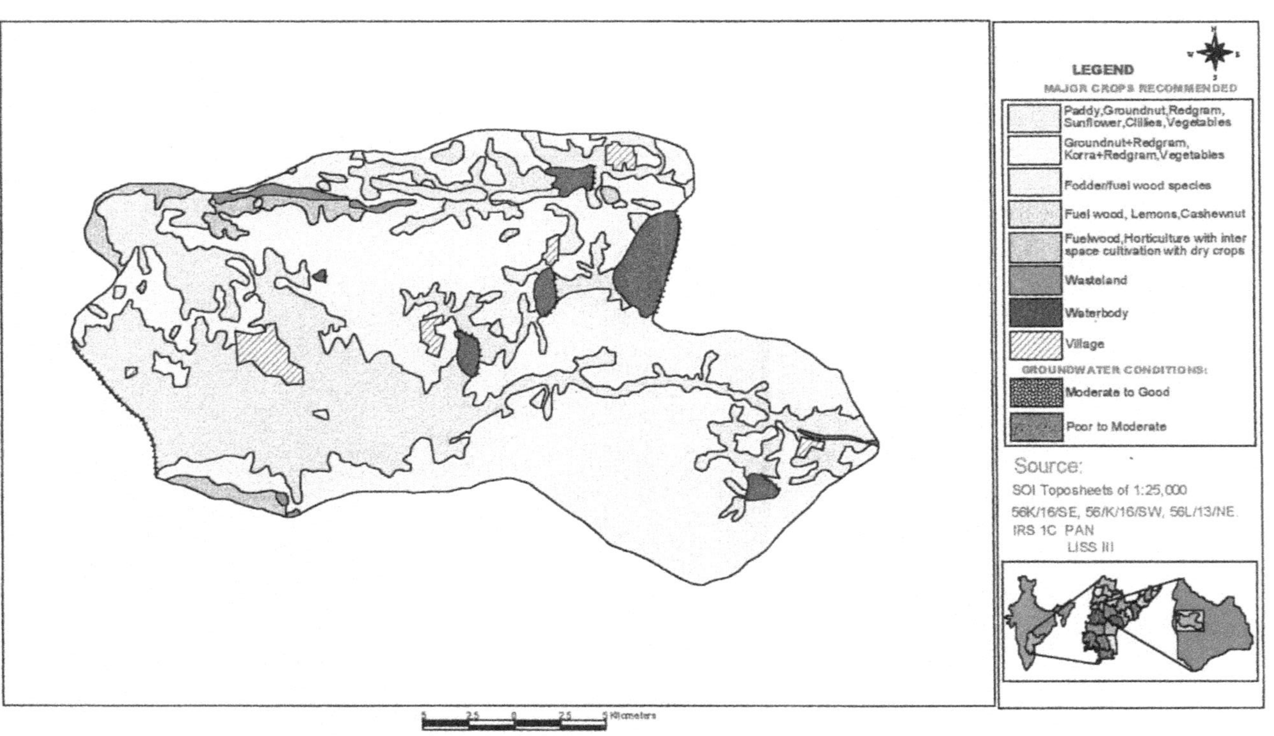

Plate 22 Action plan for Land Resources : Major crops recommended for Shivannagudem watershed, Nalgonda Dt., (A.P.) INDIA

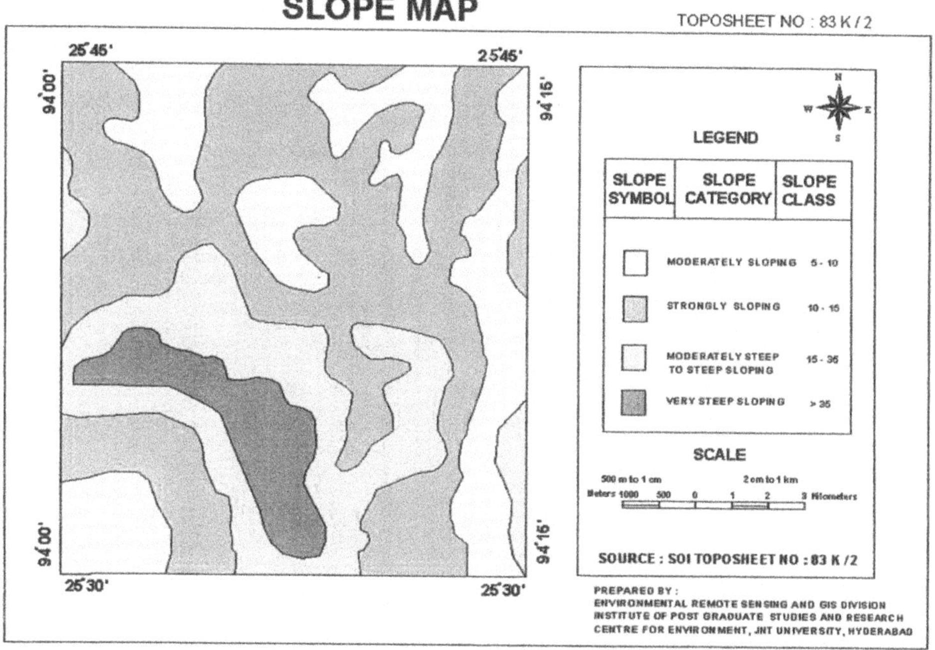

Plate 23 Slope stability map of Kohima area Nagaland

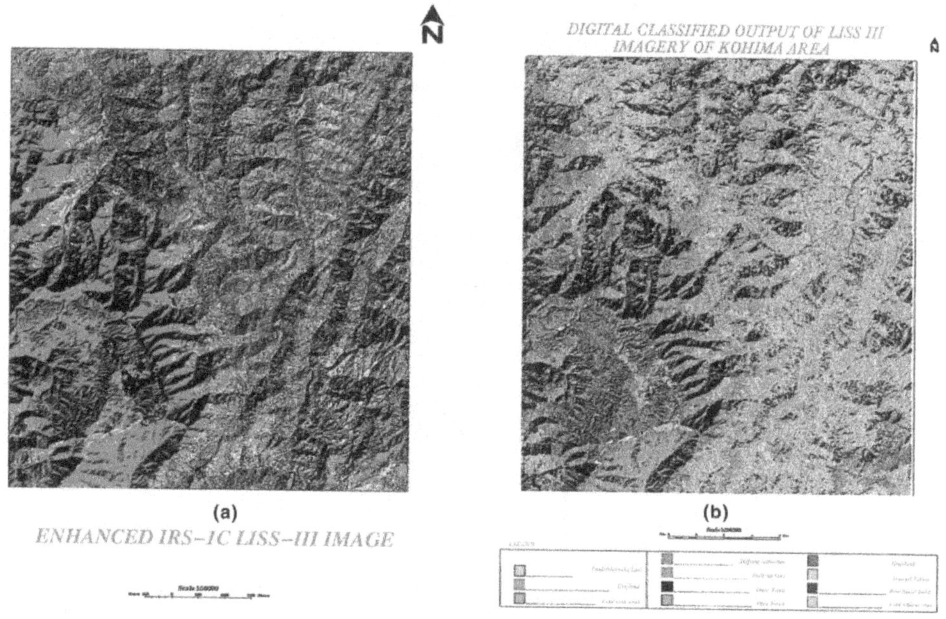

Plate 24 (a) Shows the enhanced satellite image of Kohima area
(b) Color coded output of classified image showing Land use/Land cover of Kohima area Nagaland

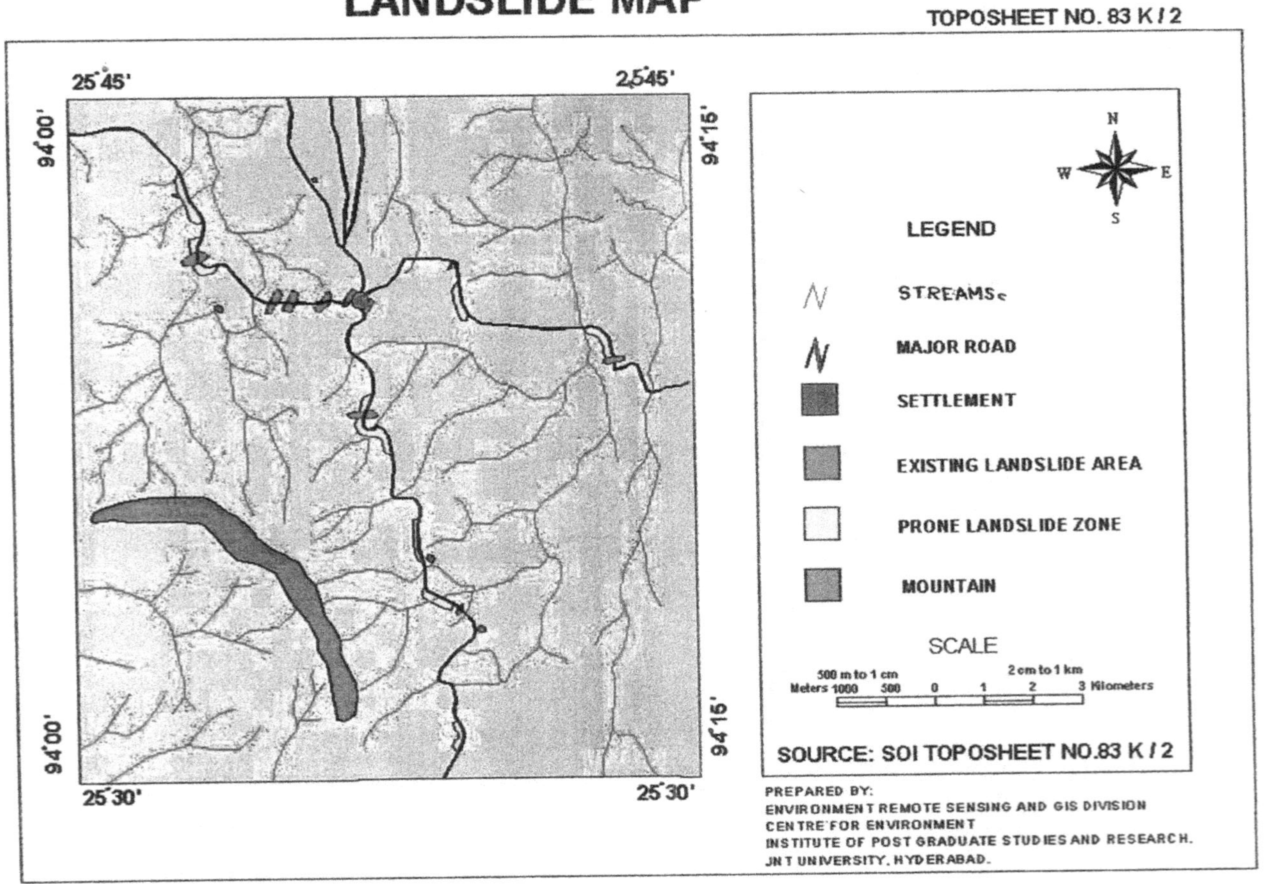

Plate 24 Map showing major Land slide activities of Kohima area in Nagaland

The human transition to urban settlement is a continuing phenomenon creating a complex environmental problem on global climate, the intricate web of forest, ecology, and diversity of living beings. It is also manifested in the form of soil erosion, climatic changes, landslide and so on which are a common phenomenon, possessing a great menace to the economic development of the study area. The study has been carried out to identify various changes in landuse/landcover pattern over a selected period of time using IRS IC LISS III digital data. The spatial information on landuse/landcover and their pattern of change are essential for planning, utilisation and management of the area because they permit better understanding of land utilisation aspects on cropping pattern, fallow land, forest, grazing land, wasteland, and surface water bodies which are very vital for developmental planning.

Landslide which is a recurrent phenomenon, is mainly found below settlement areas, terrace fields, rolling Jhumland slopes and road construction sites. Landslide occurrence around settlement areas is mainly due to lack of proper drainage to convey high runoff water during the rainy days and this accounts for forty percent of landslides. Road construction is also quite significant and it accounts for forty percent. Jhumland areas and deforestation are also responsible for about twenty percent of the landslide occurrence. Slope stability (Plate 22) for the various types of down slope movement is generally explained by examining the forces on slopes and their interrelated, variable types of earth materials, slope, topography, climate, vegetation, water and time. Climatic condition is aslo significant in influencing the down slope movement because it controls the nature and extent of the precipitation and thus the moisture content of slope materials. The role of vegetation in landslides and related phenomena is complex because the vegetation in an area is a function of several factors including climate, soil type and topography.

Because of expanding population, urbanisation, and construction, we have been rapidly modifying the landscape on a massive scale and have triggered through our own carelessness. The nearly universal constructions of buildings, immense highway constructions and hillside development systems have added toll to increase landslides. Most man made landslides are caused by changes in the slope configuration or by activities that permit additional moisture to enter the substratic material.

19.10.1 *Objective of the study*

(i) To study the existing land use pattern of Kohima area, Nagaland, with special reference to the evaluation of landslides.

(ii) To find the changes in the land use pattern since 1989 with the help of remote sensing data.

(iii) To map the zone of landslides using GIS and to suggest remedial measures for prevention of landslides.

19.10.2 Slope Analysis

Slope plays a major role in destabilizing the slope besides other factors, so slope analysis is one of the factor used in hazard analysis. Four categories of slope representing Escarpment/Cliff, very step slope, moderately step slope, and gentle slope have been used.

Slope in the study area ranges in between 15^0 to 45^0. There are few fairly flat areas on the top of the knolls and ridges.

Slope Classes	Slope in Degrees
Gently slope	$15\text{-}25^0$
Moderately steep sloping	$25\text{-}35^0$
Very Steep Sloping	$35\text{-}45^0$
Escarpment/Cliff	$>45^0$

19.10.3 Drainage and its Density

Drainage is the life line of any given area which determines the productive capability of the region in terms of its water resources. The main drainage systems present in the study area are Dzuna, Sidzu, Dzurii, Dzulakie and Kehorii. The drainage pattern observed in the area is dendritic type with varying density. A number of Perennial and Ephemeral streams with 'V' shaped valleys and gorges are common.

Drainage density is an indicator of linear scale landform elements that reflect permeability, rock types and local relief. The drainage density in the study area ranges between 0.5 to 3.2 Km/Sq Km.

19.10.4 Land use/Land cover Analysis

Digital image processing involves the manipulation and interpretation of images obtained in the digital form using various modules like GCP'S, geo-referencing and registration with the aid of computers. One of the most often used methods of information extraction is multispectral classification. Most of the information extraction techniques rely on analysis of the spectral reflectance properties of imagery and employ special algorithms designed to perform various types of "spectral analysis". The process of multispectral digital classification is performed using supervised maximum likelihood Classifier.

The total master plan area of Kohima is around 3855.62 hectares out of which 1080.52 Ha is vacant land. The total developed area is 2436.47 Ha. Table 19.2 shows the existing landuse of Kohima. Plate 23 (a) shows the enhanced satellite digital image and Plate 23 (b) depicts the corresponding colour coded output of classified image for landuse/landcover of a part of Kohima.

Table 19.2 Existing land use of Kohima

Land use	Area in hectare	% of the total land
Residential		
Government	496.25	12.87
Private	746.93	19.37
Commercial	232.83	6.03
Administration	454.42	11.78
Industrial	13.56	0.35
Recreational	8.70	0.22
Transportation & Communication	93.24	2.41
Security uses	56.65	1.46
Agricultural uses	338.63	8.78
Other uses	11.56	0.29
Vacant land	1080.52	28.02
Total	**3855.62**	**100.00**

Source : Land use survey, town-planning organisation. Kohima, 1998

Based on the land use/land cover classified system, the different landuse/ landcover patterns are classified for the years 1989 and 1999 as shown below in the Table 16.3 and pie chart. (Fig. 19.1)

Table 19.3 Land use/ land cover of Kohima area for 1989 and 1999

Land use	Areal extent (Ha)				Change Percent from '89 to '99
Classes	1989	Percent (total area)	1999	Percent (total area)	
Built up	3242.72	5.93	3933.05	7.19	21.28
Dense Forest	14020.08	25.65	11424.25	20.90	18.5-
Open Forest	5925.08	10.84	5299.37	9.69	10.57-
Land with scrub	8020.30	14.67	7298.48	13.35	8.99-
Land without scrub	10115.90	18.51	9399.93	17.20	7.07-
Grassland	7561.57	13.83	6898.48	12.62	8.76-
Crop land	1194.03	2.18	1328.84	2.43	11.29
Shifting Cultivation	3716.19	6.80	4993.89	9.13	34.38
Seasonal follow	1820.11	3.33	1783.11	3.26	2.02-
River	N.A.	498.91	0.91	0.00	
Landslide	N.A.	574.07	1.05	0.00	
Total	**54645.36**		**54645.36**		

* -ve: indicate decrease in landuse classes.
Source : NRSA, Hyderabad.

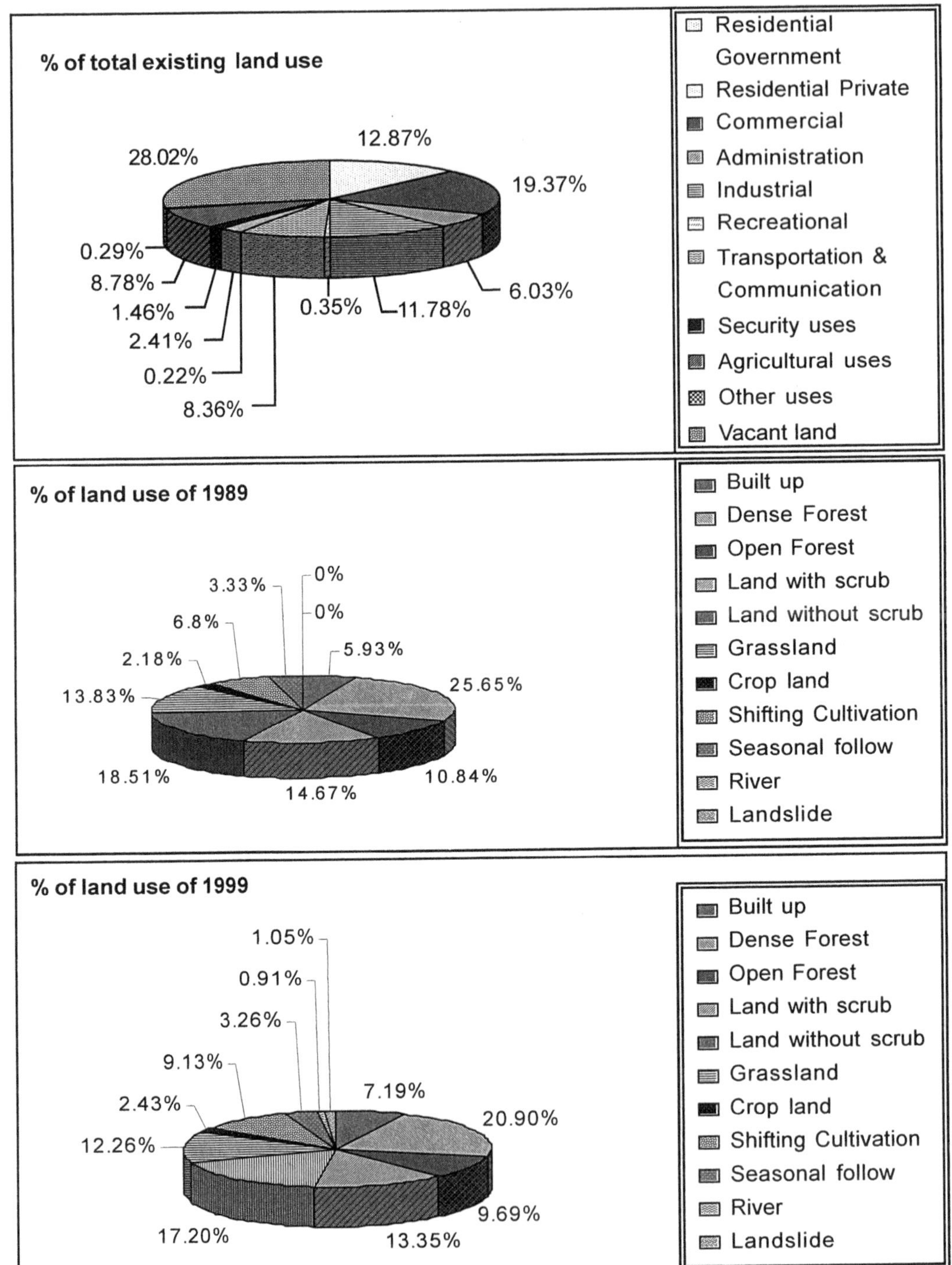

Fig. 19.1 Pie-Chart Showing Existing land use of Kohima, Nagaland.

19.10.5 Land Use Change Analysis

From the above analysis it has been shown that the land use pattern has been changing. It is observed that the dominant change which occur in the study area, is the decrease of forest area by 3944.02 hectares. This change may be due to tremendous increase in shifting cultivation and felling of trees. Increase in built-up land, which increases to 690.33 ha, is due to increase in population and economic activities. Since the people depend mostly upon agriculture, and the land not being very suitable to terrace cultivation, shifting cultivation is adopted as a major practice which increases to 1277.7 ha. The decrease in the area under wasteland, which is about 715.77 ha, may be due to transformation of land to built-up and waste management. Major landslide activities (Plate 24) seen along the road cutting and settlement areas covering an area of 574.07 ha are due to man's activity, lack of proper drainage and geological instability of land.

19.10.6 Results of the GIS Analysis System

The best viable solution is complete avoidance of the area and this strategy has the advantage of being the safest alternative. However, as urbanisation is in its development, prevention and management of the area have been a key factor. So some of the alternative prevention practices are slope reduction, retention structures like ground covers or other vegetation development, retaining walls or other stabilised structures and hillside benching. Surface water may be diverted around the slope by a series of gutters, roadcuts, and surface ditches, and collector systems may also be used to drain ponds and remove standing water. Groundwater may be inhibited from entering a slope by excavating a cut of trench, galleries, and tunnels. Control of toe and surface erosion are caused by using concrete or crib walls, or spur dykes. Control surface erosion due to runoff by use of surface drains, vegetation cover and gabion. People should be made aware of consequences of shifting cultivation.

As observed from the study undertaken, there have been rapid changes in landuse pattern like decrease in forest area, increasing shifting cultivation and so on which have serious consequences on the environment as well as on climatic condition. Hence awareness should be infused into the people as to the sustainable management of the land. Prevention of large natural landslides is nearly impossible but good geological engineering design could be identified and practised for preventive measures and control exercises to minimise the losses. Furthermore GIS software can be used to creat operational database for taking remedial measures to protect the environment in which the buffer zone can also be created around sensitive areas for future planing.

20

Creation of Information System: A Case Study

General

Over 75% of population in India live in rural areas and depend mainly on agricultural activities. The Father of the Nation late Mahatma Gandhi has enunciated that, any development in the country should be centered on the development of a village. Though several developmental programmes have been initiated and executed on various scales, by both Central and State Govt. agencies in Warangal dustrict, microlevel planning taking an individual mandal as a centre has not yet been formulated with full thrust. It is therefore proposed to develop a Mandal Information System and model action plan for sustainable development of Maripeda Mandal of Warangal district in Andhra Pradesh.

20.1 Objectives

1. To study the present status of water resources, land resources, natural resources, soil fertility/productivity, cropping patterns, crop water requirements, water quality, ground water potential using satellite data, collateral data and field data.

2. To prepare the digital thematic maps namely, land use/ land cover, hydrogeomorphology, slope, physiography, soil, geology, drainage etc., using satellite imageries on ARC / INFO GIS platform. This constitutes the spatial database.
3. To compute environmental quality index (EQI) based on the water quality index (WQI) and soil quality index (SQI) maps.
4. Preparation of action plan for land resources and water resources for the Mandal development towards sustainability of the Mandal.
5. Creation of Maripeda MIS (Mandal Information System)

20.2 Methodology

The broad methodology adopted and followed to achieve the objectives of the present study involves the following steps : (Fig. 20.1)

20.2.1 Work Flow

1. Collection of source data like satellite data of two seasons, SOI toposheets, village maps and tentative soil erosion maps. These are the main inputs for the preparation of thematic layers.
2. Two seasons satellite data of PAN and LISS III are geometrically corrected and enhanced. Then both PAN and LISS III data are merged using principal component method and Cubic Convolution resampling technique. Finally after map composition satellite imagery is printed in FCC in 1: 50,000 scale.
3. Preparation of basic themes like base map, transport and settlement map, village map, contour map, drainage map and soil erosion map from the source data. Then updation of base map, transport map and drainage map from the satellite image by visual interpretation.
4. Watershed map and slope map are prepared from the drainage map and contour map respectively.
5. Thematic maps (related to natural resources) like land use / land cover map, lithology map, structure map and geomorphology map are prepared by visual interpretation of the satellite imagery. Visual interpretation is carried out based on the image characteristics like tone, size, shape, pattern, texture, location, association, background etc., in conjunction with existing maps / literature.

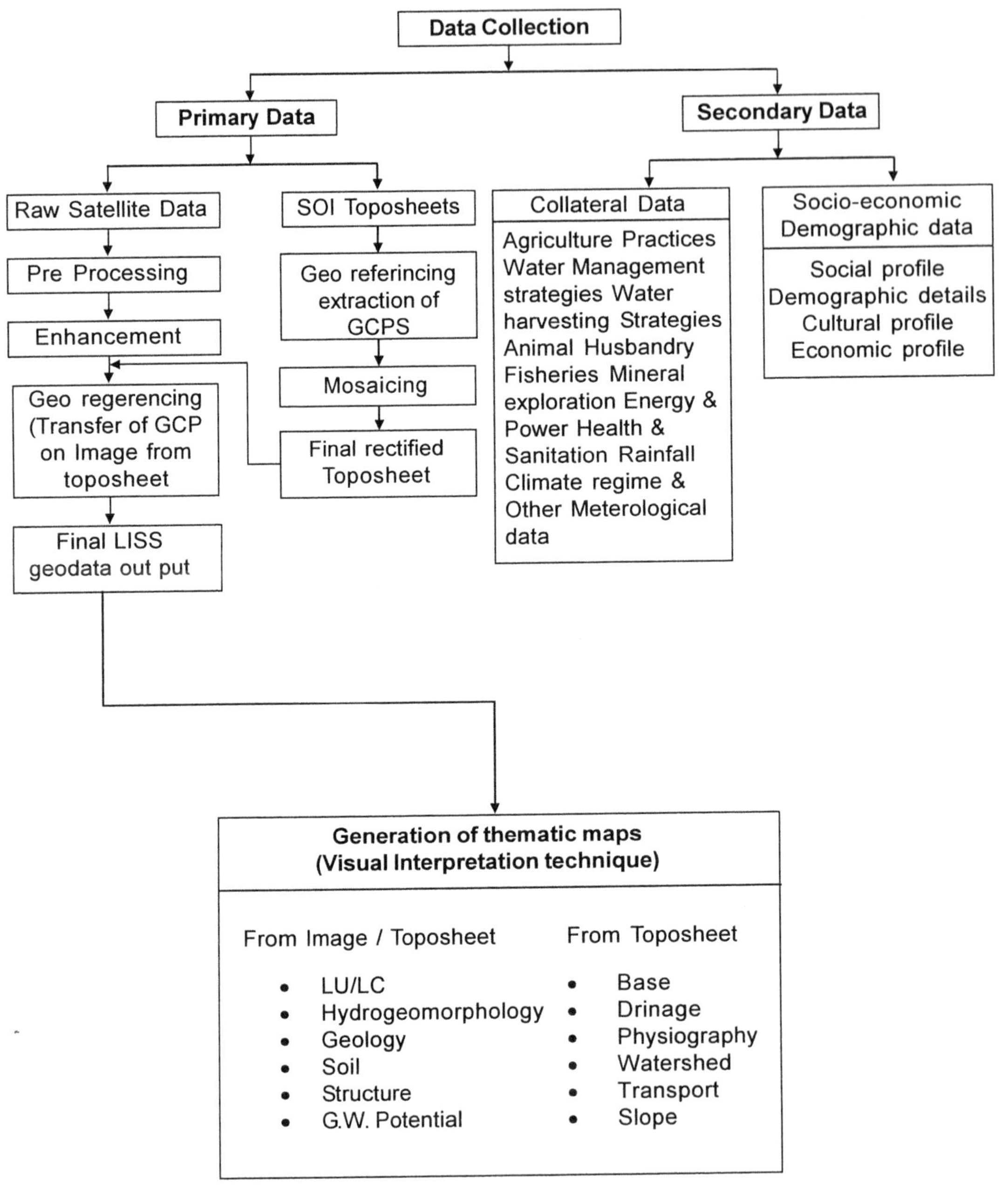

Fig. 20.1 Methodology for generation of thematic maps.

6. Preliminary quality check and necessary corrections are carried out for all the maps prepared.

7. Field visits are carried out to check the delineated units of the maps prepared by visual interpretation of satellite imageries. Photographs related to the study areas were, primary data of land use, soil, well inventory and secondary data related to irrigation, agriculture, land use and ground water are collected.

8. Field observations are incorporated in to the related thematic layers. Well status map is prepared by plotting the well inventory data on the village maps.

9. Ground water prospects map is prepared by the combination of lithology map, geomorphology map, structure map, soil thickness (depth) map and well inventory map. Command area map is prepared by the combination of land use / land cover map, drainage map, contour map, primary and secondary data related to irrigation and tanks.

10. Final quality check and necessary corrections are carried out for all the maps prepared.

11. All the maps prepared are converted into soft copy by digitization. In that process editing, labeling, mosaicing, quality checking, data integration etc., are carried out.

12. Land use / land cover map, drainage map, ground water prospects map, well status map, slope map and command area map are integrated with village map and analysed to get village-wise statistical findings.

13. Villages are categorized by irrigation utilization, natural resources utilization based on the village-wise statistical findings.

14. Ranking criteria is prepared for prioritization of villages for the developmental activities based on the available natural resources and accordingly villages are ranked and categorized.

20.3 Data Used

To study the natural resources, different types of data are used for the preparation of different maps Table 20.1.

Table. 20.1 Details of Sources and the Maps Prepared

Serial No.	Source	Maps Prepared
1.	Survey of India's topographic maps	Contour
2.	Survey of India's topographic maps and satellite imageries	Base Transport Drainage Geomorphology
3.	Satellite imageries	Land use / Land cover Soil Thickness Structure
4.	Maps derived / prepared from the above maps by using GIS	Watershed Ground water Agricultural Area prospects Command area

20.3.1 Details and Limitations of the Data Used

20.3.1.1 Survey of India's topographic Maps

Serial No.	Topographic Map No.	Scale	Year of Survey	Year of Publication
1	56 O/14	1: 50,000	1983-84	1987
2	56 O/15	1: 50,000	1967-68	1969

20.3.1.2 Satellite data from NRSA, Hyderabad

Serial No.	Season	LISS III
1.	Kharif	14–Nov–2004
2.	Rabi	24–Feb–2001

20.4 The Study Area

The Maripeda Mandal lies geographically between latitudes 17^0 20' 00" and 17^0 35'00" and longitudes 79^0 45' 00" to 80^0 00' 00" is covered in the Survey of India toposheet numbers 56 O/14 and 56 O/15. It is one of the 51 Mandals of Warngal district,

(Table 20.2) in Andhra Pradesh. Maripeda town is at a distance of kms from Warangal (District H.Q.) and 120 kms from Hyderabad (State Capital).

Warangal district occupies an area of 12846 Km^2 with various agricultural, mining and quarrying, manufacturing and other household industries. As per the census of 1997-98, forests occupied 28.69% of the area and gross irrigated area as percentage of gross cropped area is 52.02%. The district has a total road length of 54.85 km per 100 Km^2 and a railway route of 1.19 km per 100 Km^2. The total population of the district as per 2001 census is 3231.17 thousands which comprise of 620.79 thousands of urban population and 2610.38 thousands of rural population. The decadal growth rate of the district from 1991 – 2001 is 14.63%.

The major crops grown in the study area are: Rice, Jowar, Cotton, Turmeric, Maize, Chillies and Sesame. Because of long range of dry periods and less number of rainy days, the area suffers from poor soil moisture condition, resulting in frequent drought and famines. Due to erratic nature of rainfall and impermeable nature of rocks the stream channels are formed into shallow and wider valley floors. The Drainage pattern is dendritic as a whole. Most of the stream courses are controlled by geological structures.

The upland areas of Warangal District in A.P. form parts of semi-arid zone in peninsular India with scanty and erratic rainfall. These areas have been identified as chronically drought affected areas in the state. Famines have affected the area frequently in the past and their frequency of occurrence has increased during the last five decades. It is universally recognized that the most effective way to eliminate drought and famine and to reserve the desertification process in an area is by augmenting water supplies to the area to optimally cater the basic needs of drinking and sustaining the agriculture on which the majority of the population depends.

With the advent of satellite remote sensing technology and computer based Geographic Information System (GIS), it has become possible for environmental planners to have correct overall perspective with least investment of time and money. Keeping these in view, the present project proposal envisages to employ GIS combined with remote sensing technologies for augmenting and optimal utilization of water resources and watershed management programmes which will culminate in the overall socio-economic development and poverty alleviation in the study area.

Table 20.2 List of Revenue Mandals in Warangal District.

Mandal Code	Mandal Name	Mandal Code	Mandal Name	Mandal Code	Mandal Name
1	Cheriyal	18	Thorrur	35	Duggondi
2	Maddur	19	Nellikudur	36	Geesugonda
3	Narmetta	20	Narsimhulapet	37	Atmakur
4	Bachannapeta	21	Maripeda	38	Shayampet
5	Jangaon	22	Dornakal	39	Parkal
6	Lingala Ghanpur	23	Kuravi	40	Regonda
7	Raghunatha Palle	24	Mahabubabad	41	Mogullapalle
8	Ghanpur(Stn)	25	Kesamudram	42	Chityal
9	Dharmasagar	26	Nekkonda	43	Bhupalpalle
10	Hasanparthy	27	Gudur	44	Ghanapur
11	Hanamkonda	28	Kothagudem	45	Mulug
12	Wardhannapet	29	Khanapur	46	Venkatapur
13	Zaffergadh	30	Narsampet	47	Govindaraopet
14	Palakurthi	31	Chennaraopet	48	Tadvai
15	Devaruppula	32	Parvathagiri	49	Eturnagaram
16	Kodakandla	33	Sangam	50	Mangapet
17	Raiparthy	34	Nallabelly	51	Warangal

20.5 Basic Themes

20.5.1 Base Map

Base Map is prepared by using Survey of India topographic maps on 1: 50,000 scale. All the settlements, road network, water bodies and forest areas are taken into consideration. By comparing the Survey of India topographic maps with that of the satellite image, the size of all the settlements are increased and updated. The aerial extent of the study area is 287.9 sq. km.

20.5.2 Transportation Map

In the study area all the settlements are connected either by metalled road or un-metalled road. Where as, State Highway connects Maripeda, Railway network does not exist in the Maripeda Mandal. The nearest railway station is Khammam, which is at a distance of 18 kms SouthEast of Maripeda.

Fig. 20.2 Road Network of Maripeda Mandal.

20.5.3 Village Map

This map is prepared by digitization of the maps of Central Survey Office. Revenue boundaries of all the villages are plotted in this map. The entire data of the village available is attached to this map as database using GIS. This database is very useful to know the present scenario of a village. This map is used for analyzing village wise land and water resources.

In the study area there are 23 revenue villages, out of these Maripeda, which is the Mandal head quarter.

Village wise findings of the study area are given below (Table 20.3):

Table 20.3 Total Village Area Findings

S.No	VILLAGE NAME	VILLAGE AREA In Sq.Kms	VILLAGE AREA in Ha	Built-up AREA in Ha
1	GUDUR	14.3338	1433.38	13.95266
2	UGGAMPALLI	6.2661	626.61	5.035589
3	YELLAMPET	18.4576	1845.76	14.68401
4	VISAMPALLI	8.7504	875.04	5.072229
5	JAYYARAM	20.2711	2027.11	11.21656
6	RAMPURAM	11.8124	1181.24	8.955088
7	CHILLAM CHERLA	12.7141	1271.41	5.387084
8	NILKURTHI	9.1091	910.91	9.027011
9	ANEPURAM	12.7375	1273.75	4.852912
10	VIRARAM	16.2316	1623.16	3.891979
11	ULLEPALLI	7.4106	741.06	3.0135

Contd...

S.No	VILLAGE NAME	VILLAGE AREA In Sq.Kms	VILLAGE AREA in Ha	Built-up AREA in Ha
12	ERACHERLA	9.3641	936.41	5.555669
13	DHARMARAM	14.3409	1434.09	4.214278
14	PURUSHOTTAMGUDA	6.4151	641.51	2.204947
15	BICHRAJPALLI	8.2254	822.54	1.895173
16	PURUSHOTTAMGUDAM	4.8182	481.82	2.280693
17	ABBAIPALEM	8.5823	858.23	3.181055
18	MARIPEDA	32.4529	3245.29	28.51486
19	GUNDIPUDI	9.7557	975.57	11.36234
20	BURANAPURAM	5.7150	571.50	7.115339
21	TANAMCHERLA	19.8126	1981.26	12.357064
22	TALLAUKKAL	17.1023	1710.23	4.209495
23	GIRIPURAM	15.2834	1528.34	2.845766

20.5.4 Physiography Map

Contour is a line joining the points of equal elevation. Contour interval used while preparing this map is 20 m spot heights are also included in this map. Elevation of the study area varies from 160 to 356 m above mean sea level (MSL).

The maximum of 356 m. height exists between Viraram and Ullepalli and the minimum is 160 m. This map is used in the preparation of slope map.

20.5.5 Land Use / Land Cover Map

Land use / land cover map is prepared by visual interpretation of high-resolution satellite data with the help of Survey of India Topographic maps on 1: 50,000 scale. Two seasons' data (Kharif and Rabi of year 2004) is used for the delineation of different units. The units are confirmed by the ground truth/field visits.

Level-II classification of National (Natural) Resources Information System (NRIS) has been followed for the delineation of units.

Land use/ Land cover map of the study area is integrated with village map and analyzed with the help of GIS to get the village wise findings of the present land use of the study area, which is given elaborately in the following tables (Table 20.4 and 20.5):

Table 20.4 Category wise present land use/land cover findings of total Mandal

S.NO	DESCRIPTION	AREA in Ha
1	Tank with Water	966.4265
2	Land for Plotting	268.8404
3	Single Crop	10341.3584
4	Double Crop	8739.9779
5	Fallow Land	1439.981
6	Dry Tank	215.1961
7	Barren Rock/Sheet Rock/Stony Waste	1842.7488
8	Land with Scrub	4307.1229
9	Land without Scrub	9.2313
10	River with Water	416.4555
11	Built-up-Area	180.1635

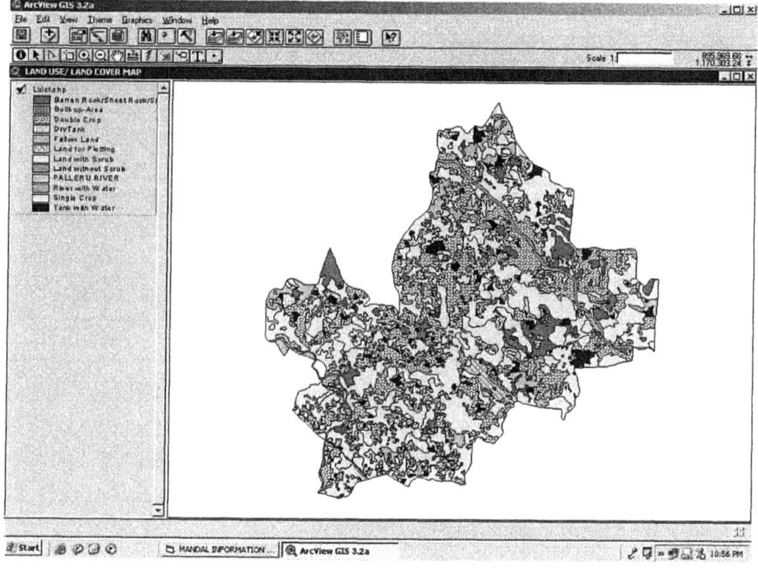

Fig. 20.3

Table 20.5 Land Use Category Details of the Study Area -Village wise in Ha.

S.No	Village Name	Waste land in Ha	Agricultural Land in Ha	Total water Area in Ha	Built-up Land
1	UGGAMPALLI	31.7402	562.8604	59.6841	13.95266
2	GUDUDR	358.4504	983.8141	130.3174	5.035589
3	JAYYARAM	447.8290	1370.7268	137.8425	14.68401
4	VISAMPALLI	104.3194	707.5095	61.2245	5.072229
5	YELLAMPET	229.4026	1895.8728	107.0032	11.21656
6	ANEPURAM	244.8893	979.7989	49.8155	8.955088
7	NILKURTHI	113.5917	624.4617	30.9209	5.387084
8	CHILLAM CHERLA	342.4216	735.7583	85.6710	9.027011
9	RAMPURAM	215.1019	902.9790	117.6542	4.852912
10	TALLAUKKAL	21202648	501.258	15.87	3.891979
11	TANAMCHERLA	224.0061	1670.2899	63.0913	3.0135
12	BURANAPURAM	104.9995	458.3002	8.6255	5.555669
13	GUNDIPUDI	91.6076	846.8924	64.3937	4.214278
14	GIRIPURAM	124.5925	526.0617	45.0077	2.204947
15	MARIPEDA	615.3195	2242.1072	106.7905	1.895173
16	VIRARAM	371.7756	1191.8894	56.6615	2.280693
17	ULLEPALLI	223.7203	498.7230	15.7869	3.181055
18	ERACHERLA	328.6808	570.0239	54.1155	28.51486
19	BALADHARMARAM	669.9388	812.0928	104.7559	11.36234
20	BICHURAJPALLI	555.8496	224.0822	43.0663	7.115339
21	PURUSHOTTAMGUDAM	202.2594	482.537	12.753	8.357064
22	ABBAIPALEM	240.0533	586.9453	29.6637	4.209495
23	GALIVARIGUDEM	115.4636	360.5177	6.6537	2.845766

20.5.6 Agriculture

In the study area, less than 2% of double crop area observed in Galivari gudem, Bichrajpalli, Gundipudi, Buranapuram and the maximum double crop area observed in Yellampet (12%) and Maripeda (10%).

Less than 2% of single crop area observed in Bichurajpalli, Gundipudi and Abbaipalem and maximum of 15% area single crop area observed in Maripeda(15%), and Jayyaram (10%). And more than 27% area of fallow land area is in Gudur and Nilkurthi; Giripuram and Viraram with 0% area.

20.5.7 Wasteland

In the study area, more than 10% of wasteland to net village area is in Dharmaram, 669.93Ha. and less than 1% of wasteland to net village area is in Uggampalli, i.e. 0.98% (31.7Ha.) (Table 20.7).

Table 20.7 Village wise wasteland findings

S.No	Village name	B.R/S.R/S.W In Ha	Land with Scrub in Ha	Land with out Scrub	waste land in Ha
1	UGGAMPALLI	0	31.7402	0	31.7402
2	GUDUDR	73.9429	284.5076	0	358.4504
3	JAYYARAM	260.9441	186.8850	0	447.8290
4	VISAMPALLI	79.5950	24.7245	0	104.3194
5	YELLAMPET	33.9202	195.4824	0	229.4026
6	ANEPURAM	116.8428	128.0466	0	244.8893
7	NILKURTHI	30.6478	82.9440	0	113.5917
8	CHILLAM CHERLA	186.4725	155.9491	0	342.4216
9	RAMPURAM	23.2275	191.8745	0	215.1019
10	TALLAUKKAL				
11	TANAMCHERLA	36.4759	179.6508	7.8795	224.0061
12	BURANAPURAM	10.9201	94.0795	0	104.9995
13	GUNDIPUDI	91.5968			
14	GIRIPURAM	8.2551	116.3375	0	124.5925
15	MARIPEDA	126.6039	488.7156	0	615.3195
16	VIRARAM	28.8390	342.9367	0	371.7756
17	ULLEPALLI	2.5173	221.2030	0	223.7203
18	ERACHERLA	126.6810	201.9998	0	328.6808
19	BALADHARMARAM	135.8153	534.1235	0	669.9388
20	BICHURAJPALLI	263.6072	292.2424	0	555.8496
21	PURUSHOTTAM GUDAM	*******	*******	********	********
22	ABBAIPALEM	148.8516	91.2018	0	240.0533
23	GALIVARIGUDEM	11.2836	104.1800	0	115.4636

B.R/S.R/S.W Barren Rock / Sheet Rock / Stony Waste

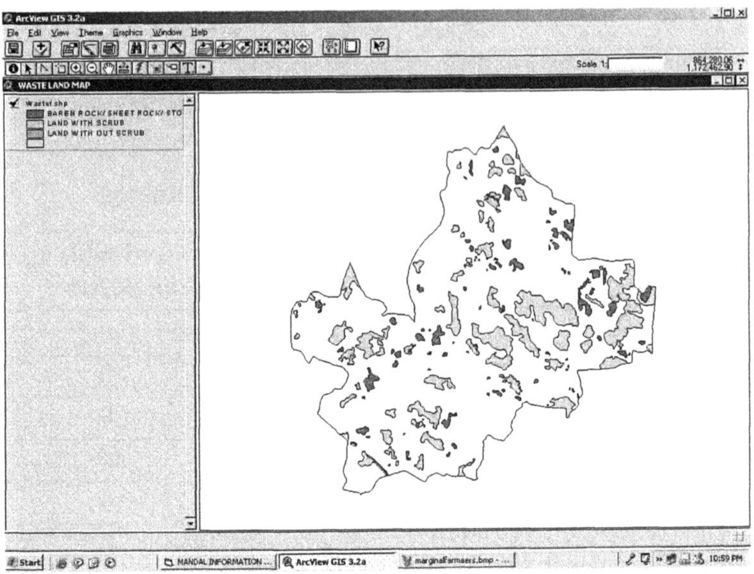

Fig. 20.4

20.6 Water Resources

Water is essential for life. It is also a key resource in all-economic activities ranging from agriculture to industry. There is severe stress on water resources with increasing pressure of human population. Lands are getting either degraded or wasteland due to inadequate management of water resources and over exploitation of available water. Water resources are divided into surface water resources and ground water resources.

20.6.1 Drainage

Drainage map is prepared by using Survey of India Topographic maps on 1: 50,000. All the streams and tanks existing in the study area are marked in this map. These streams further classified based on stream ordering. Up to fourth order streams exist in the study area.

Most of the streams right side of the state highway flow NorthEast to SouthWest in the study area and left side of the state highway SouthWast to NorthEast. Canals, reservoirs and lakes do not exist in the study area. Only two minor rivers namely Palleru and Akeru exists. The drainage system existing is dendritic. Tank bunds are also marked in the map. All the water bodies are divided into dry and wet areas. These wet (water spread) areas changes from time to time and some new tanks are found in the satellite images. For this reason, the drainage map is updated from the satellite images.

20.6.2 Ground Water Resources

Ground water is a precious and the most widely distributed resource of the earth and unlike any other mineral resource, it gets its annual replenishment from the meteoric precipitation. It is the largest source of fresh water on the planet excluding the polar icecaps and glaciers.

The only recharge of ground water is precipitation, which is less and erratic. As such often drought conditions persist. It is essential that each drop of rain water fallen on ground is utilized properly for better water management. Recharge can be effectively done through the construction of mini percolation tanks, sub surface dams and other recharge structures at feasible sites.

20.6.3 Ground Water Prospects in the Study Area

Ground water prospects of an area depend mainly on lithology unit (rock type) occurring at that area. However, within each lithology unit, the ground water conditions vary significantly depending upon the geomorphology, structure, slope, soil thickness, depth and nature of weathered material, presence of fractures / lineaments, surface water bodies, canals, irrigated areas, etc. All the parameters of the study area are studied and integrated to arrive at the ground water prospects.

20.7 Ground Water Prospects Map

All the above parameters are integrated to arrive at the ground water prospects in each geomorphic-cum-lithology units, designated as hydrogeomorphic unit. Except dyke ridge all the geomorphic units present in the study area belongs to the granites and gneisses (lithology unit) of Peninsular gneissic complex, where as dyke ridge belongs to dolerite dyke (lithology unit).

Hydrogeomorphic units occurred in the study area are described below:

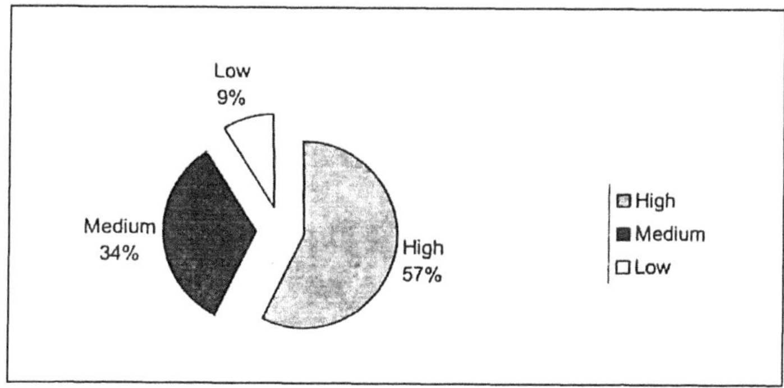

Fig. 20.5

Ground water prospects findings

S No.	GW Potential Category	Area in %
1	LOW	09
2	MEDIUM	34
3	HIGH	57

20.7.1 Pediplain shallow weathered in granitic gneiss landscape

Pediplain is a gently undulating plain developed by the coalescence of several pediments. Since it is shallow weathered, the thickness of the weathered zone is less than 10 m. Pediplains form good aquifers. Therefore this unit is categorized as good for ground water prospects. Along the lineaments the ground water prospects are even better.

20.7.2 Valley fill shallow in Granitic Gneiss landscape

These are the valleys of different shapes and sizes occupied by fill material of partly detrital and partly weathered. Since it is shallow filled, the thickness of the filled material is less than 10m. These valley fills form good productive shallow aquifers. Topographically the valleys have more recharge potential if check dams and percolation tanks are constructed across the valley. Therefore this unit is categorized as good in ground water prospects.

20.7.3 Pediment in granitic gneiss landscape

These are gently undulating plain dotted with rock outcrops with or without thin veneer of soil cover. In general the ground water prospects are moderate except along the lineaments, where it is even better.

20.7.4 Pediment-inselberg complex in granitic gneiss landscape

Pediment dotted with a number of inselbergs, which cannot be separated and mapped as individual units. Inselberg as such is a runoff zone, which does not permit infiltration. Because this unit includes pediments, it is characterized as low in ground water prospects. Along the lineaments the ground water prospects are even better.

20.7.5 Inselberg in granitic gneiss landscape

Inselbergs is an isolated small hill of massive type, rising abruptly from the surrounding plain. The inselbergs are devoid of any structure and forms a rapid runoff zone. Hence, there are no prospects for ground water occurrence and it is categorized as nil in ground water prospects.

20.7.6 Residual hill in granitic gneiss landscape

These are isolated hills formed due to differential weathering. The hills form a rapid run-off zone, eliminating the chances of ground water occurrence. Therefore the overall groundwater prospects of this unit are categorized as nil.

20.7.7 Denudational hill in granitic gneiss landscape

These hills are formed due to differential erosion and weathering, so that more resistant formation stand as hills. These hills act as a run-off zone, with no groundwater occurrences. But along the narrow valleys of these hills, there is a chance for ground water occurance. This unit is categorized as nil in ground water prospects.

20.7.8 Dyke ridge in Dolerite landscape

Dyke ridge is formed by narrow ridge of dolerite. The dyke ridges are encountered in the southern part of the study area. The general direction of the dyke ridges is East-West to NW-SE. The dyke has no ground water prospects, as it is a run-off zone. It also acts as a barrier for ground water movement and improves the potential in the upstream side. The overall groundwater prospects of this unit are categorized as nil.

20.8 Socio Economic Conditions

20.8.1 Population

The population of Maripeda Mandal is 79,107 (as per 2001 census). The population of Maripeda is 14,786 persons (as per 2001 census). The village wise population details given in Table 20.8.

20.8.2 Details of Village wise Farmers Categories:

The farmer data classified as marginal, small, medium and large according to Acrs of agricultural land each farmer is having.

Marginal Farmers	:	0 – 4.92 Acrs
Small Farmers	:	4.93 - 9.84 Acrs
Medium Farmers	:	9.85 - 24.60 Acrs
Large Farmes	:	above 24.61 Acrs

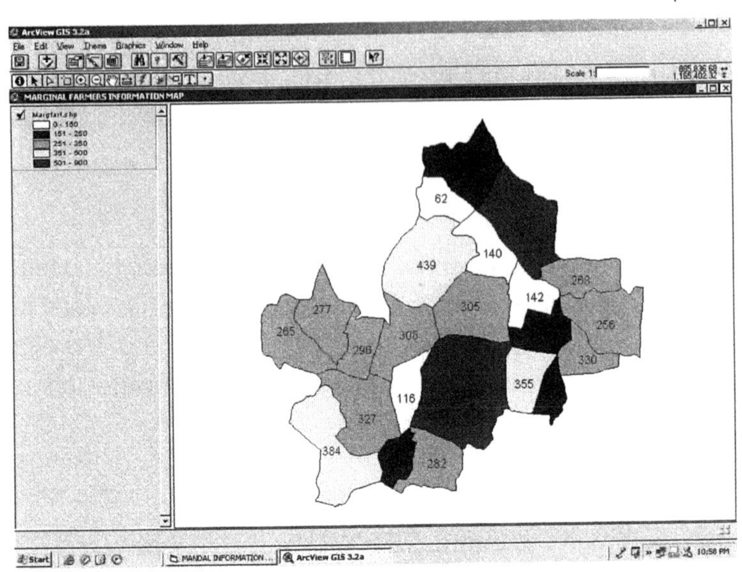

Fig. 20.6

Table 20.8 Showing Categories of Farmers in Maripeda Mandal

S.No	Revenue village Name	Marginal Farmers	Small Farmers	Medium Farmers	Large Farmers
1	UGGAMPALLI	62	80	34	7
2	GUDURU	161	80	49	9
3	JAYYARAM	581	129	49	21
4	VISAMPALLI	140	56	49	12
5	YELLAMPETA	439	187	69	5
6	ANEPURAM	308	148	43	4
7	NILKURTI	296	105	33	3
8	CHILLAM CHERLA	277	58	40	14
9	RAMPURAM	265	68	42	17
10	TALLAUKAL	327	133	72	16
11	TANAMCHERLA	384	174	34	14
12	BURANAPURAM	198	49	20	8
13	GUNDIPUDI	282	82	25	4
14	GIRIPURAM	116	24	23	9
15	MARIPAD	853	209	127	37

S.No	Revenue village Name	Marginal Farmers	Small Farmers	Medium Farmers	Large Farmers
16	VIRARAM	305	192	51	12
17	ULLEPALLI	142	55	40	5
18	ERACHERLA	268	70	32	13
19	DHARMARAM	256	78	67	25
20	BICHRAJPALLI	181	30	15	20
21	PURUSHOTTAMGUDAM	330	30	17	12
22	ABBAIPALEM	355	89	15	6
23	GALIVARIGUDEM	165	28	20	0

Source : census data 2001

20.8.3 Transportation

All the major roads, district roads and other tracks connecting different places of study area are digitized as line themes. This road network is used to develop a model for road information system using different softwares like ArcInfo, ArcView, AutoCad, Visual Basic and Map objects. This information system will help in making decisions like finding shortest path, distance between different villages, total road length, major roads that are crossing streams and the villages that are not having proper road facility etc. This is a model study for small Mandal of Maripeda, Warangal District. This model study can be used for other major cities, districts, states and etc., the road network map is shown in Fig 20.3 and distance from maripeda to each village in mandal is shown in Table 20.9.

Table 20.9 Showing the Distance from Maripeda to each Village in the Mandal

S.NO	VILLAGE NAME	DISTANCE FROM MARIPEAD In Kms.
1	UGGAMPALLI	15
2	GUDURU	17
3	JAYYARAM	21
4	VISAMPALLI	14
5	YELLAMPETA	10
6	ANEPURAM	10
7	NILKURTI	12
8	CHILLAM CHERLA	14
9	RAMPURAM	16
10	TALLAUKAL	19

S.NO	VILLAGE NAME	DISTANCE FROM MARIPEAD In Kms.
11	TANAMCHERLA	24
12	BURANAPURAM	10
13	GUNDIPUDI	3
14	GIRIPURAM	8
15	MARIPAD	0
16	VIRARAM	7
17	ULLEPALLI	11
18	ERACHERLA	14
19	DHARMARAM	9
20	BICHRAJPALLI	8
21	PURUSHOTTAMGUDAM	7
22	ABBAIPALEM	3
23	GALIVARIGUDEM	4

20.8.4 Livestock

In the drought prone areas, like Maripeda area, the villagers are mainly based on the livestock ranching, which is one of the earning sources to the farmers. The Government of Andhra Pradesh is presently concentrated in this aspect to improve the livelihood of the farmers. People of this area keep large herd of sheeps and cows. The animals are let loose in the adjoining hills for grazing, leading to a large-scale denudation of tree and grass cover.

20.8.5 Education Facilities

In the rural areas, upto high school education facilities are available in Uggampalli, Gudur, Jayyaram, Visampalli, Yellampet, Anepuram, Nilkurti, Chillamcherla, Rampuram, Tallaukal, Tanamcherla, Buranapuram, gundipudi, Giripuram, Maripad, Viraram, Ullepalli, Eracherla, Bichrajpalli, purushottamgudem, Abbaipalem,Galivarigudem villages and only primary education is available in the rest of the villages.

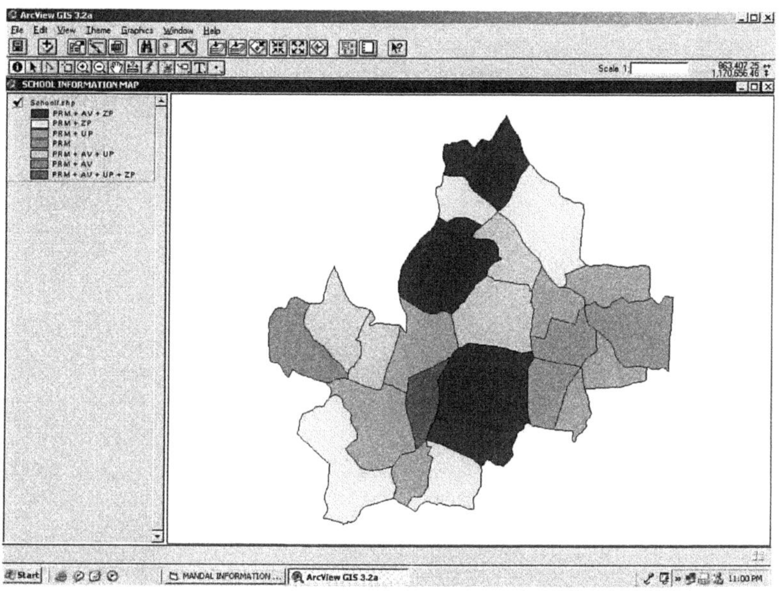
ArcView GIS 3.2a
File Edit View Theme Graphics Window Help
SCHOOL INFORMATION MAP
PRM + AV + ZP
PRM + ZP
PRM + UP
PRM
PRM + AV + UP
PRM + AV
PRM + AV + UP + ZP
Start
ArcView GIS 3.2a

Fig. 20.7

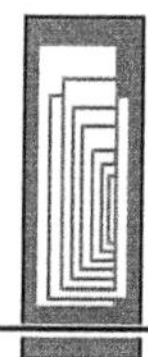

REFERENCES

Abel., D. J. 1989, SIRO-DBMS : A Database Tool-Kit for Geographical Information System, International Journal of Geographical Information System, Vol. 3, No.2, pp 103-116.

Adler., H. H. 1968, Infrared Spectra of Phosphate Minerals: Splitting and Frequency Shifts Associated with Substitution of PO_4^3 for AS O_4^{-3} in mimetite American Mineralogist, vol. 53, pp. 1740-1744.

Anderson., J. M. and E. M. Mikhail., 1985, Introduction to Surveying, New York, Mc Graw Hill Book Co.

Anji Reddy., M. and K. M. Reddy, 1996, Performance Analysis of IRS Bands for Land Use/Land Cover Classification System using Maximum likelihood Classifier, International Journal of Remote Sensing, Vol. 17, No. 13, pp. 2505-2515

Anji Reddy., M. 1995, Ph. D Thesis, Jawaharlal Nehru Technological University, Hyderabad, India

Anji Reddy., M. and S. Srinivasulu, 1994, Comparison of IRS-1B LISS-11A Pixel Array Sizes for Estimating Suspended Solids Concentration in Hussain Sager Lake, Hyderabad, India-A Statistical Approach, International Journal of Remote Sensing, Vol. 15, No. 18, pp. 3693-3706.

Anji Reddy., M. 1997, A Detailed Statistical Study on Selection of Optimum IRS LISS Pixel Configuration for Development of Water Quality Models, International Journal of Remote Sensing, Vol.18, No. 12, pp. 2559-2570.

Aronoff., S. 1989, Geographic Information Systems : A Management perspective, WDL Publication, Ottawa, Canada.

Aronoff., S.1985, The Minimum Accuracy Value as an Index of Classification Accuracy, Photogrammetric Engineering and Remote Sensing, 51, pp. 99-111.

Bannari., A., Morin. D., Bonn, F, and Hutch. A., 1995a, A Review of Vegetation Indices, Remote Sensing Reviews, 13, pp. 95-120.

Bannari., A., Morin. D., Benie, G. B., and Bonn, F. J., 1995 b, A Theoritical Review of Different Mathematical Models of Geometric Corrections Applied to Remote Sensing Images, Remote Sensing Reviews, 13, pp. 27-47.

Barrow., G. M.,1962, Introduction to Molecular Spectroscopy, New York, Mc Graw-Hill, 318p.

Bernhardsen., T., 1992, Geographic Information Systems, VIAK IT, Norway.

Bernstein and Ferneyhough., D. G. 1975, Digital Image Processing. Photogrammetric Engineering and Remote Sensing, Vol. 41, pp 1465-76.

Bernstein., R. 1983, Image Geometry and Rectificaion, Chapter 21 : In R. N. Colwett (ed.), Manual of Remote Sensing, Falls Church VQ: American Society of Photogrammetry and Remote Sensing, vol. I, pp 8775-81.

Berry., J. K., 1982, Cartographic Modelling Procedures for Extending the utility of Remotly Sensed data: In B. F. Richason (ed.) Remote Sensing : An Input to Geographic Information Systems in the 1980's, Proceedings of the Pecora 7 Symposium.

Berry., J. K., 1987, Fundamental Operations in Computer-Assisted Map Analysis, International Journal of GIS, 2, pp. 119-136.

Berry., J. K., 1993, Beyond Mapping : Concepts, Algorithms and Issues in GIS. GIS World Inc. Colorado.

Blackemore., M., 1984, Generalisation and Error in Spatial Data Base, Cartographic, 21, pp. 131-9.

Bracken., I., and Webster, C., 1989, Information Technology for Geography and Planning, London, Routeledge.

Brassel., K., Bucher., F., Stephan., E-M, and VcKovski., A, 1995, Completeness: In Guptill., S.C, Morrison., J.L., (eds) Elements of Spatial Data Quality, Oxford, Elsevier Science. pp. 81-108.

Burrough., P.A., 1983, Multiscale Sources of Spatial Variation in Soil 1. The Application of Fractal Concepts to Nested Levels of Soil Nariation, Journal of. Soil Science, 34, 577-97.

Burrough, P. A., and Frank., A. V., 1996, Geographical Objects and Indeterminate Boundaries, Taylor and Francis, London, 345p.

Burrough, P. A., 1986, Principles of Geographical Information Systems for Land Resources Assessment, Clarendon Press, Oxford.

Burrough., P. A., 1983, Geographical Information, Systems for Natural Resources Assessment, New York, Oxford University Press.

Burrough., P. A., and Masser., F. I., 1998, European Geographical Infrastructures, Opportunities and pit falls, Taylor and Francis, London.

Burrough., P. A., 1989, Fuzzy Mathematical Methods for Soil Survey and land Evaluation, Journal of Soil Science, Vol. 40., pp. 477-92.

Kardoulas., N. G., Bird, A. C., and Lawman, A. I., 1996, Geometric Correction of Spot and Landsat Imagery. A Comparison of a Map and GPS-Derived Control Points, Photogrammetric Engineering and Remote Sensing, 62, pp.1173-1177.

Mcharg., I. L., 1996, A Quest for Life, John Wiley and sons Inc. New York.

Chen., P. P-S., 1976, The Entity-Relationship Model-Toward a Unified View of Data, ACM Transactions on Database Systems, 4, pp. 397-434.

Chrisman., N. R., 1982, A Theory of Cartographic Error and its Measurement in Digital Databases, Proceedings Auto Carto, 5, pp. 159-68.

Chrisman., N. R., 1994, Beyond the Snapshot, Changing the Approach to Change, Error and Process, Unpublished Manuscript.

Chrisman., N.R., 1997, Exploring Geographic Information Systems, Wiley, New York.

Codd., E.F., 1970, A Relational Model of Data for Large Shared Data Banks. Communication of the ACM, vol. 13, No.6, pp 377-387.

Codd., E. F., 1982, Relational Database: a Practical Foundation for Productivity, Communication of the ACM, Vol. 25, No.2, pp 109-117.

Cooke, D. F. and W. F. Maxfield, 1967, The Development of a Geographic Base File and Uses for Mapping, Urban and Regional Information System for Social Programs, Proceedings of the 5th Annual Conference of the Urban and Regional Information Systems Association, pp 207-218.

Clarke., K. C., 1990, Analytical and Computer Cartography, Prentice Hall, Englewood Cliffs N J.

Cooke., D.F., 1987, Map Storage on CD-ROM Byte Magazine, vol, 12, No. 8, pp.129-138.

Crippen, R. W., 1989, A Simple Spatial Filtering Routine for the Cosmetic Removal of Scan Line Noise from Landsat TM Tape Imagery. Photogrammetric Engineering and Remote Sensing, 55, pp. 327-331.

Curran., P. J., 1988, Principles of Remote Sensing, ELBS., pp 282.

Curran., P., 1989, Principles of Remote Sensing, Longman, London.

Dale., P. F, McLaughlin. J. D. 1998, Land Information Management : An Introduction with Special Reference to Cadastral Problems in Third World Countries, Clarendon Press, Oxford.

Dale., P. E. R., Chandica., A. L., and Evans, M., 1996, Using Image Subtraction and Classification to Evaluate Change in Sub-Tropical Inter-Tidal Wetlands, International Journal of Remote Sensing, 17, pp. 703-719.

Date., C.J., 1995, An Introduction to Database systems, 6th ed. Addison-Wesley, Reading, Mass.

Date., C. J., 1982, An Introduction to Database Systems, 2nd ed. Addison-Wesley, Reading, Mass.

Davis., P. A., Major., E. R., and Jacobowitz, H, 1984, An Assessment of NIMBUS-7 ERB Short Wave Scanner Data by Correlative Analysis with Narrow Band CZCS Data, Journal of Geophysical Research., 89, pp. 5077-88.

Demers., M. N., 1999, Fundamentals of Geographic Information Systems, Second Edition, John Wiley and Sons, Inc.

Elachi., C., 1987, Introduction to the Physics and Techniques of Remote Sensing, Wiely, New York.

Eliason, E. M., and McEwen, A. S., 1990, Adaptive Box Filters for the Removal of Random Noise from Digital Images. Photogrammetric Engineering and Remote Sensing, 56, pp. 453-458.

ESRI., 1993, ESRI Build National GIS for the Real Estate Industry, ARC News 17 (4). pp. 1-2.

Estes., J., 1995, What GIS is, where it came from and what it does, Proc. Cambridge Conference for National Mapping Organizations, Workshop paper 2. 1\18.

Fisher., J., 1989, The pixel, a snare and a delusion, International Journal of Remote Sensing, 18, pp. 679-685.

Forshaw., M. R. B., Haskell, A., Miller, P. F., Stanley, D. J, and Townshend J, R, G., 1983, Spatial Resolution of Remotely Sensed Imagery : A Review paper, International Journal of Remote Sensing, 4, pp. 497-520.

Frolov., Y. S., and Maling, D. H., 1969, The Accuracy of Area Measurements by Point Counting Techniques, Cartographic J., 6, pp. 21-35.

Fungy., A.K., 1976, Remote Sensing for Environmental Sciences (Ed. E. Scandal) Springs Verlag, Beden.

Galton., F., 1884, Art of Travel, or, Shifts and Contrivances Available in Wild Countries, New York, David or Charles Reprints.

Galbaiti, L. J. Jr. 1990, Machine Vision and Digital Image Processing Fundamentals. Prentice Hall, Englewood, New Jersey. Chapter 6 : pp 105-26.

Gonzalez, R. C. and Wintz, P. 1977, Digital Image Processing, and Recognition, Addison-Wesley Publishing Company. Reading Massachusetts, pp 123-6.

Gonzalez. R.C. and Wintz, P. 1983, Digital Image Processing, Addison-Wesley Publishing Company. California, Chapter 4, pp 140-205.

Goodchild, M. F., 1991 a, Issues of Quality and Uncertainty: In. Miller, J. C., (ed), Advances in Cartography, Oxford, Elsevier Science, pp. 111-139.

Goodchild., M. F., 1980, Fractals and the Accuracy of Geographical Measures, Mathematical Geology, 12, pp. 85-98.

Goodchild., M. F. Parks., B.O. and Stayaert., L. T., (ed) 1993, Environmental Modelling with GIS., Oxford University Press, Oxford and New York.

Goodchild., M. F., 1992, Geographical Information Science, International Journal of Geographical Information Systems 6 (1), pp 31-45.

Grimshaw., D. J., 1994, Bringing Geographical Information Systems into Business, Longman, London.

Guptill., C., 1987, Desirable Characteritics of a Spatial Database Management System: In Proceedings of AUTOCARTO 8. Falls Church, VA. ASPRS, pp 278-28.

Hazelton., N. W. J., 1991, Integrating Time, Dynamic Modelling and Geographical Information Systems, Development of 4-Dimensional GIS, Unpublished Ph. D Dissrtation, Dept., of Surveying and Land Information, University of Melbourne.

Hall, E. L., 1978, Computer Image Processing and Recongnition, Academic Press, New York. Chapter 4, pp 159-84.

Healy, R. G., 1991, Geographical Information System Principles and Applications, Maguire, D.J., et al (eds), Longman Scientific and Technical, England, pp 239-267.

Henderson, S. T., 1970, Daylight and its Spectrum, Adam Hilger Ltd., London, pp 363.

Hodgkiss, A.G., 1961, Understanding Maps, Dawson, Folkestone, UK.

Heraldick., R. M., Shanmugam., K. S. and Dinstein, I, 1973, Textural Features for Image Classification, IEEE Transactions on Systems, Man and Cybernetics, 3, pp 610-622.

Heywood., D. I., Oliver J. Tomlinson., S., 1995, Building an Exploratory Multi-Criteria Modelling Environment for Spatial Decision Support : In Fisher P (ed.) Innovations in GIS, Taylor and Francis, London, pp 127-36.

Hunt., G. R., Salisbury, J. W., and Lenyoff, C. J., 1973, Visible and Near Infrared Spectra of Minerals and Rocks. V11. Acidic Igneous Rocks, Modern Geology, Vol. 4, pp 217-224.

Hurn., J., 1989, GPS : A Guide to the Next Utility, Sunnywale , Trimble Navigation Ltd.

Ian Heywood, Sarah Cornelins and Steve Carver, 1998, An Introduction to Geographical Information Systems, Longman, England.

Jenson, J.R. 1983, Education Image Processing: An overview. Photogrammetric Engineering and Remote Sensing, vol. 49, pp 1151-7.

Jenson, J .R. 1986, Introduction to Digital Image Processing; Prentice- Hall, Englewood, Cliff, New Jersey, Chapter 7, pp 117-76.

Kainz., W., 1995, Logical Consistency: In Guptill., S. C, Morreson., J. L., (eds), Elements of Spatial Data Quality, Oxford, Elsevier Science. pp 109-37.

Keefer., B. K., Smith., J. L. and Gregorie., T. G. 1988, Simulating Manual Digitizing Error with Statistical Models. Proceedings of GIS\ LIS' 88. ACSM, ASPRS, AAG and URISA, San Antonio, Texas, pp 475-83.

Kennedy, M. and C. R. Meyers, 1997, Spatial Information Systems: Introduction, Louisville, Kentucky: Urban Studies Center.

Kelmelis., J., 1991, Time and Space in Geographic Information : Toward a 4-Dimensional Spatio-Temporal Data Model, Unpublished Ph. D. Dissertation, Department of Geography, Pennsylvania State University, USA.

Kennedy., M. and C. Guinn., 1975, Automated Spatial Data Information Systems, Avoiding Failure, Louisville, Kentucky, Urban Studies Center.

Kent., W., 1983, A Simple Guide to Five Normal Forms in Relational Database Theory, Communications of the Association for Computing Machinery, 26 (2), pp 120-125.

Lange., A. F. and Gilbert., C., 1999, Using GPS for GIS Data Capture, In Geographical Information Systems : Principles and Technical Issues Edited by Paul A. Longley, M. F. Goodchild, David J. Maguire and David W. Rlind, Second Edision, John Wiley and Sons, Inc.

Langran., G. 1990, Temporal GIS Design Tradeoffs, Journal of the Urban and Regional Information Systems Association, 2, pp16-25.

Langran., G., 1989 b, Time in Geographic Information Systems, Ph. D Dissertation, University of Washington.

Langran., G., 1992, Time in Geographic Information Systems, London, Taylor and Francis.

Laurini., R. Thompson., D 1992, Fundamentals of Spatial Information Systems, Academic Press, London.

Lawrini., R. and Thompson. D., 1992, Fundamentals of Spatial Information Systems, Academic Press, London.

Leachy, C., 1987, Introduction to the Physics and Techniques of Remote Sensing, New York, Wiley.

Lillesand., T. M. and R. W. Keifer 1979, 1994, and 2000, Remote Sensing and Image Interpretation, John Wiley and Sons, New York.

MacEahren., A. M., 1994, Some Truth with Maps, Association of American Geographers, Washington D.C.

Madhusudan Reddy., P. 2000, Ground Water Pollution Studies using Remote Sensing and GIS. A Case Study from Vishakapatnan, A. P India, Proceedings of Hydro Informatics, IOWA, USA.

Maling, D. H., 1980, Coordinate Systems and Map Projections, George Phillip, London.

Matthews., E., and Rossow, W.B., 1987, Regional and Seasonal Variations of Surface Reflectance from Satellite Observations at 0.6 μm. J. Clim. APPL. Meteor., 26, pp 170-202.

Mandelfort., B. B., 1982, The Fractal Geometry of Nature, Freeman., New York.

Marble., D. F. and D. J. Peuquet, 1983, Computer Software for Spatial Data Handling, Current Status and Future Development Needs, Buffalo, New York, Geographic Information System laboratory, State University of New York.

Mark., D. M., 1986, The use of Quadtrees in Geographic Information Systems and Spatial Data Handling : In M. Blakemore (ed.) Proceedings, Auto Carto London, vol. 1, pp 517-526.

Martin., D., 1996, An Assessment of Surface and Zonal Models of Population, International Journal of Geographical Information Systems, 10, pp. 973-89.

Mather, P. M., 1987, Computer Processing of Remotely Sensed Images: An Introduction, John Wiley & Sons, 352 p.

Monmonier., M. and G. A. Schnell., 1988, Map Appreciation. Englewood Cliffs, Prentice Hall, New Jersey.

McDonnell., R. and Kemp., K., 1995, International GIS Dictionary, Geoinformation International, London.

Mead., R. A. and Szajgin., J.,1982, Landsat Classification Accuracy Assessment Procedures. Photogrammetric Engineering and Remote Sensing, 4, pp 497-501.

Nag., P. and Kudrat, M. 1998, Digital Image Processing, Concept Publishing Company, New Delhi.

NRSA, 1995,1999, IRS-Handbook, Hyderabad, India.

Oxborrow., E. P., 1989, Databases and Database systems : Concepts and Issues, 2 ed. Chartwell.

Peuckar, T. K., and N. Chrisman, 1975, "Cartographic Data Structures". The American Cartographer, Vol. 2, No.1, pp 55-69.

Pratt, W. K., 1978, Digital Image Processing. John Wiley & Sons, New York, pp 201-344.

Puequet, D. J., 1981, An Examination of Techniques for Reformatting Digital Cartographic Data. Part I: The Raster-to-Vector Process, Cartographica, Vol. 18, No. I, pp 34-48.

Perkal., J. 1966, On the Length of Empirical Curves. Discussion Paper 10, Ann Arbor, Michigan Inter-University Community of Mathematical Geographers.

Peueker., T. and N. Chrisman, 1975, Cartographic Data Structures, American Cartographer, 2, pp 25-69.

Peuquet., D. J., and Duan , N., 1995, An Event-Based Spatio-Temporal Data model (ESTDM) for temporal analysis of geographic data, International Journal of GIS, 9, pp 2-24.

Peuquet., D. J., 1984, A Conceptual Frame Work and Comparison of Spatial Data Models, Cartographica 21, pp 66-113.

Peuquet., D. J., and Qian ., L., 1996, An Integrated Database Design for Temporal GIS, Seventh International Symposium on Spatial Data, Handlind, Delft, Holland, International Gegraphical Union, pp 2-1 to 2-11.

Peuquet., D. J., 1994, It is About Time. A Conceptual Frame Work for the Representation of Temporal Dynamic in Geographic Information Systems, Annals of the Association of American Geographers, 84, pp 441-61.

Prasad., L., and Iyengar., S.S., 1997, Wavelet Analysis with Applications to Image Processing, Boca Raton, FL. CRC press.

Rampal., K. K., 1982, Textbook of Photogrammetry, Oxford and IBH Publishing, India.

Robinson, A., R. Sale and J. Morrison, 1984, Elements of Cartography, 4th ed. John Wiley and Sons, New York.

Redman., T. C. 1992, Data Quality, New York, Bantam.

Rees., W.,1990, The Physical Principles of Remote Sensing, Cambridge University press, Cambridge. U. K.

Reeves., R. G., 1979, Manual of Remote Sensing, American Society of Photogrammertry.

Rhind, D. 1977, Computer-Aided Cartography, Trans. Inst, British Geographers, NS2. pp 71-96.

Rhind., D. W., 1988, Personality as a Factor in the Development of a Discipline, American Cartographer, 15, pp 3277-89.

Robert Masom., 1991, Satellite Remote Sensing of Polar Regions, CBS Publishers, New Delhi.

Robertson., B., 1988, How to Draw Charts and Diagrams, North Light, Cincinnati, Ohio.

Robinove., C. J. 1982, Computation with Physical Values from Landsat Digital Data, Photogrammetric Engineering and Remote Sensing , Vol.48, No.5, pp 781-784.

Robinson., A. H., J. L Morrison, P. C. Muchrcke, A. C. Kimerling, and S. C. Guptill, 1995, Elements of Cartography, 6th ed. New York. John Wiley and Sons.

Rogers., W.F., and Singh 1986 Some Geomorphic Relationships and Hydrograph Analysis, Water Resources Research, 22, pp 777-784.

Ruby, F., Moore, R. and Fungi, A. 1981, Microwave Remote Sensing, Vol.1. Fundamentals and Radiometry, Addison-Wesley Publishing Company, Reading, Massachusetts.

Ruby, F., Moore, R. and Fungi, A., 1982, Microwave Remote Sensing. Volume2, Scattering and Emission Theory, Addison-Wesley Publishing Company, Reading, Massachusetts.

Salge., F., 1995, Semantic Accuracy : In Guptill., S. C., Morrison., J. L. (eds), Elements of Spatial Data Quality, Oxford, Elsevier Science. pp 139-51.

Sabins, F. F., 1987, Remote Sensing-Principle and Interpretation, Freeman and Co., New York, Chapter 7, pp 246-60.

Seigal, Barry., S. and Alan R. Gillespie (eds) 1980, Remote Sensing in Geology, John Wiley and Sons, New York.

Seaborn., D., 1995, Database Management in GIS-Past, Present, Future : An Enterprise Perspeetive for Executives, Seaborn Associates, 475, Clover Dale Road, Ottawa, Canada.

Seshagiri Rao., K.V. 2000, Watershed Management, BS Publications, Hyderabad, India.

Sioux Falls., S. Dakota, 1981, Falls Church VA : American Society of Photogrammetry.

Snyder., J. P., 1987, Map Projections-A Working Manual. U.S. Geological Survey, Professional Paper 1935. Washington, D.C. U.S. Government printing office.

Star, J. L., Estes., J. E., McGore, K. C., 1997, Integration of Geographic Information Systems and Remote Sensing, Cambridge University Press, New York.

Star., J. and Estes John, 1999, Geographic Information Systems An Introduction, Prentice Hall, New Jersey, USA.

Stern, N., and Stern., R. A., 1993, Computing in the Information Age, Wiley, New York.

Stewart., R., 1985, Methods of Satellite Oceanography, University of California Press, Berkeley California.

Tarpley., J. D., Schneider., S. R., and Money., R. L., 1984, Global Vegetation Indices from the NOAA-7 Meteorological Satellite, J. Clim. Appl. Meteor., 23, pp 491-4.

Thapa., K. and Bossler., J. 1992, Accuracy of Spatial Data Used in Geographic Information Systems, Photogrammetric Engineering and Remote Sensing, 58. pp 835-841.

Thomas., I. L. and Allcock., G. M.,1984, Determining the Confidence Level for a Classification, Photogrammetric Engineering and Remote Sensing, 5, pp 1491-1496.

Thomas, I. L., 1980, Spatial Post Processing of Spectrally-Classified Landsat Data, Photogrammetric Engineering and Remote Sensing, 46, pp 1201-1206.

Thomas, I. L., Howarth., R., Eggers., A. and Fowler, A.D.W., 1981, Textural, Enhancement of a Circular Geologic Feature, Photogrammetric Engineering and Remote Sensing, 47, pp 89-91.

Toll., D. L., 1984, An Evaluation of Simulated Thematic Mapper Data and Landsat MSS Data for Discriminating Suburban and Regional Land Use and Land Cover, Photogrammetric Engineering and Remote Sensing , 5, pp 1713-1724.

Tomlin., C. D., and Berry., J. K., 1979, A Mathematical Structure for Cartographic Modelling in Environmental Analysis., Proceedings of the American Congress on Surveying and Mapping, Falls Church VA. American Congress on Surveying and Mapping.

Tomlin., C. D., 1990, Geographic Information Systems and Cartographic Modelling, Englewood Cliffs, N. J, Prentice Hall.

Tomlin., C. D., 1991, Cartographic Modelling : In D. J. Maguire, M. F. Good Child and D.W. Rhind (eds) Geographical Information Systems : Principles and Applications, Harlow, Long man.

Tufte., E. R., 1983 The Visual Display of Quantitative Information, Cheshire, Connecticut, Graphic Press.

Turner., A. K. (ed.) 1992 Three-Dimensional Modelling with Geoscientific Information Systems, Dordrecht : Kluwer Academic Publishers, 443 pp.

Vincent, R.K., 1973, A Thermal Infrared Ratio Imaging Method for Mapping Compositional Variations Unpublished. Ph. D. Thesis, Univ. of Mich., Ann Arbor, Mich., 102 p.

Vincent., R. K., Rowan, L. C. Gillespie, R. E., and Knapp, C., 1975, Thermal Infrared Spectra and Chemical Analysis of Twenty-Six Igneous Rock Samples. Remote Sensing of Environment, 4.4, pp 199-210.

Walsh., S. J., 1987, Comparison of NOAA AVHRR Data to Meteorologic Drought Indices, Photogrammetric Engineering and Remote Sensing, 53, pp1069-74.

Weibel., R., and Heller., M., 1991, Digital Terrain Modelling : In Maguire D. J., Goodchild., M. F., Rhind D. W., (eds), Geographical Information Systems. Principles and Applications, Long man, London, Vol. 1, pp 269-97.

Witthuhn., B.O., D. P. Brandt, and G. J. Demo, 1974, Discovery in Geography, Dubuque, IA, Kendall/Hunt Publishing Company.

Wolfe., W. L, and Zissis, G. J., 1989, The Infrared Handbook. The Infrared Information Analysis Center, Environmental Research Institute of Michigan, USA.

Wolf, P. R., 1983, Elements of Photogrammetry, 2nd ed. McGraw Hill Book Co. New York.

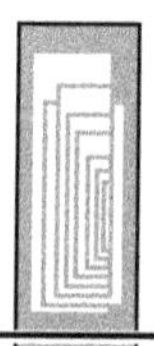

SUBJECT INDEX

A

B

C

D

E

F

G

H

I

K

L

T

U

V

W

www.ingramcontent.com/pod-product-compliance
Ingram Content Group UK Ltd.
Pitfield, Milton Keynes, MK11 3LW, UK
UKHW061535190726
13845UKWH00005B/108

9 789385 433351